Handbook of
Air Pollution Analysis

Handbook of
Air Pollution Analysis

SECOND EDITION

Edited by

Roy M. Harrison

Department of Chemistry
University of Essex
Colchester CO4 3SQ
UK

and

Roger Perry

Department of Civil Engineering
Imperial College of Science and Technology
London SW7 2BU
UK

LONDON   NEW YORK

CHAPMAN and HALL

First published in 1977 by
Chapman and Hall Ltd
11 New Fetter Lane, London EC4P 4EE
Second edition 1986

Published in the USA by
Chapman and Hall
29 West 35th Street, New York NY 10001

© 1977, 1986 Chapman and Hall Ltd

Printed in Great Britain at the
University Press, Cambridge

ISBN 0 412 24410 1

British Library Cataloguing in Publication Data

Handbook of air pollution analysis.—2nd ed.
  1. Air—Pollution   2. Air—Analysis
  I. Harrison, Roy M.   II. Perry, Roger, *1940–*
  363.7'39264   TD883.1
  ISBN 0–412–24410–1

Library of Congress Cataloging in Publication Data

Handbook of air pollution analysis.
  Bibliography: p.
  Includes index.
  1. Air—Pollution—Measurement—Handbooks, manuals, etc.
  2. Environmental chemistry—Handbooks, manuals, etc.
  I. Harrison, Roy M., 1948–   .   II. Perry, Roger, 1940–
  TD890.H36   1986   628.5'3   85–26011
  ISBN 0–412–24410–1 (U.S.)

Contents

Preface to second edition   *page* xix

Contributors   xxi

1	**General sampling techniques**	**1**
	C. A. Pio	
1.1	Sampling goals and requirements	1
	1.1.1 Ambient sampling	1
	1.1.1.1 *General objectives*	1
	1.1.1.2 *Meteorological considerations*	2
	1.1.1.3 *Sampling site criteria*	4
	1.1.1.4 *Sample scheduling*	5
	1.1.2 Source sampling	8
	1.1.2.1 *General objectives*	8
	1.1.2.2 *Stationary source sampling*	8
	1.1.2.3 *Mobile source sampling*	18
1.2	Sampling methods	21
	1.2.1 General sampling system considerations	21
	1.2.1.1 *Intake and transfer component*	21
	1.2.1.2 *Collection component*	22
	1.2.1.3 *Flow measurement component*	23
	1.2.1.4 *Air moving component*	25
	1.2.2 Aerosols	25
	1.2.2.1 *Aerosol sampling considerations*	25
	1.2.2.2 *Aerosol sampling collection components*	34
	1.2.2.3 *Ambient aerosol sampling applications*	41
	1.2.2.4 *Emission source aerosol sampling applications*	55
	1.2.3 Gases	65
	1.2.3.1 *Gas sampling considerations*	65
	1.2.3.2 *Gas sampling collection components*	66
	1.2.3.3 *Ambient gas sampling applications*	71
	1.2.3.4 *Emission source gas sampling applications*	78
	1.2.4 Sampling of rainwater and fog	82
	References	85

2	**Air pollution meteorology**	95
	D. J. Moore	
2.1	Introduction	95
	2.1.1 Wind and the turbulent mixing layer	95
	2.1.2 The effect of source height	97
	2.1.3 Plant design to achieve maximum atmospheric dispersion	99
	2.1.4 Factors affecting long-range transport of pollutants	100
2.2	Meteorological measurements	100
	2.2.1 Parameters affecting transport and dispersion of pollutants	100
	2.2.2 Wind velocity measurements	101
	2.2.3 Measurements to determine the atmospheric stability	102
	2.2.3.1 *Parameterizing the stability*	102
	2.2.3.2 *Measurements of temperature at a fixed height*	102
	2.2.3.3 *Measurements of vertical temperature gradient*	103
	2.2.3.4 *Measurements of thermal radiation*	103
	2.2.4 Turbulence measurements	105
	2.2.4.1 *Turbulent energy*	105
	2.2.4.2 *Turbulence spectra*	105
	2.2.4.3 *Turbulent fluxes*	105
	2.2.5 Measurements of mixing depth	106
	2.2.6 Precipitation measurements	107
2.3	Outline of the more important features of the atmospheric transport and dispersion of pollutants	107
	2.3.1 Transport and dispersion in different types of air mass or air stream	107
	2.3.1.1 *Air stream characteristics*	107
	2.3.1.2 *Air mass origins*	108
	2.3.2 Diurnal variations in air stream characteristics	108
	2.3.2.1 *Settled anticyclonic*	108
	2.3.2.2 *Warm advection*	111
	2.3.2.3 *Cold advection*	112
	2.3.2.4 *Unsettled cyclonic*	112
	2.3.3 Frequency of occurrence of different air streams	112
	2.3.4 Land and sea breezes	113
	2.3.5 Upslope and downslope winds	114
	2.3.6 Urban areas and elevated sources	114
2.4	Calculation of the atmospheric transmission of pollutants	115
	2.4.1 Introduction	115
	2.4.2 Calculation of plume rise	115
	2.4.2.1 *Selection of equation*	115
	2.4.2.2 *Plume rise formulae*	116
	2.4.3 The effect of particle fall velocity on plume height	118
	2.4.4 Calculation of dispersion	119

2.4.4.1 *Eddy diffusivity (or K-type) models*   120
2.4.4.2 *Gaussian models*   120
2.4.4.3 *Second and higher order closure models*   121
2.4.5   Box and cell models   121
2.4.6   Calculation of trajectories   124
2.4.7   The effects of deposition   124
2.4.7.1 *General*   124
2.4.7.2 *Dry deposition*   125
2.4.7.3 *Wet deposition*   125
2.4.7.4 *Occult deposition*   125
2.5   Examples of calculations using Gaussian models   127
         References   131

3   Air pollution chemistry   133
         A. M. Winer
3.1   Introduction   133
3.2   Inorganic reactions   134
3.2.1   The NO–NO_2–O_3 cycle   134
3.2.2   Formation of radical intermediates   135
3.2.2.1 *Hydroxyl and hydroperoxyl radicals*   135
3.2.2.2 *The NO_3 radical*   136
3.2.3   Termination reactions   137
3.2.4   Other important inorganic reactions   137
3.2.4.1 *HONO*   137
3.2.4.2 *HNO_3, N_2O_5 and acid deposition*   138
3.2.5   Peak concentrations of selected inorganic pollutants
         observed or expected in polluted atmospheres   139
3.3   Reactions involving organic compounds   140
3.3.1   Reactions of OH radicals with organics   140
3.3.1.1 *Alkanes*   140
3.3.1.2 *Alkenes*   141
3.3.1.3 *Aromatics*   142
3.3.1.4 *Aldehydes*   144
3.3.2   Reactions of O_3 with organics   144
3.3.3   Reactions of NO_3 radicals with organics   145
3.3.3.1 *Alkanes*   147
3.3.3.2 *Alkenes*   147
3.3.3.3 *Aldehydes*   147
3.3.3.4 *Aromatics*   148
3.4   Gas-to-particle conversion   149
3.4.1 SO_2 photo-oxidation and formation of sulphate
         particulate   149
3.4.2   Formation of secondary nitrate and organic particulate   149
3.5   Conclusion   150
         References   150

4   Analysis of particulate pollutants   155
     R. M. Harrison
4.1 Introduction   155
     4.1.1 Emission of particulate matter   157
     4.1.2 Emission factors for particulate matter   157
     4.1.3 Dispersion of atmospheric pollutants from a point
            source   157
     4.1.3.1 *Problems of short-term sampling*   161
4.2 Suspended material   162
     4.2.1 Sampling techniques   162
     4.2.1.1 *Filter paper techniques*   162
     4.2.2 Determination of total particulate pollutant
            concentrations   166
     4.2.2.1 *Light reflectance method*   166
     4.2.2.2 *Gravimetric techniques*   168
     4.2.2.3 *Other filter paper devices*   169
     4.2.2.4 *Piezoelectric mass monitors*   171
     4.2.3 Cascade impactors   172
     4.2.4 Light scattering techniques   175
     4.2.4.1 *The integrating nephelometer*   175
     4.2.4.2 *Aerosol particle counters*   176
     4.2.5 The directional sampler   177
4.3 Dustfall sampling   179
     4.3.1 Introduction   179
     4.3.2 Designs of national deposit gauges   180
     4.3.2.1 *The British Standard deposit gauge*   181
     4.3.2.2 *French Standard deposit gauge (Ref. NF, X43-006
              (1972))*   182
     4.3.2.3 *Norwegian NILU deposit gauge*   183
     4.3.3 Short-term surveys   183
     4.3.3.1 *Single bowl surveys*   185
     4.3.3.2 *Larger surveys*   185
     4.3.4 British Standard directional deposit gauge   187
4.4 Physical techniques for classification of particulates   189
     4.4.1 Density gradient separation   189
     4.4.1.1 *Density gradient liquids*   189
     4.4.1.2 *Recovery and cleaning of liquids*   190
     4.4.1.3 *Preparation of the gradient*   191
     4.4.2 Dispersion staining   195
     4.4.3 Microscopic techniques   197
     4.4.3.1 *Mounting samples*   198
     4.4.3.2 *Identification of dusts and reference library*   199
     4.4.3.3 *Description of dusts from different combustion and
              industrial sources*   199
     4.4.3.4 *Dust identification table*   203

4.4.4   Determination of asbestos   205
4.4.4.1 *Membrane filter method*   205
4.4.4.2 *Infrared technique for ambient atmospheres*   206
4.4.4.3 *Transmission electron microscope methods for ambient*
         atmospheres   207
4.4.5   Determination of particle size distribution   207
4.4.5.1 *Sieve techniques*   208
4.4.5.2 *Microscope techniques*   208
References   212

5   Metal analysis   215
         R. M. Harrison
5.1   Introduction   215
5.2   Analysis of particulate matter   216
5.2.1   General sampling considerations   216
5.2.2   Analytical methods involving no pretreatment of the
         sample   219
5.2.2.1 *X-ray emission analysis*   219
5.2.2.2 *Radioactivation methods*   223
5.2.3   Methods involving pretreatment of the samples   228
5.2.3.1 *Emission spectrography*   228
5.2.3.2 *Ring oven methods*   231
5.2.3.3 *Polarography*   235
5.2.3.4 *Anodic stripping voltammetry*   237
5.2.3.5 *Spark source mass spectrometry*   239
5.2.3.6 *Spectrophotometry and fluorometry*   241
5.2.3.7 *Atomic spectroscopy*   246
5.2.3.8 *Other analytical methods*   258
5.3   Gases and vapours   260
5.3.1   General sampling considerations   260
5.3.2.1 *Metal carbonyls*   261
5.3.2.2 *Hg and its compounds*   264
5.3.2.3 *Volatile Pb compounds*   267
References   271

6   Nitrogen and sulphur compounds   279
         R. M. Harrison
6.1   Introduction   279
6.2   Basic analytical techniques   282
6.2.1   Sampling techniques   282
6.2.2   Analytical methods – chemical   284
6.2.2.1 *Acidimetric methods*   284
6.2.2.2 *Colorimetric methods*   284
6.2.2.3 *Coulometric methods*   286

6.2.2.4   *Miscellaneous chemical methods*   287
6.2.3   Physical methods   288
6.2.3.1   *Chemiluminescence*   288
6.2.3.2   *Fluorescence*   290
6.2.3.3   *Absorption spectroscopy*   290
6.2.3.4   *Gas chromatography*   292
6.2.3.5   *Other physical methods*   292
6.3   Experimental section   293
6.3.1   Analysis of SO_2 293
6.3.1.1   *Chemical methods*   293
6.3.1.2   *Physical analysis of SO_2*   302
6.3.2   Analysis of SO_3 305
6.3.3   Analysis of H_2S 306
6.3.3.1   *Chemical methods*   306
6.3.3.2   *Physical methods*   309
6.3.4   Analysis of organic S compounds   310
6.3.4.1   *Chemical methods*   310
6.3.4.2   *Physical methods*   312
6.3.5   Analysis of oxides of nitrogen – NO and NO_2 313
6.3.5.1   *Chemical methods*   313
6.3.5.2   *Physical methods*   321
6.3.6   Analysis of NH_3 324
6.3.6.1   *Chemical methods*   324
6.3.6.2   *Physical methods*   327
6.3.7   Miscellaneous N_2 compounds   328
6.3.8   Preparation of standard gas mixtures for calibration   329
6.3.8.1   *Preparation of standard mixtures by static methods*   329
6.3.8.2   *Preparation of standard mixtures by dynamic methods*   330
6.4   Particulate compounds of S and N   333
6.4.1   Analysis of SO_4^{2-} 333
6.4.1.1   *Experimental procedure for SO_4^{2-} (turbidimetric)* 334
6.4.2   Analysis of particulate NO_3^- 335
6.4.2.1   *Experimental procedure for nitrate (colorimetric)*   335
6.4.3   Analysis of NH_4^+ salts   337
6.4.3.1   *Experimental procedure for NH_4^+ (colorimetric)*   337
References   338

7   Secondary pollutants   343
        R. M. Harrison

7.1   Introduction   343
7.2   Basic analytical techniques for the analysis of gaseous
            secondary pollutants   347
7.2.1   Sampling methods   347
7.2.2   Analytical techniques   348
7.2.2.1   *Chemical methods*   348

	7.2.2.2 *Physical methods*	349
7.3	Experimental section	353
	7.3.1 Analysis of 'total oxidants'	353
	7.3.1.1 *Discussion of analytical methods*	353
	7.3.1.2 *Neutral KI method for manual analyses of 'total oxidants'*	355
	7.3.1.3 *Instruments for measurements of total oxidants*	357
	7.3.2 Analysis of O_3	359
	7.3.2.1 *Chemical methods*	359
	7.3.2.2 *Physical methods*	361
	7.3.2.3 *Measurement of O_3 by the C_2H_4-chemiluminescence method*	364
	7.3.2.4 *Preparation of O_3/air mixtures for calibration purposes*	367
	7.3.3 Analysis of H_2O_2	369
	7.3.3.1 *Chemical methods*	369
	7.3.3.2 *Physical methods*	370
	7.3.4 Analysis of aliphatic aldehydes and oxygenated compounds	370
	7.3.4.1 *Chemical methods*	370
	7.3.4.2 *A colorimetric analysis of total aliphatic aldehydes in air (MBTH method)*	371
	7.3.4.3 *Colorimetric analysis of HCHO (chromotropic acid method)*	373
	7.3.4.4 *Physical methods*	375
	7.3.5 Analysis of PAN and related compounds	375
	7.3.5.1 *Chemical methods*	375
	7.3.5.2 *Physical methods*	376
	7.3.5.3 *Analysis of PAN by electron capture GC*	378
	7.3.6 Analysis of oxyacids of N	381
	7.3.6.1 *Chemical methods*	382
	7.3.6.2 *Physical methods*	383
	References	383
8	**Hydrocarbons and carbon monoxide**	**387**
	A. E. McIntyre and J. N. Lester	
8.1	Introduction	387
8.2	Volatile hydrocarbons	391
	8.2.1 Sampling procedures	391
	8.2.1.1 *Cryogenic systems*	392
	8.2.1.2 *Solid adsorption systems*	394
	8.2.1.3 *Gas sampling systems*	395
	8.2.2 Analytical methods	396
	8.2.2.1 *Continuous instrumental analysers*	396
	8.2.2.2 *Gas–liquid chromatography (GLC)*	399

8.2.2.3   *Mass spectrometry and gas chromatography/mass spectrometry*   407
8.2.2.4   *Calibration methods*   409
8.2.3   Methods for specific compounds   410
8.2.3.1 C_2–C_5 *hydrocarbons*   410
8.2.3.2 C_6–C_9 *hydrocarbons*   411
8.3   Hydrocarbon fraction of airborne particulate matter   413
8.3.1   Sampling procedures   413
8.3.2   Extraction and clean-up procedures   414
8.3.3   Analytical methods   415
8.3.3.1   *Thin-layer chromatography*   415
8.3.3.2   *High-performance liquid chromatography*   416
8.3.3.3   *Gas–liquid chromatography*   417
8.3.3.4   *Gas–liquid chromatography/mass spectrometry*   418
8.4   Carbon monoxide   418
References   419

9   Halogen compounds   425
P. W. W. Kirk and J. N. Lester

9.1   Fluorides   425
9.1.1   Sampling procedures   426
9.1.2   Analytical procedures   429
9.1.2.1   *Pretreatment for particulates*   429
9.1.2.2   *Ion-selective electrode determination*   430
9.1.2.3   *Colorimetric determination*   432
9.1.3   Recommended experimental procedures   433
9.1.3.1   *Sampling*   433
9.1.3.2   *Pretreatment and clean-up*   434
9.1.3.3   *Analytical methods*   437
9.2   Chlorine   441
9.2.1   Sampling procedures   443
9.2.2   Analytical procedures   443
9.2.3   Recommended experimental procedure   444
9.3   HCl and particulate chloride   445
9.3.1   Sampling procedures   445
9.3.2   Analytical procedures   446
9.3.3   Recommended experimental procedures   449
9.4   Bromides   450
9.4.1   Sampling procedures   451
9.4.2   Analytical procedures   451
9.5   Halogenated hydrocarbons   452
9.5.1   Fluorocarbons   453
9.5.2   Chlorinated hydrocarbons   455
9.5.3   Brominated hydrocarbons   456
References   457

10   Remote monitoring techniques   463
      R. H. Varey

10.1 Introduction   463
10.2 Correlation spectroscopy   465
      10.2.1 Mode of operation   465
      10.2.2 Baseline drift, sensitivity and multiple scattering   467
10.3 Single wavelength lidar   469
      10.3.1 Principles of lidar   469
      10.3.2 Essentials of a practical system   471
      10.3.3 Signal processing   473
10.4 Differential lidar   478
      10.4.1 Basic methods   478
      10.4.2 Examples of practical systems   480
10.5 Laser safety   488
10.6 Long pathlength absorption spectroscopy (this section by A. M.
              Winer)   489
      10.6.1 Differential ultraviolet and visible absorption
              spectroscopy   489
      10.6.2 Fourier transform infrared spectroscopy   494
10.7 Meteorological measurements   495
      10.7.1 Meteorological measurements for pollution surveys   495
      10.7.2 Sodar   498
      10.7.3 Lidar measurements of the mixing layer   502
      10.7.4 Temperature profiles   503
10.8 The use of remote sensing in field studies   505
      10.8.1 Plume rise and dispersion   505
      10.8.2 Measurement of emission fluxes from point sources   510
      10.8.3 Multisource monitoring in an industrial area   515
10.9 Conclusions   518
      Acknowledgements   519
      References   519

11   Physico-chemical speciation techniques for atmospheric
              particles   523
      R. M. Harrison

11.1 Introduction   523
11.2 Speciation methods   524
      11.2.1 X-ray diffraction (XRD)   524
      11.2.1.1 *Phases identified in air by XRD*   525
      11.2.2 Method for sampling and XRD analysis of atmospheric
              particles   526
      11.2.3 Single particle techniques   529
      11.2.3.1 *Transmission electron microscope method for
              atmospheric particles*   529

11.2.4   Speciation of sulphuric acid and other particulate
sulphates   530
11.2.4.1   *Solvent extraction method for speciation of H_2SO_4,*
NH_4HSO_4 and $(NH_4)_2SO_4$ in ambient air   531
References   532

12   **Analysis of precipitation**   535
J. R. Kramer

12.1   Introduction   535
12.2   Sampling   536
    12.2.1   Siting   537
    12.2.2   Samplers   538
    12.2.2.1   *Container material*   541
    12.2.2.2   *Sample preservatives*   543
    12.2.3   Field procedures   544
12.3   Analysis   545
    12.3.1   Filtration   545
    12.3.2   Major ions   546
    12.3.2.1   *pH and other protolytes*   547
    12.3.2.2   *Calcium, magnesium, potassium and sodium*   549
    12.3.2.3   *Ammonia*   549
    12.3.2.4   *Sulphate*   550
    12.3.2.5   *Nitrate*   550
    12.3.2.6   *Chloride*   551
    12.3.2.7   *Specific conductance*   551
    12.3.2.8   *Consistency checks for major ions*   552
    12.3.3   Trace metals   553
    12.3.3.1   *Atomic absorption analysis*   555
    12.3.3.2   *Instrumental neutron activation analyses*   556
    12.3.4   Organics   557
    12.3.5   Other analyses   557
12.4   Concluding comment   557
References   558

13   **Low-cost methods for air pollution analysis**   563
H. W. de Koning

13.1   Introduction   563
13.2   General considerations   564
    13.2.1   An air monitoring network   564
    13.2.2   Operating conditions   564
    13.2.3   What to look for when selecting a method   566
    13.2.3.1   *Sensitivity*   566
    13.2.3.2   *Specificity*   566
    13.2.3.3   *Precision*   566
    13.2.3.4   *Stability of reagents*   566

	13.2.3.5 *Calibration*	567
	13.2.4 Sampling train	567
	13.2.5 Total volume of air to be measured	569
13.3	Selected methods for measuring air pollutants	570
	13.3.1 Sulphur dioxide	570
	13.3.1.1 *Lead sulphation candle or plate*	570
	13.3.1.2 *Acidimetric method*	570
	13.3.2 Nitrogen dioxide	571
	13.3.2.1 *Sodium arsenite method*	571
	13.3.3 Carbon monoxide	571
	13.3.3.1 *Detector tube method*	571
	13.3.3.2 *Instrumental method*	572
	13.3.4 Oxidant	572
	13.3.4.1 *Neutral buffered potassium iodide method*	572
	13.3.5 Suspended particulate matter	573
	13.3.5.1 *Dustfall*	573
	13.3.5.2 *High-volume (High-Vol) sampling method*	574
13.4	Additional considerations for selecting a low-cost air pollution measurement method	575
	13.4.1 Equipment requirement	575
	13.4.2 Calibration	576
	13.4.3 Record keeping	577
	References	578

14	**Planning and execution of an air pollution study**	**579**
	D. J. Moore	
14.1	Introduction	579
14.2	Objectives of the monitoring programme	581
	14.2.1 General	581
	14.2.2 Pollutant identification	581
	14.2.3 Source identification	581
	14.2.4 Economic assessment of damage versus control	582
	14.2.5 On-line plant control	582
	14.2.6 Control of future developments	582
	14.2.7 Receptor protection	583
	14.2.8 Detection of long-term trends	583
	14.2.9 Monitoring control	583
14.3	Effluent history from source to receptor	584
	14.3.1 General	584
	14.3.2 Source network	584
	14.3.3 Effluent control processes	584
	14.3.4 Effluent transport control procedures	585
	14.3.5 Atmospheric transmission	585
	14.3.6 The receptor/sink network	585

14.4   The monitoring network   586
      14.4.1   General   586
      14.4.2   Function monitoring   586
      14.4.3   Emission monitoring   587
      14.4.4   Plume monitoring   587
      14.4.5   Meteorological monitoring   588
      14.4.6   Damage monitoring   588
      14.4.7   Dose monitoring   589
      14.4.8   Monitoring ground-level concentration (GLC)   589
      14.4.9   Optical effects of pollutants   590
      14.4.10 Monitoring wet deposition   590
14.5   The design of pollution monitoring systems   591
      14.5.1   General   591
      14.5.2   Choice of minimum averaging period   591
      14.5.3   Choice of instruments   592
      14.5.4   Choice of mobile, fixed, transportable or combined sampling system   593
      14.5.5   How many pollutants should be monitored?   594
      14.5.6   Height and exposure of samplers   594
      14.5.7   Layout and spacing of instruments in fixed surveys   595
      14.5.7.1 *General*   595
      14.5.7.2 *Discrete source surveys – verification of dispersion models*   596
      14.5.7.3 *Discrete source surveys – statistics of incidence of various levels of pollution*   597
      14.5.7.4 *Area surveys (i.e. sites within the source area)*   597
      14.5.7.5 *Distant source surveys*   599
      14.5.7.6 *Global effects surveys*   599
      14.5.7.7 *Multi-purpose surveys*   599
      14.5.8   Mobile monitoring   600
      14.5.8.1 *Surface systems*   600
      14.5.8.2 *Airborne systems*   600
14.6   Data handling   602
      14.6.1   Data transmission   602
      14.6.2   Data storage   603
      14.6.3   On-line alarm/display systems   604
      14.6.4   On-line recognition of defective readings   604
14.7   Analysis of results   604
      14.7.1   General   604
      14.7.2   Availability of historical data   605
      14.7.2.1 *Source inventories and characteristics*   605
      14.7.2.2 *Climatological data*   605
      14.7.2.3 *Topographical information*   605
      14.7.2.4 *Experience of similar source networks or other information (e.g. epidemiological) on damage/dosage relations for the pollutants under investigation*   606

14.7.3	Elimination of erroneous readings	606
14.7.4	Statistics	606
14.7.4.1	*General*	606
14.7.4.2	*Mean values over specified averaging periods*	607
14.7.4.3	*Frequency distributions*	607
14.7.4.4	*Diurnal or annual variations*	607
14.7.4.5	*Mean values in different weather situations and/or wind directions*	608
14.7.4.6	*Correlations and regression analysis*	608
14.7.5	Evaluation of physical models	609
14.7.5.1	*General*	609
14.7.5.2	*Near field (up to 25 km)*	609
14.7.5.3	*Medium range (20–250 km)*	610
14.7.5.4	*Long range (>250 km)*	610
14.8	Examples of monitoring networks and data presentations	611
14.8.1	General	611
14.8.2	Discrete source surveys	611
14.8.2.1	*US power plant studies*	611
14.8.2.2	*CEGB Midlands Region studies of SO_2 around power stations*	611
14.8.3	Area surveys	615
14.8.3.1	*Urban surveys*	615
14.8.3.2	*Regional surveys*	615
14.8.4	Distant sources and global effects	617
	Acknowledgements	619
	References	619
15	**Quality assurance in air pollution monitoring**	621
	A. Apling	
15.1	Quality and quality assurance	621
15.2	Definitions	621
15.3	Elements of the monitoring chain	622
15.4	Site location and character	622
15.5	Sampling line integrity	623
15.6	Instrument performance	623
15.7	Calibration	624
15.8	Discussion and further checks	625
	References	626
Index		628

Preface to second edition

When the first edition of the *Handbook of Air Pollution Analysis* was published in 1977 it was probably unique as a manual providing a comprehensive working knowledge of the theory and practice of air pollution analysis. The level of sales and frequency of citation in the literature suggest that the *Handbook* has fulfilled this role successfully. Air pollution analysis is a highly active subject area and inevitably there have been many developments since preparation of the first edition. In the second edition, we have attempted both to update the literature reviews and methods of analysis in the light of recent advances in knowledge, and also to restructure and extend the book to reflect current trends in the subject such as, for example, the present intense interest in rainwater chemistry. In doing so, we have maintained the policy set by the first edition of reviewing up-to-date developments and providing literature references for those wishing to read further, while also giving detailed descriptions of well-established analytical methods that may be used with confidence.

Chapters on general sampling methods, particulate pollutants, metal analysis, nitrogen and sulphur compounds, secondary pollutants and planning and execution of an air pollution study have been updated from the first edition, and edited to remove overlaps caused by restructuring the book. In most instances the updating has been carried out by the original author, but chapters written by Steve Hrudey and Tony Cox in the first edition have been updated and restructured by other authors and we are happy to record our thanks to the original authors for their most valuable contributions upon which the chapters in this book are based. The chapters on hydrocarbons and carbon monoxide, halogen compounds and remote sensing, although bearing the same titles as chapters in the first edition, have been entirely rewritten and provide a fresh and thoroughly up-to-date view of these topics. Chapters on air pollution meteorology and air pollution chemistry are entirely new to the book and reflect our belief that good air pollution analysis must always be based upon a sound understanding of atmospheric science, of which relevant aspects are described by expert authors in these chapters. Also new are chapters on physico-chemical speciation methods and rainwater analysis, introduced as an acknowledgement of current focuses of interest in the air pollution field, and low cost methods of air pollution analysis in which basic but reliable procedures, applicable especially in Third World countries where resources are strictly limited, are described. Finally, a chapter on quality assurance in air pollution monitoring provides a guide to those who need to establish the reliability of their methodology. Thus a substantial proportion of the second edition is comprised of entirely new material.

The contributing authors have been chosen because of their depth of knowledge and experience in air pollution work, and we are confident that this is reflected in a *Handbook* which will find very wide application wherever air pollution analysis is practised.

Roy M. Harrison
Roger Perry
February 1985

Readers are recommended to follow all the usual laboratory safety precautions. While care has been taken to ensure that the information in this book is correct, neither the authors nor the publisher can accept responsibility for any outcome of the application of methods and procedures outlined in this book.

Contributors

A. Apling   BSc, PhD
Air Pollution Division
Warren Spring Laboratory
Gunnels Wood Road
Stevenage
Hertfordshire SG1 2BX
UK

H. W. de Koning   DSc
Environmental Pollution
Division of Environmental Health
World Health Organization
Geneva
Switzerland

R. M. Harrison   PhD
Department of Chemistry
University of Essex
Wivenhoe Park
Colchester CO4 3SQ
UK

P. W. W. Kirk   BSc, MSc, PhD, DIC, C Chem, MRSC
Department of Civil Engineering
Imperial College
London SW7 2BU
UK

J. R. Kramer
Professor in Geochemistry
Department of Geology
McMaster University
Hamilton
Ontario L8S 4M1
Canada

J. N. Lester   B.Tech, MSc, DIC, PhD, MIPHE,
Department of Civil Engineering
Imperial College
London SW7 2BU
UK

A. E. McIntyre BSc, PhD, DIC, MIWES
Consultants in Environmental Sciences Ltd
Yeoman House
63 Croydon Road
London SW20 7TW
UK

D. J. Moore BSc, PhD
Central Electricity Research Laboratories
Kelvin Avenue
Leatherhead
Surrey KT22 7SE
UK

R. Perry BSc, PhD, FIPHE, MICOPC, FRSC
Department of Civil Engineering
Imperial College
London SW7 2BU
UK

C. A. Pio PhD
Departamento de Ambiente
Universidade de Aveiro
3800 Aveiro
Portugal

R. H. Varey BSc, PhD
Central Electricity Research Laboratories
Kelvin Avenue
Leatherhead
Surrey
KT22 7SE
UK

A. M. Winer PhD
Statewide Air Pollution Research Centre
University of California
Riverside
CA 92521
USA

1 General sampling techniques*

C. A. PIO

1.1 Sampling goals and requirements

1.1.1 Ambient sampling

1.1.1.1 General objectives

Ambient air sampling may be considered the collection of air samples in any unconfined location exposed to the atmosphere. Within this broad classification an infinite variety of air sampling locations and air mass quality will exist, from samples collected 0.5 m from the ground in a busy parking lot to samples collected from a free floating balloon over the Atlantic Ocean. Because of the variety of sampling schemes which may be used to obtain 'ambient' air samples, the objectives to be achieved by the sampling programme and subsequent analysis must be thoroughly and clearly defined before proceeding further.

Among the objectives which may commonly form the basis for ambient air sampling programmes are [1–3]:

(a) The determination of community air quality as related to local health, social and environmental effects
(b) The determination of the influence of specific emission sources or groups of sources on local air quality
(c) The generation of information to aid in planning overall pollution control and industrial and municipal zoning strategies
(d) Research into topics such as procedures for identification of emission source contributions or mechanisms of air pollutant reaction and dispersion.

In order to achieve one or more of the stated objectives, the sampling system must be considered as the 'receptor' for the influence or effect to be measured. All subsequent planning of the sampling system to be used must be performed with constant referral, at each decision step, back to the consequence that the decision will have on the success of the sampling system as a valid 'receptor'.

It is often a useful exercise to seek data from past or current air monitoring systems [4–7] and determine the advantages and shortcomings of such data when manipulated in the manner necessary to achieve the study objectives. The

* A revision of the chapter by S. E. Hrudey, *Handbook of Air Pollution*, first edition.

shortcomings of a current data source may often suggest shortcomings in the proposed sampling system which can be rectified.

In conjunction with satisfying the study objectives, the definitive planning of ambient air sampling systems requires consideration of meteorological factors, sampling site criteria and sample scheduling. These considerations will be dealt with in turn.

1.1.1.2   *Meteorological considerations*

Meteorology is the study of the physics and geography of atmospheric phenomena. The nature and changes of the atmosphere constitute what we perceive as weather [8].

Atmospheric phenomena play an important role in the determination of ambient air quality. Diurnal and seasonal fluctuations in source emissions tend to be reflected in ambient air pollution levels but the diurnal and seasonal variations in meteorological conditions superimpose an effect on those due to emission variations. Clearly, the objectives of a sampling programme will not normally be attained if due consideration is not given to the assessment of meteorological information.

Meteorological parameters have varying degrees of influence on ambient air quality. The parameters of major significance are atmospheric stability, wind speed and direction and precipitation. Other influences may be attributed to temperature, humidity and solar radiation.

The understanding of atmospheric stability is based on the knowledge that air, being a compressible gas, will cool upon expansion and will heat upon compression. Because of the downward pressure of the air mass above the earth's surface, a parcel of air at the earth's surface experiences a greater atmospheric pressure than a parcel of air experiences at some height above that point. Thus, if a parcel of air at a relatively high pressure at the earth's surface were moved to a higher altitude, it would experience a lower pressure which would allow it to expand. As the parcel of air expands, it cools. As a result of this condition and the fact that the atmosphere is heated from the surface of the earth outwards, the air temperature should be expected to decrease with increasing altitude. The rate of decrease of temperature with altitude is called the lapse rate. The dry adiabatic lapse rate is the rate of decrease in temperature of a parcel of unsaturated air as it moves upward without exchanging heat with the surrounding air and is equal to 9.8° C per 1000 m. If the air parcel is saturated, water will tend to condense as cooler temperatures are encountered. Since the condensation will release latent heat, the effective lapse rate due to expansion cooling will be lessened. The saturated adiabatic lapse rate is variable but is consistently less than 9.8° C per 1000 m.

Atmospheric stability depends on the lapse rate in that under clear sunny conditions the air near the earth's surface is heated, as the surface is warmed by solar radiation, whereupon air expands, becomes buoyant and rises. The air will continue to rise until it reaches air at the same density, which for a given altitude, will be at the same temperature. Under these conditions, the lapse rate would normally be positive so that the rising air parcel will continue to encounter cooler air and will continue to rise until it reaches equilibrium with its surroundings. If, however, the lapse rate is less than the adiabatic rate or is inverted (temperature increase with altitude), the air parcel will not rise above its current level because it would tend to lose temperature with altitude faster than the temperature change with altitude in the surrounding air. As a result, there would be no buoyant force to encourage further rising. This condition is stable, in that the air mass at the surface is prevented from mixing vertically.

The implications of stability conditions to the quality of surface air receiving consistent emission sources are clearly important. The lapse rate itself is very dependent upon the thermal capacity and physical properties of the ground [3]. Lapse rates will be found to vary from urban to suburban to rural environments. Details of the causes and air pollution effects of atmospheric stability have been discussed by Scorer [9].

Wind speed and direction will directly affect the movement and dispersion of pollutants from emission sources within a given study area. Ambient pollution levels have been found to be inversely related to wind speed [10]. Wind speed and direction are normally detected by various types of wind vane, which rotate to face into the wind. The axial rotation of the wind vane is transmitted to a recording device by one of two basic methods. A potentiometer may be used in which the angular position of the wind vane corresponds to a contact on a variable resistance. As the vane turns, the contact moves and varies the resistance to the recording device. The other system uses position motors. These are two or more small motors electrically interconnected so that the rotation imposed by the vane on one motor activates another motor to drive the recorder.

Directional devices are useful in concert with monitoring instruments in that a control system can be designed such that the sampling system samples only when the wind is coming from a specified direction. In practice, an intake sector must be defined, allowing sampling only when the wind direction is within a range of so many degrees either side of the direction of interest. Also, some damping of time delay must usually be built into the system to avoid frequent starts and stops due to wind direction fluctuation [11]. An alternative approach to directional sampling involves the positioning of fixed and continuous collection openings facing four or more directions. This method is discussed for dustfall collection in Chapter 4.

Precipitation can take the form of rain, sleet, snow, hail and various combinations thereof. The general effect of precipitation is the scavenging of particulates and gases from the atmosphere [12–15]. The net result of precipitation may be the removal of pollutants from the atmosphere before dispersion takes place. In such cases, a high concentration of the pollutant might be measured during the initial stages of rainfall.

Other meteorological factors which should be considered are temperature, humidity and sunlight. Temperature tends to have its main effect due to the resulting changes in domestic heating requirements during colder weather. Shorter term effects might be determined by continuous temperature monitoring. For example [3], turbulence within an inversion layer might bring higher level pollutants down to ground level overnight. A corresponding rise of temperature due to the higher temperature of the inverted air mass along with a rise in pollutant concentration would indicate such an effect.

Humidity can affect air quality in a variety of ways [3]. Low humidity can result in increased suspended particulate concentrations due to suspension of surface dust. High humidity, as exhibited in fog conditions, can block solar heating of the ground surface and thereby prolong the life of inversion layers. Air pollution incidents with increased morbidity and mortality were generally associated with low temperature and fog [16].

Photochemical reactions and generation of secondary pollutants are generally dependent on the solar radiation exposure available [17]. The degree of cloud cover, which will affect surface solar radiation is normally recorded based on observation and may often be obtained from local weather offices on a daily basis.

1.1.1.3   *Sampling site criteria*

Sampling site criteria generally fall into two classes. Firstly, there are criteria necessary for the proper siting of individual sampling systems in order that each site should provide a true representation of the receptor defined in the study objectives. The other class of criteria refer to the location of sites relative to one another to form sampling networks which will provide the area-wide data required to achieve the study objectives.

Many sources have discussed proper siting criteria for individual sampling systems [1–3, 18–20]. However, no comprehensive set of rules has been adopted as standard procedure, because of the infinite diversity that will be encountered in choosing new sampling sites for a specific purpose.

Some general rules of thumb include the following points.

Sampling inlets should generally be more than 2 m above ground level. The maximum height is determined by the objective of the survey but should be consistent from one site to another within a given network designed for a common objective.

Sampling sites should not be located in the lee of major obstructions such as tall buildings. A general rule is that the top of obstructions should subtend less than a 30° angle with the horizontal at the sampling point.

The sample intake should not be exposed to contamination from specific localized sources (i.e. a chimney on a roof top). For general area air quality monitoring, the site should not be directly downwind from major emission sources, such as motorways, parking lots or industrial stacks.

The site must be accessible and yet secure from tampering.

The site must provide an adequate, reliable, power supply to run the sampling equipment as necessary.

Within these general rules, wide variations in sampling system siting and set-up are possible [19, 21]. Therefore, once the general siting criteria are satisfied, consistent with the objectives of the study, all sites should be made to conform with one another in as much detail as possible, if valid comparisons are to be made using the collected data.

Location criteria for spacing network sites are very dependent on the specific objectives of the study. Statistical methods based on diffusion models and historical meteorological data have frequently been recommended [22–24]. Various network configurations have been used including [13, 20, 25, 26]:

(a) Location of sites on concentric circular lines centred on the area of interest
(b) Location of sites on typical trajectories of surface winds
(c) Location of a random heavy density of sites in the core of interest with random open spacing further out
(d) Location of sites on an equally spaced grid pattern.

The actual choice of a sampling network pattern is often dictated by financial constraints and local conditions. Where flexibility exists and there is scope for putting the data into an air quality simulation model, preliminary work with the model using artificial data from possible sampling locations will often indicate efficient choices.

Meteorological and topographical factors may play an important role in defining the optimum sampling network. An emission source inventory, even in rough form, will often be required to locate sampling sites so that they will generate the information necessary to satisfy the study objectives.

Most networks require background locations for comparative purposes. These sites should normally be free from the influence of any major sources. Enough sites should be provided so that when the wind blows from the central core to a background site, there are still other background sites located upwind of the central core, so that upwind–downwind comparisons can be made.

1.1.1.4 *Sample scheduling*

Upon locating ambient air sampling sites which will provide representative samples, free from undesirable interference, the choice remains of how frequently and for what duration to obtain samples. The frequency and duration of sampling are very much dependent upon the objectives of the study and the nature of the pollutants and their emission sources.

The variation in requirements for sampling duration is evidenced by the variety of time periods specified for various pollutant criteria in the US National Air Quality Objectives (Table 1.1). The variability is due in part to the nature of a pollutant's effect upon the receptor. For example, semi-annual averages provide useful information for lead concentrations because the effect of lead on receptor organisms is cumulative and long term. Sulphur dioxide (SO_2), on the other hand, can wreak its harmful effects upon receptor organisms over short term fumigation periods and so mainly short period, high frequency information is relevant.

Saltzman [27] considered a model for air pollutant concentration fluctuations in an attempt to rationalize sampling time considerations. The model is based on a series of sine functions of varying periodicity superimposed upon an arithmetic

Table 1.1   US National air quality standard reference methods

Pollutant	Average time	Reference method	Principle of detection
SO_2	3 h, 24 h, annual	Pararosaniline	Colorimetric
Particulate matter	24 h, annual	Hi-Vol sampler	Gravimetric
CO	1 h, 8 h	Non-dispersive infrared spectrometry	Infrared
Ozone	1 h	Gas-phase O_3–ethylene reaction (calibrated against uv Photometry)	Chemiluminescence
Hydrocarbons (non-methane)	3 h	Gas-chromatography	Flame ionization
NO_2	annual	Gas-phase chemiluminescence	Chemiluminescence
Pb	90 days	Hi-Vol/Atomic Abs. spectrometry	Atomic absorption

mean concentration. Varying periodicity for pollutant fluctuations would represent factors such as hourly fluctuations due to traffic emissions, daily variations due to industrial emissions and seasonal fluctuations due to domestic heating sources. For any sampling period other than continuous sampling, the values obtained from individual samples will not fully reflect the variance of the true concentration fluctuations about the long term mean value. Application of Saltzman's model indicated that concentration fluctuations with a cycle period shorter than the sampling period were almost completely attenuated (averaged out) by looking strictly at the periodic sample results. However, when the concentration fluctuation cycle period becomes 2.25 times the sample period, 50% of the true concentration variance is observed and when the concentration fluctuation cycle period becomes 5.6 times the sample period, 90% of the true concentration variance is observed. The important application of the Saltzman analysis is in providing an estimate of the degree to which a particular sampling programme will reflect the actual pollutant concentration variations.

One further step in this analysis is the application of the pollutant fluctuation model to a biological receptor model. Since a biological receptor will tend to eliminate a given pollutant from its system at the same time it is ingesting the pollutant, the fluctuations in the effective pollutant concentration within the receptor organ will be attenuated in comparison with the ambient concentration fluctuations. Saltzman proposes that if the required information could be gathered to define the factors needed for a valid biological model, the model could be used to choose rationally the sampling period. Provided that the chosen sampling period consistently provided less attenuation of the true pollutant concentration than would be provided by the biological receptor, then the maximum useful biological information will be obtained using the chosen sampling period. As an example, based on a biological half life of 20 min for SO_2 and a pollutant retention function value of 0.5, a sampling period of 20 min would be expected from the model, to give the maximum useful biological information.

Sampling schedule possibilities have been discussed by Akland [28]. The modified random system, commonly used in air sampling networks, calls for sampling intervals of fixed length (i.e. weekly). During the sampling interval one day is randomly chosen for sampling. The random choice may be restricted to provide for approximately the same number of sampling occurrences on each day of the week over the year. The systematic system calls for starting the sampling programme on a day picked at random, followed by sampling at fixed intervals, other than 7 days, from that day onward. Comparing the relative precision of the two methods for hypothetical sampling programmes drawn from an existing data base which had daily samples, Akland found the systematic approach to be consistently more precise.

The precision of air pollutant sampling that may be expected, given that pollutant concentrations usually follow a log normal distribution, can be calculated [29–31]. Hunt [32] provides the formulae to calculate the confidence interval about the geometric mean.

Confidence interval bounds at $1 - \alpha$ confidence level

Lower bound $= \bar{X}_{geo} - m_1 \bar{X}$

Upper bound $= \bar{X}_{geo} + m_2 \bar{X}$

where

$$m_1 = 1 - \exp\left[-t_{1-\alpha/2} \frac{S_{log}}{n^{1/2}} \left(1 - \frac{n}{N}\right)^{1/2} \right]$$

$$m_2 = \exp\left[t_{1-\alpha/2} \frac{S_{log}}{n^{1/2}} \left(1 - \frac{n}{N}\right)^{1/2} \right]$$

$\bar{X}_{geo}$ = geometric mean

and

$1 - \alpha$ = level of confidence (i.e. 95%, 99%)

n = number of samples taken during given interval

N = maximum possible number of samples during given interval (i.e. $N = 30$ daily samples in average month)

S_{log} = Standard deviation (S.D.) of the logarithms of air pollutant measurements (given by the log of the geometric S.D.)

$t_{1-\alpha/2}$ = the t statistic with $n-1$ degrees of freedom.

Thus if the overall study objectives require a certain level of precision to ensure the detection of subtle effects, provided an estimate of the geometric S.D. of the pollutant concentrations can be obtained, it is possible to calculate the number of samples that are necessary, within a given time interval, to provide the required precision.

1.1.2   Source sampling

1.1.2.1   *General objectives*

Source sampling in air pollution work may be considered as the collection of airborne pollutants before emission to and dilution by the atmosphere. Emission sources are usually categorized as stationary or mobile sources. Stationary sources include various emissions from industrial plants such as steel mills, pulp mills, chemical plants and oil refineries, from municipal sources such as electricity generating plants and refuse incinerators and from domestic sources such as house chimneys. Mobile sources include emissions from petrol and diesel engined vehicles and from aircraft.

Emission sources are studied for several reasons, including:

(a) Determination of the mass emission rate of particular pollutants from a particular source and how it is affected by process variations
(b) Evaluation of control devices for the reduction of pollutant emissions
(c) Data gathering of emissions from several sources for input to air quality management models.

The approaches and requirements for the two source categories are basically different and will be discussed in turn.

1.1.2.2   *Stationary source sampling*

Planning and preparation

The requirements for source sampling often necessitate long sampling periods. As a result, source sampling requirements can often be met only by careful advance planning so that the time spent at the sampling location is efficiently used.

Proper planning requires a thorough knowledge of the process to be sampled. The nature of process conditions and their effect upon emission parameters should be determined by discussion with a knowledgeable person in charge of the facility. The frequency of any process cycles should be considered and the reported level of emissions from similar processes studied [33, 34].

The nature of the emission source for the purposes of planning source testing can be considered by applying the classification scheme of Achinger and Shigehara [35]. These authors specify two requirements for valid source testing. Firstly, the sample should accurately reflect the true magnitude of the pollutant emission at a specific point in a stack at a specific instant of time. This requirement is determined by the design of the sampling instrument and is considered in Section 1.2. The second requirement is to obtain enough measurements varied in space and in time such that their combined results will accurately represent the entire source emission. Satisfying the latter requires consideration of the fluctuations of the source emission both in space, across the stack diameter, and in time. Achinger and Shigehara classify sources into categories with the four possible combinations of steady conditions (no variation with time) and uniform conditions (no variations in space).

Class I

Sources in this classification are both steady and uniform. Theoretically, only one measurement need be taken as this will represent the whole stack cross-section for the whole period of the time of steady operation. The example specified for this condition is sampling for a gaseous pollutant from a turbulent gas stream.

Class II

Sources in this classification are steady but non-uniform. For this type of operation the collection of a composite sample at several locations on the stack cross-section will produce results representing the whole stack for the period of steady operation. The example specified for this condition is sampling for particulates at a large continuous feed coal fired power station.

Class III

Sources in this classification are unsteady but uniform. In this case, only one sample need be used, but sampling must take place for the entire cycle of a cyclic operation or for as long as possible in a non-cyclic operation. The example specified for this condition is sampling a gaseous pollutant from a cyclic operation with turbulent stack flow.

Class IV

Sources in this classification are both unsteady and non-uniform. The sampling approach to be taken depends upon the nature of the source variability with time. All measured variables may vary proportionally or non-proportionally with time. Furthermore, the non-proportional variations may be reproducible or non-reproducible.

If all the parameters vary proportionally with time, individual rather than composite samples must be taken. Simultaneous sampling at a reference location and at the specified sampling points is recommended. The results for the individual sample points can then be adjusted back to the original point in time by applying a factor determined by comparing the simultaneously measured reference point value with the initial reference point value.

If stack parameters vary non-proportionally with time, but vary over reproducible short cycles, then complete cycles may be sampled at each sample location. Further, composite samples are valid provided the same number of cycles are sampled at each location.

If stack parameters vary non-proportionally with time and over long or erratic cycles (non-cyclic), then all measurements must be made simultaneously over the entire cycle. This requirement becomes impossible for an adequate number of sample points in most cases. A recommended compromise is the use of two to four statistically random sites for simultaneous measurement and corresponding caution in determining statistical confidence in the results.

Sampling site selection is the next major step in conjunction with classifying the nature of the source emission. The criteria used in site selection must include:

(a) Safety of the location for the test personnel
(b) Relationship to the points of particular interest (i.e. at the pollution control device, for device efficiency testing)
(c) Availability of a platform for men and equipment
(d) Accessibility of the platform to the men and equipment
(e) Access to the stack interior from a suitable port
(f) Provision of power supply for sampling equipment, and
(g) Satisfaction of flow disturbance criteria.

Many of the above criteria will have to be compromised as the ideal sampling location seldom exists. However, safety considerations cannot be compromised as stack sampling is a hazardous undertaking even under good conditions. Inability to satisfy safety criteria is sufficient reason to abandon a given site until modifications to allow safe operations can be made.

Various criteria have been specified for avoiding flow disturbances [36–41]. Generally, a sampling location in a vertical flue sited eight flue equivalent diameters* downstream and two flue equivalent diameters* upstream from a flow disturbance such as a bend, inlet or outlet, is considered good. Since many emission sources will not provide a site meeting this criteria, Fig. 1.1 [36, 38] has been developed to compensate for non-ideal sampling locations with increased sample points across the stack cross-section.

The location of sampling points at the centroids of equal areas across the stack cross-section is recommended. For a rectangle, sample points are located at the centroids of smaller equal area rectangles (Fig. 1.2) and for a circular cross-section, at the centroids of equal area annular segments (Fig. 1.3). The location of the centroids for the annular segments of Fig. 1.3 are summarized in Table 1.2 for various numbers of sample points on a stack diameter.

Data requirements

Upon locating a suitable sampling site and classifying the source as to the sampling approach required, certain basic physical measurements must be performed on the stack gas [36, 38]. These are:

(a) A stack gas composition analysis by Orsat apparatus (primarily for combustion sources)
(b) A stack gas moisture determination
(c) A stack gas temperature determination
(d) A stack gas pressure determination
(e) A stack gas velocity determination.

The stack gas Orsat analysis [36–38, 42–45] may be obtained by drawing a

---

* Flue (stack) equivalent diameter = 4 [area of flue cross-section/perimeter of flue cross-section] = actual diameter for circular flue.

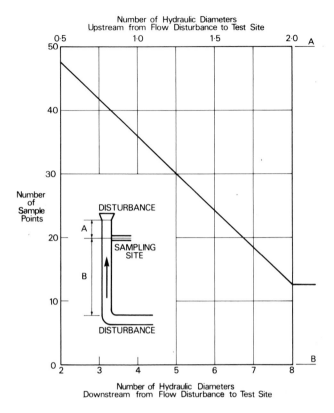

Fig. 1.1   Number of sample points. (After [36].)

small gas sample through a sample probe into the Orsat analyser using a hand operated or hydraulic aspirator, or a small electric pump. The stack gas is bubbled through the distilled water in the Orsat levelling bottle (Fig. 1.4) for 10 min. The stack gas sample is then taken with the Orsat apparatus and analysed for CO_2, O_2, CO, and N_2 by difference. The dry molecular weight of the stack gas is calculated from the Orsat volumetric analysis with the formula

$$M_D = (0.44)(\%CO_2) + (0.28)(\%CO) + (0.32)(\%O_2) + (0.28)(\%N_2)$$

where M_D = dry molecular weight of stack gas.

The stack gas moisture determination may be obtained by sampling a known volume of air and condensing the water vapour in ice cooled condensers or by measurement of the wet and dry bulb temperatures of the stack gas [36–38, 42–45]. In practice, the former method is performed in conjunction with the actual sampling for many of the available source sampling procedures, while the latter method may be performed before testing to provide data for necessary calculations to control the sampling rate.

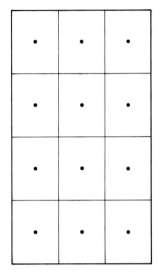

Fig. 1.2   Cross-section of rectangular stack divided into twelve equal areas showing location of traverse points at centroid of each area. (After [36].)

KEY

○   Information sample point for velocity and temperature traverses only

●   Regular sample points

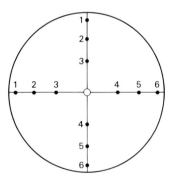

Fig. 1.3   Cross-section of circular stack divided into twelve equal areas showing location of traverse points at centroid of each area. (After [36].)

Table 1.2 Percentage of circular stack diameter from inside wall to traverse point (after [36])

Traverse point number along diameters[a]	Number of traverse points on a single diameter[b]											
	2	4	6	8	10	12	14	16	18	20	22	24
1	14.6	6.7	4.4	3.3	2.5	2.1	1.8	1.6	1.4	1.3	1.1	1.1
2	85.4	25.0	14.7	10.5	8.2	6.7	5.7	4.9	4.4	3.9	3.5	3.2
3	–	75.0	29.5	19.4	14.6	11.8	9.9	8.5	7.5	6.7	6.0	5.5
4	–	93.3	70.5	32.3	22.6	17.7	14.6	12.5	10.9	9.7	8.7	7.9
5	–	–	85.3	67.7	34.2	25.0	20.1	16.9	14.6	12.9	11.6	10.5
6	–	–	95.6	80.6	65.8	35.5	26.9	22.0	18.8	16.5	14.6	13.2
7	–	–	–	89.5	77.4	64.5	36.6	28.3	23.6	20.4	18.0	16.1
8	–	–	–	96.7	85.4	75.0	63.4	37.5	29.6	25.0	21.8	19.4
9	–	–	–	–	91.8	82.3	73.1	62.5	38.2	30.6	26.1	23.0
10	–	–	–	–	97.5	88.2	79.9	71.7	61.8	38.8	31.5	27.2
11	–	–	–	–	–	93.3	85.4	78.0	70.4	61.2	39.3	32.3
12	–	–	–	–	–	97.9	90.1	83.1	76.4	69.4	60.7	39.8
13	–	–	–	–	–	–	94.3	87.5	81.2	75.0	68.5	60.2
14	–	–	–	–	–	–	98.2	91.5	85.4	79.6	73.9	67.7
15	–	–	–	–	–	–	–	95.1	89.1	83.5	78.2	72.8
16	–	–	–	–	–	–	–	98.4	92.5	87.1	82.0	77.0
17	–	–	–	–	–	–	–	–	95.6	90.3	85.4	80.6
18	–	–	–	–	–	–	–	–	98.6	93.3	88.4	83.9
19	–	–	–	–	–	–	–	–	–	96.1	91.3	86.8
20	–	–	–	–	–	–	–	–	–	98.7	94.0	89.5
21	–	–	–	–	–	–	–	–	–	–	96.5	92.1
22	–	–	–	–	–	–	–	–	–	–	98.9	94.5
23	–	–	–	–	–	–	–	–	–	–	–	96.8
24	–	–	–	–	–	–	–	–	–	–	–	98.8

[a] Points numbered from inside wall toward opposite wall.
[b] The total number of points along two diameters would be twice the number of points along a single diameter.

The wet and dry bulb method requires withdrawing the stack gas through a sample system (Fig. 1.5) provided with a psychrometer. By determining the wet and dry bulb temperatures (the wet bulb must be lower), it is possible to use a hygrometric (psychrometric) chart [46] to obtain the specific humidity ω in kilograms of H_2O per kilogram of dry air. The volumetric humidity θ in m^3 of water vapour per m^3 of dry air can be obtained from the formula

$$\theta = \frac{28.8 \text{ kg (kmol dry air)}^{-1}}{18.0 \text{ kg (kmol H}_2\text{O)}^{-1}} \omega$$

$$= (1.6)\omega \frac{\text{kmol H}_2\text{O}}{\text{kmol dry air}}$$

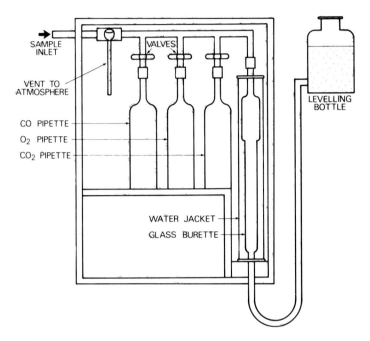

Fig. 1.4   Orsat apparatus.

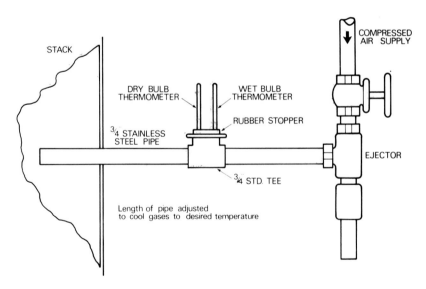

Fig. 1.5   Moisture measurement. (After [36].)

and because a molar ratio is equivalent to a volumetric ratio using Amagat's law

$$= (1.6)\omega \frac{\text{m}^3 \text{ H}_2\text{O}}{\text{m}^3 \text{ dry air}}$$

Therefore, the moisture content of the stack gas as a volume fraction is given by

$$B_{wo} = \frac{1.0\omega}{1 + 1.6\omega}$$

For the condensation method, the volume of liquid collected is measured and converted into the equivalent volume of water vapour at standard conditions using

$$V_w = V_c \frac{\rho_{H_2O}}{M_{H_2O}} \frac{RT}{P}$$

where

V_c = volume of liquid collected, litre

ρ_{H_2O} = density of water, 1 kg litre^{-1}

M_{H_2O} = molecular weight of water, 18 kg kmol^{-1}

R = ideal gas constant 8.31×10^{-2} bar m^3 (kmol K)$^{-1}$

T = standard temperature, 293 K, and

P = standard pressure, 1.01 bar

$V_w = (1.35 \text{ m}^3 \text{ litre}^{-1})V_c$

The moisture content of the stack gas as a volume fraction is then given by

$$B_{wo} = \frac{(1.35 \text{ m}^3 \text{ litre}^{-1})V_c}{(1.35 \text{ m}^3 \text{ litre}^{-1})V_c + V_m}$$

where V_m = volume of dry gas through the meter, at standard conditions.

The stack gas temperature determination can be relatively easy depending upon temperature range encountered. For stack temperatures up to 400° C mercury-in-glass thermometers may be used directly. Maximum reading thermometers which retain the stack temperature while the thermometer is pulled from the stack offer an advantage. For higher temperatures or for conducting temperature traverses across the stack to assure uniformity, direct reading thermocouples or thermistors may be used.

The relative stack gas pressure determination can be included in the determination of stack gas velocity by determining the static pressure in the stack with the standard pitot tube. A barometer should be used to establish the atmospheric pressure at the stack sampling location if conditions are likely to vary markedly from standard conditions.

The average stack gas velocity determination is obtained by performing a pitot tube traverse [36, 38, 42–44, 47]. The velocity head is measured at several points

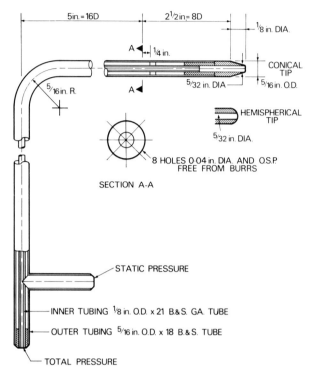

Fig. 1.6   Standard pitot tube details. (After [37].)

in the stack cross-section, chosen according to the discussion on location of sampling points (Figs. 1.2, 1.3). Either a standard pitot tube (Fig. 1.6) or a Staubscheibe, or type S pitot tube (Fig. 1.7) may be used, but the latter must be calibrated against a standard pitot tube.

The velocity head at each point must be determined by directing the pitot tube opening along the axis of the stack into the oncoming flow. The velocity head is read from an inclined gauge manometer indicating the difference between the dynamic pressure of the flow and the static stack gas pressure. If the velocity head is found to be negative at one or more sampling points, the pitot tube should be checked. If it is found to be functioning reliably and the negative reading is correct, the site is unsuitable for further sampling [38]. Likewise, if the direction of flow is found to be more than 30° from the axis of the stack, as determined by rotating the standard pitot tube in 5° increments to determine the maximum and minimum velocity heads, the site is unsuitable and another site consistent with the other specified criteria should be chosen.

The average velocity is obtained by calculating the square roots of the velocity pressures at the various sample points and applying the formula

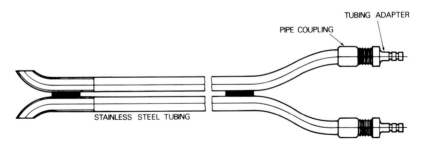

TUBING ADAPTER

PIPE COUPLING

STAINLESS STEEL TUBING

Fig. 1.7 Type S pitot tube (special). (After [37].)

$$(V_s)_{av} = K_p C_p (\sqrt{\Delta p})_{av} \sqrt{\left(\frac{(T_s)_{av}}{P_s M_s}\right)}$$

where $(V_s)_{av}$ = average stack gas velocity, m s^{-1};

C_p = pitot tube coefficient, determined by calibration for an S type or = 1 for a standard pitot tube;

K_p = 4.08 m s^{-1} (kg kmol^{-1} K^{-1})$^{\frac{1}{2}}$, a unit conversion factor for the units specified herein;

$(\sqrt{\Delta p})_{av}$ = the average of the roots of the velocity heads in mb (not equal to the root of the average velocity head);

$(T_s)_{av}$ = average absolute stack gas temperature, K;

P_s = absolute stack gas pressure, bar

and M_s = molecular weight of the stack gas on a wet basis, kg kmol^{-1}.

The molecular weight of the stack gas on a wet basis is determined from

$$M_s = M_D(1 - B_{wo}) + 18 B_{wo}$$

where M_D = dry molecular weight of the stack gas (kg kmol^{-1}) calculated from the Orsat analysis;

and B_{wo} = volumetric moisture fraction calculated from the moisture determination.

Finally, the volumetric emission rate for the stack may be calculated by applying the formula

$$Q_s = 3600 \, (V_s)_{av} A_s$$

where Q_s = volumetric emission rate at wet stack conditions in m^3 h^{-1};

$(V_s)_{av}$ = average stack gas velocity in m s^{-1};

A_s = stack area at the location of the velocity traverse, in m^2.

The volumetric emission rate corrected to standard conditions on a dry basis may be calculated from

$$Q_{std} = 3600 \ (1 - B_{wo})(V_s)_{av} A_s \left[\left(\frac{293 \ K}{(T_s)_{av}} \right) \left(\frac{P_s}{1.01 \ b} \right) \right]$$

with the symbols as defined previously.

After obtaining the required physical information, several sampling procedures may be used for aerosols and gases. Some of these techniques are discussed in Sections 1.2.2.4 and 1.2.3.4.

When sampling has been successfully completed, the analysis will provide the information to calculate the concentration of a given pollutant in the stack emission. Given the pollutant concentration and the volume emission rate determined during sampling, the total mass emission rate can be calculated as follows

$$M_s = Q_{std} C_s$$

where

M_s = pollutant mass emission rate, g h^{-1}

Q_{std} = volumetric emission rate corrected to dry standard conditions, m^3 h^{-1}

C_s = concentration of pollutant, measured at dry standard conditions, g m^{-3}.

Finally, it is often useful in assessing industrial sources to normalize the pollutant emission rates with production units so that comparison may be maintained between various plants. Thus, for example, where a refuse incinerator burns 5 t h^{-1}, and emits particulates at a rate of 20×10^3 g h^{-1}, the normalized emission rate would be

$$\frac{20 \times 10^3 \ g \ h^{-1}}{5 \ t \ h^{-1}} = 4.0 \ kg \ particulates \ (t \ of \ refuse \ burned)^{-1}.$$

The value of this procedure for comparing different operations with similar processes is self evident.

1.1.2.3   Mobile source sampling

Testing conditions and requirements

Exhaust gas sampling from vehicles and aircraft is an involved process requiring sophisticated equipment. Vehicle and aircraft emissions are heavily dependent upon the engine operating mode and therefore the operating cycle used during exhaust sampling is fundamental to the interpretation of the results.

Soltau and Larbey [48] have reviewed driving cycles which are available for vehicle testing. Exhaust emission tests are usually performed with the vehicle on a dynamometer equipped with inertia flywheels to represent the vehicle weight and brake loading to reproduce the level road load at a given speed.

The California 7-mode driving cycle is illustrated in Fig. 1.8. The emissions during the shaded portion of the curve are not used in the calculation of the total vehicle emission. Tests with the California 7-mode cycle require the use of seven consecutive complete cycles with only the first four and last two monitored and emissions recorded.

The US Federal cycle [49] for the 1972 model year is shown in Fig. 1.9. The vehicle is set up on a dynamometer as for the 7-mode California cycle and is

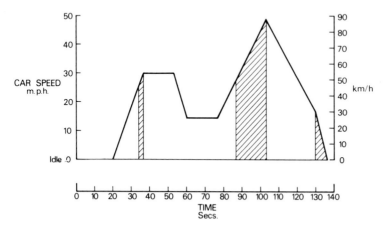

Fig. 1.8 California driving cycle. (After [48].)

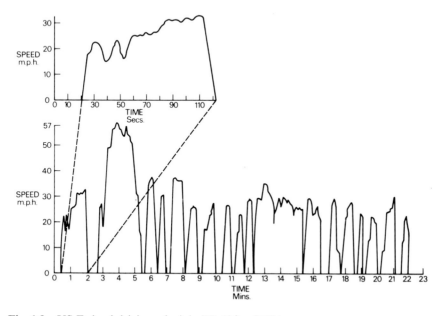

Fig. 1.9 US Federal driving schedule (II). (After [48].)

operated from a cold start through the entire non-repetitive cycle. The total emissions throughout the whole cycle are collected for the calculation of the total vehicle emission.

The Economic Commission for Europe driving cycle is shown in Fig. 1.10. This cycle is similar to the California 7-mode cycle, in that four repetitive cycles are run from a cold start.

The Japanese driving cycle is shown in Fig. 1.11 with the shaded portions being neglected from the emission calculation. Three such cycles are driven from a warm start. Weighting factors for idling (0.11), accelerating (0.35), cruising (0.52) and decelerating (0.02) are applied to the emission calculation for each segment by averaging the results of three successive test cycles.

The test procedures for aircraft engines [50] tend to follow the approach of the California and Japanese operating cycles in that emissions are measured during each phase of the cycle with the results averaged and weighted to produce a calculated total emission. Aircraft emission cycles tend to be based on steady-state operating conditions only. The approach of the 1972 US Federal cycle and

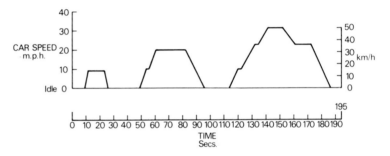

Fig. 1.10   European ECE driving cycle. (After [48].)

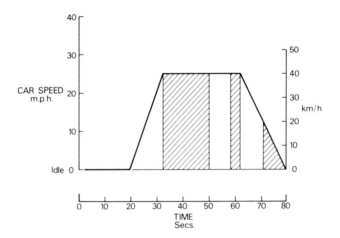

Fig. 1.11   Japanese driving cycle. (After [48].)

the ECE cycle where sampling is composited through a representative complete operating cycle has not been applied to aircraft engine testing [50].

Some actual sampling system applications to mobile source exhaust particulate and gaseous emissions are considered in Sections 1.2.2.4 and 1.2.3.4.

1.2 Sampling methods

1.2.1 General sampling system considerations

Sampling systems for airborne pollutants generally consist of four component subsystems:

(a) Intake and transfer component
(b) Collection component
(c) Flow measuring component
(d) Air moving component.

Malfunction by any one component will hinder the successful performance of the whole system. Therefore, the performance of each component must be evaluated when synthesizing the overall sampling system.

1.2.1.1 *Intake and transfer component*

The nature of the intake is basically determined by the objective of the sampling programme, varying from thin-walled probes used for aerosol source sampling to free vertical access for dustfall deposit gauges. Specific types of intake will be discussed, as necessary, with the various types of sampling, but certain general considerations apply to the intake and transfer function.

The primary consideration in evaluating the intake for a given sampling system is the ability of the device to inhale faithfully the total quantity or a reproducible representative portion of the airborne constituent being studied from a given volume of air sampled. Although fundamental to the interpretation of air pollution measurements, the difficulties encountered in verifying intake performance for a given constituent have resulted in limited evaluation of many devices commonly used in air sampling. Therefore, serious consideration should be given to a preliminary evaluation of intake performance, particularly when embarking on a sampling programme for airborne constituents not commonly reported.

Upon satisfaction of the above constraint on the intake, the transfer system must then transport the constituent of interest to the collection device without modifying the quantity or any properties of the constituent. Phenomena such as adhesion of aerosols to tube walls, condensation of volatile components within the transfer lines, reaction of gaseous components with transfer system materials and adsorption and reaction with collected particulates are some of those which have been reported [1, 18, 51–56]. Many specific problems have been recognized for given sampling systems and these will be included in the discussion of specific procedures.

1.2.1.2   *Collection component*

The collection component chosen for the sampling system is determined by the airborne constituent or constituents which are sought. Aerosol sampling may incorporate filtration, impingement, thermal or electrostatic precipitation, and gravity or centrifugal collection. Gaseous sampling may incorporate adsorption, absorption, condensation or grab sampling. Descriptions of these components are provided in Sections 1.2.2.2 and 1.2.3.2. However, the need to determine the collection efficiency of the device for the specific airborne constituent or constituents sought is common to all devices. Although 100% collection efficiency would be desirable, in practice, lower efficiencies can be used, provided the efficiency can be precisely and reproducibly measured for the specific constituent being sampled. The lowest collection efficiency that can be tolerated is determined by the importance attached to quantitative results, but collection efficiencies of 90% or better are generally acceptable [54] for quantitative analysis. Verification of collection efficiency has received attention with most commonly used sampling systems but specific verification of the collection component selected for a given constituent is a recommended practice.

Evaluation of the collection component requires the preparation of known test atmospheres [54]. For gaseous constituents this may be performed by adding measured quantities of the test gas into a container providing a known dilution volume. Collapsible containers provide the advantage of preventing dilution or pressure changes within the test atmosphere as the sample is drawn from the container. Gaseous pollutants with convenient vapour pressures can be obtained in known concentrations, using permeation devices (see Chapter 6). The pollutant enclosed in the liquid form in a plastic tube is placed in an oven maintained at precise constant temperature. The pollutant permeates at a constant rate through the plastic membrane to the outside where by the passage of a known air flow rate predetermined concentrations can be produced [57].

For aerosol testing, the preparation of an aerosol dispersion of known size distribution is required [58, 59]. Aerosol standard generators usually function by dispersing a material dissolved in a liquid solvent in a manner to create aerosol droplets. The solvent subsequently evaporates from the small particles leaving relatively non-volatile particles. Examples of such materials include polystyrene latex, dioctyl phthalate and methylene blue.

For either gaseous or aerosol constituents, losses within the test generating equipment must be considered as well as the possible inefficiency of a given collector. Frequently, the use of two collectors in series has been used in the past to evaluate the collection efficiency of a given collector [54]. This method, however, will only highlight overloading or poor efficiency on the part of the first collector, but it cannot, by itself, guarantee that both collectors are not, in fact, inefficient, if no collection is seen on the second collector [54]. A second collector may be of more value where it is proved to be efficient for the constituent sought while the first collector is an unknown which requires testing.

When collection components are used in series for regular sampling, specific evaluation of the effect of upstream components on the constituents reaching downstream components is usually necessary. For example, particulates collected on an upstream filter may adsorb a gaseous constituent intended for collection in a downstream collector. In some cases, it is necessary to place intentionally a screening collection component upstream of the primary collector, in order to remove a constituent that will interfere with the collection or analysis of the constituent of interest.

While some of these problems may be unavoidable for the system required, a knowledge of such interferences will allow modifications to the analytical schemes chosen or the quantitative interpretation placed on the results.

1.2.1.3   *Flow measurement component*

Any attempt to measure accurately the concentration of a given airborne constituent is fundamentally dependent upon the accurate knowledge of the original air volume relationship with the sample being analysed. Unfortunately, the flow measurement component of an air sampling system often receives less attention than other components.

Flow measuring components generally fall into two classes: volume meters and rate meters [18, 57, 60, 61]. Volume meters measure the total integrated volume which has passed through them for a given period of time. As such, they have the advantage of providing a direct record of the volume of air sampled. The common types of volume meters that are available are the dry test gas meter, the wet test gas meter and the cycloid gas meter. Of these, the dry test meter is most commonly used because of its relative sturdiness, low cost and weight, in comparison with the latter two meters.

The dry test gas meter measures volume flow by mechanical displacement of internal bellows by the air flow. The displacement is recorded on a mechanical counter via a series of levers. A properly maintained and calibrated unit can give an accuracy of 2–4% [57].

Rate meters measure the instantaneous volume flow rate through the sampling system and, therefore, have the disadvantage that frequent checks are required to ensure accurate calculation of the total volume sampled. In some cases, permanent flow rate recording can be provided, which will solve this problem. The advantage of rate meters is their relatively small size in comparison with the bulky dry test gas meter. Some sampling devices need to work at precise flow rates which can be measured directly with rate meters. The common types of rate meters that are available include: venturi meters, orifice meters, flow nozzle meters, rotameters, pitot tubes, turbine and hot wire anemometers.

Venturi, orifice and flow nozzle meters all depend on the air stream flowing through a constriction. The measured flow through the constriction is a function of the static pressure before and after (or at) the constriction. Thus, the measurement of static pressure at two selected points allows the calculation of flow, given the appropriate constants for the constriction. The venturi meter provides the highest accuracy with the smallest pressure loss, but is more expensive and requires more installation space than either the orifice or flow nozzle meters. These devices may all be adapted to provide continuous flow recording.

The rotameter consists of an expanding conical flow section which is mounted vertically and contains a pointed float which rides the upward flow in the flow section and effectively provides a variable annular flow area between itself and the flow section walls. The downward gravity force on the float is counteracted by the upward pressure, buoyant and drag forces due to the upward air flow. The flow section is graduated in arbitrary units. By means of calibration, the flow value of the arbitrary scale is established and the flow is then indicated by noting the level of the float on the scale.

Pitot tubes may be used to establish flow by determining the velocity profile of a flow section as discussed in Section 1.1.2.2. Turbine and hotwire anemometers are becoming increasingly popular with the development of more automated sampling equipment. The former utilizes the rate of rotation of various types of propeller devices within the flow stream to determine the volume flow rate. The latter utilizes the dependence of the convective cooling rate of a hot wire on the velocity of air flow past the wire to determine flow velocity and thereby flow rate. As the cooling rate is dependent only on mass velocity the precision is not affected by changes in air flow temperature.

The flow devices discussed tend to be inaccurate at low flow rates, i.e. less than 10 ml min^{-1} [1], but sampling rates for most sampling systems are usually adequate to maintain accurate flow measurements. Frequent calibration of all measuring devices is recommended and can be performed according to the manufacturer's recommendations or standard calibration procedures [1, 62, 63]. Calibration should be performed at anticipated ambient temperature conditions with the intake and collection components in place so that the calibration will be valid for field conditions.

Flow measuring device maintenance must be aimed at possible fouling of meter flow restrictions with particulates or reactive gas corrosion products. Ideally, however, the meter should be protected from upstream contamination by an efficient particulate collector and an adsorber for reactive gases. Provided that suitable calibration, maintenance and protection of the flow measuring component are used, the primary remaining source of error will be leaks in the sampling system. The latter may be significant and must be evaluated by measuring inflow to the system when the intake is sealed. A leakage rate of less than 0.6 litre min^{-1} at 0.51 b of vacuum is considered acceptable for source testing [38] while ambient gas sampling containers should maintain a vacuum of 0.97 ± 0.0013 b for 1 h without losing more than 0.0013 b [54].

Finally, because air volume is both pressure and temperature dependent, the measurement of these two parameters at the inlet to the volume or rate measuring device is an essential part of the flow measuring component. The determination of these parameters allows the conversion of the measured air volume to chosen standard conditions. Without this information, concentration determinations under different conditions could not be compared and would be relatively meaningless.

1.2.1.4 *Air moving component*

The final component of the sampling system is necessary to draw or force the air to be sampled through the overall system. In practice, a device creating a vacuum to draw the air through the collection component is preferable to devices which would have to intake the air and then push it through the collection component. The latter scheme obviously provides greater scope for sample contamination and change as well as greater likelihood of damage by the air sample to the air moving component itself.

Types of vacuum source include mechanical blowers, aspirators and hand operated pumps. The major determining factors in choosing the air moving component are the volumetric flow rate required and availability and type of power source.

For continuous operation at medium to high flow rates, mechanical blowers are usually chosen. Electrically powered pumps are preferable, but where a power supply is not available petrol driven pumps may be necessary. However, the possibility of sample contamination by the motor exhaust is a serious limitation to the use of petrol driven pumps. Under no circumstances should they be used without conducting the exhaust gases to a discharge point remote from the sampling intake and in all cases the degree of contamination possible must be evaluated by sampling in a clean laboratory atmosphere. The brushes on electric motors for vacuum blowers have been reported to produce aerosols [64–66] which may contaminate aerosol sampling systems. Proper maintenance of air moving equipment is required to minimize such problems. Aspirators may be used where low to medium flow rates are required and a flow of water, air or steam under pressure is available. Head loss build up in the collection component must be avoided as water or steam may be drawn back into the sampling system from the aspirator if the water or steam flow through the aspirator should decrease because of a drop in supply line pressure.

Hand pumps and syringes may be used for grab sampling of gases, but these devices are of little use where continuous sampling is required.

1.2.2 Aerosols

1.2.2.1 *Aerosol sampling considerations*

Aerosol is defined [67] as a dispersion of solid or liquid particles in a gaseous media, where particle means a small discrete mass of solid or liquid matter. As a result of the discrete mass nature of the particles of an aerosol, the collection of a representative sample of the aerosol requires special consideration.

The flow pattern of a homogeneous gas is described by a pattern of streamlines (lines drawn tangentially to the direction of instantaneous velocity for all points in the gas flow). If the flow is steady, then the actual path followed by the flow will correspond to the streamline pattern.

If, for example, a uniform horizontal flow exists in the gas stream (Fig. 1.12a),

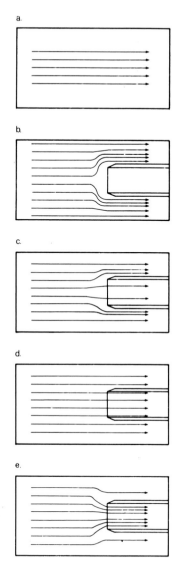

Fig. 1.12   Streamline pattern around a sampling probe in a uniform flow field.

and a sampling intake probe, connected to a closed valve, is inserted in the flow, the streamlines will be distorted, as the flow diverts around the obstruction (Fig. 1.12b). If the valve to the sampling probe is opened and the vacuum source activated, then a certain flow into the probe will result (Fig. 1.12c). As the intake flow increases, until the velocity at the face of the probe matches the velocity immediately upstream from that point in the uniform flow, then the air will flow into the probe with minimal disturbance of the streamline pattern (Fig. 1.12d). Although some distortion of the streamline pattern is inevitable, even at this sampling condition, distortion can be minimized by the use of a sharp edged, thin-walled probe. Further increases in the intake flow rate will result in a distorted pattern with streamlines converging into the probe (Fig. 1.12e). The various sampling conditions which distort the streamline pattern do not interfere with obtaining a representative sample when sampling only gaseous constituents of the air mass, since gaseous constituents, provided they are thoroughly mixed with the air mass, will follow the streamlines.

However, because components of an aerosol are discrete mass particles, they will be subject to inertial effects relative to the gas stream. Particles will possess inertia determined by the product of their mass and velocity. Their inertia is a vector quantity and as such has a directional component, which coincides with the directional component to the particle's velocity. Therefore, the particles travelling in the uniform gas stream of Fig. 1.13a will have inertia directed along the streamlines. If as in Figs 1.13b, c, the streamlines are forced to change direction abruptly in order to avoid the obstruction presented by the sampling probe, the interactive forces of the molecular gas stream may be insufficient to deflect the particles to follow the streamline. As a result, the particle may depart from the streamline and continue its path into the sampling probe. The net result will be the collection in the probe of a greater proportion of particles per unit volume of gas than exists in the actual gas flow. Only when the velocity at the face of the probe is equal to the approach velocity of the gas stream, will the streamline pattern remain undisturbed, with the result being the correct proportion of particles per unit volume of the gas being sampled into the probe. This condition is referred to as isokinetic (equal velocity) sampling (Fig. 1.13d). If the intake flow rate is increased beyond isokinetic conditions, as in Fig. 1.13e, the inertia of particles originally in the outer streamlines which are now being drawn into the probe will cause these particles to continue on their original paths and miss the probe. The net result will be a lower concentration of particles per unit volume of gas sampled than exists in the external gas flow. Clearly the attainment of isokinetic sampling conditions is fundamental to the collection of quantitatively representative aerosol samples, for which particle inertia is significant.

The errors incurred by anisokinetic sampling have been studied by several workers including Zenker [68], Hemeon and Haines [69], Watson [70], Sehmel [71], Badzioch [72] and Vitols [73] and various correction formulae have been developed. A plot based on the work of Watson is presented in Fig. 1.14.

In addition to the collection of an inaccurate total mass concentration of particles when sampling anisokinetically, an inaccurate particle size distribution will be obtained because of the relation of particle size to inertial effects.

Particle inertial effects will also cause sampling errors if the probe is not properly aligned with the flow direction, as this will also result in a disturbance of the streamline pattern at the probe intake. The errors due to misalignment of the probe are demonstrated in Fig. 1.15 [70].

Inaccurate particle distributions in the flow will be caused by inertial effects

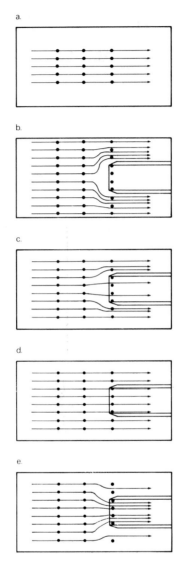

Fig. 1.13   Suspended particulate trajectories around a sampling probe in a uniform flow field.

when the flow is diverted around bends. Sansone [74] studied sampling for particulates following a 90° bend. As would be expected, the particulate concentration was found to be higher at the wall further from the centre of curvature of the bend. It was also found that, under turbulent flow conditions, particle concentrations were higher around the periphery of the flow duct.

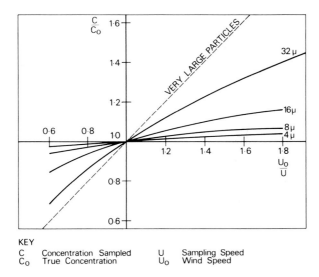

KEY

C	Concentration Sampled	U	Sampling Speed
C_O	True Concentration	U_O	Wind Speed

Fig. 1.14 Relationship of concentration sampled/true concentration to wind speed/sampling speed for spheres of unit density [70].

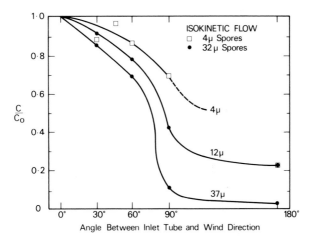

Fig. 1.15 Isokinetic sampling factors with the inlet tube aligned at various angles to the windstream for 4, 12 and 37 MMD diethyl phthalate clouds. (After Mayhood and Langstroth, from Watson [70].)

The requirements for isokinetic sampling can generally be relaxed when sampling particles exclusively less than 3 to 5 microns in diameter, as the inertial effect on these smaller particles is insufficient to cause significant sampling error. Likewise, the need for isokinetic sampling is lessened by sampling from stagnant or low velocity air-masses.

Sampling of ambient air masses is seldom made in isokinetic conditions due to the high variability of wind intensity and direction. In these conditions isokinetic sampling is difficult to achieve, being possible only with complicated equipment. The inlet must be directed to face into the wind and the suction speed must be adjusted to match wind speed. Sehmel [75] developed the 'rotating coal' sampler in which a wind vane was used to direct the inlet of a high volume cascade impactor against the wind. The suction speed was kept fixed and the sampling was therefore generally sub-isokinetic.

In anisokinetic conditions gravitational and wind effects can interfere with quantitative collection of aerosols from ambient air. The effectiveness of ambient aerosol sampling devices has been shown to be dependent on particle size, inlet geometry and wind conditions [76–78]. Entrance efficiencies for large particles (dp $> 5\ \mu$m) are quite sensitive to small changes in wind speed and angle of approach. Entrance efficiency generally decreases with increasing particle size and wind speed. The sampling effectiveness for two ambient aerosol samplers are shown in Table 1.3 and Fig. 1.16 as a function of wind condition and particle diameters.

Table 1.3   Sampling effectiveness of high volume sampler at 4.6 m/s (after [77])

Particle diameter (μm)	*Sampler orientation* (degrees)	*Sampling effectiveness* (%)
5	0	97
5	45	100
15	0	35
15	45	55
30	0	18
30	45	41
50	0	7
50	45	34

The gravitational effect is simply described here for the case where the collecting surface is pointed upwards [1] as

$$C_c = \left[1 + \frac{V_s}{V_c}\right] C_a$$

where C_c = collected particulate concentration; V_s = settling velocity of particles; V_c = collecting velocity of sampler; C_a = ambient particulate concentration and where the collecting surface is pointed downwards

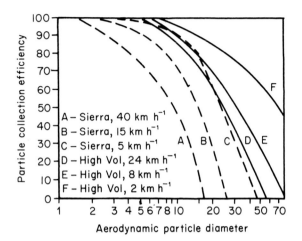

Fig. 1.16 Collection effectiveness of standard Hi-Vol and Sierra 244 dichotomous samplers at several wind speeds. (After [78].)

$$C_c = \left[1 - \frac{V_s}{V_c} \right] C_a$$

Since the particle settling velocity is directly dependent upon the size and density of the particle, the correction factor is small for smaller particles. The deviation from accurate sampling has been calculated for the standard high volume sampler [1] (Section 1.2.2.3) in Table 1.4.

Table 1.4 Gravitational effect on collected particles[a] (after [13])

Particle size (μm)	Percentage reduction of collected particles
11	1
36	10
85	50
135	100

[a] Assumes 645 cm² of opening; 1.42 m³ min⁻¹ sampling rate.

Davies [79] has studied the theoretical limitations on the entrance efficiency due to inertia and gravitational effects. In order to avoid significant inertial effects, Davies suggests that the radius of the sampling intake must be much larger than the stopping distance* for the particle concerned. In turn, for

---

* The stopping distance is the distance a particle would require to stop if injected into a stagnant air mass with the velocity it possessed in the air stream. Thus, the stopping distance is a function of the particle momentum.

gravitational effects, the inlet velocity at the sampling intake must be large in comparison with the settling velocity for the particle concerned. The inlet velocity for a given flow rate is a function of the inlet radius and, thus, Davies derived an expression for the upper and lower bounds of the sample inlet radius at a given flow rate for a given particle size. A tabulation of these bounds for sample inlets is presented in Table 1.5.

Bien and Corn [80] applied Davies' criteria to several commercially available samplers. Their findings, summarized in Table 1.6 and Fig. 1.17, indicate that many commercially available samplers fail to conform to the criteria of Davies' theoretical model for unit density spheres in calm ambient air.

Table 1.5   Permissible radii of tubes (cm) for sampling aerosols in calm conditions (after [79])

Particle diameter (µm)	Rate of suction F (cm^3 s^{-1})					
	1	10	10^2	10^3	10^4	10^5
1	0.033–1.9	0.071–6.0	0.15–19	0.33–60	0.71–190	1.5–600
2	0.051–1.0	0.11–3.2	0.23–10	0.51–32	1.1–100	2.3–320
5	0.093–0.41	0.20–1.3	0.43–4.1	0.93–13	2.0–41	4.3–130
10	0.15–0.21	0.31–0.65	0.68–2.1	1.5–6.5	3.1–21	6.8–65
20	(0.23~0.10)	(0.50~0.33)	(1.1~1.0)	2.3–3.1	5.0–10.3	11.0–31
50	(0.42~0.042)	(0.90~0.13)	(1.9~0.42)	(4.2~1.33)	(9.0~4.2)	(19~13.3)
100	(0.63~0.023)	(1.4~0.071)	(2.9~0.23)	(6.3~0.71)	(14~2.3)	(29~7.1)
200	(0.89~0.014)	(1.9~0.037)	(4.1~0.14)	(8.9~0.37)	(19~1.4)	(41~3.7)
500	(1.26~0.008)	(2.7~0.025)	(5.8~0.08)	(12.6~0.25)	(27~0.80)	(58~2.5)

Table 1.6   Specified sampling rates and probe inlet dimensions of some commonly used sampling instruments (after [80])

Sampler	Sampling curve and sampling designation	Rate (litres min^{-1})	Actual diameter of sampler (cm)
DEL electrostatic precipitator	A	750	6.1
MSA electrostatic precipitator	B	85	3.6
Anderson cascade impactor	C	28	1.8
Standard G-S impinger	D	28	1.2
Half-inch cyclone	E	10	0.41
Half-inch cyclone	F	8	0.41
Midget impinger	G	2.8	0.36
MRE elutriator	H	2.4	0.10
10 mm cyclone	J	2.0	0.21
10 mm cyclone	K	1.7	0.21
10 mm cyclone	L	1.4	0.21
Micro impinger	M	0.56	0.36

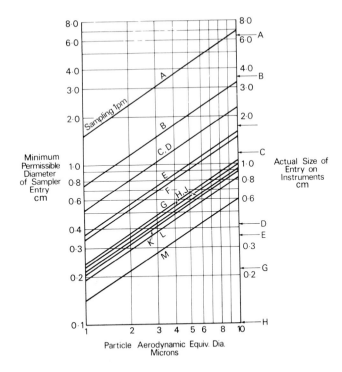

Fig. 1.17 Particle size efficiently captured by selected aerosol samplers, according to the criteria of Davies. The letter designations A to M refer to the instruments and conditions as listed in Table 1.6 (After [80].)

Although the need for considering the inertial and gravitational effects on the collection of aerosols requires primary attention, several other sources of error in aerosol sampling must also be considered, such as:

(a) Aerosols may collect on sample transfer tubing walls, particularly after bends or where condensation is allowed
(b) Aerosols may be attracted to tubing surfaces by electrostatic forces particularly where non-conducting glass and plastic tubing is used
(c) Aerosols may evaporate or decompose after collection, or react with other particles or gases.
(d) Aerosols may be hygroscopic and adsorb water vapour.

Some changes in the aerosol may be tolerated depending on the analysis to be performed. For example, agglomeration of particles without weight change would hinder particle size analysis but would not hinder a mass concentration or chemical analysis. However, a complete and thorough evaluation of possible sources of loss or change to aerosols within a sampling system is necessary in order to ensure that the aerosol which reaches the collection component quantitatively and qualitatively resembles, for the purposes of the subsequent analysis, the original aerosol sampled.

1.2.2.2   Aerosol sampling collection components

Filtration

Filtration for the collection of aerosols depends upon drawing the air sample through a network of small openings which may be formed from:

(a) The overlap of fibres as in fibre glass or cellulose filters
(b) Random pores with a fixed medium as in sintered glass filters
(c) Controlled size pores in an organic medium as in membrane filters
(d) Random pathways through a granular medium.

Filters remove the particles from the air flow as a result of the operation of various mechanisms such as direct interception, impaction, diffusion and electric and gravitational forces. Direct interception occurs for particles with dimensions larger than the distance between the elements of the filter medium. Diffusion is the most efficient mechanism for the collection of very small particles ($dp < 0.4\ \mu m$); these particles are strongly affected by Brownian motion and diffuse through the air flow to the surface of the filter material where they adhere due to attraction forces. Particles with intermediate dimensions are preferably collected by impaction; as the air flow traverses the tortuous paths of the filter, the frequent directional changes encourage the deviation of the particles from the air stream with consequent impingement on internal surfaces. Electrostatic forces may contribute to a higher filtration efficiency if the filter and the aerosol are capable of acquiring static electrical charges.

The type of filter medium chosen for the sampling of a given aerosol depends upon several factors, such as price, availability, efficiency of filtration, the requirements of analytical techniques and the capacity of the filter to maintain its characteristics of filtration and integrity in the sampling conditions.

The efficiency of the filter in the capturing of the various sized particles present in the sampled air is frequently one of the most important factors in the selection of a filter medium. As the deposition of the particles over the filter substrate is dependent on several mechanisms it becomes obvious that the efficiency of sampling of an aerosol is variable with the velocity of filtration, the size of the particles and the amount of material already collected. Low sampling velocities improve the collection of smaller particles due to the action of diffusion processes. An increase in the sampling velocity decreases the efficacy of diffusion but increases inertial forces with consequent collection by impaction. At very high flow rates the efficiency of collection may decrease due to the re-entrainment of particles previously deposited.

Theoretical models evaluating the performance of filter media in the collection of aerosols are not sufficiently refined to describe real filter capabilities which need to be studied experimentally [81–85]. Table 1.7 presents experimental

Table 1.7   Efficiencies of some common filters, expressed as a percentage, for particles with 0.1 and 1.0 μm diameter and flow rates equivalent to head pressures of 1.0 and 10.0 cm Hg (after [86])

Filter	Composition	$D_p = 0.1$ μm Δp 1.0 cm Hg	$D_p = 0.1$ μm Δp 10.0 cm Hg	$D_p = 1.0$ μm Δp 1.0 cm Hg	$D_p = 1.0$ μm Δp 10.0 cm Hg
Whatman No. 1	Cellulose fibre	56	87	98.1	99.7
Whatman No. 41	Cellulose fibre	44	84	92.0	99.1
Whatman No. 42	Cellulose fibre	94.0	97.1	99.4	99.91
Gelman A	Glass fibre	>99.99	99.97	>99.99	>99.99
Gelman A/E	Glass fibre	99.94	99.6	>99.99	>99.99
Gelman Spectrograde	Glass fibre	99.92	99.7	99.99	99.99
Whatman GF/A	Glass fibre	>99.95	99.90	>99.99	99.98
Whatman EPM 1000	Glass fibre	99.94	99.9	>99.99	99.98
Pallflex 2500 QAO	Quartz fibre	92.6	–	99.6	–
Millipore MF-HA 0.45 μm pore	Cellulose acetate/ nitrate membrane	>99.99	>99.99	>99.99	>99.9
Millipore MF/SC 8.0 μm pore	Cellulose acetate/ nitrate membrane	98.8	96.5	>99.9	>99.9
Millipore FH 0.5 μm pore	PTFE membrane	>99.99	>99.99	>99.99	>99.99
Millipore FS 3.0 μm pore	PTFE membrane	99.5	99.7	99.96	99.8
Nuclepore N010 0.1 μm pore	Polycarbonate membrane	>99.99	>99.99	>99.99	>99.99
Nuclepore N100 1.0 μm pore	Polycarbonate membrane	49	43	87	98.1
Nuclepore N500 5.0 μm pore	Polycarbonate membrane	8	–	56	–

efficiencies of some of the 76 different filter media measured by Liu *et al.* [86] as a function of air flow for particles with diameters in the range 0.035–1.3 μm.

The values of filtration efficiency presented in the literature are frequently based on the study of clean filters. Generally the efficiency of sampling increases with the accumulation of particles over the filtering surface. The resistance to flow also increases with deposited mass, but at a rate smaller than for collection efficiency. For some purposes it may not be desirable to utilize filters with the highest efficiency characteristics. These frequently present higher head losses and are only utilizable at low flow rates, requiring more powerful sampling pumps. Head loss characteristics of some common filter types are shown in Fig. 1.18.

Every filter material contains various chemical species, as major, minor or trace constituents, which are variable with each individual filter [87, 88]. The filter medium used for the sampling and analysis of a certain particulate

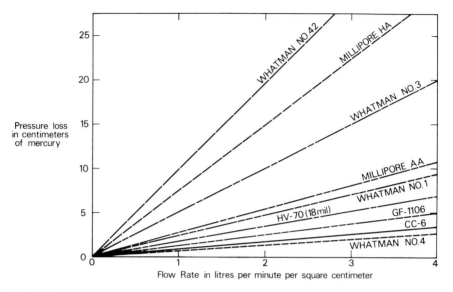

Fig. 1.18 Head-loss characteristics for various types of common filter media. (After [18].)

compound should be free from this compound or contain it in quantities much smaller than the amounts collected. The concentrations of chemical species of interest in the blank filter substrate becomes the main factor of choice when they attain values which may interfere with the determination of this species in sampled aerosols. When the blank concentrations are of the same level as the amounts collected, variations in blank values from filter to filter result in high imprecisions. Blank concentrations are frequently critical during sampling of trace elements in the atmosphere.

Possible solutions for blank filter contamination include the sampling of larger volumes of air, either by extending the duration of sampling or increasing sampling flow rates, and the cleaning of filter media. Washing of filter material with water, acids or organic solvents, or firing at high temperatures can reduce the concentrations of soluble or volatile compounds to much lower levels, dependent on the intensity and duration of the treatment [89–92]. See Table 1.8.

The filter material must be inert to atmospheric compounds of interest to prevent spurious results. An example is the formation of artifact particulate matter on the surface of alkaline filters, such as glass fibre, by oxidation and adsorption of acid gases present in sampled air [93, 94]. This is a surface-limited reaction, usually occurring early in the sampling period and is dependent on factors such as length of sampling interval, filter pH and presence of acidic gases in the sampled air. The majority of artifact mass is the sum of sulphates and nitrates formed by reactions of ambient sulphur dioxide and nitric acid. Basic

Table 1.8 Blank concentrations of untreated and acid treated and washed filters (after [92]). In the second batch washing was more intense than in the first batch

| | *Blank concentrations* ($\mu g\ cm^{-2}$) | | | | | | |
	Na^+	Mg^{2+}	Ca^{2+}	K^+	SO_4^{2-}	NO_3^-	Cl^-
Untreated Whatman GF/A glass fibre filter	17.8	0.16	0.74	3.68	0.58	0.17	0.48
Treated GF/A							
1st batch	2.52	0.02	0.15	0.58	0.20	0.09	–
	2.71	0.03	0.12	0.73	0.30	0.05	–
2nd batch	0.87	0.03	0.19	0.05	0.19	0.03	2.80
	0.84	0.06	0.18	0.06	0.21	0.03	0.30
Non-treated 2500 QA0 Pallflex quartz filter	–	–	–	–	0.48	0.39	–
Treated quartz filter							
1st batch	0.40	0.019	0.12	0.018	0.19	0.008	0.19
	0.28	0.019	0.12	0.022	0.11	0.002	0.05
2nd batch	0.04	0.002	0.02	0.014	0.02	0.03	0.08

sites on glass fibre filters substrate have also been found responsible for the neutralization of collected acidic particulate sulphates.

Witz and McPhee [95] have observed differences in concentration of total suspended particulates as high as 20% with the use of four different glass fibre filters. The variation in nitrate and sulphate concentrations measured were respectively 219 and 49%. Variable catalytic conversion of NO_2 and SO_2 was believed responsible.

Roberts and Friedlander [91] have reported a filter treatment consisting of soaking in dilute HCl, washing excess acid with distilled water and firing at 350° C, which made glass fibre filters inert to reaction with SO_2. Neutral quartz filters treated with phosphoric or hydrochloric acid and fired at 750° C have been shown to be inert in the sampling of acidic sulphates [96, 97].

During sampling, collected particles may react mutually or with ambient gaseous pollutants. Co-collected basic coarse particles have been observed to neutralize acidic sulphates [98, 99] and to adsorb and oxidize SO_2 [100, 101]. Neutralization of acidic particles by gaseous ammonia has also been reported [101]. The inter-reaction of particulates can be reduced by using size-selective sampling. Adsorption of gaseous compounds by collected aerosols can be overcome with the extraction of these compounds with a 'denuder' prior to the particulate filtration. The denuder is an apparatus formed by an assembly of long tubes, internally covered with a specific sorbent, through which the sample passes before filtration. The gas molecules have high diffusivities and are transported to the tube walls where they react chemically with the sorbent. Aerosols present much lower diffusivities and, if the flow is laminar, are not appreciably removed from the air flow. Denuder tubes have been used for removal of SO_2 [102, 103], NH_3 [104, 105] and HNO_3 [106].

Impingement

Impingement for the collection of aerosols consists of forcing the air flow through a jet which increases the velocity of the air stream and thereby the momentum of associated particles, followed by collision of the airstream with an abrupt obstruction. Because of their high momentum and the abrupt change in flow direction necessitated by the obstruction, the particles will tend to collect on the obstruction.

Impingers have been developed to operate either wet or dry. Wet impingers are used with the jet and collecting obstruction below the surface of a liquid (in which the particles must be insoluble). Dry impingers, as the name implies, have the jet and collecting obstruction dry, although an adhesive may be utilized to prevent the particles from becoming resuspended.

Two common types of collection impingers, the midget impinger and the full sized Greenburg–Smith impinger are illustrated in Fig. 1.19. The midget impinger is commonly used wet, while the Greenburg–Smith may be used either wet or dry.

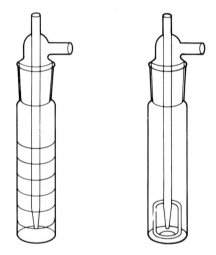

Fig. 1.19   Impingers. Left: Midget; right: G eenburg–Smith.

Because the impinger uses particle inertia for collection and that, in turn, depends upon the mass and velocity of the particles, the size collection efficiency of impingers depends on the flow rate. The midget impinger is commonly operated at 3 litres min^{-1} while the Greenburg–Smith impinger is operated at 30 litres min^{-1}.

The dependence of the particle size collection efficiency on jet velocity has been used to advantage in the design of cascade impactors. These are devices which use a series of sequential jets and collection plates with increasing jet velocities and/or decreasing gaps between the jet and the collecting plate (Fig. 1.20). As the air

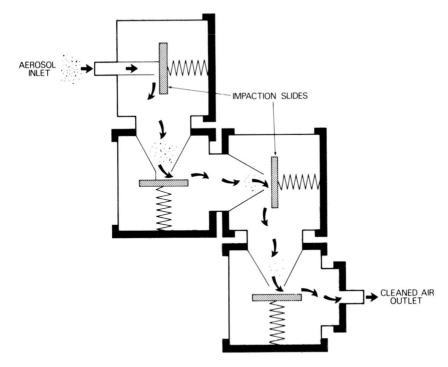

Fig. 1.20   Cascade impactor. (After [3].)

stream progresses through the device smaller particles will be collected more efficiently. As a result, the cascade impactor fractionates the air sample to obtain a particle size distribution. Unfortunately, in some cases large agglomerated particles will shatter upon impinging on the collecting plate and may resuspend giving an inaccurate size distribution. Cascade impactors are considered in more detail in Chapter 4.

Precipitation

Precipitation for the collection of aerosols depends upon exposing the air stream to either a thermal or electrostatic gradient.

In the case of the former, a sharp temperature gradient is created by a high temperature wire suspended across a cylindrical collection chamber (Fig. 1.21) [107]. The temperature gradient causes the particles to migrate towards the cooler collecting surface, apparently because gas molecules in contact with the hot wire acquire higher kinetic energy and by impacting on particulates transmit part of their momentum which has a larger component in the direction of cooler regions. The device is extremely efficient for particle sizes from 5 μm down to 0.01 μm and possibly smaller [25], but the basic instrument is limited to a relatively low flow rate (0.007 to 0.02 litre min^{-1}). Although higher flow rate

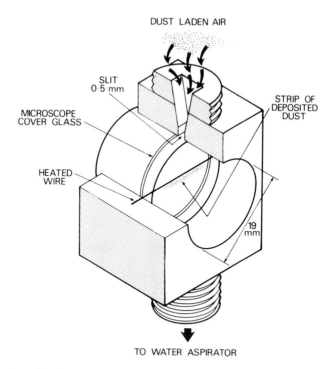

DUST LADEN AIR

SLIT
0·5 mm

MICROSCOPE
COVER GLASS

STRIP OF
DEPOSITED
DUST

HEATED
WIRE

19
mm

TO WATER ASPIRATOR

Fig. 1.21   Sampling head of standard thermal precipitator. Crown Copyright. Reproduced by permission of the Controller of Her Majesty's Stationery Office,
London.

samplers have been developed (0.5 litres min^{-1}), the thermal precipitator is
recommended primarily for specialized research studies where efficient collection
of particles for microscopic investigations is required, but not for general aerosol
sampling usage.

In the case of electrostatic precipitation, a high voltage potential (12 kV, d.c. or
20 kV, a.c.) is maintained across the plates of a capacitor with the air to be
sampled flowing between the capacitor plates. The ionizing potential created by
the electrostatic field produces ions in the gas stream which, in turn, tend to
collide with particles and charge them. The charged particles then migrate under
the influence of the electrostatic field to the collecting electrode. This device has
been found to be virtually 100% efficient for a wide range of particle sizes at flow
rates up to 85 litre min^{-1} [107]. Precautions must be observed to avoid excessive
voltage, as direct arcing across the plates may occur and sampling efficiency will
drop to nil. Also, the device is clearly unsuitable for use in explosive atmospheres
[25]. However, other than possible agglomeration of particles due to charge
effects, this device collects at a high efficiency with relatively little physical
disruption of the particles.

Gravity and centrifugal collection

Gravity and centrifugal collection depend upon the action of an external remote force on the aerosol particles to cause their movement toward a collecting surface. In the case of the former, the gravitational attraction of the particle mass towards the earth causes particles to settle at a rate dependent on the fluid drag forces counteracting gravity. For smaller particles the settling rate is small in relation to air movements with the effect that small (less than $7 \mu m$) particles remain suspended.

In the case of centrifugal collection, the air stream is made to follow a spiral pathway in a cyclone, with the resulting centrifugal action on the particles causing them to move outward to a collecting surface. Both collection methods are limited to larger particles than are collected by other collection components, with their use aimed primarily at preliminary separation of larger particles before a secondary collector.

Diffusion

The motion of suspended particles in the submicrometer range is strongly affected by random collisions with gas molecules. These particles traverse a random irregular path, their position at any given time being dependent on their most recent collisions with molecules. This movement is called Brownian diffusion and is, at a given temperature, dependent on particle dimensions. Smaller particles with less momentum are more strongly affected by collisions with molecules and present higher diffusion velocities than larger particles.

Brownian diffusion is the mechanism used for the fractionation and collection of aerosols smaller than $0.3 \mu m$, particles for which gravitational, centrifugal and impaction mechanisms are inefficient.

The fractionation of particles smaller than $0.3 \mu m$ is achieved with the diffusion battery which is an assembly of equally spaced circular tubes or holes, through which the air flow containing the aerosol passes. Due to Brownian motion the particles diffuse to the walls where they stick and are removed from the airborne state since, upon physical contact, surface adhesive forces dominate. At the boundary of every tube internal surface the particle concentration is nil and a gradient of aerosol concentration exists along the tube radius and tube axis. Smaller particles, having larger diffusion coefficients, are collected preferentially, the sampled air becoming enriched in larger particles. Fractionation of aerosol mass according to diameter can be achieved with a cut-off diameter dependent on the length of the diffusion tubes. Several lengths can be obtained by an assemblage of smaller units in series.

1.2.2.3   *Ambient aerosol sampling applications*

British suspended particulate sampler

The suspended particulate or smoke sampler is commonly used in the United

Kingdom National Survey of Air Pollution. The method is the subject of a British Standard [108].

Although most commonly used as a relatively low volume sampler (1 to 1.6 litres min^{-1}) for estimating smoke density, a system is also described for sampling at 25 to 100 litres min^{-1}. This system is believed to sample only particles less than 20 to 30 μm.

The whole system is shown in Fig. 1.22. The inlet consists of an inverted glass funnel located more than 2.5 m above the ground and 1 m from the adjacent wall. The inlet may be protected by a mesh guard, but this must be constructed of a material which will not flake or corrode in the atmosphere.

A 100 mm diameter glass fibre filter is recommended and the method calls for the use of a control filter to determine weight changes due to handling. The system employs a volume meter, capable of being read to 0.025 m^3 or less. For the determination of suspended particulate mass concentration, a sample of at least 10 mg should be obtained, which will normally require a sampling period of 24 h. For a mass concentration of 10 mg or greater, the precision of the procedure has been found to be within 10%.

American high volume sampler for suspended particulates

The high volume sampler is used in the United States National Air Sampling Network, being the current Environmental Protection Agency (EPA) reference method for total suspended particles (TSP) measurement [1, 62].

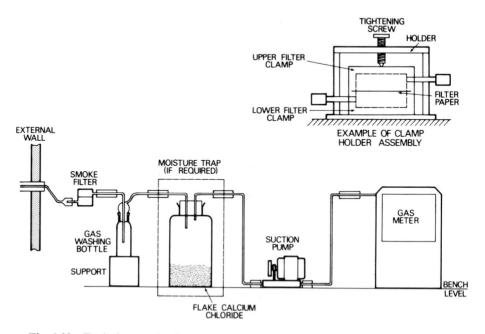

Fig. 1.22   Typical example of aerosol sampling apparatus assembly. (After [108].)

A high flow rate blower (1.1 to 1.7 m³ min⁻¹) draws the air sample into a covered housing and through a 20×25 cm rectangular glass fibre filter (Fig. 1.23). The mass of particles collected on the filter is determined by difference between gravimetric measurements of the filter material before and after exposure. The method is also frequently applied in the analysis of aerosol chemical species, which are measured by appropriate methods after extraction from the filters with convenient solvents.

The most commonly used filter medium is glass fibre because of its high filtering efficiency, and non-hygroscopic and low head loss properties. As we have seen, this material is not inert towards acidic compounds, resulting in the formation of artifacts (Section 1.2.2.2).

The air is drawn in through a free space between the roof and shelter body. In order to reach the filter the air has first to move upwards through the surrounding rectangular aperture, the dimensions of which are such that particles with diameters larger than 100 μm have a sedimentation speed that prevents them from entering and being collected.

This value for the sampling cut-off only ..as any significance in conditions of complete calm. It has been observed that samples collected on rainy windy days

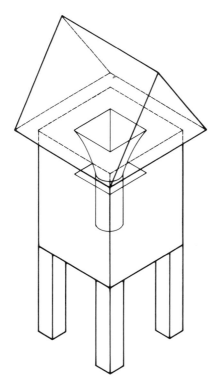

Fig. 1.23　High volume sampler and shelter.

have been impacted by rain drops much larger than the theoretical 100 μm. Wedding *et al.* [77] (Table 1.3) observed from wind tunnel experiments that the sampling efficiency of the high volume sampler was only 55% for a 15 μm diameter aerosol, at 4.6 ms^{-1} wind speed; the efficiency was variable with intensity and direction of wind relative to the high volume shelter, and started to become important for particles larger than $\sim$ 7 μm. Hotschreuder *et al.* [109], in experiments made in the open with 17 μm diameter aerosols, observed that, at conditions of atmospheric stability class D (neutral), the collection efficiency of the Hi-Vol sampler was only 33 $\pm$ 7%.

On the other hand it seems that the efficiency of the Hi-Vol sampler on particle collection is relatively unaffected by changes in sampling flow rate. Variations of sampling flow from 0.5 to 2 m^3 min^{-1} gave, for 29 μm diameter particles, a change in sampling efficiency of only $\pm$ 2.5% [110].

The Hi-Vol sampler is commonly operated for 24 h, although other sampling periods may be used, depending on the expected concentrations of suspended particles. Frequently the sampling units are provided with a timer which permits automatic sampling for a predetermined time, during a seven-day period. A usual operation process consists of sampling for 24 h in each six days, with the changing of the filter once each week. The presence of the filter on the shelter during the non-operating five-day period, both before and after sampling, can introduce positive errors in the determination of suspended particulate mass. The shelter roof provides a settling chamber for larger particles blown by the wind, which are then deposited over the filter. The length of time for which the filter is exposed, the wind speed and TSP concentrations in the ambient air seem to be the main factors influencing passive deposition [111]. Passive loaded particles are mainly of dimensions 10–60 μm diameter [112]. The process is more important in summer and may introduce a positive bias in total collected mass from 3 to 46%, with a mean value of $\sim$ 10% [113–116].

Wind action across the top of the Hi-Vol filter can also blow off the deposited particles, principally in conditions of high wind and dust concentration. Particles removed are the larger ones ($>$ 100 μm), such that in dust storm conditions results may be only 50% of the true values [117].

Errors due to passive sampling or wind blowing can be reduced by equipping the sampler with one of the commercially available automatic mechanical devices that keep the filter covered during the non-sampling periods. Alternatively, installation and retrieval of filters immediately before and after sampling will also minimize the problem.

During sampling, the filtered air traverses the motor, giving rise to the formation of copper and carbon aerosols due to disintegration of motor brushes; nitrate and sulphate particles are also produced by reaction with NO_x, SO_x and NH_3. To prevent the contamination of the sample through exhaust recirculation, deviation to some distance, or preferably air exhaust filtration, have been recommended [118, 119].

Contamination or damage to the filters is possible when inserting or removing them from the sampler, particularly on windy rainy days. There are commercially available filter-holding cassettes which allow the filter to be inserted and sealed in the laboratory, with subsequent rapid and trouble-free installation in the field [120].

The volume of sampled air is commonly measured with flow measurement devices such as rotameters, or using flow recorders. The rotameter is used to measure only the initial and final flow values, from which an average is calculated. The flow recorder provides a continuous trace that can be integrated for a more accurate measurement. Smith *et al.* [121], using Hi-Vol samplers with both devices, noticed that the flow recorder produced smaller errors (2–4%), than the rotameter (6–11%), when compared with a reference flow device. Operating the Hi-Vol sampler at lower flow rates (~ 0.85 m^3 min^{-1}) permits maintenance of a more constant flow over the 24 h sampling period, besides prolonging electric motor brush life [122].

The flow rate can be maintained constant, against variations in voltage input and pressure drops on the filter, by the utilization of flow controllers. Avera [123] has developed an electronic flow regulation system using a hot wire anemometer as the feedback sensing device to a flow controller which will adjust the blower to compensate for the increasing head loss at the filter. The anemometer is set for a null signal at a given sampling rate. Fluctuations from the set rate produce a feedback signal to compensate the flow produced by the blower to bring the anemometer back to the null setting. A similar device was presented by Wolf and Carpenter based on a turbine flow meter/control pressure system [124]. Electronic flow controllers for Hi-Vol sampling are available commercially although at relatively high prices.

The sampler must be calibrated regularly and at least every six months. Standard calibration procedures are available [1, 62]. Lynam *et al.* [125] have criticized calibration procedures which are based on varying the flow rate only with the use of a variac transformer on the vacuum blower power supply. They recommended calibration by modifying the flow resistance by placing extra filters on the sampler. Tebbens [126] explains the discrepancies noted by Lynam *et al.* as due to inward leakage between the filter and the flow recorder, but he does recommend calibration with the normal flow resistance in place.

The precision of high volume samplers, as determined from parallel sampling under field conditions, have been reported by several researchers [127, 128]. Collaborative tests showed that the relative standard deviation for the repeatability of the method (parallel samples by one analyst) was 3.0% and the corresponding value for reproducibility (parallel samples by several analysts) 3.7% [129].

The Hi-Vol blower produces an intense noise which can become annoying when sampling in residential areas [130]. In these cases the utilization of special shelters and mufflers has proved to reduce the annoyance [131, 132].

Sampling of inhalable particles

The EPA Hi-Vol sampler has been shown to give large uncertainties in the characterization of total suspended particles in the atmosphere now that the upper cut-off limit is known to be highly variable.

The effect of atmospheric aerosols on human health is strongly dependent on their capability to penetrate the respiratory tract. Generally, the smaller the particles are, the further and more profoundly they penetrate the respiratory system. Coarse particles may deposit in the pharynx and larynx, causing dryness of the nose and throat, but have no effect on mucociliary clearance.

The most satisfactory means of calculating health effects of atmospheric aerosols would be by multistage measurements of particle distributions, which must be cost effective. Recent EPA emphasis is aimed at developing air quality standards based on the specific size fraction of particles which can reach the trachea (inhalable particles) [133, 134]. The definition of inhalable particle matter (IPM) is still a matter for discussion and upper cut-off sizes of 15 μm or 10 μm diameter have been proposed. The choice of the 15 μm cut-off point is based on the worst case situation of mouth breathers, because in nose breathers particles larger than 10 μm are either rejected by the nose or restricted to the nasopharyngeal region. Although initially 15 μm seemed the most acceptable cut-off size [133–135], recently a strong possibility has arisen that EPA IP standards may be based upon a cut-off point which eliminates 50% of the particles larger than 10 μm diameter [136, 137]. The choice of 10 μm as the cut-off point is based in part on the fact that this value permits the construction of more reproducible and less wind-dependent size selective inlets.

The International Standards Organisation has proposed considering a standard for suspended particles based upon those particles depositing in the tracheo-bronchial and alveolar regions of the human respiratory tract and has termed them the thoracic particle fraction (TP). The sampler inlet should have a 50% cut-off size for 10 μm diameter particles, and the fractional penetration to the filter a log normal function with geometric standard deviation $\sigma_g = 1.5 \ \mu m$ [138].

Having in mind that only particles smaller than 3.5 μm are capable of depositing in the gas exchanging regions of the respiratory system, a second cut-off point of 2.5 μm has been recommended [133]. Although it gives a slight under-estimation of particles penetrating gas exchanging areas, this value has been chosen because it is already the cut-off diameter of several size selective samplers. Furthermore, atmospheric aerosol mass distributions generally have a minimum in this size region. Particles smaller than 2.5 μm are designated Fine particles (FP) [134].

For measurement of IP and FP, high volume samplers and dichotomous samplers have been provided with size selective inlets which give a single fraction of 0–10 μm (IP 10), or 0–15 μm (IP 15), for the Hi-Vol and two fractions (< 2.5 μm and 2.5–10 μm) for the dichotomous sampler.

Various inlets have been proposed and tested [77, 139–141]. Some inlets have variable efficiencies with wind conditions. Wedding [141] presents a table comparing the performance data of nine aerosol sampling inlets showing that most of them have cut-off efficiencies which are highly variable with wind speed. Furthermore, some of the inlets do not give a sufficiently sharp cut-off curve to present an acceptable performance, in accordance with the criteria proposed by Miller *et al.* [133] for inhalable particle samplers.

Sierra Instruments market a size-selective inlet which can be attached to a

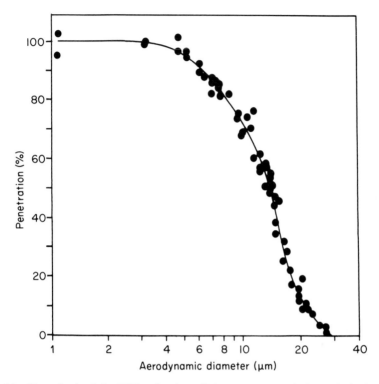

Fig. 1.24   Size selective inlet (SSI) collection efficiency curve at wind speed of 2 km h^{-1} and flow rate of 40 ft^3 min^{-1}. (From the Sierra Bulletin No. 320–281, Sierra Instruments Inc., California.)

Hi-Vol sampler and which, working a 1.13 m^3 min^{-1}, can separate particles with a 50% cut-off size of 15 ± 2 μm over a wind range of 2–24 km h^{-1} (see Fig. 1.24). The inlet is omnidirectional and maintains stable cut-off points for a range of wind speeds and flow rates (see Figs 1.25 and 1.26).

The Sierra 244E dichotomous sampler is provided with a size-selective inlet which presents a 50% cut-off point for 15 μm diameter particles. Wedding *et al.* [142] studied the apparatus in wind tunnel experiments, at wind speeds from 5 to 40 km h^{-1}, and observed the 50% cut-off point to vary from 10 to 22 μm with a maximum at 25 km h^{-1}. The variation was attributed to the opposing effects of sedimentation and impaction with wind velocity. The author proposed an inlet for the dichotomous sampler, based on a cyclone effect, which is omnidirectional and, contrary to inlets based on impaction, exhibits negligible solid particle bouncing [141]. The apparatus has a 50% cut-off size of 13.7–14.4 μm, for wind speeds between 0.5 and 24 km h^{-1}.

A similar inlet was proposed by Wedding and co-workers [142] which presents instead a 10 μm 50% cut-off point.

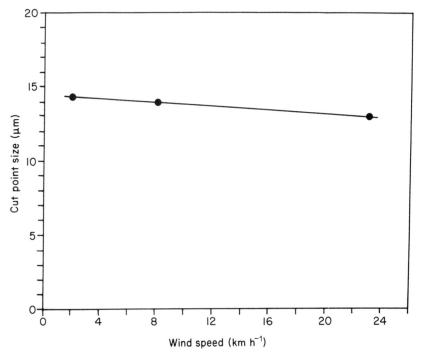

Fig. 1.25   Wind speed dependence of SSI cut-point at 40 ft^3 min^{-1}. (From the Sierra Bulletin No. 320–281, Sierra Instruments Inc., California.)

Inertial impactors

Often referred to as cascade impactors when used in multiple stages, inertial impactors provide a means of collecting ambient particles divided into sub-fractions of specific particle sizes. In cascade impactors the stages are designed and arranged so that the largest particles are aerodynamically impacted on the first stage and successively smaller particles are aerodynamically impacted on successive stages. The smallest particles are collected on a back-up filter (see Fig. 1.27).

The range of particle diameters which is collected on each stage can be determined using either theoretical calculations [143, 144], or with laboratory calibration techniques. Commercially available cascade impactors typically have 2 to 6 stages and are designed either for sampling at high flow rates (0.6–1.1 m^3 min^{-1}), or low flow rates (0.01–0.04 m^3 min^{-1}). High volume cascade impactors are designed to be mounted on high volume samplers.

The particle size cut-off points are dependent on geometrical factors and air flow velocity and, therefore, the sampling flow rate should be maintained constant during the entire sampling period.

Impaction sampling is only efficient for particles with aerodynamic equivalent

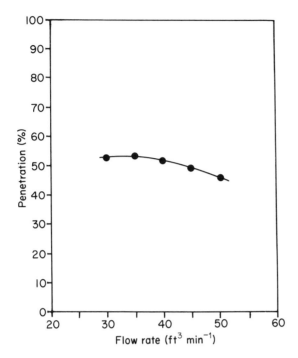

Fig. 1.26 Flow rate dependence of SSI cut-point at a particle size of 14.1 μm and a wind speed of 2 km h^{-1}. (From the Sierra Bulletin No. 320–281, Sierra Instruments Inc., California.)

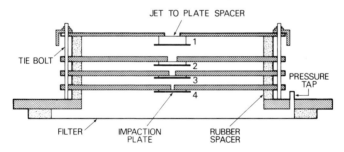

Fig. 1.27 Schematic cutaway view of high volume cascade impactor. (After [145].)

diameters larger than $\sim$0.3 μm. Below this size, particles can only be impacted and collected using very high jet velocities. The high suctions and small nozzle diameters required to attain the necessary jet velocities can result in changes in particle physical characteristics and rapid clogging of the smaller stage nozzles.

For collection, the particles are impacted on surfaces made commonly of materials such as aluminium foil, glass fibre and cellulose filter paper. Impaction surfaces can be removed after sampling and weighed for the determination of deposited mass fractions. The chemical composition of each fraction can also be determined by chemical methods after extraction from the impaction substrate.

Each stage fraction is formed by particles having a range of dimensions between a maximum and minimum. The larger and smaller cut-off sizes are not sharp, there existing a region of dimensions where particles are collected with an efficiency between zero and one hundred per cent. The response functions of impactors are generally sigmoidal and sometimes not symmetrical. Frequently, impactors present response functions which are well approximated by cumulative log-normal distributions [146]. Generally the cut-off size for each stage is given as the diameter for which 50% of the mass of particles of that size is collected and deposited on that stage.

From the determination of the size fractions collected on each stage it is possible to draw size distribution curves for the ambient aerosol concentrations. These size distributions can provide information valuable in the determination of emission sources and the understanding of transformation, transport and deposition mechanisms of atmospheric aerosols. Size distribution curves are frequently calculated and drawn considering the impactors as presenting a step function response with a value equal, either to the 50% mass diameter cut-off size, or to the equal area cut-off size [147]. Numerical methods have been developed, taking into account the real cut-off size characteristics of impactors, with the aim of obtaining less biased aerosol size distribution curves [147–152].

The collection efficiency of each impaction stage is frequently not identical for all particles having the same aerodynamic diameter. While liquid particles stick generally to the impaction surface, solid particles impacting on dry surfaces frequently bounce or are blown off [153–155]. The problem can be reduced by coating the surfaces with an adhesive material, such as paraffin or Vaseline [156–159], although this is not always possible, because the process may interfere with later chemical analyses. Furthermore, the technique is effective only during the deposition of the first layer of particles [153].

The surface substrate seems to influence the impaction characteristics. Smooth surfaces like aluminium foil or stainless steel give higher losses, due to particle re-entrainment, than porous materials like glass fibre or cellulose filter paper. Porous substrates can, however, give less sharp cut-off size values and result in the distortion of impaction response functions, as a result of the penetration of particles between the filter fibres and filtration effects for smaller particles in the horizontal flow over the impaction surface [159].

Cascade impactors suffer also from inlet and wall losses, which can distort further the determination of the true aerosol size distribution curve. Inlet losses occur particularly for larger particles and are the consequence of anisokinetic sampling. The deposition of large particles on the inlet is variable with inlet geometry and wind conditions. The addition to the impactor of special inlet devices has been used for pre-separation of large particles, with a constant efficiency independent of ambient conditions [160, 161]. Pre-separation of larger particles before impaction can help in reducing bouncing problems.

Wall losses (or inter-stage losses) are a result of particle deposition on surfaces other than the impaction plate. These losses are higher for large particles and happen principally on the inlet and first stages [162]. Wall losses tend to decrease with particle size, being negligible for particles with dimensions smaller than 1–2 μm [163]. Wall losses are more important in conditions of particle re-entrainment, being generally of the order of 5 to 10% of particle mass [164].

The problems of particle bouncing and blowing off can be overcome by use of the virtual impaction principle, in which the solid impaction surface is substituted by impaction into a void [165]. The particles are accelerated through a nozzle, after which the flow stream is drawn off at right angles (see Fig. 4.9). The small particles follow the right angle flow stream, while the larger ones, because of their inertia, continue towards the collection nozzle. The aerosol mass is separated into two fractions (hence the apparatus is also known as a dichotomous sampler).

To minimize wall losses it is necessary to maintain a small flow of air into the coarse particle receptor (generally 10% of inlet flow). Inherent in this separation technique is the contamination of the coarse particle fraction with a small percentage of fine particles. This is not considered a substantial problem in mass measurements, as a simple mathematical correction can be applied. To reduce coarse particle fraction contamination, two single stage dichotomous samplers can be used in series. Dzubay and Stevens [166] describe a virtual impactor with two stages, each stage presenting similar cut-off point characteristics and having a 7:1 (large fraction/fine fraction) flow rate ratio. Therefore an overall concentration ratio of 49:1, for the inlet and second stage coarse particle stream, was obtained, and only 2% of fine particle mass was collected with the coarse particle fraction. The authors verified the existence of important wall losses ($\sim 25\%$) for particles in the 2–3 μm and > 10 μm ranges.

In dichotomous sampling both particle fractions are collected by filtration on membrane filters. The particles are deposited uniformly over the entire surface of the collection substrate, which can be very advantageous for certain analytical techniques such as X-ray fluorescence. The commercially available equipment generally separates the aerosol mass into two fractions characterized by their capability either to penetrate the alveolar region, or to be retained in the upper respiratory tract of the human body.

Centrifugal impactors

The empirical nature of cyclone design may be a handicap as compared with conventional cascade impactors, as the fractionation curves cannot be explained by existing models.

Cyclones are generally used for separating the aerosol mass into two fractions: a coarse and fine fraction. Collection of larger particles excluded by the cyclone on a removable substrate is difficult. When only the fine fraction is to be collected the sampling process is much more simple. The coarse particles fraction can be determined by sampling of total aerosol mass in parallel.

John and Reischl [167] designed and calibrated a cyclone which presented a sharpness comparable to that of cascade impactors and which was the same for solid and liquid aerosols. The 50% cut-off point could be changed by varying the flow rate and ranged between 2.1 and 4.5 μm for flow rates between 26.6 and

11.8 litres min $^{-1}$. Re-entrainment of deposited particles was small, always less than 1% of mass loading.

Buchanan *et al.* [168] have developed a liquid scrubber which uses centrifugal action to impinge aerosol micro-organisms on to a thin liquid film. Collecting fluid is aspirated into the sampled air stream and the combined flow enters a cylindrical tube tangentially, creating a spiral flow pattern. The aspirated mist is forced out to the tube walls by centrifugal effect and forms a continuous thin film which then moves in a spiral pattern along the tube to a collection flask under slight vacuum. The micro-organisms are collected on the liquid film as a result of centrifugal impingement and subsequently flow with the collection liquid into the collection flask.

The system operates at flow rates up to 950 litres min $^{-1}$ and is found to be effective for sampling viable micro-organisms intact. The relatively gentle collection mechanism may have application to the collection of other aerosols.

Sequential filtration with Nuclepore filters

Nuclepore filters are thin (6–10 μm), homogeneous, non-hygroscopic poly-carbonate membranes, with a mass density of ~ 0.95 μg cm^{-2}, penetrated by circular pores normal to the surface. These pores are distributed randomly over the surface, being very uniform in size, with geometric standard deviations of the order of 1.1, or less.

The special characteristic of Nuclepore filters, with pores of well-defined geometry, makes possible the development of theoretical models which can describe, to a good approximation, the true sampling efficiency of this filter type [169–171]. The collection of particles in filtered air flow is the result of impaction, interception and diffusion mechanisms, the efficiency of filtration being variable with filter pore size, particle dimension and face velocity.

Sequential filtration with Nuclepore filters is sometimes used for size fractionation of atmospheric aerosols [172]. The method exploits the well-defined geometry of Nuclepore filters, which collect the large particles and let the smaller ones pass to a sequence of Nuclepore filters of decreasing pore sizes. The smallest particles are collected by a back-up filter of high efficiency.

Stacked filter units containing two filters were used by Parker *et al.* [173] for sampling separately respirable and non-respirable aerosols from the atmosphere. The larger particles were collected on a 12 μm pore size filter and the smaller ones in a subsequent 0.2 μm pore size filter. Cahill *et al.* [174] used a stacked filter unit with 8.0 μm and 0.4 μm pore size Nuclepore filters for the same purpose. Twomey and Zalabsky [175] employed a set of Nuclepore filters (0.95, 2.7, 4.5 and 7.0 μm) and flow rate combinations to obtain data for the determination of natural aerosol size distributions.

As sampling progresses, deposition of particles at pore edges and on pore walls can cause clogging, with a gradual decrease in effective pore size, which affects separation characteristics. Clogging manifests itself by an increase in pressure drop over the filter. It seems that clogging is not important with the amounts of particles collected in usual air quality measurements. Parker *et al.* [173] verified that with 12 μm pore size Nuclepore filters, loadings as high as 90 μg cm^{-2} did not alter collection characteristics. For sampling periods of 24 h this loading is equivalent to an atmospheric coarse particle concentration of the order of 100 μg m^{-3}. Jensen and Kemp [176] used a Nuclepore 6 μm pore size filter in an urban area and verified that clogging did not take place for filter loadings of up to 130 μg cm^{-2}.

Bouncing problems may affect the efficiency of collection, principally at face velocities exceeding 80 cm s^{-1}. For solid particles bounce effects may appear even at low face velocities, which give different cut-off size characteristics for solid and liquid particles [177]. The coating of the filter surface with a thin layer of grease reduces bouncing and particle fall-out during handling and transport, without affecting the separation characteristics of the filters [178].

Diffusion battery

The diffusion battery has long been used for the dynamic measurement of aerosol particles with diameters smaller than 0.2 μm [179]. Sinclair [180] has described a battery with a collimated holes structure which was formed by an assembly of stainless steel discs of several different thicknesses, containing a large number of nearly circular holes. The discs were connected in series, having between each group of discs a side tube for sampling of processed aerosol mass. Experimental tests with monodisperse aerosols showed a good agreement between theoretical predictions and experimental data [181].

The collimated holes structure diffusion battery was used by Sinclair in conjunction with a Condensation Nucleus Counter for the determination of the numerical size distribution of particles in the sub-micrometer range. By sampling in turn from various ports of the battery, each of which involves a different path length travelled by the air stream within the battery, a size distribution could be obtained.

Sinclair and Hoopes [182] developed a screen type diffusion battery, in which the perforated steel plates were substituted by groups of circular stainless steel screens mounted in sequential groups containing increasing numbers of screen units. The apparatus has been marketed commercially (TSI 3040 diffusion battery), and works at flow rates of 4 or 6 litres min^{-1}. The battery consists of ten stages having 1,2,3 . . . 10 stainless steel screens in respectively the first, second . . . tenth stages. At the end of each stage there is a sampling port from which the air can be alternatively sampled. The port selected on the battery determines the size range of particles collected.

The utilization of the diffusion battery for the determination of the chemical composition of size-differentiated aerosols has become attractive in recent years, in consequence of the realization that an important fraction of atmospheric sulphates, nitrates [183] and lead [184] may exist as particles smaller than 0.3 μm, where cascade impactors are inefficient.

Tanner and co-workers [183, 185, 186] have used the collimated structure diffusion battery for the size-fractionated sampling and analysis of sulphate aerosols in the New York atmosphere. Ambient samples were collected directly and after passing through the battery. The battery-processed sample contained only particles larger than 0.035 or 0.15 μm, depending on the number of stages used. A flow of 29 litres min^{-1} was drawn through the first seven segments of the battery and 11.4 litres min^{-1} through the remaining eleven segments. The processed samples, collected on quartz filters, contained particles respectively larger than 0.035 μm and 0.15 μm. Analysis of unprocessed samples permitted the determination, by difference, of the mass of smaller particles.

Harrison and Pio [187, 188] employed a TSI 3040 screen type diffusion battery, associated with a Hi-Vol cascade impactor, for simultaneous size differentiated sampling of optical and sub-optical aerosols which were analysed for sulphates, nitrates and ammonium. The apparatus is shown schematically in Fig. 1.28. Air enters the system into an Andersen 4-stage Hi-Vol impactor initially, where particles are separated into fractions of aerodynamic diameter >7 μm; 3.3–7 μm; 2–3.3 μm; 1.1–2 μm and <1.1 μm. Fractionation of particles smaller than 1.1 μm was achieved by drawing air, from immediately before the back-up filter, through the diffusion battery, at a flow rate of 6 litres min^{-1}. Air could be sampled from any port of the battery and particles penetrating the battery were collected on a 47 mm diameter, 0.45 μm, Teflon membrane filter. The port selected on the diffusion battery determined the size range which was collected. Expressed in terms of 50% penetration, the size cut-off for ports 1 to 10 was respectively 0.005; 0.014; 0.026; 0.042; 0.060; 0.080; 0.105; 0.130; 0.158; 0.190 μm. Hence if port 5 was selected the particles collected on the Teflon filters

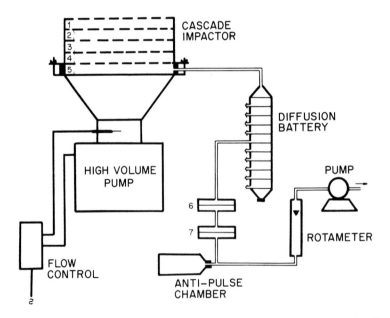

Fig. 1.28   Schematic diagram of apparatus for size-differentiated sampling of optical and sub-optical aerosols. 1–4 impactor stages; 5 impactor back-up filter; 6 Teflon filter; 7 ammonia-impregnated filter [187].

represent the size range 0.06–1.1 μm. The fraction <0.06 μm could then be determined from the analysis of the back-up and Teflon filters.

This design has the advantage over the apparatus of Tanner and co-workers [183, 185, 186] that very large particles are pre-separated by the cascade impactor and do not enter the diffusion battery where they may be intercepted, giving a misleading picture of size distributions.

1.2.2.4   *Emission source aerosol sampling applications*

Stationary sources

The sampling of aerosol emissions from stationary sources has received a great deal of attention over the years. As a result, there exists a wealth of detailed literature on overall source sampling procedures [18, 35, 42, 45, 189, 190] and standardized source sampling techniques [36–41, 44, 191–193]. Some of these will be discussed in order to illustrate the techniques available.

British Standard, BS 3405: 1971 [39, 40]

This standard provides for sampling from stacks for particulates larger than 1 μm with an expected accuracy of $\pm 25\%$. The equipment for the method includes: a pitot static tube, inclined gauge manometer, thermometer, probe tube, flow meter, flow rate control valve, connecting tubing, vacuum blower and a collection component. The latter should be capable of filtering 98% of particulates over 1 μm. Some different collection component assemblies have been approved as complying with the requirements of the standard.

Testing is performed at a minimum of 4 sampling points at the centroids of equal area sectors of the stack cross-section for a minimum of 2 min at each location as illustrated in Fig. 1.29. For stacks in excess of 2.5 m^2 cross-sectional area or where velocity pressures at sampling points vary from one another by more than 4 to 1, 8 sample locations are used.

Following a gas velocity traverse, as described in Section 1.1.2.1, the sample probe is positioned at the sampling location. The vacuum blower is started with the control valve closed. The control is opened until a precalculated flow rate for isokinetic conditions is reached. The probe is moved from location to location, providing the same sampling period at each. After completion of sampling at all locations, the probe may be withdrawn, or improved accuracy may be achieved by repeating sampling at each location.

After removal of the probe from the stack, the collection device is detached, deposits from the probe and sample lines are added and the collected particulate matter is sent to the laboratory for gravimetric and/or chemical analysis.

American Society for Testing and Materials (ASTM) [37]

This standard provides a procedure for sampling stacks utilizing an in-stack particulate collector. An example of one assembled sampling train which is consistent with this standard is shown in Fig. 1.30. Three types of filter media are specified for the in-stack filter holder. Alundum thimbles are useful for

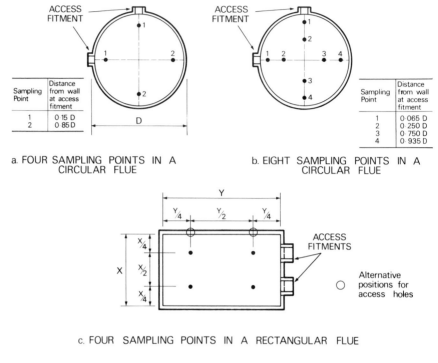

Sampling Point	Distance from wall at access fitment
1	0·15 D
2	0·85 D

a. FOUR SAMPLING POINTS IN A CIRCULAR FLUE

Sampling Point	Distance from wall at access fitment
1	0·065 D
2	0·250 D
3	0·750 D
4	0·935 D

b. EIGHT SAMPLING POINTS IN A CIRCULAR FLUE

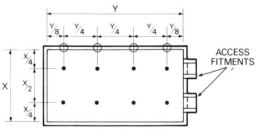

c. FOUR SAMPLING POINTS IN A RECTANGULAR FLUE

d. EIGHT SAMPLING POINTS IN A RECTANGULAR FLUE

Fig. 1.29 Sampling points in circular and rectangular flues. (After [40].)

applications at temperatures up to 550° C. Filter holders providing wire mesh or sintered stainless steel support for various round sheet filter media are included, but these are generally not suitable for high temperature operation.

The condenser system consists of two 500 ml modified Greenburg–Smith impingers, each filled with a minimum of 100 ml of distilled water, a third

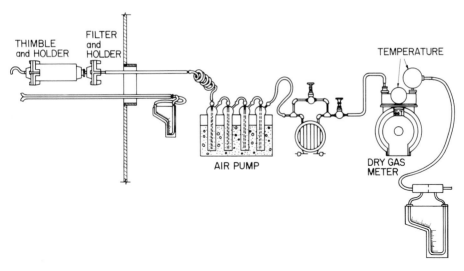

Fig. 1.30   Schematic diagram of sampling train for dry particulate matter using an in-stack sampling arrangement. (After [37].)

modified impinger empty and a fourth modified impinger containing 300 g of desiccant (indicating type). All impingers should be connected in series and immersed in an ice bath. Flow metering is provided by a calibrated orifice meter. A dry test meter provides an independent means of verifying isokinetics and gas volume. The vacuum source is a pump capable of maintaining sufficient flow through the sampling train resistance to provide for isokinetic sampling conditions.

The method calls for a minimum of 4 sample locations for rectangular stacks and a minimum of 8 for circular stacks. These may increase to 24 or more for larger stack cross-sections. Furthermore, the number of sample points should be doubled when only 4 to 6 diameters of straight stack are available from the closest upstream disturbance rather than the required 8 stack diameters. When less than 4 stack diameters are available, isokinetic sampling is considered to be difficult to achieve.

Following a gas velocity traverse and prior to sampling, the equipment is leak tested by plugging the probe and drawing a vacuum of 500 mbar. If the leakage rate is not in excess of 0.56 litres min^{-1}, the test is started with the probe in position and nozzle facing downstream. When the test is started the vacuum source is actuated, the control valve opened and the nozzle turned to face directly upstream. The sampling time at each location should be at least 2 min under steady conditions with a total sampling time of at least 1 h when possible. Sampling is maintained isokinetically during the sampling period by adjusting the flow through the sample train to precalculated values determined from the velocity traverse and stack gas conditions.

Upon completion of the test, the control valve is closed, the vacuum pump shut off, the nozzle turned downstream and the probe removed from the stack. The whole apparatus, or the collection component only (where it is feasible to change the component in the field), is removed and sent to the laboratory for analysis.

United States Environmental Protection Agency (EPA) procedure [38]

This standard provides a procedure for sampling stacks utilizing an out-of-stack collection component. The relative accuracy of out-of-stack versus in-stack collectors has been discussed by Hemeon and Black [194] who maintain that out-of-stack collectors are subject to serious error.

The EPA procedure described is currently the most restrictive and specific with regard to optional equipment as it is used for regulatory purposes. Details of construction [195] and operation, maintenance and calibration [47] have been published for this sampling method.

The sampling equipment required includes the same general components as the ASTM procedure except that an orifice flow meter is mandatory. The assembled EPA sampling train is illustrated in Fig. 1.31. A sharp edged stainless steel nozzle is used with a heated Pyrex or corrosion resistant alloy sample probe. A type S pitot tube is attached directly to the sample probe to allow simultaneous sample collection and velocity measurement.

The particulate collection medium is specified as a glass fibre filter without organic binder, exhibiting at least 99.95% efficiency on 0.3 μm dicotyl phthalate smoke particles. The filter is mounted against a sintered glass plate which is

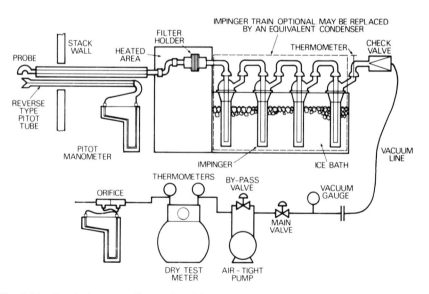

Fig. 1.31   Particulate-sampling train. (After [38].)

mounted in a heated Pyrex filter holder. This is followed by an impinger train or a measuring condenser, then an air tight pump, dry test meter and orifice flow meter.

The number and positions of sample locations are determined from Figs 1.1–1.3. Following a gas velocity traverse and before sampling, the equipment is leak tested by plugging the probe. A leakage rate of less than 0.6 litre min^{-1} at 0.51 b vacuum is considered acceptable.

The sampling procedure is similar to the ASTM and BS procedures. A minimum of 2 min is required at each sample location. During sampling, isokinetic flow rates are maintained by adjusting the flow control valve to achieve calculated isokinetic flow rates through the orifice, using the simultaneous stack gas velocity measurement from the pitot tube. Field nomographs are available and necessary to perform the flow rate calculations, since adjustment must be made for the different gas conditions in the stack and at the orifice.

At the completion of the test, the blower is turned off and the probe and nozzle removed from the stack. The equipment is removed to a clean location and disassembled. Particulate matter collected in the nozzle or probe is washed out with acetone and collected in a sealed container. The water volume collected in the condenser is measured for the moisture determination and the filter holder with filter in place and probe washings are returned to the laboratory for analysis.

The high volume stack sampler [189]

The stack sampling devices described in the foregoing procedures are suitable for high velocity, confined stack emissions. Boubel [189] has developed a stack sampling system for use with low flow rate and unconfined emission sources.

Boubel considered the disadvantages of the common standard sampling techniques. Among those listed are: expensive equipment, complicated use and analysis, inability to use collected sample for particle size analysis, and inability to sample at flow rates greater than 30 litres min^{-1}.

The high volume stack sampler developed, is constructed entirely from aluminium, to provide the convenience of light weight. The device incorporates a nozzle connected to a sample probe which in turn is connected to a filter holder designed to hold a standard 20×25 cm high volume glass fibre filter. From the filter holder, the flow runs through a valve and a sharp edged orifice, connected to a dial reading manometer (magnehelic gauge). The dial is located in a control panel positioned on the sampler body with a magnehelic gauge reading from a pitot tube attached to the sampling probe. The unit is then connected to a high volume blower by means of flexible tubing.

After a preliminary velocity traverse with the pitot tube and a preliminary sampling run to determine sample temperature through the filter and orifice, the isokinetic sampling rate is determined by reference to precalculated operating curves. Sampling may then proceed at selected sampling locations as long as necessary to collect a measurable sample. For particulate emissions of 0.25 g m^{-3}, a 1 min sample period would produce an adequate sample. At the completion of testing, the particulates collected in the nozzle and probe are collected by rinsing with methyl alcohol. The filter and probe rinsings are then returned to the laboratory for gravimetric, chemical and/or particle size analysis.

Mobile sources

Aerosol sampling from mobile sources provides a difficult challenge. Vehicle aerosol emissions cover a wide spectrum from particles several millimetres in diameter down to submicron particles. The wide range in particle sizes, high temperature and humidity of exhaust gases and the unsteady and non-uniform vehicle exhaust flow all contribute to make exhaust aerosol sampling difficult.

Hirschler *et al.* [196] pioneered vehicle exhaust aerosol sampling in a study aimed at determining vehicle lead emissions. The entire exhaust stream was passed through an electrostatic precipitator after approximately a 10 to 1 dilution with filtered air. The system efficiency was determined to be 90 to 95% for lead particulates based on collection with a back-up membrane filter. For this study, the precipitator was lined with a polyvinyl acetate liner which was dissolved in trichloroethylene after sampling. The inorganic particulates were then collected from the solvent by filtration.

McKee and McMahon [197] collected particulate samples on 20 cm × 25 cm sheets of glass fibre filter, using a vacuum blower to draw the exhaust sample through the filter. Samples for electron microscopy were collected from the exhaust line, downstream from the exhaust manifold, and diluted with 2 to 3 volumes of dry air to prevent condensation. The diluted air was sampled with a thermal precipitator which collected particulates directly on to a perforated brass screen mounted on a glass slide for electron microscopy.

Mueller *et al.* [198] studied isokinetic probe sampling for particulates directly from the vehicle tailpipe. The collected exhaust was diluted with 4 volumes of filtered, dry air. The flow stream passed through a baffle into a sampling chamber from which aerosol samples were drawn for particle size fractionation.

Habibi [199, 200] studied the requirements for vehicle particulate sampling and developed criteria for a suitable partial (proportional) collection system. Included in the criteria were a suitable vehicle operating cycle, prevention of condensation by dilution or heating, equal proportion collection of all particle sizes, isokinetic sampling, no loss or change of particles in the sampling system and capability of long duration sampling. Habibi developed the system illustrated in Fig. 1.32. With this system, the exhaust stream is led into a large tunnel and diluted with filtered ambient air. A dew point calculation showed that a 4 to 1 dilution was required to avoid condensation, but the system provided much higher dilutions (23 to 1 for 70 km h^{-1} road load). The tunnel was operated on the variable dilution principle whereby a fixed outflow from the tunnel is maintained by the exhaust blower while the total fixed inflow is provided by the variable engine exhaust flow and a compensating dilution air inflow. A tunnel mixing length of 12.2 m was provided along with an inlet orifice to promote turbulent mixing. The combination of tunnel length and inlet orifice was a compromise to produce good mixing along with a steady velocity profile at the

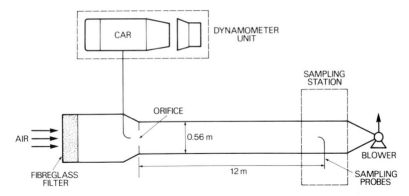

Fig. 1.32 Details of proportional vehicle exhaust sampling system. (After [199].)

sampling point. Isokinetic sampling was undertaken with 0.95 and 1.9 cm diameter sampling probes. The system performed satisfactorily except for slightly anisokinetic sampling conditions due to temperature fluctuations and the inevitable loss of particulates which settled to the bottom of the dilution tunnel.

Habibi [199] also developed a total flow filter which mounts directly on the exhaust pipe and provides less than 5 mb pressure drop at 110 km h^{-1}, making the system suitable for road testing. The device (Fig. 1.33) consists of a cylinder, packed with a glass fibre filter medium. The exhaust flows into the cylinder and passes radially out through the medium which is supported on a stainless steel grid. The unit is sealed by internal springs and stainless steel strip over the seams. When used for extraction and recovery of exhausted lead particulates, the total filter was found to be more than 99% efficient.

Campbell and Dartnell [201] worked with a total filter (Fig. 1.34) and found difficulties with condensation and temperature effects when trying to determine total emitted particulate concentration. However, they found the system was useful for the determination of metallic particulates. Exhaust gas flow would collect on the glass filter medium which was removed, solvent extracted to remove hydrocarbons and then macerated prior to metals analysis.

Ter Haar *et al.* [202] developed a system to run the total exhaust into a 70 m^3 black polyethylene bag along with a minimum of 8 to 1 filtered, dry (less than 10% relative humidity) dilution air, to minimize the effects of humidity on particle agglomeration and fallout. Upon completion of a driving cycle, sampling commenced within the bag using a 0.6 cm diameter stainless steel probe inserted 1.8 m into the bag. The sample was drawn through a 47 mm Millipore type AA, 84 μm membrane filter at a flow rate of 7 litres min^{-1}. Sampling was also performed with the bag using an open faced filter holder with a 4.25 cm glass fibre filter.

Ter Haar *et al.* [202] also developed a total filter for connection directly to the

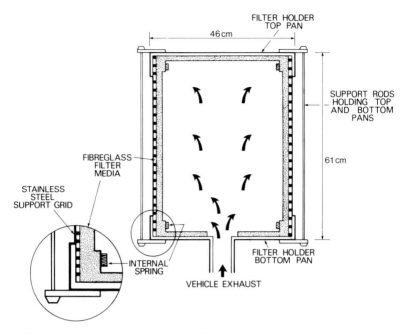

Fig. 1.33   Total exhaust filter. (After [199].)

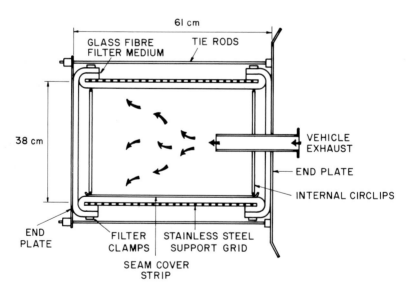

Fig. 1.34   Alternate total exhaust filter. (After [201].)

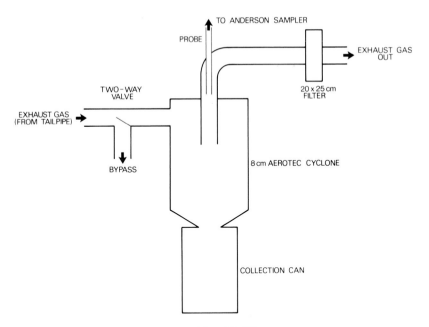

Fig. 1.35 Cyclone total exhaust filter. (After [202].)

exhaust pipe (Fig. 1.35). The unit uses a cyclone having a 50% cut-off for 0.5 to 5 μm particulates backed up by a glass fibre filter with a 99.9% efficiency for 0.3 μm particulates.

Sampson and Springer [203] conducted particulate sampling at several points along a simulated exhaust system shown in Fig. 1.36, using three separate collection components (Fig. 1.37). Each component was used in turn with a sharp nosed probe for isokinetic sampling. The first collection component was a high temperature cascade impactor with back-up and bypass filters. This component provided particle size fractionation at exhaust gas temperatures up to 450° C or total particulate mass counts depending on which valve was open. The second collection system consisted of two filters in series with a heat exchanger between them, to effect exhaust gas cooling. The third system provided metered ambient dilution air flow into the system to cool the exhaust before the filter. In all cases, glass fibre filters were used and mounted in filter holders capable of withstanding 450° C. After passing through the collection component in use, the sampled air flow was passed through heat exchangers to bring exhaust gas temperatures to 21° C. Total gas flow was measured with a wet test gas meter.

Sampson and Springer [203] evaluated several of the procedures for their system. They found that 250 cm downstream, from the exhaust port, the unsteady flow had evened out sufficiently that relatively small sampling errors were incurred while attempting isokinetic sampling. However, they added a surge tank to smooth further the unsteady exhaust flow before the sampling points.

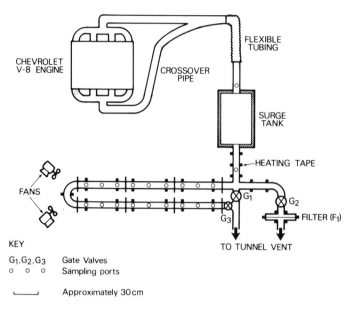

Fig. 1.36 Schematic diagram of exhaust system. (After [203].)

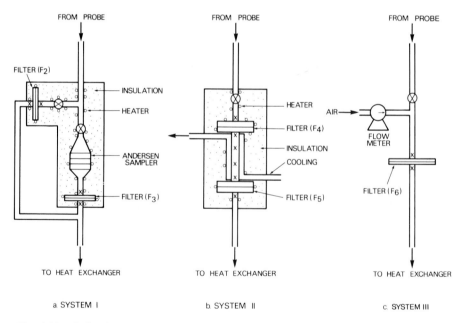

Fig. 1.37 Collection components for a particulate sampling of automobile exhaust. (After [203].)

Variations in sampling flow rate around the isokinetic rate were found to have limited effect on measured particle concentrations, but isokinetic sampling was used anyway. The effect of sampling probe diameter was evaluated and found to have no consistent effect on measured particle concentrations. The particle distribution across the exhaust pipe was evaluated and found to vary by less than 4% so that sampling in the centre of the pipe was adopted, along with a 10 min sampling period.

1.2.3 Gases

1.2.3.1 *Gas sampling considerations*

Gaseous sampling from the atmosphere requires careful attention to avoid the presentation of a modified or diminished sample to the analysis step. Although gaseous constituents of the atmosphere are not subject to the inertial and electrostatic effects which were discussed for aerosols, special precautions are necessary for accurate sampling. Several general considerations in gaseous sampling will be mentioned.

Where heavy concentrations of a gas may be present in a stagnant or slow moving air mass, a density differential between the gas and air may result in an uneven distribution of the gas within the air mass. This possibility is generally overcome by sampling from a turbulent flow or in a well mixed air mass.

Gaseous constituents of the atmosphere are in a relatively high energy state and may be reactive with components of the sampling system or other gaseous constituents. It is essential to evaluate all components of the sampling system carefully for their possible reactivity with sampled gaseous constituents. In many cases, it may be necessary to pretreat the sampled air stream before the gas reaches the collection component of the sampling system. In such cases, for example, prefiltration of particulates or moisture removal, the pretreating units may remove significant quantities of the gaseous constituent to be sampled. Where pretreatment is used, the pretreatment units should be analysed for their content of the relevant gaseous constituents.

All possible precautions should be taken against inward or outward leakage, particularly for stored samples. However, sealing lubricants may collect gaseous constituents and their use should be avoided or carefully evaluated. Self lubricating PTFE joints and stopcocks are preferable and these can generally provide relatively air tight seals. Diffusion of gases through apparently impermeable materials such as rubber, Neoprene and plasticized PVC is possible [1] and should be evaluated where samples will be stored for periods in excess of a few hours.

Condensation of atmospheric moisture within sample lines can create reactive conditions for some combinations of gaseous constituents, or the moisture may simply collect gaseous constituents before they reach the collection component. Many gaseous constituents are photochemically reactive and may undergo reactive changes within the sampling system if they remain exposed to light energy. Light sealed containers for sampling systems would simply eliminate this problem.

The sampling system must be designed to draw continually a fresh sample of the atmosphere being studied and should not allow the resampling of air which has passed through the system. An extended exhaust line to a location remote from the sample intake will normally prevent resampling from occurring.

1.2.3.2   *Gas sampling collection components*

Adsorption

Adsorption of gases from the atmosphere is a surface phenomenon whereby gas molecules are concentrated and bound by intermolecular attraction to the surface layer of a collection phase. Under equilibrium conditions at constant temperature the volume of gas adsorbed on the collection phase is proportional to a positive power of the partial pressure of the gas (Freundich's adsorption isotherm). As well, adsorption, being a surface phenomenon, is dependent upon the relative surface area of adsorbent material. Various gaseous constituents of a mixture will be adsorbed in amounts approximately inversely proportional to their volatility.

Adsorption is useful in a collection component for gaseous constituents of the atmosphere because of its ability to concentrate trace concentrations by sampling large volumes of air. Materials which are commonly used as adsorbents include activated carbon, silica gel, alumina and various gas chromatographic support phases.

Several factors need to be considered in selecting an adsorbent. A high relative surface area is important to maintain a large contact area for adsorption, while retaining the maximum space between adsorbent granules for maximum air flow rates. Silica gel and alumina offer surface areas of $200-600 \text{ m}^2 \text{ g}^{-1}$, activated carbon offers $500-2000 \text{ m}^2 \text{ g}^{-1}$, whereas GC phases can offer considerably higher surface areas per g.

Relative affinity for polar or non-polar compounds is important for selecting the adsorbent to sample a given gaseous constituent. Activated carbon is non-polar and therefore has affinity for organic compounds to the exclusion of polar gases, including water vapour. Silica gel and alumina are polar and have increasing affinity for higher polarity gases. This may cause desorption of lower polarity gases in the presence of higher polarity gases. In addition, water vapour does collect on these adsorbents and may lower adsorption efficiency as well as clogging the granular bed, resulting in reduced flow rates.

The adsorbents must not be chemically reactive with the gases to be collected, unless chemisorption is used intentionally. In this instance, adsorbents are specially treated to provide a reactive surface for chemical interaction between a molecular layer of the gas and the chemisorption media. Where this technique is used, the quantitative and qualitative results will be evaluated with full awareness of the reactions which have occurred.

The adsorbent material should not be prone to fracturing, crushing or flaking which may result in carrying over fine particulates with the sampled air stream. Most adsorbents are sieved to remove fines before use.

The retention capacity of the adsorbent should be predictable and high in order to avoid non-quantitative recovery of gaseous constituents. For activated carbon, the adsorption of gaseous constituents is complete initially, but after the retention capacity has been reached, the gases will be incompletely adsorbed until the carbon becomes saturated and no further adsorption takes place.

The desorption properties of adsorbents must be amenable to quantitative recovery of the collected sample, preferably with regeneration of the adsorbent for subsequent use. Activated carbon, although a very efficient adsorber, is very difficult to desorb quantitatively. Steam stripping, with the attendant risks of hydrolysis reactions with collected constituents, is often necessary for assurance of quantitative recovery of adsorbed materials. Alternatively, heating under vacuum and distilling components to another collection medium is often possible. In contrast, support bonded GC phases have the advantage of allowing desorption by solvent extraction or by heating and flushing with an inert carrier gas.

Relative selectivity for atmospheric gases will be important depending upon the objectives of the sampling programme. Activated carbon is relatively unselective and as such is useful for screening atmospheric constituents while silica gel is somewhat more selective and particular GC phases may be very selective.

Adsorption is also a temperature dependent phenomenon and efficiency will be improved at lower temperatures, particularly for gases which have lower boiling points. Gases such as hydrogen, nitrogen, oxygen, carbon monoxide and methane are not normally adsorbable without resorting to chemisorption. Ammonia, ethylene, formaldehyde and hydrogen sulphide are not completely adsorbed at normal ambient temperatures, but adsorption can be increased by cooling the adsorption device.

Adsorption systems provide a flexible means of collecting diverse gaseous constituents from the atmosphere and concentrating them prior to analysis, but the system chosen must be carefully evaluated to avoid non-quantitative adsorption, breakthrough effects due to exceeding the retention capacity or non-quantitative desorption of gaseous constituents.

Absorption

Absorption of gases from the atmosphere is a solubility phenomenon whereby gas molecules are preferentially dissolved in a liquid collecting phase. The degree of absorption of a given gaseous constituent in a particular solvent is limited by the equilibrium partial pressure of the vapour over its liquid solution. This limitation on absorption of gases into solution can be overcome where the gaseous constituent undergoes a relatively irreversible chemical reaction with the absorbent.

The efficiency of absorption of gases into a liquid phase is also very dependent on contact surface area and absorbing devices are designed to achieve the maximum area contact between gas bubbles and the absorbent. This is normally achieved by transforming the air stream to small, finely dispersed bubbles with a relatively long travel time through the absorbent.

As with adsorption, absorption is temperature dependent with lower temperatures improving efficiency for particularly volatile constituents. The efficiency of absorption on various gaseous constituents was studied by Sexton [204] and is summarized in Table 1.9. Various absorption devices have been devised to provide efficient absorption. Several absorber designs are illustrated in Fig. 1.38.

Table 1.9   Gas washing bottle efficiency tests (after [204])

Chemical	Solution in gas washing bottle	Sampling rate (litres min^{-1})	Gas washing bottle efficiency (%)	Stack concentration ppm (by weight)
Sulphuric acid	Water	3.4	99.8	72
Hydrochloric acid	Water	28.3	97.1	47
Hydrochloric acid	Water	3.4	100	47
Perchloric acid	Water	3.4	99.8	41
Acetic acid	Water	3.4	99.5	1065
Nitric acid	Water	3.4	100	182
Ammonia	Water	3.4	84.3	103
Chlorine	Water	3.4	Very low[a]	103
Chlorine	2% Na_2CO_3	3.4	77.0	103
Ethylenediamine	Water	3.4	99.6	211

[a] Chlorine is only 1% soluble in water.

The simple Dreschel bottle may be used in some applications. It has the advantage of high flow rate capacity, up to 30 litres min^{-1}, but is relatively inefficient in comparison with other designs. Standard impingers, as used for aerosol sampling, have been used as absorbing devices, but they are not markedly more efficient in many cases, than Dreschel bottles.

Absorbers with fritted glass diffusers have been constructed in various configurations. These have the advantage of creating a small bubble size to enhance efficient absorption. Fritted glass absorbers are the best choice in many applications, but they must be protected from blockage by particulates and adequate freeboard with the collecting fluid must be provided to avoid foaming over. As well, adsorption of gaseous constituent on the glass frit is possible and must be evaluated. Spiral absorbers have been designed to achieve longer detention times for the rising air bubbles, but these devices are flow rate limited and may be unable to sample adequate air volumes.

Bead packed absorbers operate on the principle of buffeting the gas flow through bead packing to provide mixing and dispersion of the gas throughout the absorbent. This principle may be used to create long pathways for the gas flow to ensure complete absorption. However, the beads themselves must be evaluated to ensure that they do not adsorb or react with the gaseous constituents. It is advantageous with most absorber configurations to concentrate the gaseous

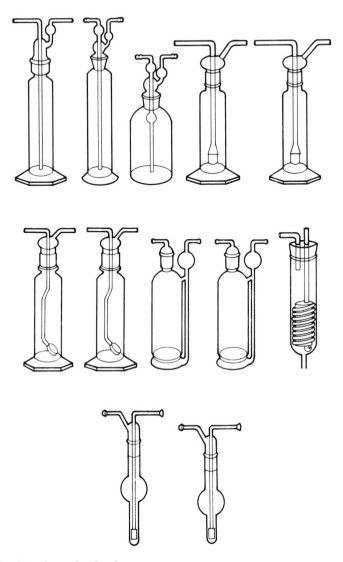

Fig. 1.38 Gas absorption bottles.

constituent into a small volume of absorbing solution in order to enhance subsequent analytical procedures.

Absorbers may provide effective concentration of gaseous constituents, but, in all cases, the actual efficiency of collection must be established and saturation values determined to avoid oversampling and subsequent misinterpretation of results.

Condensation

Condensation of gases from the atmosphere depends upon cooling the gas stream to temperatures below the boiling or freezing points of the gases to be collected. The temperatures used in a condensation device are selected to ensure adequate condensation of the gas by lowering its vapour pressure to less than 1.33 mb generally [1].

Various refrigerants are possible and a list of refrigerant temperatures is provided in Table 1.10. Because of the low temperatures involved, good insulation must be provided and, therefore, the equipment involved in condensation devices is rather bulky. Dewar flasks are commonly used to contain the refrigerant and collecting trap.

Table 1.10 Summary of cold bath solutions (after [18])

Coolant	*Temperature* ($^\circ$ C)
Ice and water[a]	0
Ice and NaCl	−21
Carbon tetrachloride slush[a,b]	−22.9
Chlorobenzene slush[a,b]	−45.2
Chloroform slush[a,b]	−63.5
Dry ice and acetone[a]	−78.5
Dry ice and cellosolve[a]	−78.5
Dry ice and isopropanol[a]	−78.5
Ethyl acetate slush[a,b]	−83.6
Toluene slush[b]	−95
Carbon disulphide slush[a,b]	−111.6
Methyl cyclohexane slush[a,b]	−126.3
N-Pentane slush[b]	−130
Liquid air	−147
Isopentane slush[b]	−160.5
Liquid oxygen	−183
Liquid nitrogen	−196

[a] Adequate for secondary temperature standard.
[b] The slushes may be prepared by placing solvent in a Dewar vessel and adding small increments of liquid nitrogen with rapid stirring until the consistency of a thick milkshake is obtained.

The primary problem encountered in condensation devices is the large volume of water vapour in the air which will freeze out. This is often overcome by providing sequential traps at progressively lower temperatures, with the first trap designed to collect primarily water vapour. The lowest temperature system which is recommended is liquid oxygen, because lower temperature systems will condense atmospheric oxygen into the collecting device which will present a serious combustion hazard.

Relatively low flow rates (less than 1 litre min^{-1}) are generally applicable with condensation systems, but the concentration efficiency of these devices somewhat compensates. Higher flow rates would provide less refrigerant contact time and would tend to enhance the loss of condensation aerosols from the system. The latter should be recovered by filtration, in any case, to ensure quantitative sampling. Although condensation systems are not suitable for unattended operation, they have the advantage of providing preserved bulk atmospheric samples which are readily amenable to further analysis.

Grab sampling

Grab sampling of gases from the atmosphere involves the direct collection and isolation of the test atmosphere in an impermeable container. The sample may be returned to the laboratory for direct analysis. Clearly, this sampling technique is limited to those gaseous constituents for which sensitive analytical techniques are available, or for which there is a high concentration of a given constituent in the test atmosphere. Due to the fact that the sample is not concentrated, it is much more sensitive to contamination by leaks.

Many devices including evacuated bottles, syringes and various types of plastic bags have been used for this type of sampling. In all cases, the containers must be evaluated for their reactivity with, or adsorption of, gaseous constituents. In some cases, where adsorption is reproducible, preconditioning of the container with large concentrations of the gaseous constituent to be sampled may be used. However, care must be exercised to avoid exposing conditioned containers to temperature and pressure conditions different from those at the time of conditioning, because desorption of the gaseous constituents may occur.

An advantage of most grab sampling techniques is that the volume of air sampled need not be accurately determined, if a small sample of air is to be withdrawn from the grab for direct analysis. However, the temperature and pressure conditions of the subsample taken must be known in order to relate measured quantities of gaseous constituents back to reference conditions. Grab sampling techniques generally require relatively simple equipment and offer variety and flexibility to many air sampling applications.

1.2.3.3 *Ambient gas sampling applications*

Charcoal adsorption tubes

Charcoal adsorption tubes are commonly used to collect many gaseous organic constituents. Reid and Halpin [205] report the successful use of activated charcoal for sampling halogenated and aromatic hydrocarbons while a standard method [206] uses charcoal for the collection of organic vapours. Sampling efficiencies for adsorption of aliphatic hydrocarbons were reported by Fraust and Hermann [207] to be independent of mass sampling rate and to have an efficiency greater than 95% until breakthrough occurred.

Specifications for activated carbon used in adsorption sampling have been reported by the American Society for Testing and Materials [54] and are summarized in Table 1.11. The carbon fines may be removed by sieving or washing with distilled water and drying above 100° C.

The carbon granules are usually packed in a glass U tube, providing 5 to 10 cm length of carbon. The carbon packing is capped at either end by glass wool plugs.

Table 1.11   Specifications for air-purification activated carbon. (After [54].)

Property	Specification
Activity for CCl₄[a]	at least 50%
Retentivity for CCl₄[b]	at least 30%
Apparent density	at least 0.42 g ml⁻¹
Hardness (ball abrasion)[c]	at least 80%
Particle size	passing on No. 6 (3.35-mm) sieve retained on No. 14 (1.40-mm) sieve

[a] Maximum saturation of carbon, at 20° C and 760 mm Hg, in air stream equilibrated with CCl_4 at 0° C.
[b] Maximum weight of adsorbed CCl_4 retained by carbon on exposure to pure air at 20° C and 760 mm Hg.
[c] % of carbon passing a No. 6 (3.35-mm) sieve and largely retained on a No. 8 (2.36-mm) sieve, that remains on a No. 14 (1.40-mm) sieve after vibrating with 30 steel balls of 0.25 to 0.37 in. diameter/50 g carbon, for 30 min.

Reid and Halpin [205] recommend adjusting the plugs to provide a 46–53 mb head loss through the adsorption tube at a flow rate of 2 litres min⁻¹. Sampling rates of 1 to 2 litres min⁻¹ can normally be maintained with carbon adsorption tubes. The required sampling time will normally be determined by experimentation in order to provide adequate sample quantity for subsequent analysis.

Activated carbon desorption is generally done by steam stripping or vacuum distillation [54]. For the former, superheated steam at 300° C or higher is flushed through the carbon bed and the steam effluent is condensed for subsequent analysis. With vacuum distillation the carbon is heated to 200 to 250° C and distilled into a condensation train maintained under vacuum.

Gas chromatographic phase adsorption media

A wide variety of gas chromatographic phases have been reported in use for gaseous sampling [208–219]. Conventional GC support materials coated with a liquid phase may be used for sampling, but these materials are limited by their relatively low volumetric retention capacity. Cantuti and Cartoni [210] utilized the relationship of the gas phase concentration to the partition coefficient between the gas and liquid phases and the relationship for specific retention volume to the partition coefficient to calculate sampled concentrations of lead alkyls. The GC phase material was placed in a syringe within an upper leg of a U tube located in the sampling line. Sampling was either done at ambient conditions or at 0° C with the excess moisture collecting at the bottom of the U tube. Flow rates of 1.5 litres min⁻¹ for 10 to 15 min were used. Desorption was performed by flushing the syringe with the GC carrier gas.

Porous polymer GC phases have shown utility for sampling organic gases. Jeltes [208] found virtually 100% recovery of trichloroethylene at sampling rates of 0.05 litre min^{-1}. Mann *et al.* [209] used 10 cm lengths of Chromosorb 101 in 10 mm internal diameter glass tubing to achieve efficient recoveries of hexachlorobenzene at sampling rates of 2 litres min^{-1} for 3 h. Desorption of the sample from the Chromosorb was achieved by shaking the sample with hexane.

Aue and Teli [211] have evaluated the utility of support bonded Chromosorb GC phases for air sampling and found them to be convenient for the collection of high molecular weight organic gases. If the phase material can undergo exhaustive solvent extraction without loss of polymer materials, then solvent extraction of adsorbed material can be used in place of temperature programmed desorption.

Perry and Twibell [212] used Chromosorb 102 to collect atmospheric hydrocarbons, after preparation of the media by extensive solvent extraction clean up. The use of gas chromatographic phases for adsorption of gaseous samples is a versatile, rapidly evolving sampling technique for a variety of gaseous constituents of the atmosphere.

Impregnated filters

The sampling of atmospheric gaseous pollutants may be achieved by passage through a filter material impregnated with a chemical compound, which, by reaction, retains and concentrates it on the filter medium. The technique has several advantages over absorption methods with bubblers, which are fragile, difficult in the handling of the liquid content and present evaporative losses of liquid during sampling.

The sorbed pollutant, or the consequent chemical species formed on the impregnated filter, is extracted with water or other solvent, and analysed by appropriate chemical methods. As during sampling any aerosol present is also collected on the impregnated filter, it becomes essential, in certain cases, to remove suspended particles by pre-filtration of the aerosol with a filter non-reactive towards the gaseous pollutant to be measured.

One of the problems in impregnated filter sampling is related to the total capacity of the impregnated substrate for reaction and collection of the pollutant. If sampling is continued when the filter capacity is near exhaustion, a diminution in the efficiency of collection occurs, resulting in under-estimation of ambient concentrations. For each technique the capacity of the impregnated filter must be determined in the laboratory, utilizing atmospheres containing known concentrations of the pollutant.

The efficiency of impregnated filters for pollutant collection is usually dependent on the concentrations in the sampled air and on the air velocity across the filter. A high efficiency is desirable for a precise measurement of ambient concentrations. Intermediate efficiencies can be accepted if two or more filters are used in series. If the reactive efficiency of each filter is constant and independent of pollutant concentration in the gas phase, the true ambient levels can easily be determined using simple mathematical expressions [220].

Frequently the efficiency of sampling is highly dependent on relative humidity in the collected air stream. Lewin and Zachau-Christiansen [221] have tried the sampling of SO_2 with cellulose filters impregnated with various solutions and concluded that impregnation with 0.5 N KOH gives the best results, with efficiencies better than 90%, for humidities over 30% and concentrations of at least 0.1 ppm. For lower relative humidities the efficiency of collection decreased very rapidly. The technique was based on the reaction of SO_2 with KOH and formation of sulphate compounds, which were analysed by spectrophotometric methods, after water extraction. Interferences by particulate sulphates were overcome by pre-filtration of suspended particles on an inert filter.

Impregnated filter sampling has already been used in the measurement of several other gaseous pollutants such as HCl [222–225], NH_3 [226], volatile arsenic [227], H_2S [228, 229], gaseous fluorides [230], HNO_3 [231] and NO_x [232, 233].

The impregnated filter sampling technique can easily be automated for the sequential collection of gaseous pollutants. A commercially available apparatus (Biolafitte Capteur F115, France) can be used for sequential collection of gaseous pollutants over impregnated filters, for periods of 2 to 24 h, with automatic changes of exposed filters. A maximum of 15 samples can be collected without the intervention of the operator. Flow rates between 40 and 1200 litres h^{-1} are possible. The sampler weighs only 5 kg, occupies a very small volume and is relatively inexpensive. Impregnated filters for several pollutants are available with the apparatus.

Condensation traps

A condensation trap designed to collect gaseous pollutants was designed by Shepherd *et al.* [234]. The system takes account of the formation of frozen aerosols from water vapour by providing a glass wool filter within the condensation tube to prevent loss of the aerosol from the system. The sampler is immersed in liquid oxygen contained in a Dewar flask. The sampling system also includes a wet test meter, a control needle valve and mechanical pump capable of drawing several litres min^{-1} through the system. One variation to the basic condensation trap is the preconditioning for moisture removal, although this runs the risk of causing gas losses by adsorption to the preconditioner. Another variation is the provision of sequential condensation traps starting with one trap just cold enough to freeze the majority of the incoming water vapour, followed by sequentially colder traps to condense sequentially gaseous constituents. This latter modification has the advantage of some preliminary fractionation of gaseous components collected. However, sequential condensation traps are involved and cumbersome, making their routine usage often impractical.

Flexible container sampling

Various flexible containers have been evaluated and used for grab sampling of gaseous constituents. Bags constructed of polyester film, aluminium foil lined polyester, polyvinyl chloride film and various fluorinated plastics have been used successfully for particular applications.

Foil lined polyester bags are recommended in a standard procedure for low molecular weight atmospheric hydrocarbons [235]. These bags are reported as unsuitable for the collection of highly polar gases like SO_2, NO_2 or O_3 [1]. Desbaumes and Imhoff [236] evaluated saran bags for solvent collection by determining the diffusion curves of the solvents escaping the container. They concluded that saran bags were suited for all the common solvents tested except styrene which exhibited too high a diffusion rate through saran.

Schuette [237] reviewed the available sources and applications for several plastic films. Although the references, provided, reported various satisfactory applications, Schuette recommends specific testing of the selected materials to investigate their effectiveness in sampling and storing the specific pollutants to be collected.

For all bag sampling techniques, some means of filling is required. Conner and Nader [238] developed a bag sampling method in which a deflated bag is contained in a rigid box. The sample is taken by applying suction from a vacuum blower to the air surrounding the deflated bag in the box. The net result is the inflation of the bag and sampling rates of up to 10 litres min^{-1} were achieved using a 75-litre bag. Performance comparisons were done between Mylar and Teflon bags and the former were found to be superior. Predictable decay of NO_2, SO_2, O_3 and hydrocarbon samples occurred in the Mylar bags, but the decay rate was low enough to permit several days storage.

Oord [239] reports equipping bags with hardboard sides and handles to allow manual flushing and filling of the bag by compression and expansion of the bag. Curtis and Hendricks [240] developed a gravity operated sampling system with a collapsed bag with cardboard sides supported on a wooden frame. Sampling was activated by opening the valve to the bag which concurrently allowed one of the side plates to drop, expanding the bag and, thereby, collecting the air sample. A sampling rate of 50 litres min^{-1} for 5 min was achieved by this method. Bag sampling methods using small hand held blowers, and by holding the bag in the wind have been suggested [54].

In all cases, the bags used must be carefully checked for leaks, particularly at valves and seals. They must be flushed and completely evacuated before sampling by collapsed bag techniques or thoroughly flushed where manual sampling is done. The bags should not be filled to capacity in order to allow for temperature and pressure variations which may cause gas expansion and encourage leakage. Finally, the bags must be protected in transit against rupture or association with contaminated atmospheres and samples should be analysed as soon as possible after collection.

Syringe sampling

Syringe sampling techniques are useful for rapid and convenient sampling of atmospheres for gases with sensitive analytical techniques or high concentrations of the gas. Lang and Freedman [241] report the use of disposable 10 ml glass barrel syringes equipped with butyl rubber plungers for sampling low molecular weight gases such as CO_2, CO, O_2, N_2 and CH_4. They evaluated several glass and plastic barrel syringes with the criteria of one week's storage without significant gas loss before choosing the glass barrel–butyl rubber combination.

Losses from syringes were attributed to leakage through pinholes or around seals and permeation losses by passage of the gas through interstices on the sealing surfaces.

Meader and Bethea [242] developed the use of polypropylene syringes in preference to glass syringes for sampling reactive gases such as NO_2, Cl_2, HCl and HF. However, the syringes required preconditioning with high concentrations of the gas to be tested in order to control adsorption and reaction effects. Different syringe materials may be expected to react differently depending upon the gases to be sampled so that direct evaluation of the syringes to be used is necessary.

Syringe sampling offers a relatively simple sampling technique but applications are limited to specific small volume sampling requirements.

Rigid container sampling

Rigid containers may be used for gaseous sampling, either by pre-evacuation of the container and filling in the test atmosphere or flushing the test atmosphere through the container several times before sealing the container.

Glass tubes may be evacuated and provided with a breakable seal to allow the inward flow of the test atmosphere. After collection, the tube may be capped with a wax plug. Tubes evacuated for sampling must be structurally sound and they must be used with care after evacuation to avoid accidental implosion. Evacuation to less than 1.3 mb should be achieved. If this is not possible, the bottle should be evacuated in a clean atmosphere and the result corrected to the volume sampled according to the formula [54]

$$V_s = V_b \times \frac{P_1 - P_2}{P_1}$$

where V_s = volume sampled; V_b = volume of bottle; P_1 = pressure after sampling; P_2 = pressure after evacuation of tube.

Evacuated containers are not recommended for reactive gas sampling because of possible reactions of the gas with the sealing wax plug [54]. This can be overcome by filling the container with an absorber with specific chemical reactivity for the gaseous constituent sought.

Many types of rigid containers are possible for use in displacement sampling. Van Houten and Lee [243] report the use of 120 ml glass bottles for sampling solvents with a 50 ml hand operated aspirator bulb. The aspirator was squeezed 20 times to achieve 8 volume changes for the bottle resulting in over 99% air change. The bottles were capped and sealed with three layers of saran and a rubber gasket. Recoveries of 88 to 100% were found after 30 days storage for a variety of organic solvents. Rigid container sampling may be adapted to provide for sampling kits which are suitable for convenient sample collection and shipment.

Passive sampling

An alternative to dynamic techniques, in which an air pump is necessary to draw air through the sampling system, is passive sampling. Passive samplers are based upon the diffusion or permeation properties of the gas to be measured [244].

Most passive samplers rely on the diffusion of the pollutant through a stagnant air layer and absorption in a specific reagent, where an irreversible chemical reaction takes place. If an effective collection medium is employed, the concentration at the boundaries of the reagent surface is nil and a gradient of concentration exists in the stagnant air layer. Each pollutant has a unique diffusion coefficient in air and, due to the gradient of concentration, is transported to the surface with a velocity dependent upon ambient concentration.

Some passive sampling techniques involve the adsorption and subsequent permeation of pollutant molecules through a solid membrane. Passive samplers relying on the permeation principle are especially useful when a liquid medium is used as collecting reagent, or when the pollutant is mixed with interfering gases or vapours. In this case the membrane employed should be highly permeable to the pollutant in question and present a lower permeation capability to the interfering gases.

Some of the limitations of passive sampling are resultant from the much lower transference rates to the absorption medium, compared with that of dynamic techniques. Due to limitations in sensitivity of available analytical methods, passive sampling can only generally be used either for long-term sampling, or for determination of high pollutant levels, as may occur in industrial environments. In these conditions the method has several advantages due to the simplicity and low volume of equipment used and it is becoming an alternative means of determination of personal exposure in factory environments [245].

The limitations and sources of errors in passive sampling are related to the determination of the velocity of diffusion, dependent on sampler geometry and diffusion coefficient, which, being specific for each pollutant, is variable with atmospheric pressure and temperature. If the geometry of the sampling device is not optimum, the thickness of the stagnant air layer near the collection surface may change with wind turbulence and intensity, influencing the velocity of pollutant transference. Also, in conditions of very light winds, an external resistance to pollutant transference may appear; ideally all the resistance to the pollutant transference should be within the stagnant layer inside the device.

Although it is possible to calculate theoretically the velocity of collection in relation to air pollutant concentration, the best way to assess the rate of passive collection is to run the apparatus in parallel with an active device, or to expose it to atmospheres containing known concentrations of the pollutant and possible interferents.

Passive sampling has long been used for measurement of ambient sulphur dioxide with the lead peroxide candle method [246]. In this technique a stick containing lead peroxide is exposed within a louvred box to ambient air over a month. The SO_2 gases in the air react with the lead peroxide, with formation of sulphate, which is measured by turbidimetric methods as barium sulphate. The results are presented as mg SO_2, per day and per 100 cm^2 of candle surface. Determination of true concentrations in the atmosphere with this method are difficult, due to interferences from H_2SO_4 and H_2S and the influence of wind and humidity conditions in the velocity of SO_2 transference to the collection surface. The precision and accuracy of the method are also affected by the reproducibility in preparation of the absorption candles [247].

Huey [248] used an alternative method for SO_2 sampling, in which PbO_2 slurry was added to a 48 mm disposable Petri dish and allowed to dry. The prepared plate was mounted horizontally in a holder and exposed to ambient air for a specific period. Lynch *et al.* [249] have adapted the Huey technique to the sampling of gaseous fluorides, using CaO as absorbent; calibration factors, obtained by running in parallel an impinger, allowed the measurement of atmospheric concentrations on a semi-quantitative basis, with a precision of $\pm 25\%$.

McDermott *et al.* [250] have measured long-term SO_2 ambient concentrations using a permeation device. The method involved the permeation of the gas through a dimethyl silicone polymer membrane, into a manganese salt solution which catalysed oxidation of SO_2 to H_2SO_4. This process, unlike the lead peroxide candle method, provides the possibility of quantitative determination of SO_2 ambient concentrations, because the permeation rate is, in the range 0–80% R.H., independent of atmospheric humidity and functional over a wide range of temperatures. Exposure times as great as 3 months are possible. The permeation rate was linear with atmospheric SO_2 concentrations in the range 0.1–10 ppm.

Hardy *et al.* [251] described a permeation system for collection of chlorine in atmospheres with 0.1–2.0 ppm concentrations. The method was not affected by changes in relative humidity and temperature, nor suffered from interferences from strong oxidants and HCl. The samples were collected, after permeation through a silicone membrane, into a fluorescein-bromide solution. The resultant eosin was measured spectrophotometrically. The method can be used for sampling periods of 30 min to 12 h.

1.2.3.4.   *Emission source gas sampling applications*
Mobile sources

A detailed standardized sampling apparatus has been developed for regulatory vehicle emission testing in the USA [49]. The system (Fig. 1.39) operates on the principle of variable dilution similar to that described for Habibi [199] in Section 1.2.2.4 on aerosol sampling from vehicle emissions. The entire exhaust cycle is sampled and diluted to prevent condensation.

The system provides for continuous constant flow sampling of filtered inlet dilution air and filtered, diluted and cooled exhaust into sample bags for either transient or stabilized vehicle operation modes. The minimum sampling flow rate required is 5 litres min^{-1}. The collected samples can then be run through analysing instruments for gaseous constituents including hydrocarbons, carbon monoxide and oxides of nitrogen. The total diluted exhaust volume is measured by recording revolution counts and pressure drop across a positive displacement blower. Total mass emissions can be calculated using the total diluted exhaust volume and the measured concentrations in the sample bags. The measured

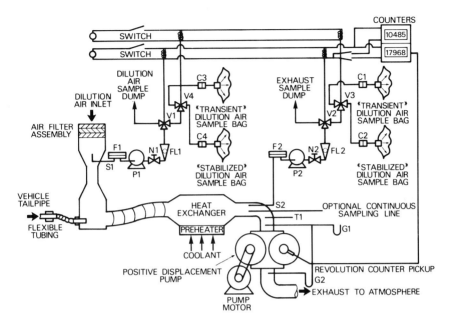

Fig. 1.39 Exhaust gas sampling system. (After [49].)

concentrations for the diluted exhaust are corrected for background concentrations measured on the inlet dilution air.

Soltau and Larbey [48] describe the advantages of variable dilution constant volume sampling in terms of reduction of reaction rates between hydrocarbons and oxides of nitrogen and prevention of water vapour and heavy hydrocarbon condensation.

A sampling system is also described [49] for measurement of evaporative losses from the carburettor and fuel tank. The collection device consists of a canister with a sealing screw cap, inlet and outlet tubes and capacity for 150 g of activated carbon (Fig. 1.40). The canister is connected to evaporative emission sources, as shown in Fig. 1.41. Connections must be sealed and the canister must be protected from heating by radiant or conductive heat from the engine during vehicle testing.

Siegel [50] reports that procedures for aircraft emissions must account for variations due to engine power level: time and spatial variations of exhaust composition; sampling line diameter, length, material and temperature; ambient temperature and humidity and ambient pollution levels.

Probes constructed of stainless steel are currently being used for sample intake. These need not be constructed for isokinetic sampling since particulates in jet exhausts are generally less than 1 μm. No variation in smoke density was found between facing the proble into the exhaust stream or facing 180° in the opposite direction.

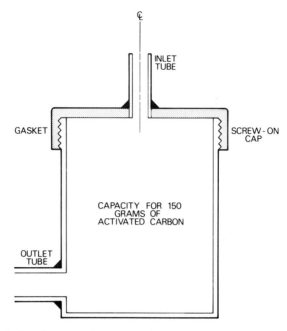

Fig. 1.40   Typical activated carbon trap (schematic diagram). (After [49].)

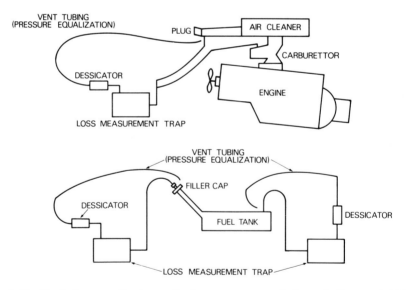

Fig. 1.41   Typical evaporative loss collection arrangements (schematic diagrams). (After [49].)

Multipoint sampling is necessary and more than 12 locations (3 radial positions in each of the 4 quadrants) are recommended. The vertical sampling plane should be located no more than one nozzle diameter downstream of the exit nozzle. The sampling lines should be constructed of stainless steel and maintained at greater than 175° C to prevent condensation of water vapour and high boiling point organics.

Stationary sources

Sampling systems for the collection of gaseous constituents from emission sources are generally tailored to the specific constituent sought. Since there is no requirement for isokinetic conditions when sampling for gaseous constituents, operating procedures are somewhat simplified. However, where prolonged sampling is undertaken in variable flow emission sources, it is desirable to obtain a flow proportional sample. Alternatively, where variable flow conditions are encountered, a series of grab samples may be taken with the flow velocity recorded. The results for the various samples may then be proportioned to flow.

Specific standard procedures for sampling for oxides of nitrogen, SO_2 and SO_3 are available [38, 252]. The sampling trains of the United States Environmental Protection Agency methods for SO_2 and NO_x are shown in Figs 1.42 and 143. These procedures use sampling trains similar to those employed for particulate sampling except that in addition to a filter to collect aerosols, absorbers or grab vessels are provided to collect the gaseous sample for subsequent analysis. Sampling equipment for gaseous emissions other than oxides of sulphur and nitrogen may be developed by using specific absorbers for other gases. Likewise, low temperature condenser or adsorber collection components may be used,

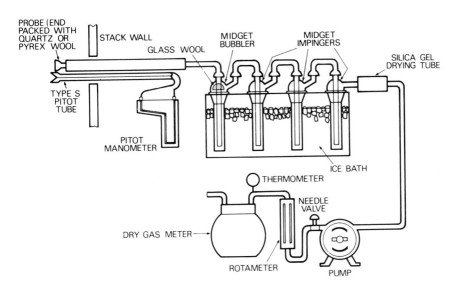

Fig. 1.42 SO_2 sampling train. (After [38].)

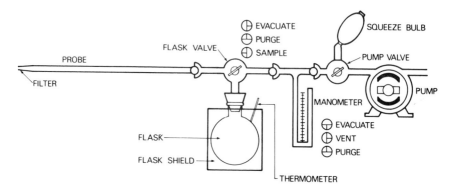

Fig. 1.43 Sampling train, flask valve and flask. (After [38].)

depending on the properties of the gaseous constituents to be measured. Sampling train possibilities are very flexible provided that provision is made for accurate flow measurement, adequate sampling rate and protection of the sample from loss or contamination in sampling lines.

The actual sampling procedures generally follow the same pattern used for aerosol sampling with the exception of the acceptability of proportional sampling in place of isokinetic sampling.

Berger *et al.* [253] reviewed the precision of common stack gas sampling procedures for oxides of sulphur and nitrogen and found the results exhibited relatively good precision. In general, accurate results may be obtained, as well, provided that care is taken to avoid leaks in the actual sampling system.

The sampling points should be selected to take advantage of flow turbulence to provide a homogeneous gas mixture rather than a stratified flow. Flow restrictions that provide for increased gas flow and turbulence, provided that a consistent velocity profile is available across the stack cross-section, are useful. Sample points should also be evaluated to avoid sampling in the proximity of flow stagnation or significant inward leakage of dilution air.

1.2.4 Sampling of rainwater and fog

Rainwater scavenging is one of the prime mechanisms by which gaseous and particulate pollutants are removed from the atmosphere. Thus the measurement of pollutant concentrations in rainwater is crucial to the full understanding of pollutant pathways in the atmosphere.

The types of rainwater samplers include simple bulk collectors, open both to wet and dry deposition, and wet-only collectors, having a sliding lid which opens only when rainfall begins. Some samplers have two collectors one for wet and the other for dry deposition.

Wet-only samplers are generally preferable to bulk samplers which are prone to contamination from dry deposited particles.* However, as bulk collectors are much simpler to build and install, they are frequently used in precipitation sampling. When a wet-only collector is not available, a narrow-neck bottle and funnel collector is recommended, rather than a wide-mouth bottle collector; the former minimizes dry deposition by the effect of the wind on the funnel and reduces evaporation of collected rain.

Rainwater collected with bulk samplers can suffer large alterations in the concentrations of some ionic compounds as a result of reaction with co-deposited particles. Coarse basic particles dry deposited on the sampler have been known to increase markedly the pH of collected rain [254]. Changes in concentration of Ca^{2+}, K^+, Mg^{2+} and Na^+ have also been detected in stored rainwater samples resulting from bulk collection [255]. The ionic stability of precipitation samples can be improved if the samples are filtered immediately after the precipitation event and stored at $4°$ C. Refrigeration only, although retarding alteration processes, does not seem to be sufficient to maintain sample integrity [255].

Rainwater samples collected with wet-only samplers are much more stable. Most of the possible degradation of ionic composition occurs rapidly during the sampling period, before collection of the daily sample can occur [256]. However, even in stored samples of wet-only collected rain, alterations of the order of 15–20% in the concentration of some chemical species (Ca^{2+} and Mg^{2+}) have been detected, after a 6 week period [255].

The material of collection and storage bottles can interfere with the determination of rainwater compounds, due to liberation of soluble species from the container walls and to adsorption of compounds from the liquid. Galloway and Likens [257, 258] concluded that polyethylene is a material that, at low cost, gives good results in the determination of inorganic compounds in precipitation. Samant and Vaidya [259] recommend the use of Pyrex bottles for the sampling of heavy metals. These authors studied the capability of the Sangamo precipitation collector (see Fig. 1.44) for the measurement of heavy metals in rainwater. The apparatus is an event sampler with two hard plastic buckets designed to collect dry or wet deposition. In the instrument a moisture sensing grid controls the motor-driven cover to collect precipitation, or dry deposition, depending on weather conditions. The plastic buckets were found to be suitable for sampling of Cd, Cu, Fe, Pb and Zn, but to introduce errors in Hg measurement, in consequence of leaching from container walls and sorption from the solution.

The amounts and composition of pollutants in precipitation are dependent on the intensity of rainfall and vary temporally throughout each precipitation event. The study of rainwater composition, in relation to precipitation intensity and meteorological parameters, has been carried out with sequential samplers [260, 261]. Raynor and McNeil [262] describe an automatic sequential sampler for collection of wet and frozen precipitation, with exclusion of dry fall-out between precipitation events. The sampler is formed by a box containing a turntable holding 30 sampling bottles. The bottles are changed automatically

---

* Sampling of dry deposition is dealt with in Chapter 4.

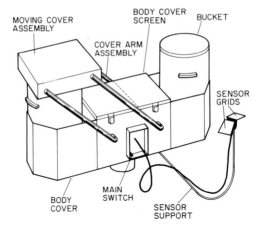

Fig. 1.44 Sangamo precipitation collector, Type A (courtesy Sangamo Co. Ltd).

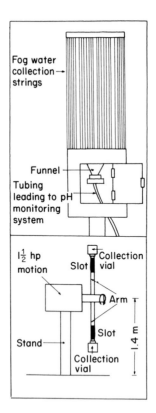

Fig. 1.45 Fog water collectors. (After [264].)

and periodically from the moment the cover opens. Times of cover opening/closing and bottle changing are recorded. A similar commercially available sequential rain sampler is described by Coscio *et al.* [263], in which the samples are sealed upon collection, to prevent gas exchanges, and refrigerated on site at $4 \pm 2°$ C.

The collection and analysis of fog droplets may become rather important for the understanding of atmospheric acidity processes; recent studies revealed that fog droplets may be 100 times more acidic than precipitation in the same region [264].

The techniques for the sampling of fog are still under development. Fuzzi *et al.* [265] collected fog water by impacting the fog droplets in an impinger designed by May [266] and also used by Garland *et al.* [267] in their fog droplet sampling networks. The impinger allowed the droplets to be captured with an efficiency of about 100%, for $d_p > 10$ μm.

Hileman [264] describes two fog samplers used in Californian studies (see Fig. 1.45). One is a rotating device consisting of an arm with slits along each side to gather fog water. The arm rotates at 1700 r.p.m., picking up fog water droplets larger than ~ 8 μm, with a good efficiency; the water is pushed by centrifugal force into collection bottles inserted in each end of the arm. The other is a passive device formed by vertical plastic strings disposed cylindrically, which, through impaction, collects fog droplets transported by the wind.

References

1. Intersociety Committee (1972) *Methods of Air Sampling and Analysis*, American Public Health Association, Washington.
2. ASTM (1981) Standard Recommended Practice for Planning the Sampling of the Atmosphere, D1357–57, Part 26. In *ASTM Annnual Book of Standards*, American Society for Testing and Materials, Philadelphia, Pa.
3. American Industrial Hygiene Association (1972) *Air Pollution Manual. Part I Evaluation*, 2nd edn, AIHA, Detroit.
4. National Air Sampling Network, US Environmental Protection Agency, Washington, DC.
5. National Air Pollution Surveillance, Annual Summaries, Air Pollution Control Directorate, Environmental Protection Service Environment Canada, Ottawa, Canada.
6. Continuous Air Monitoring Program, US Environmental Protection Agency, Washington, DC.
7. National Survey of Air Pollution in the United Kingdom, Warren Spring Laboratory, Stevenage.
8. McIntosh, D. H. and Thom, A. S. (1978) *Essentials of Meteorology*, Wykeham, London.
9. Scorer, R. (1968) *Air Pollution*, Pergamon Press, Oxford.
10. Brief, R. S. and Confer, R. G. (1972) Air quality monitoring: procedures; data analysis. *Heating, Piping, Air Conditioning*, pp. 103–10.
11. Deyo, J., Tomas, J. and King, R. B. (1977) *J. Air Pollut. Control Assoc.*, **27**, 142–4.

12. Hales, J. M. (1972) *Atmos. Environ.*, **6**, 635–59.
13. Bielbe, S. and Georgii, H. W. (1968) *Tellus*, **20**, 435–48.
14. Slinn, W. G. B. (1977) *J. Water Air Soil Pollut.*, **7**, 513–43.
15. Marsh, A. R. W. (1978) *Atmos. Environ.*, **12**, 401–6.
16. First, M. W. (1969) *Environ. Res.*, **2**, 88–92.
17. Seinfeld, J. H. (1975) *Air Pollution. Physical and Chemical Fundamentals*, McGraw-Hill, New York.
18. Stern, A. C. (1976) *Air Pollution*, 3rd edn, Academic Press, New York.
19. Charlson, R. J. (1969) *J. Air Pollut. Control Assoc.* **9**, 802.
20. Pooler, F. (1974) *J. Air Pollut. Control Assoc.*, **24**, 228–31.
21. Yamada, V. M. (1970) *J. Air Pollut. Control Assoc.*, **20**, 209–13.
22. Noll, K. E., Miller, T. L., Norco, J. E. and Raufer, R. K. (1977) *Atmos. Environ.*, **11**, 1051–9.
23. Munshi, U. and Patil, R. S. (1982) *Atmos. Environ.*, **16**, 1915–18.
24. Seinfeld, J. H. (1972) *Atmos. Environ.*, **6**, 847–58.
25. Katz, M. (1969) *Measurement of Air Pollutants: Guide to the Selection of Methods*, World Health Organisation, Geneva.
26. Hougland, E. S. and Stephens, N. T. (1976) *J. Air Pollut. Control Assoc.*, **26**, 51–3.
27. Saltzman, B. E. (1970) *J. Air Pollut. Control Assoc.*, **20**, 660–5.
28. Akland, G. G. (1972) *J. Air Pollut. Control Assoc.*, **22**, 264–6.
29. Hale, W. E. (1972) *Atmos. Environ.*, **6**, 419–22.
30. Tong, E. Y. and De Pietro, S. A. (1977) *J. Air Pollut. Control Assoc.*, **27**, 1008–10.
31. Hirtzel, C. S., Quon, J. E. and Corotis, R. B. (1982) *J. Environ. Eng. Div., Am. Soc. Civ. Eng.*, **108**, 488–501.
32. Hunt, W. F. (1972) *J. Air Pollut. Control Assoc.*, **22**, 687–91.
33. Ministry of Health and Environmental Protection (1980) *Handbook of Emission Factors*, Government Publishing Office, The Hague, Netherlands.
34. US Environmental Protection Agency (1977–1984) *Compilation of Air Pollutant Emission Factors*, AP42, 3rd Edn (including supplements 1–15) Office of Air Quality Planning and Standards. Research Triangle Park, NC 27711 USA.
35. Achinger, W. C. and Shigehara, R. T. (1968) *J. Air Pollut. Control Assoc.*, **18**, 605–9.
36. Pollution Control Branch, BC Water Resources Service (1974). *Source Testing Manual for the Determination of Discharges to the Atmosphere*, 3rd edn, Victoria, BC.
37. ASTM (1981) Standard Test Method for Particulates Independently or for Particulates and Collected Residue Simultaneously in Stack Gases. D3685–78, Part 26. In *ASTM Annual Book of Standards*, American Society for Testing and Materials, Philadelphia, Pa.
38. Environmental Protection Agency (18 Aug. 1977) Standards for Performance for New Stationary Sources, US Federal Register, Part II, Vol. 42, No. 160, 41753–89.
39. BSI (1961) British Standard Simplified Methods for Measurement of Grit and Dust Emissions from Chimneys (Metric Units), BS 3405: 1961, British Standards Institution, London.
40. BSI (1971) British Standard Simplified Methods for Measurement of Grit and Dust Emissions (Metric Units), BS 3405: 1971, British Standards Institution, London.
41. American Society of Mechanical Engineers (1957) Test Code for Determining Dust Concentrations in a Gas Stream, Power Test Code 27.
42. Brenchley, D. L., Turley, C. D. and Yarmac, R. F. (1973) *Industrial Source Sampling*, Ann Arbor Science, Ann Arbor.
43. ASTM (1981) Standard Method of Test for Average Velocity in a Duct (Pitot Tube

Method), D3154-72, Part 26. In *ASTM Annual Book of Standards*, American Society for Testing and Materials, Philadelphia, Pa.

44. Harlerd, H. H. (ed.) *Methods for Determination of Velocity, Volume, Dust and Mist Content of Gases*, Bulletin WP-50, 7th edn, Joy mfg Co., Los Angeles.

45. Cooper, H. P. H. and Rossano, A. T. (1971) *Source Testing for Air Pollution Control*, Environmental Science Services.

46. Rogers, G. F. C. and Mayhew, Y. R. (1967) *Engineering Thermodynamics. Work and Heat Transfer* (SI Units), Longmans, London.

47. Rom, J. J. (1972) Maintenance, Calibrations and Operation of Isokinetic Source Sampling Equipment, Publication APTD-0567, Environmental Protection Agency, Office of Air Programs.

48. Soltau, J. P. and Larbey, R. J. (1971) *Symp. of Institution of Mechanical Engineers*, 218, London.

49. Environmental Protection Agency, Protection of Environment, *Title 40*, Chap. 1, Part 85 Control of Air Pollution from New Motor Vehicles and New Motor Vehicle Engines, US Federal Register, 606.

50. Siegel, R. D. (1972) *J. Air Pollut. Control Assoc.*, **22**, 845–53.

51. Slowik, A. A. and Sansome, E. B. (1974) *J. Air Pollut. Control Assoc.*, **24**, 245–7.

52. Wohlers, H. C., Newstein, H. and Daunis, D. (1967) *J. Air Pollut. Control Assoc.*, **17**, 753.

53. Byers, R. J. and Davis, J. W. (1970) *J. Air Pollut. Control Assoc.*, **20**, 236.

54. ASTM (1981) Standard Recommended Practises for Sampling Atmospheres for Analysis of Gases and Vapours, D1605-60, Part 26. In *ASTM Annual Book of Standards*, American Society for Testing and Materials, Philadelphia, Pa.

55. Crittenden, P. D. and Read, D. J. (1976) *Atmos. Environ.*, **10**, 897–8.

56. Mamane, Y. and Donagi, A. E. (1976) *J. Air Pollut. Control Assoc.*, **26**, 991–2.

57. Nelson, G. O. (1971) *Controlled Test Atmospheres*, Ann Arbor Science, London.

58. Mercer, T. T. (1973) *Aerosol Technology in Hazard Evaluation*, Academic Press, London.

59. Bergland, R. M and Liu, B. Y. H. (1973) *Environ. Sci. Technol.*, **7**, 147–53.

60. Beckwith, T. G. and Buck, N. L. (1961) *Mechanical Measurements*, Addison Wesley, London.

61. Baker, W. C. and Pouchot, J. F. (1983) *J. Air Pollut. Control Assoc.*, **33**, 66–72, 156–62.

62. Environmental Protection Agency (1971) Reference Method for the Determination of Suspended Particulates in the Atmosphere (High Volume Method), US Federal Register 36, No. 84.

63. *ASTM Annual Book of Standards* (1975) Recommended Practice for Rotameter calibration D3195-73, Part 26, American Society for Testing and Materials, Philadelphia, Pa.

64. Countess, R. J. (1974) *J. Air Pollut. Control Assoc.*, **24**, 605.

65. Hoffman, G. L. and Duce, R. A. (1971) *Environ. Sci. Technol.*, **5**, 1135.

66. Patterson, R. K. (1980) *J. Air Pollut. Control Assoc.*, **30**, 169–71.

67. ASTM (1981) Standard Definitions of Terms Relating to Atmospheric Sampling and Analysis, D1356-73a, Part 26. In *ASTM Annual Book of Standards*, American Society for Testing and Materials, Philadelphia, Pa.

68. Zenker, P. (1971) *Staub*, **31**, 30–5.

69. Hemeon, W. C. L. and Haines, G. F. (1954) *Air Repair (J. Air Pollut. Control Assoc.)*, **4**, 159.

70. Watson, H. H. (1954) *Am. Ind. Hyg. Assoc. Q.*, **15**, 21–5.
71. Sehmel, G. A. (1970) *Am. Ind. Hyg. Assoc. J.*, **31**, 758–71.
72. Badzioch, S. (1960) *J. Inst. Fuel*, **33**, 106–10.
73. Vitols, V. (1966) *J. Air Pollut. Control Assoc.*, **16**, 79–83.
74. Sansone, E. B. (1969) *Am. Ind. Hyg. Assoc. J.*, **30**, 487–93.
75. Sehmel, G. A. (1973) *An Evaluation for High Volume Cascade Particle Impaction System*, Proceedings of the Second Joint Conf. on Sensing of Envir. Poll., 109–15, Washington, DC, 10–12 Dec., 1973.
76. Raynor, G. S. (1968) *Am. Ind. Hyg. Assoc. J.*, **29**, 397–404.
77. Wedding, J. B., McFarland, A. R. and Cermak, J. E. (1977) *Environ. Sci. Technol.*, **11**, 387–90.
78. Watson, J. G., Chow, J. C., Shah, J. J. and Pace, T. G. (1983) *J. Air Pollut. Control Assoc.*, **33**, 114–19.
79. Davies, C. N. (1968) *Br. J. Appl. Phys.*, Ser. 2, **1**, 921–32.
80. Bien, C. T. and Corn, M. T. (1971) *Am. Ind. Hyg. Assoc. J.*, **32**, 453–6.
81. Liu, B. Y. H. and Lee, K. W. (1976) *Environ. Sci. Technol.*, **10**, 345–50.
82. John, W. and Reichle, G. (1978) *Atmos. Environ.*, **12**, 2015–19.
83. Lee, K. W. and Liu, B. Y. H. (1980) *J. Air Pollut. Control Assoc.*, **30**, 377–81.
84. Stelson, A. W. and Seinfeld, J. H. (1981) *Environ. Sci. Technol.*, **15**, 671–9.
85. Elder, J. C., Tillery, M. J. and Ettinger, H. J. (1982) *J. Air Pollut. Control Assoc.*, **32**, 66–8.
86. Liu, B. Y. H., Pui, D. Y. H. and Rubow, K. L. (1983) in *Aerosol in the Mining and Industrial Work Environments* (eds V. A. Marple and B. Y. H. Liu), Vol. 3, 898–1038, Ann Arbor Science.
87. Baudo, R. (1982) *Trends Anal. Chem.*, **1**, 393–7.
88. Krause, W. J. and Albretcht, H. (1977) *Int. J. Environ. Anal. Chem.*, **5**, 34–46.
89. Gandrud, B. W. and Lazrus, A. L. (1972) *Environ. Sci. Technol.*, **6**, 455–7.
90. Forrest, J. and Newman, L. (1973) *Atmos. Environ.*, **7**, 561–73.
91. Roberts, P. T. and Friedlander, S. K. (1976) *Atmos. Environ.*, **10**, 403–8.
92. Harrison, R. M. and Pio, C. A. (1983) *Environ. Sci. Technol.*, **17**, 169–74.
93. Pierson, W. R., Hammerle, R. H. and Brachaczek, W. W. (1976) *Anal. Chem.*, **48**, 1808–11.
94. Pierson, W. R., Brachczek, W. W., Korniski, T. J. *et al.* (1980) *J. Air Pollut. Control Assoc.*, **30**, 30–4.
95. Witz, S. and McPhee, R. D. (1977) *J. Air Pollut. Control Assoc.*, **27**, 239–41.
96. Tanner, R. L., Cederwall, R., Garber, R. *et al.* (1977) *Atmos. Environ.*, **11**, 955–66.
97. Leahy, D. F., Phillips, M. F., Garber, R. W. and Tanner, R. L. (1980) *Anal. Chem.*, **52**, 1779–80.
98. Barret, W. J., Miller, H. C., Smith, J. E. and Christina, H. C. (1977) *Development of a Portable Device to Collect Sulfuric Acid Aerosol*, Interim Report EPA-600/2-77-027, Environmental Protection Agency, Research Triangle Park, North Carolina, USA.
99. Klockow, D., Jablowski, B. and Niebner, R. (1979) *Atmos. Environ.*, **13**, 1665–76.
100. Coffer, J. W. and Charlson, R. J. (1974) SO_2 oxidation to sulphate due to particulate matter on a high volume air sample filter. *Extended Abstracts of the Division of Environmental Chemistry, Amer. Chem. Soc. Meeting, Los Angeles, April.*
101. Hitchcock, D. R., Spiler, L. L. and Wilson, W. E. (1980) *Atmos. Environ.*, **14**, 165–82.
102. Durham, J. L., Wilson, W. E. and Bailey, E. B. (1978) *Atmos. Environ.*, **12**, 883–6.
103. Kaplan, D. J., Himmelblan, D. M. and Kanaoka, C. (1981) *Environ. Sci. Technol.*, **15**, 558–62.

104. Stevens, R. K., Dzubay, T. G., Russwurm, G. and Rickel, D. (1978) *Atmos. Environ.* **12**, 55–68.
105. Hara, H., Kurita, M. and Okita, T. (1982) *Atmos. Environ.*, **16**, 1565–6.
106. Forrest, J., Spandau, D. J., Tanner, R. L. and Newman, L. (1982) *Atmos. Environ.*, **16**, 1473–85.
107. Hodkinson, J. R. (1972) *Air Sampling Instruments for Evaluation of Atmospheric Contaminants*, 4th edn, American Conference of Government Industrial Hygienists, Cincinatti.
108. British Standard Method for the Measurement of Air Pollution, Part II – Determination of Concentration of Suspended Matter, BS 1747: Part II, 1969, British Standards Institution, London.
109. Hotschreuder, P., Vrins, E. and Van Boxel, J. (1983) *J. Aerosol Sci.*, **14**, 65–8.
110. McFarland, A. R. and Rodhes, C. E. (1979) Characteristics of aerosol samplers used in ambient air monitoring, *86th National Meeting, American Inst. of Chem. Eng., Houston, April.*
111. Bruckman, L. and Rubino, R. A. (1976) *J. Air Pollut. Control Assoc.*, **26**, 881–3.
112. Sweitzer, T. W. (1980) *J. Air Pollut. Control Assoc.*, **30**, 1324–5.
113. Lizarra-Rocha, J. A. (1976) *The effects of exposure of filters during nonoperating periods in the High Volume Sampling method*, MSc Thesis, University of North Carolina, USA.
114. Chahl, H. S. and Romano, D. J. (1976) *J. Air Pollut. Control Assoc.*, **26**, 885–6.
115. Blanchard, G. E. and Romano, D. J. (1978) *J. Air Pollut. Control Assoc.*, **28**, 1142–4.
116. Swinfor, R. (1980) *J. Air Pollut. Control Assoc.*, **30**, 1322–4.
117. Thanukos, L. C., Taylor, J. A. and Kay, R. E. (1977) *J. Air Pollut. Control Assoc.*, **27**, 1013–14.
118. Patterson, R. K. (1974) *J. Air Pollut. Control Assoc.*, **24**, 605.
119. Patterson, R. K. (1980) *J. Air Pollut. Control Assoc.*, **30**, 169–71.
120. King, R. B. and Fordyce, J. S. (1971) *J. Air Pollut. Control Assoc.*, **21**, 720.
121. Smith, F., Wohlschlegel, P. S., Rogers, R. S. C. and Mulligan, D. J. (1978) *Investigation of flow rate calibration procedures associated with high volume method for determination of suspended particulates*, EPA 600/4-78-047, US Environmental Protection Agency, USA.
122. Giever, P. M. and Ruch, W. E. (1971) *Am. Ind. Hyg. Assoc. J.*, **32**, 260–6.
123. Avera, C. B. (1968) *Am. Ind. Hyg. Assoc. J.*, **29**, 397–404.
124. Wolf, I. and Carpenter, R. L. (1982) *J. Air Pollut. Control Assoc.*, **32**, 744–6.
125. Lynam, D. R., Pierce, J. O. and Cholak, J. (1969) *Am. Ind. Hyg. Assoc. J.*, **30**, 83–8.
126. Tebbens, B. (1970) *Am. Ind. Hyg. Assoc. J.*, **31**, 44–51.
127. Clements, H. A. *et al.* (1972) *J. Air Pollut. Control Assoc.*, **22**, 955–8.
128. McKee, H. C., Childers, R. E. and Saluz, O. (1971) Collaborative study of reference method for the determination of suspended particulates in the atmosphere (High Volume method), South West Research Inst., Houston, TX, June.
129. Katz, M. (1980) *J. Air Pollut. Control Assoc.*, **30**, 528–57.
130. Van Winkle, W., Putnicki, G. J. and Crowder, J. W. (1981) *J. Air Pollut. Control Assoc.*, **31**, 168–9.
131. Sacco, A. M., Rinsky, A. H. and McDonald, J. T. (1976) *J. Air Pollut. Control Assoc.*, **26**, 883–5.
132. Heckler, L. H. and Anglen, D. M. (1981) *Am. Ind. Hyg. Assoc. J.*, **38**, 650.
133. Miller, F. J., Gardner, D. E., Graham, J. A. *et al. J. Air Pollut. Control Assoc.*, **29**, 610–15.

134. Lioy, P. J., Watson, J. G. and Spengler, J. D. (1980) *J. Air Pollut. Control Assoc.*, **30**, 1126–30.
135. Miller, S. S. (1978) *Environ. Sci. Technol.*, **12**, 1353–5.
136. Farthing, W. E. (1982) *Environ. Sci. Technol.*, **16**, 237A–44A.
137. Environmental Protection Agency (20 March, 1984) *National Ambient Air Quality Standards Proposed Rules*, US Federal Register, Vol. 49, No. 55, 10408–62.
138. ISO TC-146 (1981) *Am. Ind. Hyg. Assoc. J.*, **42**, A64–A68.
139. McFarland, A. R. and Ortiz, C. A. (1982) *Atmos. Environ.*, **16**, 2959–65.
140. Watson, J. G. (1983) *J. Air Pollut. Control Assoc.*, **33**, 114–19.
141. Wedding, J. B. (1982) *Environ. Sci. Technol.*, **16**, 154–61.
142. Wedding, J. B., Weigand, M. A. and Carney, T. C. (1982) *Environ. Sci. Technol.*, **16**, 602–6.
143. Ranz, W. E. and Wong, J. B. (1952) *Ind. Eng. Chem.*, **44**, 1371–81.
144. Marple, V. A. and Liu, B. Y. H. (1974) *Environ. Sci. Technol.*, **8**, 648–54.
145. Gussman, R. A., Sacco, A. M., McMahon, N. M. (1973) *J. Air Pollut. Control Assoc.*, **23**, 778–82.
146. Kubie, G. (1971) *J. Aerosol Sci.*, **2**, 23–30.
147. Cooper, D. W. and Guttrich, G. L. (1981) *Atmos. Environ.*, **15**, 1699–1707.
148. Cooper, D. W. and Spielman, L. A. (1976) *Atmos. Environ.*, **10**, 723–9.
149. Natusch, D. F. S. and Wallace, J. R. (1976) *Atmos. Environ.*, **10**, 315–24.
150. Raabe, O. G. (1978) *Environ. Sci. Technol.*, **12**, 1162–7.
151. Puttock, J. S. (1981) *Atmos. Environ.*, **15**, 1709–16.
152. Williams, P. C. (1982) *J. Air Pollut. Control Assoc.*, **32**, 1071–2.
153. Dzubay, T. G., Hines, L. E. and Stevens, R. K. (1976) *Atmos. Environ.*, **10**, 229–34.
154. Cautreels, W. and Van Cauwenberghe, K. (1978) *Atmos. Environ.*, **12**, 1133–41.
155. Rao, A. K. and Whitby, K. T. (1978) *J. Aerosol Sci.*, **9**, 87–100.
156. Lawson, D. R. (1980) *Atmos. Environ.*, **14**, 195–9.
157. Rao, A. K. and Whitby, K. T. (1977) *Am. Ind. Hyg. Assoc. J.*, **38**, 174–9.
158. Chan, T. L. and Lawson, D. R. (1981) *Atmos. Environ.*, **15**, 1273–9.
159. Barr, E. B., Newton, G. J. and Yeh, H. (1982) *Environ. Sci. Technol.*, **16**, 633–5.
160. McFarland, A. R., Wedding, J. B. and Cenmak, J. E. (1977) *Atmos. Environ.*, **11**, 535–42.
161. McFarland, A. R., Ortiz, C. A. and Bertch, Jr, R. W. (1979) *Atmos. Environ.*, **13**, 761–5.
162. Anderson, A. A. (1965) *Am. Ind. Hyg. Assoc. J.*, **27**, 160–5.
163. Cushing, K. M., McCain, J. D. and Smith, W. B. (1979) *Environ. Sci. Technol.*, **13**, 726–31.
164. Marple, V. A. and Willeke, K. (1976) *Atmos. Environ.*, **10**, 891–6.
165. Conner, W. D. (1966) *J. Air Pollut. Control Assoc.*, **16**, 35–8.
166. Dzubay, T. G. and Stevens, R. K. (1975) *Environ. Sci. Technol.*, **9**, 663–8.
167. John, W. and Reischl, G. (1980) *J. Air Pollut. Control Assoc.*, **30**, 872–6.
168. Buchanan, L. M. *et al.* (1972) *Appl. Microbiol.*, **23**, 1140–4.
169. Melo, O. T. and Philips, C. R. (1974) *Environ. Sci. Technol.*, **8**, 67–71.
170. Manton, M. J. (1978) *Atmos. Environ.*, **12**, 1669–75.
171. Manton, M. J. (1979) *Atmos. Environ.*, **13**, 525–31.
172. Heidam, N. Z. (1981) *Atmos. Environ.*, **15**, 891–904.
173. Parker, R. D., Buzzard, C. H., Dzubay, T. G. and Bell, J. P. (1977) *Atmos. Environ.*, **11**, 617–21.
174. Cahill, T. A., Ashbaugh, L. L., Barone, J. B. *et al.* (1977) *J. Air Pollut. Control Assoc.*, **27**, 675–7.

175. Twomey, S. A. and Zalabsky, R. A. (1981) *Environ. Sci. Technol.*, **15**, 177–84.
176. Jensen, F. P. and Kemp, K. (1978) Miljostyrelsens Luftforureningslaboratorium, MST LUFT-A 14 (Danmark).
177. John, W., Reichel, G., Goren, S. and Plotkin, D. (1978) *Atmos. Environ.*, **12**, 1555–7.
178. Cahill, T. A., Eldred, R. A., Barone, J. B. and Ashbaugh, L. L. (1979) Federal Highway Administration Report FHWA-RD-178 (USA).
179. Thomas, J. W. (1955) *J. Colloid Sci.*, **10**, 246.
180. Sinclair, D. (1972) *Am. Ind. Hyg. Assoc. J.*, **33**, 729–35.
181. Sinclair, D. *et al.* (1976) *J. Air Pollut. Control Assoc.*, **26**, 661–3.
182. Sinclair, D. and Hoopes, G. S. (1975) *Am. Ind. Hyg. Assoc. J.*, **36**, 39–42.
183. Tanner, R. L., Cedarwall, R., Garber, R. *et al.* (1977) *Atmos. Environ.*, **11**, 955–66.
184. Chamberlain, A. C., Heard, M. J., Little, P. and Wiffen, R. D. (1979) *Phil. Trans. R. Soc. London, Ser. A*, **290**, 577.
185. Marlow, W. H. and Tanner, R. L. (1976) *Anal. Chem.*, **48**, 1999–2001.
186. Tanner, R. L. and Marlow, W. H. (1977) *Atmos. Environ.*, **11**, 1143–50.
187. Harrison, R. M. and Pio, C. A. (1981) *J. Air Pollut. Control Assoc.*, **31**, 784–7.
188. Harrison, R. M. and Pio, C. A. (1983) *Atmos. Environ.*, **17**, 1733–8.
189. Boubel, R. W. (1971) *J. Air Pollut. Control Assoc.*, **21**, 783–7.
190. Paulus, H. J. and Thron, R. W. (1976) Stack sampling. In *Air Pollution*, Vol. 3 (ed. A. C. Stern), Academic Press, New York, pp. 525–87.
191. AFNOR (1983) *Analyse des gas. Qualité de l'air*, Association Française de Normalization, Paris.
192. Leatherdale (1977) Isokinetic stack testing. In *Air Pollution Control and Design Handbook*, Part 1, (ed. P. N. Cheremisinoff and R. A. Young) Marcel De K. Ker Inc., New York, pp. 65–121.
193. Lewandowski, G. A. (1981) Stack sampling. In *Air/Particulate Instrumentation and Analysis* (ed. P. N. Cheremisinoff), Ann Arbor Sci., Ann Arbor, pp. 119–54.
194. Hemeon, W. C. L. and Black, A. W. (1972) *J. Air Pollut. Control Assoc.*, **22**, 516–18.
195. Martin, R. M. (1971) Environmental Protection Agency, Pub. No. APTD-0581.
196. Hirschler, D. A. *et al.* (1975) *Ind. Eng. Chem.*, **49**, 1131–42.
197. McKee, H. C. and McMahon, W. A. (1960). *J. Air Pollut. Control Assoc.*, **10**, 456–62.
198. Mueller, P. K. *et al.* (1964) Symp. on Air Poll. Measurement Methods, *Am. Soc. Test. Mater.*, Spec. Tech. Publ. no. 352.
199. Habibi, K. (1971) *Environ. Sci. Technol.*, **4**, 679.
200. Habiki, K. (1973) *Environ. Sci. Technol.*, **7**, 223–4.
201. Campbell, K. and Dartnell, P. L. (1971) *Symp. of Institution of Mechanical Engineers*, **14**, London.
202. Ter Haar, G. L. *et al.* (1972) *J. Air Pollut. Control Assoc.*, **22**, 39–46.
203. Sampson, R. F. and Springer, G. S. (1973) *Environ. Sci. Technol.* **7**, 55–60.
204. Sexton, R. W. (1964) *Am. Ind. Hyg. Assoc. J.*, **25**, 346.
205. Reid, F. H. and Halpin, W. R. (1968) *Am. Ind. Hyg. Assoc. J.*, **29**, 390–5.
206. ASTM (1981) Standard Practice for Sampling Atmospheres to Collect Organic Compound Vapors (Activated Charcoal Adsorption Method). In *ASTM Annual Book of Standards*, American Society for Testing and Materials, Philadelphia, Pa.
207. Fraust, C. L. and Hermann, E. R. (1969) *Am. Ind. Hyg. Assoc. J.*, **30**, 494–9.
208. Jeltes, R. (1969) *Atmos. Environ.*, **3**, 587–8.
209. Mann, J. B. *et al.* (1974) *Environ. Sci. Technol.*, **8**, 584–5.
210. Cantuti, V. and Cartoni, G. P. (1968) *J. Chromatogr.*, **32**, 641–7.
211. Aue, W. A. and Teli, P. M. (1971) *J. Chromatogr.*, **62**, 15–27.
212. Perry, R. and Twibell, J. D. (1973) *Atmos. Environ.*, **7**, 927.

213. Cropper, F. and Kominsky, S. (1963) *Anal. Chem.*, **35**, 735.
214. Williams, I. (1965) *Anal. Chem.*, **37**, 1723.
215. Bellar, T. A. *et al.* (1963) *Anal. Chem.*, **35**, 1924.
216. Stephens, E. R. and Burleson, F. R. (1967) *J. Air Pollut. Control Assoc.*, **17**, 147.
217. Williams, F. W. and Umstead, M. E. (1968) *Anal. Chem.*, **40**, 2232.
218. Novak, J., Vasak, V. and Janak, J. (1965) *Anal. Chem.*, **37**, 660.
219. Dravnieks, A. *et al.* (1971) *Environ. Sci. Technol.*, **5**, 1220.
220. Smith, J. R. (1979) *J. Air Pollut. Control Assoc.*, **29**, 969–70.
221. Lewin, E. and Zachau-Christiansen, B. (1977) *Atmos. Environ.*, **11**, 861–2.
222. Lazrus, A. L., Gandrud, B. W., Woodard, R. N. and Scdlacek, W. A. (1976) *J. Geophys. Res.*, **81**, 1067–70.
223. Lazrus, A. L. (1977) *Geophys Res. Lett.*, **4**, 587–9.
224. Bonelli, J. E., Greenberg, J. P. and Lazrus, A. L. (1978) *Atmos. Environ.*, **12**, 1591–4.
225. Williams, K. R. and Jacobi, S. A. (1978) *Atmos. Environ.*, **12**, 2509–10.
226. Bourbon, P. and Nivot, A. (1975) Note de laboratoire: Contribution au captage et dosage de NH_3 gaseux et sels ammoniacaux atmospheriques. *Cent. Belge Etude Doc. Eaux*, No. 385, Dec., Belgium.
227. Wash, P. R., Duce, R. A. and Fasching, J. L. (1977) *Environ. Sci. Technol.*, **11**, 163–6.
228. Natush, O. F. S., Sewell, J. R. and Tanner, R. L. (1975) *Anal. Chem.*, **46**, 410.
229. Jaeschke, W. and Herrmann, J. (1981) *Int. J. Environ. Anal. Chem.*, **10**, 107–20.
230. Cormis, L. and Cantuel, J. (1977) *Pollut. Atmos.*, **19**, 377–80.
231. Okita, L., Monimoto, S. and Izawa, M. (1976) *Atmos. Environ.*, **10**, 1085–9.
232. Bourbon, M. P., Alary, J., Esclassan, J. and Lepert, J. C. (1976) *Pollut. Atmos.*, **69**, 11–15.
233. Bourbon, M. P., Alary, J., Esclassan, J. and Lepert, J. C. (1977) *Atmos. Environ.*, **11**, 485–8.
234. Shepherd, M. *et al.* (1951) *Anal. Chem.*, **23**, 1431.
235. ASTM (1981) Test for C_1–C_5, Hydrocarbons in the Atmosphere by Gas Chromatography, D2820-72T, Part 26. In *ASTM Annual Book of Standards*, American Society for Testing and Materials, Philadelphia, Pa.
236. Desbaumes, E. and Imhoff, C. (1971) *Staub*, **31**, 36–41.
237. Schuette, F. J. (1967) *Atmos. Environ.*, **1**, 515–19.
238. Conner, W. D. and Nader, J. S. (1964) *Am. Ind. Hyg. Assoc. J.*, **25**, 291–7.
239. Oord, F. (1970) *Am. Ind. Hyg. Assoc. J.*, **31**, 532–3.
240. Curtis, E. H. and Hendriks, R. H. (1969) *Am. Ind. Hyg. Assoc. J.*, **30**, 93–4.
241. Lang, H. W. and Freedman, R. W. (1969) *Am. Ind. Hyg. Assoc. J.*, **30**, 523.
242. Meader, M. C. and Bethea, R. M. (1970) *Environ. Sci. Technol.*, **4**, 853–5.
243. Van Houten, R. V. and Lee, C. (1969) *Am. Ind. Hyg. Assoc. J.*, **30**, 465–9.
244. Pozzoli, L. (1981) *Giornale Igienisti Industriali*, year VI; (5), III–XIV; (6), IV–X; (7), III–IX.
245. Rose, V. E. and Perkins, J. L. (1982) *Am. Ind. Hyg. Assoc. J.*, **43**, 605–25.
246. BSI (1969) British Standard Method for Measurement of Air Pollution, The Lead Peroxide Method (1969), BS 1747, Part 4: 1969, British Standards Institution, London.
247. Warner, P. W. (1976) *Analysis of Air Pollutants*, John Wiley, New York.
248. Huey, N. A. (1968) *J. Air Pollut. Control Assoc.*, **18**, 610–11.
249. Lynch, A. J., McQuaker, N. R. and Gurney, M. (1978) *Environ. Sci. Technol.*, **12**, 169–72.
250. McDermott, D. L., Reiszner, K. D. and West, P. W. (1979) *Environ. Sci. Technol.*, **13**, 1087–90.

251. Hardy, J. K., Dasgupta, P. K., Reiszner, K. D. and West, P. W. (1979) *Environ. Sci. Technol.*, **13**, 1090–3.
252. BSI (1971) British Standard Methods for the Sampling and Analysis of Flue Gases, Part I, Methods of Sampling, BS 1756-1: 1971, British Standards Institution, London.
253. Berger, A. W., Driscoll, J. N. and Morgenstern, P. (1972) *Am. Ind. Hyg. Assoc. J.*, **33**, 397–404.
254. Slanina, J., Mols, J. J., Baard, J. H. *et al.* (1979) *Int. J. Environ. Anal. Chem.*, **7**, 161–76.
255. Peden, M. E. and Skowron, L. M. (1978) *Atmos. Environ.*, **12**, 2343–9.
256. Madsen, B. C. (1982) *Atmos. Environ.*, **16**, 2515–9.
257. Galloway, J. N. and Likens, G. E. (1976) *Water, Air Soil Pollut.*, **6**, 241–58.
258. Galloway, J. N. and Likens, G. E. (1978) *Tellus*, **30**, 71–82.
259. Samant, H. S. and Vaidya, O. C. (1982) *Atmos. Environ.*, **16**, 2183–6.
260. Gascoyne, M. (1977) *Atmos. Environ.*, **11**, 397–400.
261. Ronneau, C., Cara, J., Navarre, J. L. and Priest, P. (1978) *Water, Air Soil Pollut.*, **9**, 171–6.
262. Raynor, G. S. and McNeil, J. P. (1979) *Atmos. Environ.*, **13**, 149–55.
263. Coscio, M. R., Pratt, G. C. and Krupta, S. V. (1982) *Atmos. Environ.*, **16**, 1939–44, 2272–3.
264. Hileman, B. (1983) *Environ. Sci. Technol.*, **17**, 117A–23A.
265. Fuzzi, S., Orsi, G. and Mariotti, M. (1983) *J. Aerosol Sci.*, **14**, 135–8.
266. May, K. R. (1961) *Q. J. R. Meteorol. Soc.*, **87**, 535.
267. Garland, J. A., Branson, J. R. and Cox, L. C. (1973) *Atmos. Environ.*, **7**, 1079.

2 Air pollution meteorology

D. J. MOORE

2.1 Introduction

2.1.1 Wind and the turbulent mixing layer

One of the natural defences of living organisms against the effects of toxic air pollutants is the great capacity of the atmosphere to dilute and in many cases subsequently to rid itself of these materials. The mechanisms which ensure that dilution is sufficient to enable life to continue are the wind and atmospheric turbulence.

This wind removes the material from the source region and if the emission is continuous at rate Q_p units s^{-1}, then the quantity of material in length x of the plume will be equal to $Q_p x/U$ units where U is the wind speed (m s^{-1}). The wind therefore effects a dilution proportional to its speed.

The turbulence is usually restricted to a layer – the 'mixing' layer, which may be anything from a few metres to a few kilometres deep. Within this layer, the effluent material will disperse in all directions until it reaches the surface or the top of the mixing layer. The rate of dispersion depends in a complicated way on the three-dimensional spectrum of turbulent energy and the size and shape of the effluent cloud.

Turbulence is produced:

(a) By the stirring of the wind due to the drag of the surface and objects projecting from it on the air as it moves over it
(b) Thermal convection currents which rise from the surface when it is warmer than the air and may also descend from the tops of cloud layers which cool due to long wave radiation out into space.

The turbulence will be dissipated or suppressed by:

(c) Viscosity and
(d) Atmospheric stability, caused by surface cooling or mixing of warm air into the top of the mixing layer.

At a particular location the depth of the mixing layer will also be determined to a large extent by factors (a) to (d) and the way in which they vary in time or with fetch upstream.

95

Factors (b) and (d) will depend partly on current values of net incoming solar radiation and outgoing long-wave infra-red radiation from the earth's surface and from cloud layers. They will also be affected by the previous history of the air stream, including the trajectories above and below the mixing layer. If the air stream is warmer than the surface, buoyancy forces will act to reduce turbulence and heat transfer is always due to 'forced convection'. On the other hand, if there are convection currents present and the winds are light, the heat flux is attributed to 'free convection'. In practice the free stream, i.e. the air above the mixing layer, is almost always warmer. Therefore the turbulence at the top of the convective layer is due to 'overshoot' of the convection currents originating at the surface into the warmer air above. The turbulent heat transfer in the top of the layer is, therefore, downward, and the convection is forced by the upward heat flux occurring throughout the air below and the wind shear at the top of the layer. On the other hand, free convection may occur in a cloud or fog layer whose top is cooling by radiation even when the bulk of the layer is warmer than the surface. Such a situation occurs when cloud or fog moves off the sea over cold ground or if the ground lost heat by radiation before the fog formed.

In the absence of heating, cooling or condensed water, the decrease of pressure with height causes the air to cool at just under $1°$ C per 100 m of ascent. This temperature gradient is called the 'dry adiabatic lapse rate'. If the decrease of temperature with height is greater than this, any parcel of air displaced vertically will be subject to a buoyancy force tending to move it further away. The air is then said to be unstable and the lapse rate super-adiabatic. Conversely, if the decrease of temperature with height is less than the dry adiabatic rate, buoyancy forces will tend to restore any vertically displaced air parcel back towards its original level. The lapse-rate is then said to be stable.

Free convection occurs with unstable and forced convection with both unstable lapse rates in strong winds and with all stable lapse rates. If the air is saturated with water vapour, then heat is released as the air rises and the air cools at the 'saturated adiabatic' lapse rate. This is about $0.6°$ C/100 m, but varies with temperature. In the presence of cloud, vertical temperature gradients which are steeper than the saturated adiabatic lapse rate, are unstable.

Transport and dispersion of pollution is also complicated by the effect of the earth's rotation. This causes a force (the Coriolis force) proportional to the wind speed to act in a direction perpendicular to the wind direction. Therefore, in the absence of surface drag the wind blows parallel to the isobars. When the wind speed is reduced by surface friction, the balance between the Coriolis force and the pressure gradient force is upset and the surface air accelerates towards the low pressure. Since the drag decreases with height, the wind direction veers, i.e. turns in a clockwise direction, with height in the mixed layer in the Northern Hemisphere. The flow is anticlockwise and converging towards the centre in a depression and clockwise and diverging away from the centre in an anticyclone. All rotations are in the opposite sense south of the equator. As a result of this variation of wind direction with height, plumes do not travel in the direction indicated by measurements made at a single height. Furthermore, the dispersion of material in the cross-wind direction is increased as the plume depth increases.

The above discussion shows that for air streams arriving at a given location with common previous histories, dispersion within the mixing layer will be determined by:

(a) The surface roughness (z_0) (m)
(b) The Coriolis parameter (f) (s^{-1})
(c) The free stream wind speed (U_f) (m s^{-1})
(d) Some parameter or parameters to represent the thermal effects, such as the net radiation fluxes at the top and bottom of the layer
(e) Some parameter or parameters to represent the previous history of the air stream, such as the mean temperature difference between the mixed layer and the surface.

Factors (a) to (c) above may be regarded as external parameters of the layer, i.e. they are not affected by its development. (d) will not be completely independent of (a) to (c) and will clearly be affected by cloud within the layer. Of course, the presence of a flux of heat means that the properties of the layer are changing, so one would not expect the thermal characteristics of the layer to be determined entirely by external parameters. Both the thermal and mechanical properties will depend on previous history, (e), to some extent. Taking all the above factors into account, the best that we can expect is that the dispersive properties of the atmosphere downstream of a given site are determined by (a) to (c) above plus:

(d) The time of day
(e) The time of year
(f) The air mass type
(g) The free stream wind direction
(h) The cloud cover above and within the mixing layer
(i) The 'wetness' of the surface
(j) The large scale features of the weather situation, i.e. the degree of convergence or divergence of the free stream flow.

The meterological measurements required to characterize the mixing layer are discussed in Section 2.2 below. Air mass types and weather systems and their effects on the properties of the layer are described in Section 2.3.

2.1.2   The effect of source height

It follows from the nature of the atmospheric dispersion mechanisms described above that dilution of pollutants released at the surface is at a minimum at a given distance downwind (i.e. ground level concentration (GLC) is at its maximum) when the wind is light and there is little or no thermal convection. This occurs when the surface is cooler than the air and there is no cloud in the mixing layer. This situation does not produce the highest GLCs when the emission is from tall chimneys, especially when the emissions have large initial buoyancy and/or upward momentum. In conditions of reduced turbulence near the ground, when the mixing layer is usually very shallow, emissions from such releases rise above the turbulent layer into the 'free atmosphere' where turbulence levels are generally very low. Consequently, the effluent does not diffuse to the ground until something happens to make the air at the height at which the plume has levelled out become turbulent. This will not occur until either the atmospheric conditions change or the plume moves into a region where the surface topography or temperature is different from what it was at the location and time of release.

The effective source height (H) is defined as the stack height (h) plus the plume rise due to momentum and/or buoyancy (Z_r). Zero GLCs from emissions with high effective source heights will therefore be the rule when low level emissions are producing their worst effects.

There is one exception to this rule. When calm, subsiding anticyclonic conditions persist for several days and convective mixing does not develop during the day due to persistent fog, snow cover or frozen ground, the plumes may eventually descend into the shallow mixed layer and add to the high concentrations due to the low level emissions. Otherwise the greatest GLCs from sources with high effective heights will occur in strong winds (when the plume rise from buoyancy or momentum is low) or in lighter winds when there is strong convective turbulence. In both these conditions, the GLC due to low level emissions will be well below its maximum value.

If the plume rise is completed within the mixing layer, then the concentration of effluent material at the point where it first reaches the ground is inversely proportional to the vertical cross-section area of the plume. In the case where the plume material is uniformly distributed within this area and the cross-section is circular, then, provided no material has been lost in transit, we may write

$$C_{max} = Q_p/(\bar{U}H^2\pi) \tag{2.1}$$

i.e. (C_{max}) is inversely proportional to the product of the square of the effective stack height and the wind speed ($\bar{U}$) averaged over the vertical cross-section of the plume.

In general the plume shape will not be circular and will change as the material drifts downwind, e.g. consider an elliptical vertical cross-section with the horizontal cross-wind radius (R_y) and the vertical radius (R_z) given by expressions of the form

$$R_y = A_y L_y^{(1-a)} x^a = A_y L_y (x/L_y)^a \tag{2.2}$$

$$R_z = A_z L_z^{(1-b)} x^b = A_z L_z (x/L_z)^b \tag{2.3}$$

where L_y and L_z are length scales and A_y and A_z are dimensionless parameters. In this case, the expression for the concentration of effluent material in the plume is

$$C_A = Q(L_y/x)^a (L_z/x)^b / (\pi A_y A_z L_y L_z \bar{U}) \tag{2.4}$$

for values of $R_z < H$ and twice this value at ground when $R_z \geq H$ if the material is 'reflected' by the surface. $C_A = C_{max}$ when $R_z = H$. This will occur when

$$x = L_z (A_z L_z/H)^{-(1/b)} = x_{max} \tag{2.5}$$

Consequently, in dimensionless form, we have

$$(x_{max}/H) = \frac{1}{A_z}\left(\frac{H}{A_zL_z}\right)^{-(1/b-1)}$$

(2.6)

and

$$(\pi \bar{U} C_{max}H^2/Q_p) = \left(\frac{A_z}{A_y}\right)\left(\frac{L_z}{L_y}\right)^{(1-a)}\left(\frac{A_zL_z}{H}\right)^{(a/b-1)}$$

(2.7)

The presence of boundaries to dispersion of pollutants at the top and bottom of the mixing layer and the absence of such boundaries in the cross-wind direction have the effect of making $a > b$ in most meteorological situations in locations with no topographical complications. Thus Equation (2.7) tells us that C_{max} will be inversely proportional to a power of H which is >2. Since $H = h + Z_r$ the importance of both the chimney height (h) and the plume rise (Z_r) in reducing C_{max} is obvious.

Provided that the effluent is buoyant, dilution factors of order 10 000 can be achieved by good stack design for high level emissions by the time the material reaches the ground. These may be compared with reductions of source strength by a factor of 2 to 200 achievable by most emission control processes.

Equations (2.1) to (2.7) are, of course, an over-simplification of plume behaviour because the material is not uniformly distributed over the plume cross-section. The calculation of plume rise, cross-wind and vertical plume dispersion will be discussed further in Section 2.4.

2.1.3 Plant design to achieve maximum atmospheric dispersion

Ensuring that the maximum possible dilution by atmospheric dispersion is achieved before effluent reaches a location where it could cause a nuisance should be regarded as an essential requirement in the design of large emitters of pollution. This is true whether such dilution is an alternative to or an additional safeguard in parallel with emission control. Such design considerations include:

(a) Avoiding downwash (flagging) of effluent into the low pressure area in the lee of the stack by maintaining a high efflux velocity
(b) Avoiding entrainment of the plume into downdraughts in the lee of adjacent buildings or topographical features by careful site selection and by building a tall stack
(c) Avoiding emitting gases with low or negative buoyancy by emitting at a sufficiently high temperature
(d) Ensuring maximum plume rise by fulfilling requirements (a) to (b) and by emitting from as few stacks as possible
(e) Removing particulates large enough to fall a substantial fraction of the plume height in the distance travelled to the point of maximum gaseous GLC.

Methods for taking the effect of downdraughts, downwash and particle fall velocity into account in situations where it is not possible to avoid such problems by good plant design, are described in Section 2.4 or by Jones [1].

2.1.4   Factors affecting long-range transport of pollutants

The amount of pollutant material reaching distant receptors depends on the efficiency of removal processes as well as on dispersion. There are three main removal mechanisms for pollutants. These are: (a) diffusion to or impaction and sedimentation on the earth's surface or protuberances attached to it (usually referred to as dry deposition); (b) diffusion to or impaction on cloud or spray droplets and (c) diffusion to or impaction by raindrops, hailstones or snowflakes, which is usually referred to as washout. Process (b) will result in permanent removal only if the droplets are themselves impacted on the surface or swept up by hydrometeors. (This latter process is usually referred to as rainout.) Process (c) will result in permanent removal if the hydrometeors reach the ground without evaporating.

Process (a) will be most effective near low level sources, which produce high GLCs at and near the source. Thus the low GLCs near high level sources described in Section 2.1.1 will result in reduced deposition compared with the deposition which would have occurred had the same quantity of material been released at ground level. However, in most meteorological conditions the increases in long-range transport resulting from reduced near field deposition are small. There are further discussions of removal mechanisms and long-range transport of pollutants in Sections 2.4.7 and 2.4.6.

2.2   Meteorological measurements

2.2.1   Parameters affecting transport and dispersion of pollutants

The simplest approaches used to evaluate the effects of atmospheric transport and dispersion on pollutant releases require measurements of wind speed and direction at one height and location only, together with the estimated cloud cover. A standard procedure is then followed to obtain quantitative estimates of the effects of atmospheric stability, including thermal convection, on the dispersion. The mixing depth may also be estimated. The procedure involves using the latitude, longitude, time of the day and year to determine the solar radiation incident on the surface or the top of the cloud layer and an estimate of the surface roughness and thermal characteristics of the site. One of the best known of such schemes is described by Pasquill and Smith [2]. It provided the basis for the calculation methods described by Clarke [3]. Such models are appropriate for application to low level releases from single plant and may also be used for application to multiple source releases in areas of a few hundred square kilometres where complications due to such factors as topography and sea-breezes are absent, so the wind field is reasonably homogeneous.

Where high level releases are concerned it is desirable to have wind measurements at the height of release.

2.2.2 Wind velocity measurements

Wind velocity is, of course, a vector quantity with two horizontal and a vertical component. The mean vertical velocity is usually much smaller than the vector sum of the two horizontal components (the speed). Many instruments ignore the vertical component and, historically, instruments on towers have measured the speed and the direction. These are usually averaged over a period of several minutes to give the mean wind speed and direction. The mean direction derived from the vector mean of the two horizontal wind components will not be the same as the mean of the angular positions of a wind vane if there is any correlation between the wind speed and direction. In general there will be such a correlation because faster travelling air is usually descending from aloft with a direction deflected towards that of the upper wind. Measurements made with propeller or sonic anemometers are therefore marginally to be preferred to those from cup or vane instruments for the calculation of trajectories. However, the difference is very small unless the wind speed is very light and the direction is very variable. Also, the dilution resulting from injection of material into the air stream is inversely proportional to the average *speed* at the stack top level.

A summary of the various instruments used to measure wind velocity is given in Table 2.1.

Table 2.1 Instruments used to measure wind speed and/or direction and/or inclination

Instrument	*Mounting*	*Quantity measured*
Wind vane[a]	Cable/tower	Wind direction at each point
Rotating vertical shaft operates circular potentiometer or selsin unit. Recording technique has to deal with difficulties due to excursions through north		
Propeller (fixed)[ab]	Cable/tower	1 velocity component/sensor
Propeller/vane[a]	Cable/tower	Speed + direction at each point
Pressure tube (fixed)	Tower	1 velocity component per sensor
Pressure tube/vane	Tower	Speed + direction at each point
Vortex shedding/vane[a]	Cable/tower	Speed + direction at each point
Sonic anemometer[ab]	Tower	1 velocity component per sensor
Cup anemometer[a]	Tower	Speed at a point
Rotating cups drive a generator or operate a counter		
Paddle wheel	Tower	Direction
Tracked tetroon	Free air	Maintains fixed height
Tracked balloon	Ascending in free air	1 velocity/balloon at each level
Tracking by radar or theodolites. Averages over *c* 100 m deep layers		
Vane mounted vane[ab]	Tower	Wind inclination to horizontal
Kite	Tether	Speed and direction
Doppler Sodar[b]	Ground	Velocity and direction
Average velocities over layers up to range of equipment (~ 1000 m)		

[a] Suitable for high frequency (~ 1 Hz) turbulence measurements.
[b] Suitable for measuring vertical component of turbulence.

2.2.3   Measurements to determine the atmospheric stability

2.2.3.1   *Parameterizing the stability*

In general it will not be possible to measure turbulent energies and spectral densities (Section 2.2.4) throughout the depth and distance where the pollutant cloud concentrations need to be predicted. It is therefore desirable to be able to choose a parameter or parameters, whose values are relatively easy to determine or estimate, from which the relevant properties of the turbulence at all points of interest may be deduced. Such parameters include (2.2.3.2) measurements of temperature at a fixed height, (2.2.3.3) vertical temperature gradients and (2.2.3.4) radiation fluxes.

2.2.3.2   *Measurement of temperature at a fixed height*

Standard temperature measurements made at a reference height of about 1.3 m by the national meteorological service are usually available. These may be used to estimate sensible heat fluxes over the area of interest, provided that the temperature is changing at about the same rate at all heights and locations within the mixing layer. If this is so, it means that the shape of the vertical temperature profile is not changing with time (t) at a fixed location, or with distance (x) travelled in the direction of the wind. In this case we may write

$$d\theta/dt = \partial\theta/\partial t + U\partial\theta/\partial x$$
$$= -\frac{\partial((E+R_a)/(\rho C_p))/\partial z}{\rho C_p} \tag{2.8}$$

where R_a is the net radiative heat flux, E is the turbulent flux of sensible heat, θ is the potential temperature and ρ the potential density.

Estimates of $\partial\theta/\partial t$ and $\partial\theta/\partial x$ can be made from standard surface observations. Provided that radiative effects can be corrected or neglected, the rate of change of temperature at a fixed point ($\partial T/\partial t$), coupled with measurements at the same height at adjacent sites, to determine ($\partial T/\partial x$), can thus give a fair indication of the divergence of the turbulent sensible heat flux. If the mixing depth (Section 2.2.5) is also known, then the heat flux, E_0, at the surface can be deduced. For example, if ($d\theta/dt$) is not changing with height in the mixing layer, then if $d\theta/dt = 2°$ C h^{-1} and the height at which $E = 0$ is 600 m, taking $\rho C_p = 1250$ J m^{-3} ° C^{-1}

$$E_0 = 600 \times 1250 \times 2/3600 = 208 \text{ W m}^{-2}$$

If $\partial\theta/\partial t = 0$ *and* $\partial\theta/\partial x$ are independent of z, then the wind speed used in the advective part of Equation (2.8) should be an average speed through the mixing layer, rather than the speed at the measurement height. This is especially important if standard 1.3 m height screen temperature measurements are being used to estimate heat fluxes.

The potential temperature (θ) and density (ρ) are the temperature and density which air at temperature T and pressure p would have if brought adiabatically to a standard pressure of 1000 mb.

Temperature measurements are also important in their own right in air pollution studies because the temperature affects the rates of chemical reactions and also the rate of emission of pollution from combustion processes associated with space heating. It is important that the instruments used should record true air temperatures, i.e. they should be housed in ventilated radiation screens such as louvred boxes.

2.2.3.3   *Measurements of vertical temperature gradient*

The vertical temperature gradient is important both as an indicator of atmospheric stability and as a parameter which affects the plume rise (Section 2.4.2). There are some disadvantages in using the 'lapse rate' as a stability parameter. Gradients near the surface depend very much on the nature of the ground in the vicinity of the measurement site and those in the height range from 100 m to the top of the mixed layer usually give a slightly stable $(\partial\theta/\partial z) > 0$ or adiabatic $(\partial\theta/\partial z = 0)$ gradient in unstable $(\partial\theta/\partial z < 0)$ conditions near the surface.

Since the temperature differences are small, greater accuracy is required than for fixed height temperature measurements. The thermometers in the radiation screens should therefore be aspirated. A platinum resistance thermometer bridge, with the thermometer at the reference height used as the reference resistance, appears to be the best configuration for a permanent installation. Compensating leads and regular inspections of all junctions, which should be kept to the minimum practicable number, are essential.

2.2.3.4   *Measurements of thermal radiation*

Solar radiation

The energy spectrum of the radiation emitted by the sun peaks at a wavelength of 0.48 μm. In the absence of any terrestrial atmosphere, this would reach the earth's surface with a density

$$R_0 = 1400 \sin(\lambda) \text{ W m}^{-2}$$

where λ is the sun's angle of elevation.

In fact, solar radiation reaches the surface in two ways and leaves it in a third. These are: (a) direct radiation (R_d), (b) diffuse or scattered incoming radiation (R_s) and (c) reflected radiation (R_r). Both the sum and the ratio of R_d and R_s depend on such factors as solar elevation, atmospheric pollutant burden and the extent and nature of the cloud cover. R_r depends mainly on the solar elevation angle and the reflectivity or albedo of the surface. Some radiation will also be absorbed in the atmosphere.

R_d peaks at a longer wavelength than R_0 due to the more efficient scattering and absorption of the shorter wavelengths by gases and particles in the earth's atmosphere. R_s peaks at a shorter wavelength, for the same reason.

The short-wavelength energy, especially that within the ultra-violet range ($<0.4\,\mu$m), outside the visible spectrum (0.4–0.7 μm), is important in photo-chemical reactions. Consequently, measurements of total incoming radiation do not give sufficient information to enable the rates of reaction of some pollutants to be determined. Furthermore, the spectral density will change as the beam traverses the mixing layer, so that precise evaluation of the photochemically important part of the spectrum throughout the reacting pollutant cloud is only possible if a radiometer is mounted in a sampling aircraft.

The total incoming solar radiation is one term in the surface heat balance (below).

Long-wave radiation

The earth loses heat to space by emitting infrared radiation whose energy spectrum peaks at about 10 μm wavelength, corresponding to an absolute temperature of 290 K, compared with the solar surface temperature of around 6000 K. Over the whole earth's surface and long periods of time, this outgoing long-wave radiation balances the incoming solar radiation. Locally, there will be large imbalances and these result in diurnal variations in surface temperature and the transfer of heat from one region to another by large and small scale weather systems.

The atmosphere also emits and absorbs infra-red radiation. These processes are enhanced by the presence of cloud, water vapour and CO_2. Thus the infra-red flux consists of an outgoing (R_{lo}) and a returning (R_{lr}) component.

Measurement of the net radiation flux, or its two components, is usually accomplished by an array of thermocouples embedded in a surface with good absorption characteristics. These surfaces are protected by suitable transparent domes to eliminate turbulent heat transfer and various artifices prevent interference from condensation on the domes.

The surface heat balance

The surface heat balance equation may be written

$$R_d + R_s + R_{lr} = E_o + E_v + R_r + R_{lo} + E_s \tag{2.9}$$

where E_o is the turbulent sensible and E_v the turbulent latent heat flux into the air and E_s is the flux of sensible heat into the soil. The way Equation (2.9) is written, E_o, E_v and E_s represent upward heat fluxes if they are positive. In strong sunshine, E_s would be negative in all except very cold winter conditions. E_o and E_v would be negative at night when the air temperature and dew point were higher than the ground surface temperature. The infra-red radiation terms R_{lr} and R_{lo} will always be positive.

Equation (2.9) shows that it is a complicated matter to estimate the sensible heat flux to or from the mixing layer, even given accurate radiation data at one point.

2.2.4   Turbulence measurements

2.2.4.1   *Turbulent energy*

Measurements of the variance of each of the three components of the wind velocity vector are equal to twice the turbulent kinetic energy for that component while the square root of the variance gives the r.m.s. turbulent velocities σ_u, σ_v and σ_w, i.e. the KE (kinetic energy) $= \frac{1}{2}\sigma_u^2$, etc. The response time of the instrument will filter out the high frequency components of the energy, while the duration of sampling will restrict the effect of long period variations.

The types of instrument which are suitable for measuring turbulence down to frequencies ~ 1 Hz are listed in Table 2.1.

2.2.4.2   *Turbulence spectra*

The dispersive properties of the turbulence depend not only upon the turbulent energy but also on the distance over which the velocity excursions are maintained. We really need to know these distance scales following the motion of the air, i.e. the Lagrangian length scales, but what is usually available are measurements at a fixed point. These may be processed to give the proportion of the turbulent energy contained within given frequency intervals, the turbulence spectra. The frequencies at which each of these spectra peak determines the 'time scale' of that component of the turbulent velocity. However, because they are fixed point measurements, these are termed Eulerian time scales, and further hypotheses are involved in deducing the Lagrangian length scales that we need from them.

A discussion of the relationships between Lagrangian and Eulerian scales and methods of deriving them from spectral data is given by Pasquill and Smith [4]. Direct measurements of the Lagrangian fluctuations have been made by tracking tetroons or balloons, but such data would not normally be available. Furthermore, their validity depends upon the balloons faithfully following the air motions, i.e. they must be truly neutrally buoyant with respect to the surrounding air. Also a large number of balloons must be tracked to give statistically significant results.

2.2.4.3   *Turbulent fluxes*

The product of the air density (ρ_e) and the covariance of the vertical velocity and either of the other velocity components equals the vertical turbulent flux of momentum of that component. Similarly, the covariance of the vertical velocity and the potential temperature is a measure of the vertical turbulent heat flux. Therefore, if fast response thermometers are mounted adjacent to the turbulent velocity sensors, both heat and momentum fluxes can be measured. These fluxes change with height in the mixing layer. When the air is not accelerating, the divergence of the momentum fluxes balance the pressure gradient and Coriolis forces, the divergence of the turbulent heat flux determines the non-radiative component of the rate of heating or cooling of the air.

Sonic anemometers (Table 2.1) can measure the air temperature directly over the same path as they are measuring the wind velocity, because they measure the velocity of sound in air.

2.2.5   Measurements of mixing depth

Before discussing measurements of mixing depth, it is necessary to define, as precisely as possible, what we are talking about. Nieustadt and van Dop [5] define the top of the layer as follows:

(a) For unstable conditions, the height of the convective boundary layer (CBL) is the height of the lowest temperature inversion
(b) For stable conditions, the height of the stable boundary layer (SBL) is the height at which the turbulence falls to 'say 5 per cent' of the turbulence level 'near the surface'.

In fact, mixing is occurring in the stable air at the top of the CBL and fluctuations in velocity due to gravity waves occur above the height at which turbulent mixing is insignificant in the stable boundary layer. Therefore neither of these definitions is entirely satisfactory. From a practical point of view, we know exactly what we should like the mixing height to mean. It is the maximum height from which material currently being released into the atmosphere with no buoyancy or vertical momentum will diffuse to the surface. It is also the maximum height which material released in a similar fashion from the surface will reach.

In general, the interface will be 'humpy' or 'wavy' due to the presence of gravity waves and 'plumes' of hot air rising from local 'hot spots' on the surface. Therefore, although there may be a sharp change in the properties of the air (temperature, humidity, wind speed and direction and pollutant burden) between the air mass above and below the interface, this will usually show up as a gradient when measurements are averaged over a period of minutes or more. Sharp tops to the mixing layer are mainly confined to divergent airstreams (Section 2.3.1.1) because there will be less stable air, more moisture and systematic upward motion in the air aloft in convergent flows.

From the above it appears that there is no very precise definition and consequently no prospect of very precise measurement of the mixing depth.

The mid-point of the region where there is an obvious transition from the properties of the air mass in the mixing layer to the lower tropospheric air mass, is probably the best definition. In many cases this will also correspond to a minimum in the wind velocity gradient and a maximum in the potential temperature gradient.

Remote sensing equipment such as sodars, lidars and direct measurements from towers or balloons are all capable of giving estimates of the mixing depth. The height of the top of the lowest cloud layer is an indication of the maximum vertical excursion of the air mass in the mixing layer.

2.2.6 Precipitation measurements

An ideal raingauge measures the quantity of rain which would have fallen on an area of ground equal to the area of the raingauge orifice. It is therefore important to ensure that the wind flow over the gauge does not prevent drops falling into it. Splashing of water, either into or out of the collector, as a result of drop impaction on the raingauge or the adjacent surface, should also be avoided. Evaporation from the container before measurement is another potential source of error.

If the precipitation falls as snow, wind effects are more important and the container may become full unless there is some form of heating to prevent the temperature of the collector falling below freezing.

2.3 Outline of the more important features of the atmospheric transport and dispersion of pollutants

2.3.1 Transport and dispersion in different types of air mass or air stream

Air streams can be characterized as:

(i) Warmer than the underlying surface, or 'stable'
(ii) Cooler than the underlying surface or 'unstable'
(iii) At the same temperature as the underlying surface or 'neutral'.

These in turn may be:

(a) Divergent
(b) Convergent or
(c) Non-divergent.

Characteristics (a) to (c) depend mainly on the large scale weather systems, but this may be modified in the mixing layer by gradients in the surface temperature, which will cause convergence into warmer areas.

2.3.1.1 *Air stream characteristics*

In divergent air streams, clean air from above the mixing layer is being fed into it as a result of subsidence. The wind speed is usually increasing with distance travelled but not necessarily at a fixed point. Convergence may occur temporarily in the mixing layer during the day over coastal areas, due to sea-breeze effects, but will be restricted in depth to <1000 m on most occasions.

In convergent air streams, air can be lost from the mixing layer as a result of upward motion over a wide area. The wind speed may increase with distance travelled as the cyclonic vorticity increases, but will decrease as centres of low pressure are approached. Convergence destabilizes the air and the probability of deep thermal convection is consequently increased in this type of air stream.

In non-divergent air streams, the wind speed and direction can be very steady, which can lead to persistently high GLCs at sites which are downstream of multiple source areas.

2.3.1.2   *Air mass origins*

It is customary to categorize an air stream as containing air of a given air mass. Air masses begin their life in the light wind areas near the centres of the anticyclones. As the air moves away towards lower pressure, the air mass, when it reaches a given location, will have acquired a temperature and humidity which have been determined in part by the properties of the surface over which it has travelled, and partly by the properties of the air which subsided into it from the lower troposphere above during that part of its travel where the flow was divergent. The air mass is usually identified by its potential dew-point temperature. This is usually considered to be more conservative than its potential temperature, which is more subject to diurnal variation. The range of values of the potential dew-point which characterize a given air mass will depend on the location.

Air streams which originate in the quasi-permanent sub-tropical anticyclones, which are located at about latitude 30° in summer and latitude 35° in winter, arrive in middle latitudes as tropical maritime (designated mT) or tropical continental (cT) air streams, depending on the location of the anticyclone and the subsequent air trajectory between their area of origin and the area of concern. Similarly, air streams which originate in the travelling anticyclones between latitudes 40 N and 70 N, arrive in the area of concern as polar maritime (mP) or polar continental (cP) air masses. Further subdivision, into those generally warmer (mPw, or returning polar maritime) and those generally cooler (cPk or mPk) than the local daily average surface temperature, is also made. Finally, air streams which originate in the quasi-permanent arctic anticyclones, north of about 70 N, arrive in middle latitudes as arctic maritime (mA) or arctic continental (cA) air masses.

2.3.2   Diurnal variations in air stream characteristics

The dispersive characteristics and the depth of turbulent mixing in the air streams described in Section 2.3.1 will depend to a greater or lesser extent on diurnal heating or cooling of the surface. Figure 2.1a–d is a series of vertical time sections over land showing the properties of the mixing layers generated by each of the principal air stream types. Figure 2.2 shows the locations of the air streams in relation to typical features of a middle latitude synoptic weather map. Each of these air stream types will now be discussed briefly.

2.3.2.1   *Settled anticyclonic*

The principal dynamic features of these air streams are:

(a) Large-scale divergence (i.e. flow outwards from the high-pressure centres) across the isobars and corresponding subsidence (descending motion) in the lower half of the troposphere.

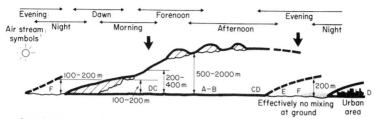

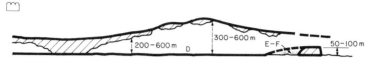

a. Settled anticyclonic – Turbulence mainly convective in origin.
With snow cover, the convective layer may not develop.

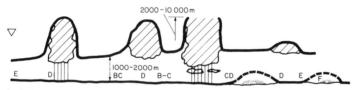

b. Warm advection – Turbulence mainly mechanical in origin.
On-shore wind at night, off-shore wind in day.

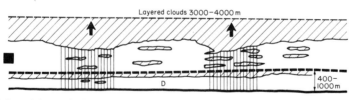

c. Cold advection – Turbulence convective mechanical. Land breeze, sea breeze.

d. Unsettled cyclonic – Turbulence mechanical. Thick layered cloud inhibits diurnal variation.

KEY

Fog or Cloud		—— Well defined	Boundary of turbulent layer
		– – Ill defined	
Precipitation		↑↓ Upward or downward motion	
		Ground fog	

Fig. 2.1 Time sections of the lower atmosphere in each of the four principal air stream types of middle latitudes. The time axis is calibrated in general terms (morning, afternoon, etc.) as the actual times will vary with location and time of year. The heights are indicated at various points along the time scale as these again will be functions of location, time of year, etc. The corresponding 'Pasquill' categories are indicated by capital letters just above the surface.

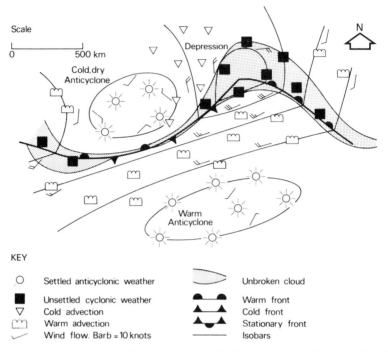

Fig. 2.2   Locations of principal air stream types in relation to typical features of a middle latitude (Northern Hemisphere) surface, synoptic weather map.

(b) The descending air above the mixing layer has a stable vertical temperature gradient and low humidity.
(c) The free stream wind speed is light or occasionally calm, but increases as the air moves away from the centre of the anticyclone.
(d) Any high turbulence intensities in the mixing layer are usually convective in origin, but relatively strong wind shears can develop and lead to outbreaks of mechanical turbulence.

A typical diurnal history of the mixing layer over land in locations where there is warm sunshine during the day, is shown in Fig. 2.1a.

Under clear skies at night a surface radiation inversion develops over rural areas. The surface wind becomes light and its speed and direction are determined at times by the slope of the surface and local horizontal temperature gradients. At other times, provided that the free stream wind speed is strong enough, the wind speed will pick up and its velocity will then be determined by the free stream velocity. In most locations there will be a minimum 10 m wind speed of 0.5 to 1 m s^{-1}, but in hollows there will be virtually no wind. This situation can lead to very low levels of atmospheric turbulence and where the surface wind is calm, no effective mixing layer. Later in the night, if the air near the surface contains sufficient moisture, fog will form. If the fog is deep enough, the maximum radiation cooling will be transferred from the surface to the top of the fog layer. Convective turbulence will then develop in the fog layer and any low-level emissions of pollutants will become well mixed through the fog layer. The top of the mixing layer is now the top of the fog.

As the sun rises, the surface will be warmed sufficiently for the fog to evaporate near the ground, low stratus cloud (base 60–300 m) will develop and eventually disperse. Further surface heating then results in the depth of the boundary layer continuing to increase and again, if there is sufficient moisture present, fair-weather cumulus cloud will form later in the forenoon, but with a much higher cloud base (400–2000 m).

The convective thermals will penetrate some distance into the stable capping layer of dry subsiding air and some of this air will be fed into the top of the mixing layer, helping the warming process (see Caughey [6] for a quantitative discussion). As the intensity of solar radiation begins to decline after noon, the supply of heat will eventually be insufficient to maintain the warming process and shortly after the maximum temperature is reached the air near the surface will begin to stabilize. The large-scale turbulence at higher levels will then dissipate, as the supply of further energy from the surface is cut off, and the diurnal cycle is then complete. This is the classic air pollution situation for low-level emissions. Modifications due to the presence of surface features and the importance of effective emission height on pollutant behaviour in these conditions will be discussed in Sections 2.3.4–2.3.6.

2.3.2.2 *Warm advection*

Warm advection occurs when the underlying surface is cooler than the surface upwind. Typical examples over north-west Europe are the south-westerly wind in the warm sector of a depression or the southerly flow ahead of warm fronts.

The principal features of this type of air stream are:

(a) A deep, stably stratified flow with high humidity
(b) Moderate or strong winds
(c) Little organized vertical motion, mainly weak subsidence
(d) Turbulence in the mixing layer, mainly mechanical; some weak convection over land during the day.

The diurnal history is illustrated in Fig. 2.1b.

Overnight there will be extensive low cloud. This may well reach the surface as sea- or hill-fog. Diurnal heating may burn off the cloud, or at least cause it to break, but the diurnal range in mixing layer height will be much smaller than in an anticyclonic situation. At sufficient distance from the coast, if the cloud has been dissipated by surface heating and the wind speed is light enough, a surface radiation inversion and fog patches may develop in the evening. However, in maritime locations low stratus cloud will usually spread over the area from the direction of the coast later in the evening, causing the surface air to warm again and a mechanically stirred mixing layer to reform.

2.3.2.3 *Cold advection*

These streams occur typically on the western sides of low-pressure areas (Fig. 2.2) and also locally as a result of sea breezes and shallow katabatic winds (see Sections 2.3.4 and 2.3.5).

The potential temperatures at heights up to several thousand metres are lower than those at the surface, except intermittently during the night and in subsiding air between clouds.

Other features of these streams are:

(a) Winds moderate or occasionally strong
(b) Convective cloud often giving showers
(c) Clear subsiding air between the clouds may give locally limited mixing depths corresponding roughly to the cloud base
(d) Turbulence – partly mechanical, but the mixing depth is controlled largely by the convective instability
(e) Little net vertical motion, but a tendency towards upward motion, especially where there are showers.

The diurnal history is shown in Fig. 2.1c. Winds and shower activity may be sufficient to prevent some stabilization of the surface air after sunset, but any fog patches will be shallow and clear as convective clouds move overhead probably at times during the night, but in any case soon after sunrise. The radiational cooling under clear skies may be very rapid because of the low humidity.

2.3.2.4 *Unsettled cyclonic*

These streams show fairly rapid changes in wind direction and generally overcast skies, often with rain. A general upward vertical motion and the 'baroclinic' nature of the flow often result in an ill-defined top to the mixing layer. However, in some situations the presence of a warm moist air mass aloft may lead to a sharp temperature inversion restricting vertical mixing. There will be a component of flow across the isobars towards the low pressure. Turbulence near the surface is mainly mechanical in origin.

There is little diurnal variation, as shown in Fig. 2.1d.

2.3.3 Frequency of occurrence of different air streams

The frequency with which the different types of air streams will occur at a given site will depend upon its geographical location. They will often be associated with a particular wind direction.

For the UK, a rough breakdown would be:

(a) Settled anticyclonic 30% (variable, often NE to E wind)

(b) Warm advection 50% SE to SW wind in summer, SW to W wind in winter
(c) Cold advection 10% NW to N wind in summer, NW to SE through E in winter
(d) Cyclonic 10% (variable).

2.3.4   Land and sea breezes

When the pressure gradient is slack, i.e. mainly in settled, anticyclonic conditions, the temperature difference between land and water leads to flow from the cooler towards the warm surface. At some height (usually < 1000 m) and at some distance (up to ~80 km) from the shore, the 'normal' flow will be maintained. Thus this situation, illustrated in Fig. 2.3a, leads to a limited mixing depth and also, in circumstances in which the land or sea breeze opposes the general flow, to the possibility of recirculation of pollution. For a recent quantitative discussion of sea breeze effects see Jones [1].

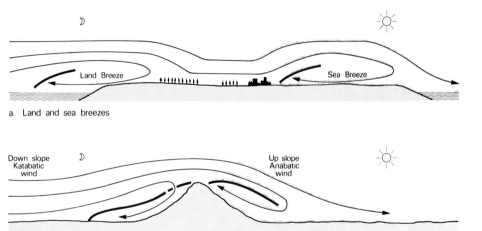

a. Land and sea breezes

b. Up slope and down slope winds

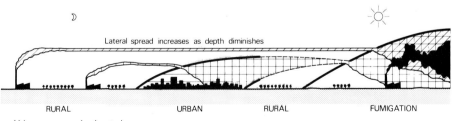

c. Urban areas and elevated sources

Fig. 2.3   Effect of surface features: (a) coasts, (b) high ground and (c) urban areas and source height on pollutant dispersion.

2.3.5   Upslope and downslope winds

Local winds also develop when air cooled by radiation at night over upland regions runs downhill to displace warmer air at lower levels (Fig. 2.3b). These katabatic winds are usually turbulent and in narrow valleys or fjords can be quite strong. Mixing within the stream will be quite efficient but there will be little exchange between the cold air stream and the warmer air above it. On the other hand, when low ground is covered with fog in the early morning, and the upper slopes clear, the differential heating may cause upslope (anabatic) winds to develop (Fig. 2.3b). The top of such circulations is generally limited by the snow-line.

2.3.6   Urban areas and elevated sources

An urban area enjoys its own microclimate and the reduced levels of turbulence associated with radiation inversions in rural areas are seldom encountered in built-up areas (see Bringfelt *et al.* [7]). Temperatures are also a few degrees higher than they are in rural areas in light winds (see WMO [8]). The consequence of these effects is that the urban area develops its own internal boundary layer and that low-level emissions which are trapped within this layer become well mixed within it (Fig. 2.3c).

In settled anticyclonic conditions at night the warmer urban air rises above the cooler rural air on the lee-side of the city, and an 'urban plume' drifts downwind. As the convective boundary layer develops over the rural area during the morning, the urban plume will eventually be brought down to the ground (right-hand side of Fig. 2.3c), leading to a rapid increase in ground-level pollution. As the mixing depth increases, the concentration will fall from this peak 'fumigation' value (see Martin and Barber [9]).

Figure 2.3c also illustrates some features of the behaviour of plumes from elevated sources. These will generally level out at some height depending on atmospheric conditions, stack height and heat emission rate. The initial dilution caused by turbulence due to relative motion will be much greater than that experienced by 'inert' tracers and will generally be comparable to that from a point source 10–20 km upwind in 'Pasquill F' (see Section 2.5) conditions by the time the plume has travelled a few hundred metres. If the air stream at plume level is stably stratified the depth of the plume will diminish (Fig. 2.3c) under the action of gravity forces as it drifts further downwind, resulting in additional lateral spread, with the result that the plume width is often greater than in a 'neutral' boundary layer. This again differs from the behaviour of inert plumes, which usually show much-reduced lateral spread (see Singer and Smith [10]).

As the plumes from elevated sources drift over urban areas at night, they may be fed into the urban boundary layer, or, if the source is large enough, they may pass over it (middle of Fig. 2.3c). It is interesting to note that there could be some descending air flow on the approach to the urban plume owing to the presence of a surface wind over the urban area and consequent downwind displacement of the streamlines. The opposite effect would occur during the day when wind speeds over the urban area would in general be lower because of the greater surface roughness there.

2.4   Calculation of the atmospheric transmission of pollutants

2.4.1   Introduction

The object of calculations of the atmospheric transmission is to determine the spatial distribution of pollutants downstream of their sources in all likely meteorological and emission conditions. To do this, it is necessary to have available methods to determine:

(a) The locus of the plume axis
(b) The dispersion of plume material about the axis
(c) The loss of material to the surface and
(d) The effects of atmospheric chemical reactions.

The plume axis may be defined as the line joining the points of maximum concentration at each distance downwind and its location will be determined by the wind field, the plume rise and the fall velocities of the plume material if this consists of large particles or droplets.

2.4.2   Calculation of plume rise

2.4.2.1   *Selection of equation*

It was shown in Section 2.1.2 that the most important parameter in determining the magnitude of the maximum concentration is the height of the plume. Early empirical formulae, such as that due to Holland [11] are now generally accepted as unsuitable for general application because they were based on observations made at various distances downwind usually before the maximum plume rise was achieved.

Two plume rise formulae are recommended by Jones [1]. These are:

(a) Briggs [12], developed from a model which assumes that the plume behaves like a horizontal cylinder containing all the effluent moving vertically through the air
(b) Moore [13], who assumes the plume comprises a series of interacting puffs.

In the stage where the growth of the plume elements is dominated by their own momentum and buoyancy, both models use conservation of vertical momentum in a plume element, as given by

$$V_t\rho_t w_t = V_0\rho_0 w_0 + Bg(\rho_t - \rho_e)V_0 t = \rho_e(M_0 + BFt)t_p \qquad (2.10)$$

where ρ_e is the air density (kg m^{-3}) and $\rho_{0,t}$ the plume density; V_0 is the volume of a plume element containing material emitted during time interval t_p; t is the time of travel from the source(s); w_0 is the vertical velocity of the plume elements (m s^{-1}); suffix 0 refers to the initial conditions at the source, and suffix t to conditions after travel time t; g is the acceleration due to gravity (m s^{-2}); B is a dimensionless numerical constant; F is the buoyancy flux (m^4 s^{-3}); M_0 is the initial momentum flux (m^4 s^{-2}).

In emissions from boiler plant, F_0 is proportional to the heat flux from the chimney, provided the gases are well ($\sim 100°$ C) above ambient temperature.

In both models the rate of entrainment of air into the plume element is taken to be proportional to the product of the relative velocity and the surface area of the element. This is equivalent to the assumption that the increase in plume radius (r_t) is proportional to the distance which the plume element has moved relative to the surrounding air during this phase of growth. In the absence of wind shear this means the w is proportional to dr/dt. In the Briggs model the plume is assumed to remain continuous in the streamwise direction and hence $V_t \propto r_t^2 \bar{U} t_p$ which means that the left-hand side of Equation (2.10) is proportional to $d(z_t^3)/dt$. In the Moore model the plume is assumed to consist of a series of puffs and hence $V_t \propto r_t^3$. The left-hand side of Equation (2.10) is then proportional to $d(z_t^4)/dt$.

If there is no gradient of potential density, i.e. potential temperature, in the ambient air, $F = F_0$ and the plume rise after a given time (i.e. distance) of travel will be proportional to the quarter power of the sum of two terms, one containing the product of t_p and M_0 and the other the product of t_p and F_0 (the initial buoyancy flux). Thus if t_p is independent of the plume radius and the flow is three-dimensional (puffs), plume rise is proportional to the 1/4 power of M_0 and F_0 as in the Moore model, but if t_p is proportional to the plume radius, even if the flow is three-dimensional, then the plume rise will be proportional to the 1/3 power of M_0 and F_0.

The plume rise formulae differ:

(a) Because of different assumptions about the nature of t_p
(b) Because of different assumptions about the plume dilution mechanism once relative motion becomes comparatively ineffective in diluting the plume.

Lamb [14] has suggested a plume model which reproduces the Briggs expressions in a non-turbulent environment but develops three-dimensional characteristics when the plume is in the mixing-layer. This type of approach appears most likely to yield improvement in practical prediction techniques.

2.4.2.2   Plume rise formulae

The expression for the rise of a positively or neutrally buoyant plume (Z_r) given by Moore [13] and which was one of those recommended in Jones [1] is

$$Z_r^4 = QmK^{1/2}(f'/U^*)^4 \, U_m x^{*3} \; (m^4) \tag{2.11}$$

Equation (2.11) may also be used to calculate the rise and droop of non-buoyant plumes (z_r'') provided a virtual excess temperature term T_g^* is introduced. Variation of the surface roughness parameter z_0 (Section 2.1.1) is also taken into account by introducing the term h_r where:

$h_r = \min (120, \max (h_s - 50(z_0 - 0.38)), 30)$ (m)
$h_s =$ the stack height in m

The other terms in the expression are:

$Q = 0.25\pi d_0^2 w_0$, the volume emission (m³ s⁻¹)
$d_0 =$ the stack diameter (m)
$K = \max (T_g^*, 12)/110$
$T_g^* = (C_{pg}T_0/C_{pa}) - (T_0 + \theta_e)(m_g - m_a)/m_g$ (K)
m_g and m_a are the molecular weights of the flue gas and air respectively
T_0 is the excess temperature of the flue gas (K)
C_{pg} and C_{pa} are the specific heats at constant pressure of the flue gas and ambient air (MJ kg⁻¹ K⁻¹)
$\theta_e =$ the ambient potential temperature at $1.5h_s$ (K)
$U_m =$ an empirical entrainment velocity (0.975 m s⁻¹)
$U^* = \max (0.2, U_{1.5h})$
$U_{1.5h} =$ the mean wind speed at $1.5h_s$ (m s⁻¹)
$f = 1$ when $S_1 \geqslant 0.0025$ and for large heat sources (>40 MW)
$\quad = \min (1, 0.16 + 0.007h_r)$ for $S_1 < 0.0025$, where
$S_1 = \Delta\theta^*/U^{*2}$ and $\Delta\theta^* = \max (\Delta\theta, 0.08)$ where
$\Delta\theta =$ the potential temperature gradient above the stack top (K/100 m)
$X_S = 120/S_1^{1/4}$ (m)
$X_N = 120(16 + 0.16h_r)$ (m)
$X_T = X_S X_N/(X_S^2 + X_N^2)^{1/2}$ (m)

$$x^* = \min \left(\frac{2000X_T}{(2000^2 + X_T^2)^{1/2}}, \frac{xX_T}{(x^2 + X_T^2)^{1/2}} \right) \text{(m)} \qquad (2.12)$$

$$m = \frac{gm_g[T_g^*(x^* + 27d_0) + \theta_e x_w(1.5 + 54d_0/x^*)]}{m_a x^*(T_0 + \theta_c)} \text{(m s}^{-2})$$

$x =$ downwind distance from stack (m)
$x_w = U^* w_0/g$ (m)

The rise from non-buoyant plumes (z_r), when the virtual temperature term $T_g^* < 0$, is equal to Z_r from Equation (2.11), provided $x^* < -\theta_e x_w/T_g^*$. If $x^* > -\theta_e x_w/T_g^*$ then

$$z_r'' = 2z' - (2z'^4 - Z_r^4)^{1/4} \qquad (2.13)$$

where z' is the value of Z_r at $x^* = -\theta_e x_w T_g^*$. At values of $x^* > -\theta_e x_w/T_g^*$, Z_r^4 will become negative so the fourth root of this term is no longer real. In this case z_r'' may become negative, indicating that the plume has returned to below the stack-top level.

The equivalent Briggs equation to (2.11) which is also recommended in Jones [1] is

$$Z_r^3 = (1.5/(\pi\beta^2))Qx^{*2}m/U_{1.5h}^3 \qquad (2.14)$$

where m now equals $gm_g[T_g^* + \theta_e(2x_w/x^*)]/m_a(T_0 + \theta_c)$ and the empirical entrainment factor β has a value of about 0.5.

Various alternatives to Equation (2.14) for determining x^* extracted from papers by Briggs are given in Jones [1].

The trajectories of plumes which are effectively hot air, in which the terms $T_g^* = T_0$, and m equals $gT_0/(T_0 + \theta_e)$, are well represented by the simpler expressions

$$Z_r = 2.4\, Q_h^{1/4}\, x^{*\,3/4} K^{1/8}/U^*$$

and

$$Z_r = 3.3\, Q_h^{1/3}\, x^{*\,2/3}/U_{1.5h}$$

where Q_h is the sensible heat emission from the chimney $Q\rho_e C_{pg} T_0/(\theta_e + T_0)$.

2.4.3   The effect of particle fall velocity on plume height

Calculating the effective height of a particle plume is difficult. The motion within a rising plume is complicated. Figure 2.4 shows the idealized motion in a cross-section of a rising line-thermal. The motion in a three-dimensional puff is qualitatively similar, but one has to rotate the cross-section about the axis of symmetry to obtain the three-dimensional flow pattern. Actual flow patterns are less regular, i.e. even more complicated, than that shown in Fig. 2.4. The problem to be solved is that initially, falling particles are within the rising envelope of the gaseous plume, indicated by the solid boundary of the 'thermal'. However, once they fall out of it, they no longer benefit from any further rise which the gaseous plume achieves. An effective plume height must therefore be calculated for each size of particle. Foster [15] has discussed this problem using a two-dimensional plume model. He showed that fine dust, with fall velocities of the order of a few centimetres per second, will achieve much the same rise as the gaseous plume. However, the centre of the particle plume will fall relative to the surrounding air at the fall velocity of the particles, so its effective height at the distance of maximum GLC may still be lower than that which would be expected for the gaseous plume.

Larger particles of interest include agglomerates in stack emissions or water-droplets carried out of cooling towers. These may have fall velocities of the order $1\ \mathrm{m\ s^{-1}}$. A very limited data base exists to evaluate theoretical predictions of this sort of particle plume (see, for example, Dunn *et al.* [16]).

A practical approach to the problem is to assume that the particles leave the plume when they have fallen a distance relative to the plume axis equal to half the plume rise. This roughly corresponds to the distance where they would have

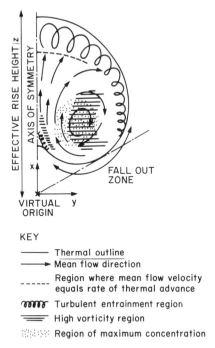

KEY

——— Thermal outline

——➤ Mean flow direction

- - - - Region where mean flow velocity equals rate of thermal advance

ᵐᵐᵐ Turbulent entrainment region

⹀ High vorticity region

⠉⠉ Region of maximum concentration

Fig. 2.4 Flow structure in a thermal. The illustration is taken from Foster [15] and describes a line thermal, but flow in a profile is very similar. (Reproduced with permission of Pergamon Press.)

fallen below the plume envelope. If this distance is x_f, then $w_f x_f / \bar{U} = Z_r(x_s)/2$ and the height of the particle plume axis is given by

$$Z_r(x) - w_f x/\bar{U} \text{ for } x < x_f \text{ and } Z_r(x_f) - w_f x/\bar{U}, \text{ for } x \geqslant x_f$$

U is the average wind speed between the gas and particle plume axes.

2.4.4 Calculation of dispersion

In a dispersion model, the ground level concentration pattern is calculated for each source by assigning an effective height to the plume (Section 2.4.2) and then using one of a number of techniques to estimate the spread of the effluent material as it travels downwind. The overall pattern is then derived by summing the contributions of the individual sources. Small sources of similar type, e.g. domestic, commercial or vehicles, may be treated as area or line sources, with the total estimated or observed emission divided by the estimated area or length of the source to give the source strength.

Broadly speaking, dispersion models fall into three types. These are eddy diffusivity, Gaussian and second or higher order closure models.

2.4.4.1 *Eddy diffusivity (or K-type) models*

A vertical eddy diffusivity (K_z, $m^2\,s^{-1}$) and a three-dimensional wind velocity vector, U, which will in general vary with location, are attributed to each part of the pollutant's path between source and receptor areas. The distribution of material is calculated by solving the diffusion equations with appropriate boundary conditions, which can include surface deposition.

These models are often simplified by treating U as one-dimensional and U and K_z as functions of height only. In this case if

$$\bar{I}(z, x) = \int_{-\infty}^{+\infty} C(x, y, z)\,\mathrm{d}y$$

$$U\frac{\partial \bar{I}}{\partial x} = \frac{\partial}{\partial z}\left[K(z)\frac{\partial \bar{I}}{\partial z}\right] \tag{2.15}$$

The diffusivity is usually equated with $\sigma_w l$, where l is a 'mixing length' which equals $0.4\,Z$ near the ground. Equation (2.15) only applies if σ_z exceeds twice l. For small distances from the source $\sigma_z = \sigma_w\,x/U$. For cross-wind dispersion, the 'mixing lengths' are very large and so a Gaussian distribution (see Section 2.4.4.2) is usually used to represent the lateral spread of the plume. In that case $\bar{I} = \sqrt{2\pi\sigma_y}\,C(x, 0, z)$.

2.4.4.2 *Gaussian models*

The distribution of material is assumed to be normally distributed about the effective height (H) of the plume so that the concentration at points (x, y, z) with the source of strength Q_p as origin is given by

$$C(x, y, z, \tau_s) = \frac{Q_p}{2\pi \bar{U}\sigma_z\sigma_y} \exp\left(\frac{-(H-z)^2}{2\sigma_z^2}\right) \exp\left(\frac{-y^2}{2\sigma_y^2}\right) \tag{2.16}$$

Additional terms in which H is replaced by $-H$ and $2nz_m \pm H$, with $n = 1, 2$, etc., are summed to take account of multiple reflections of the plume at the ground and at the top of the mixing layer (z_m), should be included. Equation (2.16) is a solution of Equation (2.15) when K_z and U do not vary with height. In that case $\sigma_z^2 = 2K_z \times /\bar{U}$.

As an alternative to the multiple reflection model Moore [26] has proposed the use of the expression

$$C(x, y, O, \tau_s) = \frac{2Q_p \exp(-H^2/(2\sigma_{z\downarrow}^3))\exp(-y^2/(2\sigma_{y\downarrow}^2))}{\pi(\bar{U}_\uparrow\sigma_{z\uparrow}\sigma_{y\uparrow} + \bar{U}_\downarrow\sigma_{z\downarrow}\sigma_{y\downarrow})} \tag{2.17}$$

where the arrows refer to upward and downward dispersion. The wind speeds $\bar{U}$ are averages above and below the plume axis. Plume reflection is taken into account by restricting $\sigma_{z\uparrow}$ and $\sigma_{z\downarrow}$ to maximum values of ($z_m - H$) and H respectively. Equation (2.17) also takes some account of the variation of wind speed and turbulence with height.

σ_z and σ_y, the vertical and lateral standard deviations of the concentration distributions, increase with distance downwind and with the sampling interval τ_s. A number of methods have been proposed for estimating them either from measurements of atmospheric turbulence or from the general properties of the mixing layer. This usually involves determining a stability category from: (i) the wind speed at a fixed height and (ii) one of the following:

(a) The vertical temperature gradient between two heights
(b) The surface sensible heat flux (measured or estimated from cloud cover and time of day and year)
(c) Fluctuations in the wind direction at a fixed height.

The vertical temperature gradient and wind speed above the stack top are also important in determining the plume rise, so these must be measured or predicted from the properties of the stability category whether or not they are used in the category definition.

All these schemes are qualitatively similar but individual occasions may be assigned to different stability categories depending on the scheme used. The variation of σ_z and σ_y with x within each stability category also varies to some extent between one scheme and another. A comprehensive comparison of these differences for twelve different schemes has been made by Kretzchmar and Mertens [17] (Fig. 2.5), which should be consulted for more detailed information.

2.4.4.3 *Second and higher order closure models*

Equation (2.15) is a first moment equation and implies that net turbulent transfer of material is always in the direction of the gradient of its mean value. This is not always the case, especially when the turbulence is mainly produced by thermal convection. Second order closure methods attempt to solve the equations for the second moments, e.g. the equation for $\partial(\overline{w'C'})/\partial t$. This contains six terms (see Pasquill and Smith [18]), involving various properties of the above covariance plus buoyancy, pressure fluctuation and viscous dissipation terms. An attempt is being made to model the Plains Experiment pollution data (Chapter 14) using such techniques (see Lewellen and Sykes [19]).

2.4.5 Box and cell models

In a box model, the effects of individual sources are ignored, and the air stream (often termed an airshed) under investigation is treated as if it were a single box or a number of boxes (cells). The depth of the boxes is normally the maximum depth of the mixing layer or internal boundary layer (e.g. the urban boundary layer, Fig. 2.3c) during the period of each time step.

The surface area of each box may correspond to some reasonably homogeneous part of the area or may be fixed arbitrarily, e.g. as the area round each pollutant sampling site or a square of given side length. Unless special precautions are taken to reduce spurious mathematical diffusion effects, the sides

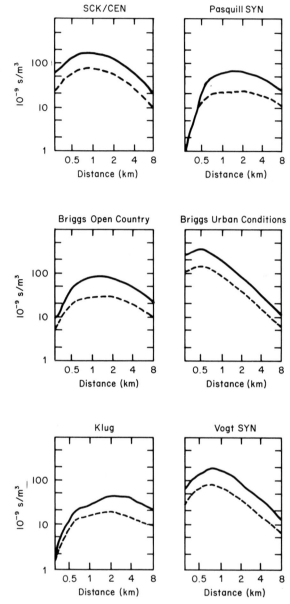

Fig. 2.5   Yearly averages for Mol, Belgium for a continuous release from a point source 100 m above ground level. Calculated by twelve different dispersion schemes (after Kretzchmar and Mertens [17]. (Reproduced with permission of Pergamon Press.)

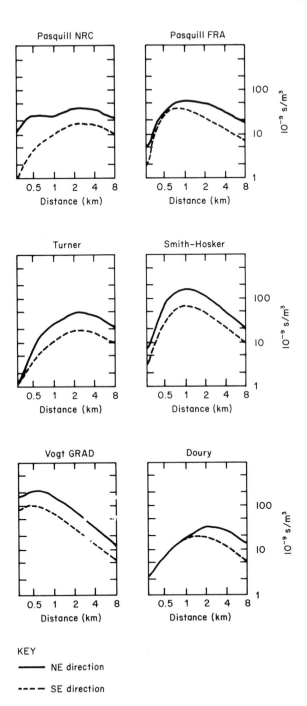

KEY

——— NE direction

- - - - SE direction

of the boxes should be kept parallel and normal to the local wind direction in multiple box or cell models.

For each of a number of time steps, whose duration is normally between one hour and one day, depending on the degree of spatial and temporal variability which the model is designed to represent, the emissions within each box are assumed to be immediately mixed uniformly throughout it. Pollutant removal or conversion processes can be taken into account if this were deemed necessary. The inflow and outflow of pollutant from each box are calculated from estimated wind speeds and directions. In its simplest form, this type of model ignores high pollution produced by nearby low level sources and the zone of zero GLC around high level sources. A detailed discussion of box models is given by Pasquill and Smith [20].

2.4.6 Calculation of trajectories

When the time of travel of a pollutant is many hours, the wind speed and direction may change appreciably between the time it leaves the source and the time it arrives at the receptor.

In this case it will be necessary to construct a trajectory, e.g. to use estimates of wind speed and direction, to work backwards from the place and time an 'incident' was observed, or forwards from the time and place of a release of pollutant, to find the probable path of the pollutant. Most trajectory modelling uses winds at a fixed height or pressure level. One difficulty is that air at different levels is often travelling in different directions and may be rising or descending. Isentropic trajectories, i.e. trajectories in which the potential temperature is conserved, are thought to give a better representation of this sort of motion than isobaric trajectories. Danielsen [21] has demonstrated that isentropic trajectories, particularly in the vicinity of a rain-producing weather system, may be very different from those estimated from two-dimensional (isobaric) flow patterns. This difficulty, which has still not been entirely solved, may be especially important in deciding the origin of dissolved material in precipitation.

A discussion of errors in trajectory estimates is given by Kallend [22]. Using current wind data, the error in speed was about 20% and in direction about 20°.

2.4.7 The effects of deposition

2.4.7.1 *General*

Deposition occurs at the surface, but removal processes operate throughout the depth of the pollutant cloud when rain, snow or hail are falling through it. Deposition affects the pollutant distribution in two ways:

(a) It decreases the total flux
(b) It alters the vertical profile.

In particular if deposition is occurring only at the surface, it results in lower GLCs than would otherwise have occurred.

It is customary to express the flux to the surface as the product of the GLC and a 'deposition velocity'. When this is done, then provided that $K(z)$ in Equation (2.15) is one of a few simple functions of z, analytical solutions of the equation may be obtained.

2.4.7.2   Dry deposition

Dry deposition occurs when gases or particles are taken up at the surface in unsaturated air. The deposition velocity (w_d) may be regarded as the inverse of the resistance of the surface to deposition. This resistance may itself be regarded as the sum of an aerodynamic (R_a), a viscous (R_v) and a chemical or biological resistance (R_c). Hence

$$w_d = 1/R_a + 1/R_v + 1/R_c \qquad (2.18)$$

If $R_c \gg R_a$ and R_v, then the deposition velocity will depend on the chemical or biological interactions between the surface and the material being deposited on it, but not on the wind speed or atmospheric stability. However, if R_c is low compared with the other resistances, the wind and stability will control the deposition rate.

Deposition velocities have been reviewed and tabulated by Jones [1]. For most gases and small particles they are within the range 1–10 mm s^{-1}.

2.4.7.3   Wet deposition

Wet deposition occurs when gases or particles are washed out of the air by rain. This happens:

(a) Within the cloud, when the pollutant material may first have been incorporated into cloud-drops, and the process is termed 'rainout', and
(b) Below the cloud, where the process is called 'washout'.

In rainout, material which is present as condensation nuclei will begin to be removed immediately it enters the rain cloud. However, if the material is present as a gas or particles which do not become centres of condensation, there is usually a delay time before such material is taken into the cloud water. This delay may depend on the presence of other chemical species.

2.4.7.4   Occult deposition

Occult deposition occurs when cloud drops impact on vegetation or other protuberances from the surface, or fall on to the ground. Such droplets have fall velocities 10 to 100 times greater and much greater inertia than the nuclei on to which the water condensed. The rate of removal of the nucleus material in cloud may therefore be an order of magnitude faster than in unsaturated air, i.e. if a site is in cloud 10% of the year, total occult deposition could be comparable with total dry deposition as far as annual deposition is concerned.

This also happens, but to a lesser extent, to soluble gaseous pollutants and smaller aerosol particles, which do not form centres of condensation which may be captured by the cloud-drops. Wind plays an important role in the impaction process, so hill-fog situations are likely to produce more deposition than radiation fogs. It is therefore important to try to determine how frequently hills, which are in the vicinity of source of pollutants likely to become centres of condensation or taken up into cloud water, are in cloud when the wind is blowing towards them from the sources.

Some measurements of occult deposition are described by Dollard and Unsworth [23].

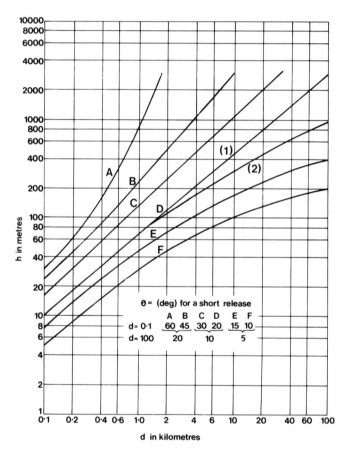

Fig. 2.6 Estimates of vertical (h) and lateral (θ) spread of a plume. The extent of h and θ are defined by pollutant concentrations one-tenth of those on the axis of the plume and equal to 2.15 σ_z and 2.15 σ_θ.

2.5 Examples of calculations using Gaussian models

Figures 2.6 and 2.7 show values of $h = 2.15\sigma_z$ and σ_y for the various stability conditions described in Table 2.2, following the scheme suggested by Pasquill [25] for ground level releases.

If the plume height is calculated by using the expressions for plume rise given in Section 2.4.2.2, the distance and magnitude of the maximum GLC and the concentration at any location (x, y, o) may be calculated. Figure 2.8 shows the results of a series of such calculations for a sampling interval τ_s of about 10 min.

Such estimates are only a rough guide to likely surface concentrations because σ_z and to a lesser extent σ_y are functions of height of release as well as the

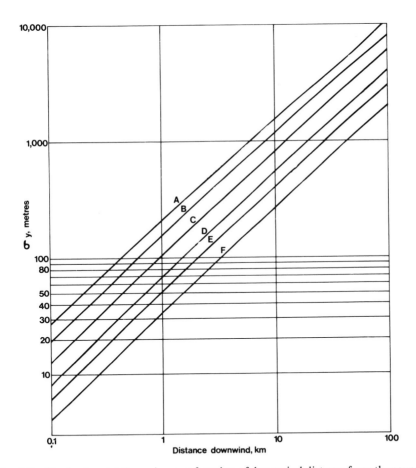

Fig. 2.7 The horizontal dispersion as a function of downwind distance from the source. The function σ_y is the standard deviation of the plume concentration distribution in the horizontal. (Turner [24].)

Table 2.2   Key to stability categories

Surface wind speed at 10 m (m s⁻¹)	Sunlight			Night	
	Strong	Moderate	Slight	≥4/8 cloud	≤3/8 cloud
<2	A	A–B	B	–	–
2–3	A–B	B	C	E	F
3–5	B	B–C	C	D	E
5–6	C	C–D	D	D	D
>6	C	D	D	D	D

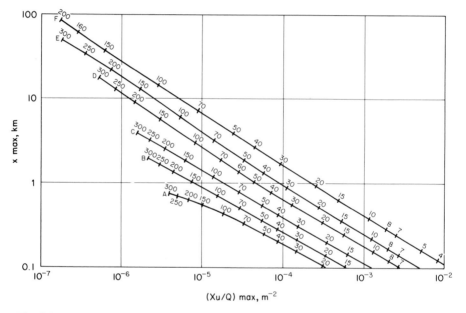

Fig. 2.8   Distance of maximum ground-level concentration and maximum $\chi u/Q$ as a function of stability (curves) and effective height (m) of emission (numbers). (After Turner [24].)

meteorological conditions. Furthermore, large fluctuations about the expected concentrations will occur and the calculated values are best regarded as ensemble means of a range of values which will be observed in the specified stability conditions.

Figure 2.9 shows a typical hourly mean SO_2 pattern observed downwind of Tilbury and Northfleet power stations and Fig. 2.10 shows the results of a

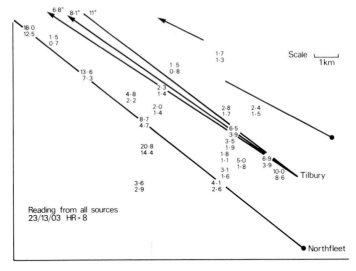

KEY

Date time group, bottom left-hand corner

03	Time of beginning of period (0330)
13	Day
23	Month (11th month of second year; November 1964)
HR = 8	8th hour of period, i.e. 10·30 - 11·30

Upper reading	Maximum during hour
Lower reading	Average during hour (p.p.h.m.)
	N.B. Decimal point of maximum reading gives meter locations on diagrammatic print-out

Line 11°	Direction of 114 m wind
8·1°	Direction from SO$_2$ distribution in near arc
6·8°	Direction from SO$_2$ distribution in second arc
	The bearing is measured in degrees off the meter axis, positive readings indicate that the axis of the plume lies east of the meter pattern axis

Fig. 2.9 Surface pollution pattern (hourly mean values and maximum 3-min reading (pphm) for the period 10.30–11.30 h 13 November 1964.

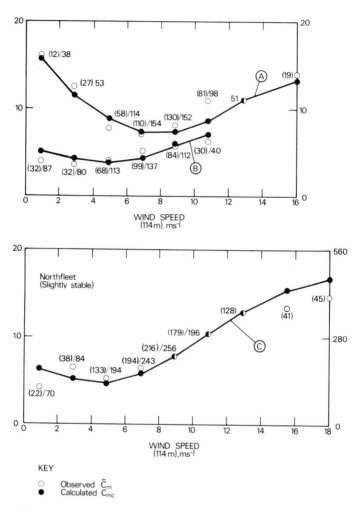

Fig. 2.10 $\bar{C}_m$ and C_{mc} as a function of wind speed in (a) unstable; (b) stable; and (c) slightly stable meteorological conditions.

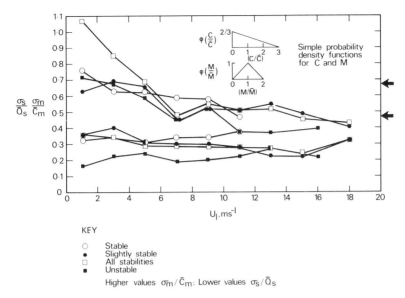

KEY

○ Stable
• Slightly stable
□ All stabilities
■ Unstable

Higher values $\sigma_m/\bar{C}_m$: Lower values $\sigma_s/\bar{Q}_s$

Fig. 2.11 Scatter of value of observed maximum GLC $(\sigma_m/\bar{C}_m)$ and SO_2 emission $(\sigma_s/\bar{Q}_s)$ as a function of wind speed of various stability groups (Northfleet plume).

calculation of the ensemble averages of ground level concentrations using a model described by Moore [26]. The scatter of the individual values within each of the categories is shown in Fig. 2.11. The dispersion parameters used in these calculations were functions of height, but in general differed by a factor of less than two from the corresponding Pasquill/Turner values.

References

1. Jones, J. A. (1983) *Models to Allow for the Effects of Coastal Sites, Plume Rise and Buildings on Dispersion of Radionuclides and Guidance on the Value of Deposition Velocity and Washout Coefficients*, NRPB-R157, National Radiological Protection Board, Chilton, Didcot, Oxford.
2. Pasquill, F. and Smith, F. B. (1983) *Atmospheric Diffusion*, Ellis Horwood, Chichester, pp. 313–41.
3. Clarke, R. H. (1979) *A Model for Short and Medium Range Dispersion of Radionuclides Released to the Atmosphere*, NRPB-R91, National Radiological Protection Board, Chilton, Didcot, Oxford.
4. Pasquill, F. and Smith, F. B. (1983) *Atmospheric Diffusion*, Ellis Horwood, Chichester, pp. 67–84.
5. Nieustadt, F. T. M. and van Dop, H. (1982) *Atmospheric Turbulence and Air Pollution Modelling*, Reidel, Dordrecht, pp. xvi, xvii.

6. Caughey, S. J. (1982) Observed characteristics of the atmospheric boundary layer. In *Atmospheric Turbulence and Air Pollution Modelling* (eds F. T. M. Nieustadt and H. van Dop), Reidel, Dordrecht, Holland.

7. Bringfelt, B., Hjorth, T. and Ring, S. (1974) *Atmos. Environ.*, **8**, 131–48.

8. WMO (1970) Urban climates, Technical Note 108, World Meteorological Organisation, Geneva.

9. Martin, A. and Barber, F. R. (1973) *Atmos. Environ.*, **7**, 17–38.

10. Singer, I. A. and Smith, M. E. (1966) *Int. J. Air Water Pollut.*, **10**, 125–36.

11. Holland, J. Z. (1953) *A Meteorological Survey of the Oak Ridge Area*, USAEC Report ORO-99, Weather Bureau, Oak Ridge, Tennessee.

12. Briggs, G. A. (1975) Plume rise predictions. In *Lectures on Air Pollution and Environmental Impact Analyses*, American Meteorological Society, Boston.

13. Moore, D. J. (1980) Lectures on plume rise. In *Atmospheric Planetary Boundary Layer Physics* (ed. A. Longhetto), Elsevier, Amsterdam, pp. 327–54.

14. Lamb, R. G. (1982) Diffusion in the convective boundary layer. In *Atmospheric Turbulence and Air Pollution Modelling* (eds F. T. M. Nieustadt and H. van Dop), Reidel, Dordrecht, pp. 219–20.

15. Foster, P. M. (1982) Particle fall-out during plume rise, *Atmos. Environ.*, **16**, 2777–84.

16. Dunn, W. E., Gavin, P., Boughton, D. *et al.* (1981) *Studies on Mathematical Models for Characterizing Plume and Drift Behaviour from Cooling Towers*, EPRI CS-1683, Vol. 3, Electric Power Research Institute, Palo Alto.

17. Kretzchmar, J. and Mertens, I. (1984) *Atmos. Environ.*, **18**, 2377–94.

18. Pasquill, F. and Smith, F. B. (1983) *Atmospheric Diffusion*, Ellis Horwood, Chichester, p. 102.

19. Lewellen, W. S. and Sykes, R. I. (1983) *Second Order Closure Model Exercise for the Plains Power Plant Plume*, EPRI EA-3079, Electric Power Research Institute, Palo Alto.

20. Pasquill, F. and Smith, F. B. (1983) *Atmospheric Diffusion*, Ellis Horwood, Chichester, p. 346 *et seq.*

21. Danielsen, E. F. (1974) *Adv. in Geophysics*, **18B**, 73–94.

22. Kallend, A. S. (1983) *The Fate of Atmospheric Emissions along Plume Trajectories over the North Sea* – Summary Report EPRI EA-3217, Electric Power Research Institute, Palo Alto, pp. 4–101.

23. Dollard, G. J. and Unsworth, M. H. (1983) *Atmos. Environ.*, **17**, 775–80.

24. Turner, J. S. (1970) *Workbook of Atmospheric Dispersion Estimates*, US Environmental Protection Agency, Research Triangle Park, N. Carolina.

25. Pasquill, F. (1961) *Meteorol. Mag.*, **90**, 33–49.

26. Moore, D. J. (1975) *Proc. Inst. Mech. Eng.*, **189**, 33–43.

3 Air pollution chemistry

A. M. WINER

3.1 Introduction

Within the past fifteen years there has been a revolution in our knowledge of the chemistry of urban atmospheres. The complexities of photochemical air pollution have gradually yielded to a combination of laboratory and environmental chamber research and computer modelling studies in which the rates and mechanisms of reactions involving organics, oxides of nitrogen (NO_x) and oxides of sulphur (SO_x) have been quantitatively characterized.

Complementing these fundamental investigations of the formation of photochemical air pollution have been the development and application of long pathlength spectroscopic techniques, and other state-of-the-art analytical methods, which have permitted the identification and measurement of many of the so-called 'trace' atmospheric pollutants in ambient atmospheres. Thus, knowledge of the time–concentration behaviour in ambient air of such species as formaldehyde (HCHO), nitric acid (HNO_3), the nitrate radical (NO_3) and nitrous acid (HONO), has provided important insights into the role these compounds play as products or intermediates in air pollution chemistry.

Finally, the development of increasingly sophisticated computer kinetic models and their application to laboratory and atmospheric data have provided a further basis for testing our understanding of the intricacies of photochemical smog formation.

The foundations of air pollution chemistry were laid down more than twenty years ago by Leighton [1]. Subsequently, in the 1960s, an empirical understanding of the $NO–NO_2–O_3$ and $NO_x–HC$ systems was obtained from environmental chamber experiments [2–10]. However, an important breakthrough, $c.$ 1969–76, was the recognition [11–19] that the hydroxyl radical (OH) is, with several important exceptions, the key reactive intermediate in the photooxidation of most organic, and many inorganic, compounds found in polluted atmospheres. Today the rates of OH radical reactions have been measured by relative or absolute techniques for more than 200 atmospherically relevant organic compounds [20, 21].

It is interesting, however, and an indication of the complexity of the polluted troposphere, that significant new reaction pathways are still being discovered at this late date. Thus, only in the last several years ($c.$ 1980–83) has it begun to be appreciated that the NO_3 radical and HONO play important roles in the night-time chemistry of urban atmospheres [22–35] and that the NO_3 radical has an important role in the natural troposphere as well [33–35]. As late as 1978 these two species had not even been observed in the atmosphere, let alone recognized as important.

Clearly, the task of describing in complete detail our present state of knowledge of air pollution chemistry is beyond the scope of this chapter. In particular no attempt has been made to treat in any detail the important area of heterogeneous chemistry (e.g. heterogeneous SO_2 oxidation). For more detailed descriptions of the relevant mechanisms and rates of reactions applicable to the topics covered here the reader is referred to the primary literature references at the end of the chapter and to a number of excellent reviews [15, 19–21, 36–39]. Here we provide a brief overview and summary of the essential elements of gas phase air pollution chemistry.

Particular emphasis is given to the role of primary air pollutants such as NO, NO_2, SO_2 and hydrocarbons, and to the formation and fates of secondary air pollutants such as ozone, peroxyacetyl nitrate (PAN), aldehydes, HNO_3 and HONO, as well as to reactive intermediates such as OH and NO_3 radicals. Particulate nitrates, sulphates and organics, which arise from gas to particle conversions in the atmosphere or from direct emissions, are also discussed briefly. In so far as possible, this treatment emphasizes the implications of these chemical processes for ambient air analysis of primary and secondary pollutants, and for the interpretation of the resulting air monitoring data.

3.2   Inorganic reactions

3.2.1   The $NO-NO_2-O_3$ cycle

Many aspects of the inorganic reaction systems in the atmosphere are now well understood. The photodissociation of NO_2 by near ultra-violet solar radiation

$$NO_2 + hv(295 \leq \lambda < 430 \text{ nm}) \rightarrow NO + O(^3P) \tag{3.1}$$

is a critical process since the subsequent reaction of the resulting $O(^3P)$ atom with oxygen

$$O(^3P) + O_2 + M \rightarrow O_3 + M \tag{3.2}$$

remains the only significant source of ozone in urban atmospheres. The rapid reaction of NO with O_3

$$NO + O_3 \rightarrow NO_2 + O_2 \tag{3.3}$$

completes this reaction cycle.

Because reaction (3.3) is rapid, ozone concentrations in urban atmospheres cannot rise until most of the NO has been converted to NO_2. This accounts in part for the fact that O_3 levels may be lower on average in city centres where high NO emissions occur, but higher in downwind suburban areas to which the resulting NO_2 is transported and then photodissociated, leading to O_3 formation.

3.2.2 Formation of radical intermediates

3.2.2.1 *Hydroxyl and hydroperoxyl radicals*
There are at least three significant formation routes leading to production of hydroxyl radicals. A pathway for OH radical formation, which becomes important later in the day as ozone concentrations rise [29], is photolysis of O_3

$$O_3 + hv(\lambda < 319 \text{ nm}) \rightarrow O(^1D) + O_2(^1\Delta g) \tag{3.4}$$

The $O(^1D)$ atoms may be quenched to $O(^3P)$ or react with water vapour to yield OH radicals

$$O(^1D) + H_2O \rightarrow 2OH \tag{3.5}$$

with approximately 20% overall efficiency at 298 K and 50% RH.

Nitrous acid, which has been shown to accumulate to concentrations of ~ 1–8 parts per billion (ppb) during the night in Los Angeles [25, 30], will photolyse at sunrise, producing a pulse of OH radicals [29].

$$HONO + hv(\lambda < 400 \text{ nm}) \rightarrow OH + NO \tag{3.6}$$

Photolysis is a major sink for HONO during daylight hours. (Other aspects of the formation and atmospheric chemistry of this important species are discussed in Section 3.2.4.1.)

A third significant source of OH radicals is the photolysis of HCHO which is both a primary (from motor vehicle exhausts) and secondary pollutant and may occur in significant concentrations in the morning hours as well as in the afternoon [40–42].

$$HCHO + hv(\lambda < 370 \text{ nm}) \begin{cases} \rightarrow H + HCO & (3.7) \\ \rightarrow H_2 + CO & (3.8) \end{cases}$$

The H atoms formed in reaction (3.7) or from reactions such as (3.9)

$$OH + CO \rightarrow CO_2 + H \tag{3.9}$$

can react with oxygen to produce hydroperoxyl radicals

$$H + O_2 \xrightarrow{M} HO_2 + M \tag{3.10}$$

which can then react with NO to form hydroxyl radicals

$$HO_2 + NO \rightarrow OH + NO_2 \tag{3.11}$$

This then completes a chain reaction involving reactions (3.9), (3.10) and (3.11).

The formyl radical in reaction (3.7) may also serve as a precursor to HO_2 radicals and hence OH radical formation

$$HCO + O_2 \rightarrow HO_2 + CO \tag{3.12}$$

Other sources of formyl radicals include higher aldehyde photolysis

$$RCHO + h\nu(\lambda < 350 \text{ nm}) \rightarrow R + HCO \tag{3.13}$$

The approximate relative importance of the three formation routes for OH radicals discussed above, are shown as a function of time for a particular day in the Los Angeles atmosphere in Fig. 3.1.

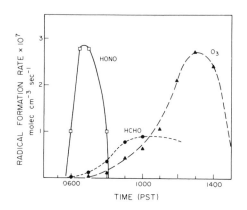

Fig. 3.1 Calculated rates of radical formation initiated by photolysis of HONO, HCHO and O_3 as a function of time for typical Los Angeles atmospheric conditions.

In addition to reactions (3.10) and (3.12), HO_2 radicals are produced by H-atom abstraction from alkoxy radicals as discussed below

$$RCH_2O^{\cdot} + O_2 \rightarrow RCHO + HO_2$$

Based upon environmental chamber data and computer modelling studies, concentrations of OH and HO_2 radicals in polluted atmospheres are believed to be in the ranges 10^6–10^7 and 10^8–10^9 radicals cm^{-3}, respectively.

3.2.2.2 The NO₃ radical

Ozone can react with NO_2 to produce the nitrate radical and an oxygen molecule

$$O_3 + NO_2 \rightarrow NO_3 + O_2 \tag{3.14}$$

However, because of its large photolytic cross-section [43, 44]

$$NO_3 + h\nu \rightarrow NO_2 + O(^3P) \tag{3.15a}$$

$$\rightarrow NO + O_2 \tag{3.15b}$$

the NO_3 radical concentration can only rise above part per trillion (ppt) levels after sunset. This important radical intermediate is discussed in more detail below.

3.2.3 Termination reactions

The inorganic chemistry system includes termination reactions for OH and HO_2 radicals with NO and NO_2 to form nitrogen acids

$$OH + NO \xrightarrow{\text{M}} HONO \tag{3.16}$$

$$OH + NO_2 \rightarrow HNO_3 \tag{3.17}$$

$$HO_2 + NO_2 \rightarrow HO_2NO_2 \tag{3.18}$$

However, HO_2NO_2 thermally back dissociates rapidly ($\tau_{1/2} \sim 10$ s at 298 K) and HONO photolyses, so $OH + NO_2$ is the major sink of reactions (3.16–3.18).

While nitrous acid [25, 29] and nitric acid [40–42] have now both been reliably measured in ambient air by longpath spectroscopic techniques, peroxynitric acid has not yet been observed in the atmosphere although it has been predicted to be present at fractional ppb concentrations [15, 45] and has been extensively studied in laboratory systems [46–50].

Although radical–radical reactions will generally not be of major importance in the atmosphere, due to the low concentrations of the radicals, an exception may be the disproportionation reaction of hydroperoxyl radicals to form hydrogen peroxide

$$HO_2 + HO_2 \rightarrow H_2O_2 + O_2 \tag{3.19}$$

However, reports of H_2O_2 detection in urban atmospheres by various wet chemical methods [51–53] have not been confirmed by unambiguous spectroscopic techniques, and one of these methods is now known to have been subject to artifacts leading to erroneously high values of H_2O_2 [54]. Indeed, reaction (3.19) may only be important in clean atmospheres containing low NO concentrations due to the rapid competing reaction $HO_2 + NO \rightarrow HNO_3$.

3.2.4 Other important inorganic reactions

3.2.4.1 *HONO*

An alternative pathway for reaction (3.18) is the formation of HONO

$$HO_2 + NO_2 \rightarrow HONO + O_2 \tag{3.20}$$

However, this reaction pathway has been shown [48, 50] to be negligible compared with reaction (18).

Still another potential source of HONO is the equilibrium between NO, NO_2 and H_2O

$$NO + NO_2 + H_2O \rightarrow 2HONO \tag{3.21a}$$

$$2HONO \rightarrow NO + NO_2 + H_2O \qquad (3.21b)$$

Reaction (3.21a) may proceed both homogeneously and heterogeneously but these reactions appear to be much too slow to be of atmospheric importance at part per million concentrations of NO_x.

Similarly the reaction

$$2NO_2 + H_2O \rightarrow HONO + HNO_3 \qquad (3.22)$$

may proceed in the gas phase or on surfaces. Thus, recent work [55] has shown that HONO is produced in environmental chambers from the reaction of NO_2 with water vapour, almost certainly via heterogeneous processes (although HNO_3 was not observed in these studies). This process can hence lead to the formation of HONO in automobile exhaust trains, power plant stacks and possibly in the atmosphere. Indeed, Pitts and co-workers have recently identified and measured nitrous acid directly emitted in auto exhaust using a 'wall-less' spectroscopic technique [56]; they have also observed HONO adjacent to major freeways using a longpath differential optical absorption spectrometer (DOAS) [29, 57]. Thus a significant portion of the HONO observed in urban atmospheres may arise from direct emissions rather than from gas phase reactions.

3.2.4.2   *HNO_3, N_2O_5 and acid deposition*

An equilibrium reaction of importance is that involving NO_2, NO_3 and N_2O_5 [56]

$$NO_2 + NO_3 \rightleftarrows N_2O_5 \qquad (3.23)$$

The equilibrium constant for this system has recently been characterized at room temperature [58, 59] by the direct *in situ* determination, using FT-IR and DOAS spectroscopy, of all three species.

Dinitrogen pentoxide is a potentially important precursor to HNO_3 (and hence acid deposition) through its reaction with water either in the gas phase [60] or on surfaces

$$N_2O_5 + H_2O \rightarrow 2HNO_3 \qquad (3.24)$$

Thus, for NO_2 and NO_3 radical concentrations of ~ 3 ppb and ~ 100 ppt, respectively, representative of receptor sites downwind from major urban areas (e.g. Los Angeles) [22], and using the $NO_2 + NO_3 \rightleftarrows N_2O_5$ equilibrium constant recently determined by Tuazon *et al.* [59], an HNO_3 formation rate of ~ 0.5 ppb h^{-1} is obtained at $\sim 50\%$ relative humidity [60].

These estimated night-time HNO_3 formation rates via reaction (3.24) can be compared to a calculated daytime formation rate of ~ 0.04 ppb h^{-1} from reaction (3.17)

$$OH + NO_2 \xrightarrow{M} HNO_3 \qquad (3.17)$$

for 1 ppb of NO_2 and 1×10^6 molecule cm^{-3} of OH radicals. Clearly reaction

(3.24) can be an important loss process for NO_x and a significant night-time pathway for HNO_3 formation in urban atmospheres.

3.2.5 Peak concentrations of selected inorganic pollutants observed or expected in polluted atmospheres

Having shown how a number of key inorganic species are believed to be formed and react in polluted atmospheres we indicate in Table 3.1 the observed or expected range of peak concentrations for the major nitrogenous compounds.

Table 3.1 Concentration ranges for selected gaseous air pollutants observed in the California South Coast Air Basin

Compound	Range of peak concentrations	Time of peak concentration	Approx. 1-day exposure[a] (ppb-h)
Ozone (O_3)	100–450 ppb	Mid-afternoon	1000
Peroxyacetyl nitrate (PAN)	5–40 ppb	Mid-afternoon	100
Nitric acid (HNO_3)	≤ 50 ppb	Mid-afternoon	250
Nitrogen dioxide (NO_2)	0.1–0.8 ppm	Morning/afternoon	1500
Nitrous acid (HONO)	1–8 ppb	Early morning	20
Nitrate radical (NO_3)	5–40 ppt	Early evening	< 0.5
Formaldehyde (HCHO)	10–70 ppb	Mid-afternoon	500
Formic acid (HCOOH)	5–20 ppb	Mid-afternoon	100

[a] Estimated for a 24-h period including a moderate air pollution event; exposures exhibit wide day-to-day and site-to-site variations within the CSCAB.

Table 3.1 also indicates the time of peak concentration and the approximate one-day exposure which might be experienced in an urban airshed experiencing moderate to severe photochemical air pollution.

One air monitoring consequence of the presence of significant ambient concentrations of species such as PAN, HNO_3 and other nitrogenous air pollutants is their interference with measurement of NO_2 by commercial chemiluminescence analysers since such analysers have been shown to respond quantitatively to these species in the 'NO$_2$ mode' [61].

3.3   Reactions involving organic compounds

We now know that all organic compounds emitted into the atmosphere, including those from automobile exhaust, evaporation of solvents and gasoline, and volatilization of herbicides and other toxic substances, may be degraded by one or more of the following four reaction pathways:

(a) Reaction with OH radicals
(b) Reaction with O_3
(c) Reaction with NO_3 radicals
(d) Photolysis.

Indeed, our knowledge of the rates and mechanisms of these processes has advanced to the point where we can predict with a reasonable degree of certainty which of those processes will predominate for a given compound. In the following sections we briefly summarize the basic features of these reaction pathways for organic compounds emitted into the atmosphere. In this section those compounds which can be detected in the atmosphere or which are stable end products occurring below present analytical detection limits are underlined in the reaction sequences leading to their formation.

3.3.1   Reactions of OH radicals with organics

We consider separately the mechanisms of reaction of OH radicals with the major classes of organic compounds including alkanes, alkenes, aromatics and oxygenated compounds.

3.3.1.1   *Alkanes*

It is now well established that the only significant atmospheric chemical loss processes for the alkanes is reaction with OH radicals. These reactions proceed by hydrogen abstraction [19, 20] to produce alkyl radicals (R) which then add O_2 to form alkyl peroxy radicals (RO_2)

$$OH + RH \rightarrow R^{\cdot} + H_2O \tag{3.25}$$

$$R^{\cdot} + O_2 \rightarrow RO_2^{\cdot} \tag{3.26}$$

In polluted atmospheres RO_2 radicals rapidly oxidize NO to NO_2 forming alkoxy radicals (RO), or add NO_2 to form alkyl peroxynitrates

$$RO_2^{\cdot} + NO \rightarrow RO^{\cdot} + NO_2 \tag{3.27}$$

$$RO_2^{\cdot} + NO_2 \rightleftarrows RO_2NO_2 \tag{3.28}$$

The latter, however, are not expected to be present in ambient air at significant concentrations due to their short (≤ 1 s at 298 K) lifetimes with respect to thermal decomposition (reaction 3.28).

Alkoxy radicals may also undergo hydrogen abstraction by molecular oxygen to form ketones

$$R'RCHO^{\cdot} + O_2 \rightarrow R'COR + HO_2^{\cdot} \tag{3.29a}$$

or decompose or isomerize to form oxygenates

$$R'RCHO^{\cdot} \rightarrow \begin{cases} RCHO + R'^{\cdot} \\ R'CHO + R^{\cdot} \end{cases} \tag{3.29b}$$

However, irrespective of the reaction sequence, HO_2 is formed, and hence OH radicals are regenerated. The carbonyl compounds thus formed may subsequently react with OH radicals or photodecompose (see Section 3.3.1.4).

A further important reaction pathway for acyl radicals is addition of O_2 followed by reaction with NO_2 to form 'PANs' (i.e. peroxyacetyl nitrates) which are eye irritants and phytotoxicants [62]

$$\underset{\substack{\| \\ RC^{\cdot}}}{O} + O_2 \rightarrow \underset{\substack{\| \\ RCOO^{\cdot}}}{O} \tag{3.30}$$

$$\underset{\substack{\| \\ RCOO^{\cdot}}}{O} + NO_2 \rightleftharpoons \underset{\substack{\| \\ RCOONOR_2}}{O} \tag{3.31}$$

The simplest of these compounds, peroxyacetyl nitrate (PAN), has been measured in a number of polluted atmospheres throughout the world [40–42, 62–69] at concentrations ranging from a few parts per billion (ppb) in the eastern US and Canada up to ~ 40 ppb in Los Angeles during severe smog episodes [40].

The reactions described above suggest the importance of simple ($\leq C_4$) alkoxy and alkylperoxy radicals in atmospheric chemistry and our knowledge of reactions involving these species has greatly increased in recent years [70–79]. In addition to these reactions for the simple alkanes, two other processes involving $RO^{\cdot}$ and $RO_2^{\cdot}$ reactions occur for larger ($\geq C_4$) alkanes, namely alkoxy radical isomerization [70–75] and alkyl nitrate formation from the reaction of $RO_2^{\cdot}$ with NO [76–79].

$$RO_2^{\cdot} + NO \xrightarrow{M} RONO_2 \tag{3.32}$$

Discussion of these processes is beyond the scope of this chapter and the reader is referred to the original literature and to appropriate reviews [20, 21, 36, 39] for summaries of these aspects of the atmospheric chemistry of longer chain alkanes.

3.3.1.2 *Alkenes*

In polluted atmosphere unsaturated hydrocarbons react primarily with OH radicals and with O_3. For ethene, and the higher methyl- and ethyl-substituted alkenes, reaction with OH radicals proceeds essentially entirely by addition to the double bond, [20, 80]. For propene, addition of the OH radical to the double bond is expected to be followed by O_2 addition with the resulting peroxy radical oxidizing NO to NO_2 to form an alkoxy radical [70].

$$\dot{OH} + CH_3CH=CH_2 \rightarrow CH_3\dot{C}HCH_2OH \tag{3.33}$$

$$CH_3\dot{C}HCH_2OH + O_2 \rightarrow CH_3\overset{\overset{OO^\cdot}{|}}{C}HCH_2OH \tag{3.34}$$

$$CH_3\overset{\overset{OO^\cdot}{|}}{C}HCH_2OH + NO \rightarrow CH_3\overset{\overset{O^\cdot}{|}}{C}HCH_2OH + NO_2 \tag{3.35}$$

Decomposition of the alkoxy radical and subsequent reactions lead to the formation of acetaldehyde and formaldehyde, both of which can be detected in polluted atmospheres.

$$CH_3\overset{\overset{O^\cdot}{|}}{C}HCH_2OH \rightarrow CH_3CHO + {}^\cdot CH_2OH \tag{3.36}$$

$${}^\cdot CH_2OH + O_2 \rightarrow HCHO + HO_2 \tag{3.37}$$

$$\downarrow NO$$

$$OH + NO_2 \tag{3.38}$$

The overall reaction resulting from reactions (3.33) to (3.37) is

$$OH + CH_3CH=CH_2 + O_2 + 3NO \rightarrow CH_3CHO + HCHO + 3NO_2 + OH$$

$$\downarrow h\nu \tag{3.39}$$

$$O_3$$

Thus, the reaction of OH radicals with alkenes increases the rate of NO to NO_2 conversion and hence increases the yield of ozone.

This illustrates a major role the organic compounds, not only alkenes but also alkanes and aromatics, play in producing photochemical air pollution, acceleration of the conversion of NO to NO_2 and ozone formation. Figure 3.2 shows typical time–concentration profiles obtained during irradiation of a propene–NO_x–air mixture in an environmental chamber. The observed behaviour occurs in a similar time scale in ambient atmospheres.

3.3.1.3 Aromatics

Reactions of OH radicals with aromatic hydrocarbons proceed by either H-atom abstraction from the substituent methyl group(s) or OH radical addition to the ring [20, 81–85]. The only other important chemical loss processes for certain aromatics (i.e. hydroxy-substituted) is reaction with NO_3 radicals [26]. For toluene, an important organic constituent in urban atmosphere, OH radical addition is expected to occur $\sim 80\%$ of the time at the ortho position [82].

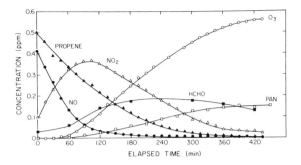

Fig. 3.2 Time–concentration profiles for selected primary and secondary pollutants during an environmental chamber irradiation of propene and NO_x in air [117].

$$\text{OH} + \quad \begin{array}{c}\text{CH}_3\\ \bigcirc\end{array} \quad \longrightarrow \quad \begin{array}{c}\text{CH}_2^{\bullet}\\ \bigcirc\end{array} + \text{H}_2\text{O} \qquad (A) \qquad (3.40a)$$

$$\begin{array}{c}\text{CH}_3\\ \text{OH}\\ \bigcirc -\text{H}\\ \bullet\end{array} \qquad (B) \qquad (3.40b)$$

The rates of reaction of OH radicals with aromatics are now well characterized [20, 21] as is the relative importance of the alternative pathways (3.40a,b) at least for selected aromatics (e.g. for toluene abstraction occurs approximately 8% of the time [85]).

Under atmospheric conditions the benzyl radical (A) will form benzaldehyde and benzyl nitrate by reactions analogous to the alkyl radical reactions discussed above and elsewhere in detail [20, 86]. However, the fate of the addition adducts (B) remains unclear although several mechanisms have been proposed [87]. Indeed, although substantial progress has been made for other aromatics as well, e.g. the xylenes and cresols, much additional research is needed before a complete understanding of the complex NO_x-photo-oxidation chemistry of these systems is obtained. It is significant, however, for analysis of trace species in polluted atmospheres that such oxygenates as biacetyl, glyoxal and methylglyoxal have been observed as products in environmental chamber studies of *o*-xylene [85, 88, 89].

3.3.1.4 *Aldehydes*

Aldehydes are consumed in the atmosphere both by photolysis and attack by OH radicals. While photolysis of, for example, acetaldehyde lead to methyl and formyl radicals

$$CH_3CHO + h\nu \rightarrow CH_3 + HCO \tag{3.41}$$

which react as discussed above, attack by OH radical forms acetyl radicals which can successively add O_2 and NO_2 to form PAN

$$CH_3CHO + OH \rightarrow CH_3\dot{C}O + H_2O \tag{3.42}$$

$$CH_3\dot{C}O + O_2 \rightarrow CH_3\overset{\overset{O}{\|}}{C}OO \cdot \xrightarrow{\cdot NO_2} CH_3\overset{\overset{O}{\|}}{C}OONO_2 \tag{3.43}$$

3.3.2 Reactions of O_3 with organics

Ozone reacts only slowly with saturated hydrocarbons and therefore its reactions are unimportant relative to attack on the alkanes by OH radicals. Under atmospheric conditions this is also the case for ozone reactions with the aromatics. For alkenes, however, reaction with ozone is important. Thus, although the rate constants for reaction of ozone with unsaturated hydrocarbons are not fast, falling in the range 10^{-15} to 10^{-18} cm^3 molecule^{-1} s^{-1} (vs. 10^{-10} to 10^{-12} cm^3 molecule^{-1} s^{-1} for OH + alkenes) ozone concentrations in polluted atmospheres may be as much as six orders of magnitude higher than OH radical concentrations (e.g. 200 ppb or $\sim 5 \times 10^{12}$ molecules cm^{-3} for O_3 vs. $\sim(1-5) \times 10^6$ radicals cm^{-3} for OH). Thus in the afternoons during air pollution episodes when O_3 concentrations may be in the range 100–300 ppb a significant portion of reaction of alkenes will occur with O_3 as indicated in Fig. 3.3.

Unfortunately, as for the case of OH radical reactions with aromatics, while the rate constants for the reactions of O_3 with alkenes are reasonably well characterized [21], substantial uncertainties remain concerning the detailed mechanisms of reaction despite a great deal of research in this important area [90–98].

It is reasonably well established [90–98] that O_3-alkene reactions proceed primarily by the formation of carbonyls and Criegee biradicals via a molozonide

$$
\begin{array}{c}
O_3 + \begin{array}{c} R_1 \quad R_2 \\ \diagdown C = C \diagup \\ \diagup \quad \diagdown \\ R_3 \quad R_4 \end{array} \longrightarrow \begin{array}{c} R_1 \diagdown \overset{O-O}{\underset{C-C}{\diagup}} \diagup R_2 \\ R_3 \qquad R_4 \end{array}
\end{array}
\begin{cases}
\rightarrow R_1COR_3 + [R_2R_4\dot{C}O\dot{O}] & (3.44a) \\
\\
\rightarrow R_2COR_4 + [R_1R_3\dot{C}O\dot{O}] & (3.44b)
\end{cases}
$$

The energy-rich biradicals may then either be thermalized or undergo rearrangement or decomposition. The thermalized biradicals have been

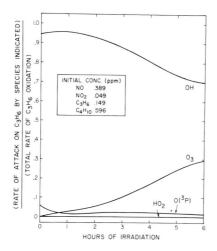

Fig. 3.3 Predicted relative importance of OH radicals, ozone, HO_2 radicals and $O(^3P)$ atoms during photo-oxidation of a propene–n–butane–NO_x mixture under simulated atmospheric conditions [18].

proposed to react with aldehydes to form secondary ozonides [92, 96], with SO_2 [92, 93, 96] and with water [93], as shown for the $CH_3\dot{C}HOO^{\cdot}$ biradical in reactions (3.45)–(3.47).

$$CH_3\dot{C}HOO^{\cdot} + RCHO \longrightarrow \text{[cyclic structure]} \qquad (3.45)$$

$$CH_3\dot{C}HOO^{\cdot} + SO_2 \xrightarrow{\ H_2O\ } CH_3CHO + H_2SO_4 \qquad (3.46)$$

$$CH_3\dot{C}HOO^{\cdot} + H_2O \longrightarrow CH_3COOH + H_2O \qquad (3.47)$$

Thermalized Criegee biradicals may also react with NO and NO_2 but the details of these and other reactions in the O_3-alkene system require further investigation. For a detailed discussion of the various decomposition and rearrangement pathways which have been proposed for 'hot' biradicals as well as further descriptions of the fate of the thermalized biradicals for various alkenes see reference [21].

3.3.3   Reactions of NO_3 radicals with organics

The pioneering work of Niki and co-workers [99, 100] showed that gaseous NO_3 radicals react with alkenes, with the rate constants increasing markedly with the degree of substitution on the double bond. Carter *et al.* [26] showed that the hydroxy-substituted aromatics (phenols and the cresols) also react rapidly with NO_3 radicals.

More recently, in a comprehensive experimental programme carried out by Atkinson and co-workers, the kinetics of the reactions of NO_3 radicals with a wide range of organics were determined at room temperature [101–105] and from these, and the previous studies, information concerning the mechanisms of these reactions has been forthcoming.

The recent kinetic data reported by Atkinson and co-workers [101–105] for the reactions of NO_3 radicals with organics allows assessment of the importance of these reactions as a sink for organics, both in an absolute sense and relative to O_3 and OH radical reactions. Table 3.2 gives the lifetimes of selected organic

Table 3.2   Calculated lifetimes of selected organics due to reaction with O_3 and OH and NO_3 radicals

	Organic lifetimes for moderately polluted[a] atmosphere		
Alkenes	τ_{O_3} *24 hour*	τ_{OH} *Daytime*	τ_{NO_3} *Night time*
Ethene	2.7 day	16 h	79 day
Propene	11 h	5.6 h	1.1 day
trans-2-Butene	35 min	2.0 h	33 min
2-Methyl-2-butene	17 min	1.6 h	1.3 min
2,3-Dimethyl-2-butene	6 min	1.3 h	0.2 min
n-Butane	> 130 year	2.2 day	172 day
Toluene	> 1.3 year	22 h	179 day
o-Cresol	19 day	3.2 h	0.6 min

[a] Assuming 200 ppb of O_3 (24 h average), $2 \times 10^6 \text{ cm}^{-3}$ (0.08 ppt) of OH during daylight hours, and 100 ppt of NO_3 during night-time hours.

pollutants for a moderately polluted atmosphere, with ~ 200 ppb of O_3, ~ 0.1 ppt of OH radicals and ~ 100 ppt of NO_3 radicals. Clearly, reaction with NO_3 radicals at night can be an important and even dominant loss process for certain alkenes and hydroxy-substituted aromatics. Winer *et al.* have proposed [34] that reaction with NO_3 radicals at night may also be an important reaction pathway for certain naturally-occurring organics such as the monoterpenes and dimethyl sulphide.

Conversely, as discussed by Winer *et al.* [34], the kinetic data recently obtained for NO_3 radical reactions also show that reactions with the more reactive alkenes (including isoprene and certain of the monoterpenes), dimethyl sulphide, and the hydroxy-substituted aromatics can be important loss processes for NO_3 radicals at night, and that in fact for certain of these organics (e.g. the monoterpenes, 2,3-dimethyl-2-butene and the hydroxy-substituted aromatics) their presence at concentrations approaching 1 ppb will lead to low NO_3 radical concentrations [34, 35].

In the remainder of this section we briefly summarize our present understanding of the mechanisms of reaction of NO_3 radicals with the various classes of organics.

3.3.3.1 *Alkanes*

The relevant kinetic data indicate [103] that these reactions proceed via H-atom abstraction from the C–H bonds, almost certainly predominantly from secondary or tertiary C–H bonds

$$NO_3 + RH \rightarrow \underline{HNO_3} + R^{\cdot} \tag{3.48}$$

Hence, these reactions lead directly to HNO_3 formation. The measured room temperature rate constants for the alkenes range between 2×10^{-17} cm^3 molecule^{-1} s^{-1} for *n*-butane to 1×10^{-16} cm^3 molecule^{-1} s^{-1} for 2,3-dimethyl-butane.

3.3.3.2 *Alkenes*

The reaction of NO_3 radicals with the alkenes have been shown from both kinetic [100, 101] and product [106] studies to proceed via initial addition of the NO_3 radical to the olefinic double bond

$$NO_3 + CH_3CH{=}CH_2 \rightarrow CH_3\overset{|}{\underset{}{CH}}\overset{\cdot}{C}H_2 + CH_3\overset{\cdot}{C}HCH_2ONO_2 \tag{3.49}$$

with addition at the terminal carbon being expected to dominate [21]. Possible reaction sequences, after this initial reaction, have been discussed [21, 106] but are highly uncertain at the present time [21]. However, thermally unstable nitro-peroxynitrates such as $CH_3CH(ONO_2)CH_2OONO_2$ and stable di-nitrates such as $CH_3CH(ONO_2)CH_2ONO_2$ have been reported as products in NO_3–NO_2–propene–air systems [106].

3.3.3.3 *Aldehydes*

Based upon the product date of Morris and Niki [99], i.e. the observed formation of HNO_3 from the reaction of NO_3 radicals with CH_3CHO, it is expected that these reactions proceed via H-atom abstraction from the relatively weak H–CO bonds

$$NO_3 + RCHO \rightarrow R\overset{\cdot}{C}O + \underline{HNO_3} \tag{3.50}$$

Thus reaction of NO_3 radicals with acetaldehyde could be a night-time source of peroxyacetyl nitrate (PAN)

$$NO_3 + CH_3CHO \rightarrow \underline{HNO_3} + CH_3\overset{\cdot}{C}O \tag{3.51}$$

$$CH_3\overset{.}{C}O + O_2 \rightarrow CH_3CO_3^{.} \tag{3.52}$$

$$CH_3CO_3^{.} + NO_2 \rightarrow CH_3\overset{\overset{\displaystyle O}{\|}}{C}OONO_2 \tag{3.53}$$

Reaction of NO_3 radicals with the higher aldehydes will lead by analogous reactions to the higher peroxyacyl nitrates RCO_3NO_2. However, reaction with formaldehyde will lead to HO_2 radical formation, since HCO reacts rapidly with O_2 [21]

$$HCO + O_2 \rightarrow HO_2 + CO \tag{3.12}$$

3.3.3.4   *Aromatics*

As discussed by Atkinson *et al.* [105] the reactions of NO_3 radicals with the monocyclic aromatic hydrocarbons and the hydroxy-substituted aromatics appear, based upon kinetic evidence, to react via H-atom abstraction from the C–H or O–H bonds on the substituent groups. This conclusion is based upon the observation that for the xylenes and the cresols the meta-isomer reacts more slowly (by a factor of ~ 2) than do the ortho- and para-isomers. This is in contrast to the addition reactions of $O(^3P)$ atoms and OH radicals, where the meta-isomer is the most reactive [20]. Furthermore, *o*-nitrophenol has been tentatively identified as a product of N_2O_5–NO_2–phenol–air reaction mixtures, presumably formed by the reaction sequence

$$\tag{3.54}$$

followed by [107]

$$\tag{3.55}$$

Thus these reactions are also a direct source of nitric acid as well as forming low volatility organic nitro compounds.

In summary it is now clear that reaction with NO_3 radicals at night is a major atmospheric reaction pathway for many organic pollutants. Thus, it must be considered, along with the reactions of OH radicals and O_3 and photolysis, as one of the dominant loss processes for organics in the atmosphere.

3.4 Gas-to-particle conversion

3.4.1 SO_2 photo-oxidation and formation of sulphate particulate

A great deal of research has been conducted concerning the atmospheric reaction pathways of sulphur dioxide and the conversion of this important gaseous species to sulphate particulate. The latter contribute significantly to both visibility degradation and potential health effects.

Although earlier it was thought that a number of homogeneous gas phase reactions of SO_2 could be significant [93], today it is generally accepted that in polluted atmospheres reaction with OH radicals is the dominant gas phase removal process for SO_2, with possibly a smaller contribution from reaction with Criegee biradical [21]. In particular reactions with peroxy radicals, including hydroperoxy radicals, can be neglected. In highly polluted atmospheres SO_2 can be oxidized by reaction with OH radicals at rates up to $\sim 5\%$ h^{-1} [93]. Calvert has recently shown [108] that a likely pathway for this process is as follows

$$OH + SO_2 + M \rightarrow HOSO_2 + M \tag{3.56}$$

$$HOSO_2 + O_2 \rightarrow HO_2 + SO_3 \tag{3.57}$$
$$\downarrow H_2O$$
$$H_2SO_4$$

Thus, the ultimate fate of atmospheric SO_2 is the formation of secondary sulphate particulate of appropriate diameter to efficiently scatter light as well as falling in the respirable range (i.e. $0.1–1$ μm). Moreover, incorporation of acidic sulphur species, including H_2SO_4 itself, into liquid droplets contributes to acid deposition, an environmental problem of increasing importance throughout the world.

Details of the mechanisms of homogeneous oxidation of SO_2 in both clean and polluted atmospheres may be found in several reviews [93, 108–111].

3.4.2 Formation of secondary nitrate and organic particulate

As discussed above there are at least two important reaction pathways leading to HNO_3 formation in polluted atmospheres. The reaction of OH radicals with NO_2 occurs during daylight hours with formation rate of HNO_3 of ~ 0.5 ppb h^{-1} and ambient levels of HNO_3 of up to 50 ppb [40–42] have been observed. If ammonia is present in the atmosphere at levels sufficient to saturate the atmosphere with ammonium nitrate by reaction with HNO_3 [112, 113], then secondary nitrate particles will form in the respirable and light scattering size range. If insufficient ammonia is present, the HNO_3 will remain in the gas phase or react with pre-existing aerosol.

A second potentially important pathway for nitrate particulate formation is the hydrolysis of N_2O_5 on wet surfaces

$$N_2O_5 + H_2O \rightarrow 2HNO_3 \text{ (surface)} \tag{3.58}$$

Thus N_2O_5 may be rapidly converted to HNO_3 in fog droplets leading to acidic fog [114, 115]. Again, if there is sufficient basic species present, when such fogs evaporate particulate nitrate will be left in the atmosphere.

A number of classes of organic compounds, including monoterpenes, alkenes ($> C_6$) and cyclic and dialkenes, have been shown to be precursors to formation of secondary organic particulate in the 0.1 to 1 μm size range [116]. Detailed discussions of the kinetics and mechanisms of organic particulate formation are beyond the scope of this chapter and may be found elsewhere [116].

3.5 Conclusion

From the foregoing discussion it should be clear that a variety of chemical processes in the atmosphere transform the relatively small number of inorganic primary pollutants, and the much larger number of organic compounds emitted from anthropogenic sources, into a diverse spectrum of secondary air pollutants. Although some of these compounds are now amenable to routine chemical analyses, many others occur in relatively low concentrations and pose great challenges to current analytical capabilities. Yet these 'trace' components may play critical roles in atmospheric chemistry or have significant impacts on visibility, vegetation and human health.

In summary, while much more remains to be learned about certain important aspects of atmospheric chemistry, the great progress made during the past decade in this field allow us to feel some confidence in directing our analytical methods and techniques toward the characterization of polluted atmospheres.

References

1. Leighton, P. A. (1961) *Photochemistry of Air Pollution*, Academic Press, New York.
2. Buckberg, H., Wilson, K. W., Jones, M. H. and Lindh, K. G. (1963) *Int. J. Air Water Pollut.*, **7**, 257–80.
3. Korth, M. W., Rose, A. H., Jr. and Stahman, R. C. (1964) *J. Air Pollut. Control Assoc.*, **14**, 168–75.
4. Nicksic, S. W., Harkins, J. and Painter, L. J. (1966) *Int. J. Air Water Pollut.*, **10**, 15–23.
5. Hamming, W. J. and Dickinson, J. E. (1966) *J. Air Pollut. Control Assoc.*, **16**, 317–23.
6. Romanovsky, J. C., Ingels, R. M. and Gordon, R. J. (1967) *J. Air Pollut. Control Assoc.*, **17**, 454–9.
7. Dimitriades, B. (1967) *J. Air Pollut. Control Assoc.*, **17**, 460–6.
8. Glasson, W. A. and Tuesday, C. S. (1970) *J. Air Pollut. Control Assoc.*, **20**, 239–43.
9. Altshuller, A. P., Kopczynski, S. L., Wilson, D. *et al.* (1969) *J. Air Pollut. Control Assoc.*, **19**, 787–90, 791–4.
10. Dimitraides, B. (1972) *Environ. Sci. Technol.*, **6**, 253–60.
11. Heicklen, J., Westberg, K. and Cohen, N. (1969) Cent. Air Environ. Stud. Pa. State Univ. Report No. 115–69.

12. Stedman, D. H., Morris, E. D., Jr., Daby, E. E. *et al.* (1970) Paper presented at the 160th national meeting of the American Chemical Society, Chicago, 14–18 September.
13. Niki, H., Daby, E. E. and Weinstock, B. (1972) *Adv. Chem. Ser.* **113**, 16–57.
14. Hecht, T. A., Seinfeld, J. H. and Dodge, M. C. (1974) *Environ. Sci. Technol.*, **8**, 327–39.
15. Demerjian, K. L., Kerr, J. A. and Calvert, J. G. (1974) *Adv. Environ. Sci. Technol.*, **4**, 1–262.
16. Levy, H., II (1974) *Adv. Photochem.*, **9**, 369–524.
17. Darnall, K. R., Lloyd, A. C., Winer, A. M. and Pitts, J. N., Jr. (1976) *Environ. Sci. Technol.*, **10**, 692–6.
18. Lloyd, A. C., Darnall, K. R., Winer, A. M. and Pitts, J. N., Jr. (1976) *J. Phys. Chem.*, **80**, 789–94.
19. Finlayson, B. J. and Pitts, J. N., Jr. (1976) *Science*, **192**, 111–19.
20. Atkinson, R., Darnall, K. R., Lloyd, A. C. *et al.* (1979) *Adv. Photochemistry*, **11**, 375–488.
21. Atkinson, R. and Lloyd, A. C. (1984) *J. Phys. Chem. Ref. Data*, **13**, 315–444.
22. Platt, U., Perner, D., Winer, A. M. *et al.* (1980) *Geophys. Res. Letts.*, **7**, 89–92.
23. Noxon, J. F., Norton, R. B. and Marovich, E. (1980) *Geophys. Res. Letts.*, **7**, 125–8.
24. Platt, U. and Perner, D. (1980) *J. Geophys. Res.*, **85**, 7453–8.
25. Platt, U., Perner, D., Harris, G. W. *et al. Nature*, **285**, 312–14.
26. Carter, W. P. L., Winer, A. M. and Pitts, J. N., Jr. (1981) *Environ. Sci. Technol.*, **15**, 829–31.
27. Pitts, J. N., Jr., Harris, G. W. and Winer, A. M. (1981) Spectroscopic Measurements of the NO_3 Radical and of HONO in the Atmosphere: Implications for Tropospheric Free Radical Chemistry. Proceedings of the 15th International Symposium on Free Radicals, Ignonish Beach, Nova Scotia, Canada, 2–7 June.
28. Platt, U., Perner, D., Schröder, J. *et al.* (1981) *J. Geophys. Res.*, **86**, 11,965–70.
29. Harris, G. W., Carter, W. P. L., Winer, A. M. *et al.* (1982) *Environ. Sci. Technol.*, **16**, 414–19.
30. Harris, G. W., Winer, A. M., Pitts, J. N., Jr. *et al.* (1983). In *Optical and Laser Remote Sensing* (eds A. Mooradian and D. K. Killinger), Springer Verlag, New York, pp. 106–13.
31. Winer, A. M., Pitts, J. N., Jr., Platt, U. F. and Biermann, H. W. (1983) Direct Measurements of Nitrate Radical Concentrations at Desert Sites in California: Implications for Nitric Acid Formation. Environmental Sciences Division, pp. 76–80, 185th National American Chemical Society Meeting, Seattle, WA, 20–25 March .
32. Stockwell, W. R. and Calvert, J. G. (1983) *J. Geophys. Res.*, **88**, 6673–82.
33. Platt, U., Winer, A. M., Biermann, H. W. *et al.* (1984) *Environ. Sci. Technol.*, **18**, 365–9.
34. Winer, A. M., Atkinson, R. and Pitts, J. N., Jr. (1984) *Science*, **224**, 156–9.
35. Atkinson, R., Winer, A. M. and Pitts, J. N., Jr. (1986) *Atmos. Environ.*, in press.
36. Finlayson-Pitts, B. J. and Pitts, J. N., Jr. (1977) *Adv. Environ. Sci. Technol.*, **7**, 75–162.
37. McCrae, G. J., Goodin, W. R. and Seinfeld, J. H. (1982) *Mathematical Modeling of Photochemical Air Pollution*, Report No. 18, 27 April, Environmental Quality Laboratory, California Institute of Technology, Pasadena, CA.

38. Logan, J. A., Prather, M. J., Wofsy, S. C. and McElroy, M. B. (1981) *J. Geophys. Res.*, **86**, 7210–54.
39. Atkinson, R. and Lloyd, A. C. (1981). In *Oxygen and Oxy-Radicals in Chemistry and Biology*, Academic Press, New York, pp. 559–92.
40. Tuazon, E. C., Winer, A. M. and Pitts, J. N., Jr., (1981) *Environ. Sci. Technol.*, **15**, 1232–7.
41. Tuazon, E. C., Graham, R. A., Winer, A. M. *et al.* (1978) *Atmos. Environ.*, **12**, 865–75.
42. Tuazon, E. C., Winer, A. M., Graham, R. A. and Pitts, J. N., Jr., *Adv. Environ. Sci. Technol.*, **10**, 259–300.
43. Graham, R. A. and Johnston, H. S. (1978) *J. Chem. Phys.*, **82**, 254–368.
44. Magnotta, F. and Johnston, H. S. (1980) *Geophys. Res. Letts.*, **7**, 769–72.
45. Pitts, J. N., Jr., Winer, A. M., Harris, G. W. *et al.* (1983) *Environ. Health Perspect.*, **52**, 153–7.
46. Niki, H., Maker, P. D., Savage, C. M. and Breitenbach, L. P. (1977) *Chem. Phys. Lett.*, **45**, 564–6.
47. Hanst, P. L. and Gay, B. W. (1977) *Environ. Sci. Technol.*, **11**, 1105–9.
48. Graham, R. A., Winer, A. M. and Pitts, J. N., Jr. (1977) *Chem. Phys. Lett.*, **51**, 215–20.
49. Graham, R. A., Winer, A. M. and Pitts, J. N., Jr. (1978) *J. Chem. Phys.*, **68**, 4505–10.
50. Howard, C. J. (1977) *J. Chem. Phys.*, **67**, 5258–63.
51. Bufalini, J. J., Gay, B. W., Jr. and Brubaker, K. L. (1972) *Environ. Sci. Technol.*, **6**, 816–21.
52. Bufalini, J. J., Gay, B. W. and Kopczynski, S. L. (1971) *Environ. Sci. Technol.*, **5**, 333–6.
53. Kok, G. L., Darnall, K. R., Winer, A. M. *et al.* (1978) *Environ. Sci. Technol.*, **12**, 1077–80.
54. Zika, R. G. and Saltzmann, E. S. (1982) *Geophys. Res. Letts.*, **9**, 231–4.
55. Pitts, J. N., Jr., Sanhueza, E., Atkinson, R. *et al.* (1984) *Int. J. Chem. Kinet.*, **16**, 919–39.
56. Pitts, J. N., Jr., Biermann, H. W., Winer, A. M. and Tuazon, E. C. (1984) *Atmos. Environ.*, **18**, 847–54.
57. Platt, U., Winer, A. M., Biermann, H. W. and Pitts, J. N., Jr. Unpublished data.
58. Malko, M. W. and Troe, J. (1982) *Int. J. Chem. Kinet.*, **14**, 399–416.
59. Tuazon, E. C., Sanhueza, E., Atkinson, R. *et al.* (1984) *J. Phys. Chem.*, **88**, 3095–8.
60. Tuazon, E. C., Atkinson, R., Plum, C. N. *et al.* (1983) *Geophys. Res. Letts.*, **10**, 953–6.
61. Winer, A. M., Peters, J. W., Smith, J. P. and Pitts, J. N., Jr. (1974) *Environ. Sci. Technol.*, **8**, 1118–21.
62. Stephens, E. R. (1969) *Adv. Environ. Sci. Technol.*, **1**, 119–46.
63. Tingey, D. T. and Hill, A. C. (1967) *Utah Acad. Proc.*, **44**, 387–95.
64. Taylor, O. C. (1969) *J. Air Pollut. Control Assoc.*, **19**, 347–51.
65. Lonneman, W. A., Bufalini, J. J. and Seila, R. L. (1976) *Environ. Sci. Technol.*, **10**, 374–80.
66. Nieboer, H. and van Ham, J. (1976) *Atmos. Environ.*, **10**, 115–20.
67. Spicer, C. W. (1977) *Atmos. Environ.*, **11**, 1089–95.
68. Lewis, T. E., Brennan, E. and Lonneman, W. A. (1983) *J. Air Pollut. Control Assoc.*, **33**, 885–6.
69. Peake, E., MacLean, M. A. and Sandhu, H. S. (1983) *J. Air Pollut. Control Assoc.*, **33**, 881–3.
70. Carter, W. P. L., Lloyd, A. C., Sprung, J. L. and Pitts, J. N., Jr. (1978) *Int. J. Chem. Kinet.*, **11**, 45–101.

71. Hendry, D. G., Baldwin, A. C., Barker, J. R. and Golden, D. M. (1978) *Computer Modeling of Simulated Photochemical Smog*, EPA-600/3-78-059, January, Environmental Protection Agency, Research Triangle Park, North Carolina, pp. 1–291.
72. Baldwin, A. C., Barker, J. R., Golden, D. M. and Hendry, D. G. (1977) *J. Phys. Chem.*, **81**, 2483–92.
73. Batt, L. (1979) *Int. J. Chem. Kinet.*, **11**, 977–93.
74. Batt, L. and Robinson, G. N. (1979) *Int. J. Chem. Kinet.*, **11**, 1045–53.
75. Carter, W. P. L., Darnall, K. R., Lloyd, A. C. *et al.* (1976) *Chem. Phys. Lett.*, **42**, 22–7.
76. Darnall, K. R., Carter, W. P. L., Winer, A. M. *et al.* (1976) *J. Phys. Chem.*, **80**, 1948–50.
77. Atkinson, R., Aschmann, S. M., Carter, W. P. L. *et al.* (1982) *J. Phys. Chem.*, **86**, 4563–9.
78. Atkinson, R., Carter, W. P. L. and Winer, A. M. (1983) *J. Phys. Chem.*, **87**, 2012–18.
79. Atkinson, R., Aschmann, S. M., Carter, W. P. L. *et al.* (1984) *Int. J. Chem. Kinet.*, **16**, 1085–101.
80. Biermann, H. W., Harris, G. W. and Pitts, J. N., Jr. (1982) *J. Phys. Chem.*, **86**, 2958–64.
81. Perry, R. A., Atkinson, R. and Pitts, J. N., Jr. (1977) *J. Phys. Chem.*, **81**, 296–304.
82. Kenley, R. A., Davenport, J. E. and Hendry, D. G. (1981) *J. Phys. Chem.*, **85**, 2740–6.
83. Tully, F. P., Ravishankara, A. R., Thompson, R. L. *et al.* (1981) *J. Phys. Chem.*, **85**, 2262–9.
84. Nicovich, J. M., Thompson, R. L. and Ravishankara, A. R. (1981) *J. Phys. Chem.*, **85**, 2913–16.
85. Atkinson, R., Carter, W. P. L. and Winer, A. M. (1983) *J. Phys. Chem.*, **87**, 1605–10.
86. Perry, R. A., Atkinson, R. and Pitts, J. N., Jr. (1977) *J. Phys. Chem.*, **81**, 1607–11.
87. Atkinson, R., Carter, W. P. L., Darnall, K. R. *et al.* (1980) *Int. J. Chem. Kinet.*, **12**, 779–836.
88. Darnall, K. R., Atkinson, R. and Pitts, J. N., Jr. (1979) *J. Phys. Chem.*, **83**, 1943–6.
89. Takagi, H., Washida, N., Akimoto, H. *et al.* (1980) *J. Phys. Chem.*, **84**, 478–83.
90. Cox, R. A. and Penkett, S. A. (1972) *J. Chem. Soc. Faraday Trans. 1*, **68**, 1735–53.
91. Herron, J. T. and Huie, R. E. (1977) *J. Am. Chem. Soc.*, **99**, 5430–5.
92. Niki, H., Maker, P. D., Savage, C. M. and Breitenbach, L. P. (1977) *Chem. Phys. Lett.*, **46**, 327–30.
93. Calvert, J. G., Su, F., Bottenheim, J. W. and Strausz, O. P. (1978) *Atmos. Environ.*, **12**, 197–226.
94. Herron, J. T. and Huie, R. E. (1978) *Int. J. Chem. Kinet.*, **10**, 1019–40.
95. Dodge, M. C. and Arnts, R. R. (1979) *Int. J. Chem. Kinet.*, **11**, 399–410.
96. Su, F., Calvert, J. G. and Shaw, J. H. (1980) *J. Phys. Chem.*, **84**, 239–46.
97. Kan, C. S., Su, F., Calvert, J. G. and Shaw, J. H. (1981) *J. Phys. Chem.*, **85**, 2359–63.
98. Niki, H., Maker, P. D., Savage, C. M. and Breitenbach, L. P. (1983) *Environ. Sci. Technol.*, **17**, 312A–22A.
99. Morris, E. D., Jr. and Niki, H. (1974) *J. Phys. Chem.*, **78**, 1337–8.
100. Japar, S. M. and Niki, H. (1975) *J. Phys. Chem.*, **79**, 1629–32.
101. Atkinson, R., Plum, C. N., Carter, W. P. L. *et al.* (1984) *J. Phys. Chem.*, **88**, 1210–15.
102. Atkinson, R., Pitts, J. N., Jr. and Aschmann, S. M. (1984) *J. Phys. Chem.*, **88**, 1584–7.
103. Atkinson, R., Plum, C. N., Carter, W. P. L. *et al.* (1984) *J. Phys. Chem.*, **88**, 2361–4.
104. Atkinson, R., Aschmann, S. M., Winer, A. M. and Pitts, J. N., Jr. (1984) *Environ. Sci. Technol.*, **18**, 370–5.
105. Atkinson, R., Carter, W. P. L., Plum, C. N. *et al.* (1984) *Int. J. Chem. Kinet.*, **16**, 887–98.

106. Bandow, H., Okuda, M. and Akimoto, H. (1980) *J. Phys. Chem.*, **84**, 3604–8.
107. Niki, H., Maker, P. D., Savage, C. M. and Breitenbach, L. P. (1979). In *Nitrogenous Air Pollutants* (ed. D. Grosjean), Ann Arbor Press, Ann Arbor, MI, pp. 1–16.
108. Stockwell, W. R. and Calvert, J. G. (1983) *Atmos. Environ.*, **17**, 2231–5.
109. Atkinson, R., Lloyd, A. C. and Winges, L. (1982) *Atmos. Environ.*, **16**, 1341–55.
110. Calvert, J. G. and Mohnen, V. (1983) The Chemistry of Acid Formation. In *N.A.S. Report, Acid Deposition; Atmospheric Processes in Eastern North America*, Appendix A, National Academy Press, Washington, DC, pp. 155–201.
111. Calvert, J. G. and Stockwell, W. R. (1983). In *Acid Precipitation: SO₂, NO and NO₂ Oxidation Mechanisms: Atmospheric Considerations*, Chap. 1, Ann Arbor Science, Ann Arbor, MI.
112. Stelson, A. W., Friedlander, S. K., Seinfeld, J. H. (1979) *Atmos. Environ.*, **13**, 369–71.
113. Doyle, G. J., Tuazon, E. C., Graham, R. A. *et al.* (1979) *Environ. Sci. Technol.*, **13**, 1416–19.
114. Waldman, J. M., Munger, J. W., Jacob, D. J. *et al.* (1982) *Science*, **218**, 677–80.
115. Munger, W. J., Jacob, D. J., Waldman, J. M. and Hoffmann, M. R. (1983) *J. Geophys. Res.*, **88**, 5109–21.
116. USNAS (1977) *Ozone and Other Photochemical Oxidants*, US National Academy of Sciences, Washington, DC.
117. Pitts, J. N., Jr., Darnall, K. R., Carter, W. P. L. *et al.* (1979) *Mechanisms of Photochemical Reactions in Urban Air*, EPA Report 600/3-79-110 (Data from EC run 216).

4 Analysis of particulate pollutants

R. M. HARRISON

4.1 Introduction

Particulate pollutants are emitted by a great many sources, both stationary and mobile. Additionally, particles are formed in the atmosphere by chemical and physical conversion from both natural and anthropogenic gaseous substances. In this chapter the methods of sampling atmospheric particulates are critically examined and techniques of physical examination are described. Analysis of the chemical composition is given comprehensive coverage in subsequent chapters.

Particulate pollutants are very diverse in character and cover a size range from <0.1 μm to >100 μm. A typical size distribution, indicating the responsible formation mechanisms, is indicated in Fig. 4.1. The major proportion of the aerosol <2 μm is generally man made, including, for example, sulphates formed from sulphur dioxide oxidation and lead from vehicle exhausts. Particles of greater than 2 μm diameter are mostly natural (e.g. marine aerosol, wind-blown soil), but the division should not be regarded as rigid, since man-made material extends in size to >2 μm and natural airborne material to below this diameter.

Particulate pollutants may be sampled either from suspension in the air, by filtration for example, or by collection of deposited particles as they fall out from the atmosphere under gravitational influence, known as dustfall. Consequently, the study of particulate pollution is simplified by division of particles into two categories: (a) suspended matter; and (b) depositable matter. This division is not clear cut and is dependent upon such factors as meteorological conditions and size, density and terminal velocity of the particles. Most workers consider a diameter of 10 μm as the dividing line between suspended and depositable matter. This division should not be taken too literally, however, and when estimates are made of deposition from chimneys, it is seen that deposition rates in the open of particles with diameter in the region of 20 μm are negligible. This is not difficult to understand when it is considered that a 20 μm particle of unit density has a falling speed of approximately 1.2 cm s^{-1}, so that a wind speed of 1 mile h^{-1} (0.45 m s^{-1} – almost calm conditions) is high enough to maintain 20 μm particles airborne for long distances. The picture is further complicated by the fact that turbulent fluctuations of the wind can maintain particles in suspension for far longer and for these reasons it is not proposed to go into greater detail on the subject of particle sizes. Another complication which will

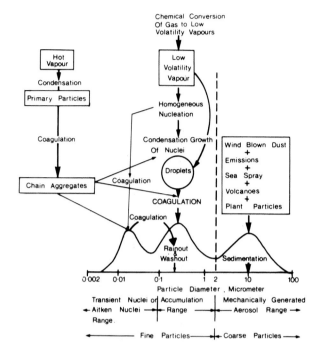

Fig. 4.1   Schematic diagram of the size distribution (expressed as surface area per increment in particle diameter) and formation mechanisms for atmospheric aerosols (reproduced from *Atmos. Environ.* **12,** 136 with permission of Pergamon Press).

arise in practice is that, at any site, pollution from different directions is likely to contain particles of radically differing densities, and consequently particles in any one size range, from one source, could well remain suspended while becoming deposited from another.

   Because of these problems, when operating in the field, sampling techniques should always be well defined. Different terms describing atmospheric particulate loadings have arisen from the different measurement techniques employed. *Smoke* is defined as fine suspended particulate air pollutants ($<15\ \mu$m) as measured by determining the staining capacity of the air. The major material responsible for staining of air filters is elemental carbon, and this arises substantially but not exclusively from the incomplete combustion of coal. Total suspended particulates (TSP) are particles (generally $<15\ \mu$m) suspended in the atmosphere, as collected by filtration with subsequent gravimetric determination, irrespective of the relative staining capacities of these emissions. In the measurement of depositable pollutants, the terms *dry deposition* and *dustfall* are both used to describe pollutant deposition, although the former term applies to gases as well as particles.

4.1.1   Emission of particulate matter

There are few industrial processes which do not emit particulate matter and the amounts emitted vary considerably from process to process as well as in different parts of any particular process. Formulae exist for estimating the behaviour of pollution emitted from discrete sources such as chimneys; but when emissions are from diffuse sources such as coke heaps, or from materials handling processes such as grabs, excavators or conveyor belts, methods of estimating overall emissions are more difficult and have to be designed for each individual case. It is true that a conveyor belt can be regarded as a line source of pollution, and a mechanical grab, unloading cargo from a ship's hold, as a generator of discrete puffs of pollution. Formulae do exist for estimating ground level concentrations from such sources but in practice it is seldom possible to apply them because such industrial sources are generally installed in heavily congested areas where nearby buildings affect dispersion of the pollution and make any accurate estimation of likely concentrations from these sources almost impossible to carry out.

4.1.2   Emission factors for particulate matter

Since different industrial processes emit differing quantities of particulate pollutants, data relating to such pollution are best presented as kilograms of particulate matter per tonne of finished product. Most industrial processes, however, employ methods, such as filtration, for reducing emissions and the efficiencies of these will, for obvious reasons, vary from factory to factory. Vandegrift *et al.* [1] looked at the particulate emission factor, i.e. the weight of particulate matter emitted per tonne of raw material, and tried to assess the average efficiency of the control techniques in use. For easy reference his estimates are quoted in Table 4.1 together with the net emission of particulate material to the atmosphere for various industries after control. These give some idea of the relative importance of some industries as emitters of particulate pollution. Although the list is not complete and substantial differences exist between different installations performing the same process, for those sampling in the field it should serve as a useful guide not only as to which sources are potential offenders but also which particulate materials it will be necessary to sample. For example, some of the particulate matter mentioned will have to be treated as suspended while other material can be treated as depositable. Factors such as these will determine the sampling technique used.

4.1.3   Dispersion of atmospheric pollutants from a point source

As a considerable amount of work has been carried out on the subject of atmospheric dispersion, it is not proposed in this book to go into full details of the various theories involved but instead to concentrate on the purely practical

Table 4.1   Emission of particulate matter from industrial processes

Source	Emission factor without control (kg tonne^{-1})	Net efficiency of applied control[a]	Net emission factor (kg tonne^{-1})	
(1) *Fuel combustion*				
Power stations – coal				
Pulverized	95	0.89	10.5	(as coal)
Stoker	73	0.70	21.9	(as coal)
Cyclone	17	0.64	6.3	(as coal)
Industrial boilers – coal				
Pulverized	85	0.81	16.2	(as coal)
Stoker	67	0.52	31.9	(as coal)
Cyclone	16	0.75	3.9	(as coal)
(2) *Agricultural operations*				
Grain elevators – grain	14	0.28	9.7	(as grain)
Cotton gins[b]	6	0.32	4.1	(kg bale^{-1})
Alfalfa dehydrators – dry meal	26	0.42	14.5	(as dry meal)
(3) *Iron and steel*				
Materials handling – steel	5	0.32	3.4	(as steel)
Sinter plant – sinter	21	0.90	2.1	(as sinter)
Blast furnace – iron	65	0.99	0.7	(as iron)
Steel furnaces – steel				
Open hearth	9	0.40	5.1	(as steel)
Basic oxygen	20	0.99	0.2	(as steel)
Electric arc	5	0.78	1.1	(as steel)
Scarfing	2	0.68	0.5	(as steel)
(4) *Cement*				
Kilns	84	0.88	10.0	(as cement)
Grinders, etc. (wet)	13	0.88	1.5	(as cement)
Grinders, etc. (dry)	34	0.88	4.0	(as cement)
(5) *Pulp mills*				
Recovery furnace	75	0.91	6.8	(as pulp)
Lime kilns	23	0.94	1.4	(as pulp)
Dissolving tanks	3	0.30	1.8	(as pulp)
(6) *Lime works*				
Crushing, screening – rock	12	0.20	9.6	(as rock)
Rotary kilns – lime	90	0.81	17.1	(as rock)
Vertical kilns – lime	4	0.39	2.2	(as rock)
Materials handling – lime	3	0.76	0.6	(as rock)
(7) *Clay*				
(a) *Ceramic*				
Grinding	38	0.60	15.2	
Drying	35	0.60	14.0	
(b) *Refractories*				
(1) *Kiln fired*				
Calcining	100	0.64	36.0	

Table 4.1   Emission of particulate matter from industrial processes – *contd*

Source	Emission factor without control (kg tonne^{-1})	Net efficiency of applied control[a]	Net emission factor (kg tonne^{-1})	
Drying	35	0.64	12.6	
Grinding	38	0.64	13.7	
(2) Castable	113	0.77	25.9	
(3) Magnesite	125	0.56	55.5	
(4) *Mortars*				
Grinding	38	0.60	15.2	
Drying	35	0.60	14.0	
(5) Mixes	38	0.60	15.2	
(8) *Primary non-ferrous*				
(a) *Aluminium*				
Grinding of bauxite	3	0.80	0.6	(as bauxite)
Calcining of hydroxide –				
alumina	100	0.90	10.0	(as alumina)
Reduction cells – aluminium				
H.S. Soderberg	72	0.40	43.2	(as aluminium)
V.S. Soderberg	42	0.64	15.1	(as aluminium)
Prebake	32	0.64	11.4	(as aluminium)
Materials handling	5	0.32	3.4	(as aluminium)
(b) *Copper*				
Ore crushing – ore	1	0	1.0	(as ore)
Roasting – copper	84	0.85	12.6	(as copper)
Reverb furnace – copper	103	0.81	19.6	(as copper)
Converters – copper	118	0.81	22.4	(as copper)
Materials handling – copper	5	0.32	3.4	(as copper)
(c) *Zinc*				
Roasting – zinc				
Fluid bed	1000	0.98	20.0	(as zinc)
Ropp, multihearth	167	0.85	25.0	(as zinc)
Sintering	90	0.95	4.5	(as zinc)
Materials handling	4	0.32	2.4	(as zinc)
(d) *Lead*				
Sintering	260	0.86	36.4	(as lead)
Blast furnace	125	0.83	21.3	(as lead)
Materials handling	3	0.32	1.7	(as lead)
(9) *Asphalt*				
Paving materials				
Dryers	16	0.96	0.7	(as material)
Secondary sources	4	0.96	0.2	(as material)

Table 4.1   Emission of particulate matter from industrial processes – *contd*

Source	Emission factor without control (kg tonne^{-1})	Net efficiency of applied control[a]	Net emission factor (kg tonne^{-1})
(10) *Ferroalloys*			
Blast furnaces	205	0.99	2.1 (as alloy)
Electric furnace	120	0.40	72.0 (as alloy)
Materials handling	5	0.32	3.4 (as alloy)
(11) *Iron foundries – metal*			
Furnaces	8	0.27	5.9 (as metal)
Materials handling (coke, limestone, etc.)	3	0.20	2.0 (as metal)
(12) *Secondary non-ferrous* (a) *Copper* *Materials preparation –* scrap			
Sweating furnace	8	0.19	6.1 (as scrap)
Blast furnace	25	0.68	8.0 (as scrap)
Smelting and refining	35	0.57	15.1 (as scrap)
(b) *Aluminium*			
Sweating furnace – scrap	16	0.19	13.0 (as scrap)
Refining furnace – scrap	2	0.57	0.9 (as scrap)
Chlorine fluxing – chlorine	500	0.25	375 (as chlorine)
(c) *Lead – scrap*			
Pot furnace	0.4	0.90	0.1 (as scrap)
Blast furnace	95	0.90	9.5 (as scrap)
Reverb furnace	50	0.90	5.0 (as scrap)
(d) *Zinc – scrap*			
Metallic scrap sweating	6	0.19	4.9 (as scrap)
Residual scrap sweating	15	0.19	12.2 (as scrap)
Distillation furnace	23	0.57	9.7 (as scrap)
(13) *Sulphuric acid*			
New acid – contact process – 100% acid	1	0.85	0.2 (as 100% acid)
Spent acid concentrators – acid	15	0.80	3.0 (as spent acid)
(14) *Phosphoric acid*			
Thermal process – P_2O_5	67	0.97	2.0 (as P_2O_5)

[a] The overall level of control in the US; the product of the application of control and the efficiency of control.
[b] kg bale^{-1}.

problems arising as a consequence of these theories, which have to be solved in order to perform effective sampling. Chapters 2 and 14 give some idea of the extent to which pollution may be dispersed, such that when sampling is undertaken sites will be selected which will give meaningful results. It must be remembered that the formulae and methods presented refer to conditions in open, level country. They will be only a very approximate guide to the dispersion of pollutants in a built-up environment.

4.1.3.1   *Problems of short-term sampling*

When short-term sampling is planned the requirements must be clearly understood before work starts otherwise the results may be of little value. An example might be to settle whether the pollution being monitored is from a particular source or whether it concerns area-wide pollution. In the latter case sites should be avoided which are under the influence of nearby sources. If on the other hand it concerns emissions from a particular source it may be desirable to locate the maximum ground-level pollution from such a source so that a suitable site may be found. In an earlier section a method was given for determining the location of peak ground-level concentrations from a specific source. On paper this is a simple problem but in practice, even in open country with no buildings to complicate dispersion, the actual location is not a simple task even using continuous recorders. The reasons are explained by the meteorological parameters which control dispersion.

If, in an attempt to locate high concentrations, sampling is carried out close to a source the pollution levels in the plume will be found to fluctuate wildly from peak values to zero in a few seconds as the plume shifts constantly in direction due to variations in wind direction, etc. The amount of these fluctuations is largely determined by the stability category and some useful figures given by Pasquill show that at a distance of 0.1 km from a source, under very stable conditions (stability category F) a plume of about $10°$ can be expected; under unstable conditions (category A) plumes can be about $60°$ in width at this distance. Further downwind concentration fluctuations will be found but they will not be quite so extreme as close to the source. Although these fluctuations will not be quite so severe it will still be difficult in practice to locate the centre and, since the crosswind concentrations of pollution in a plume are roughly Gaussian, a slight error in the location of the site could give severely reduced concentrations.

When short-term sampling has to be started it is important that any measurements should be adequately supported by the relevant meteorological data, such as wind speed, direction, stability category, etc. A fuller discussion of meteorological considerations is given in Chapters 2 and 14.

Another factor which has to be taken into consideration is the pattern of emissions. Many pollution sources emit different amounts of pollution, not only dependent on the time of day, the day of the week, but there may also be seasonal variations, e.g. from summer to winter. Some cyclic changes in pollution emissions can be very short term indeed, such as the effect on emissions from road traffic affected by traffic lights. When the effect of these variations is added to the variations due to meteorological conditions it will be seen that care will have to be taken with such short-term determinations.

4.2   Suspended material

4.2.1   Sampling techniques

4.2.1.1   *Filter paper techniques*

Probably the most common method of sampling particulate pollution (this includes 'smoke') is by collecting it on filter paper by drawing a sample of contaminated air through a filter which is held in a clamp such as is illustrated in Fig. 4.2 [2]. This figure illustrates the equipment used in the National Survey of Air Pollution in the UK. The suction pump draws approximately 2 m³ of air per day through the equipment; particulate matter is filtered out by the filter, and the Drechsel bottle which contains dilute H_2O_2 removes SO_2 contamination [2].

   It is not possible to give firm advice on the type of filter material which should be used for any particular work since this will depend on the purpose for which the sampling is being conducted. For example, cellulose and glass fibre filters of suitable porosity will be suitable for routine determination of the total mass of particulate pollution. If subsequent chemical analysis is required they may be suitable if the filter background is negligible for the particular material being

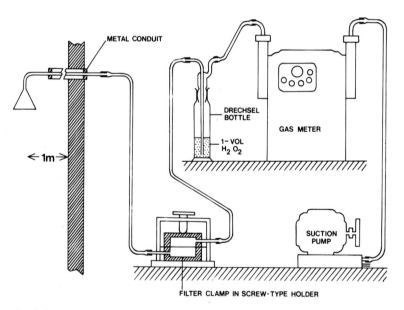

Fig. 4.2   Schematic arrangement of standard daily smoke and SO_2 sampling apparatus [2].

sought. It is always necessary to check the background contamination of the filters to be used in any analytical work. For microscopic work cellulose and glass fibre filters are unsuitable, because the atmospheric particulate matter tends to become trapped amongst the fibres of the filters, and thus membrane and Nuclepore filters are preferable. At this stage it will be useful to look at some of the more common types of filters in general use and describe the advantages of each.

Cellulose filters

Cellulose filters are frequently used for routine air pollution work. In most cases, where the density of the collected matter is determined by light reflectance methods, Whatman No. 1 is the most common, but with Whatman No. 4 in use in the USA. Whatman No. 4 offers less resistance to air flow but against this the penetration of particles into the filter is greater than with No. 1 grade. This is best illustrated by the graph shown in Fig. 4.3 which was taken from a technical bulletin published by the manufacturers [3]. This figure shows the particulate penetration of a number of the more common filters against increasing dust

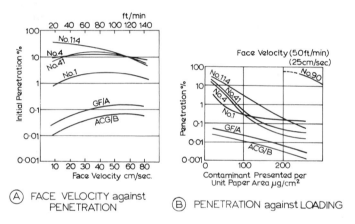

Fig. 4.3 The efficiency of Whatman cellulose and glass fibre filters. (a) The effect of face velocity upon penetration (= 100% efficiency). (b) The effect of filter loading upon penetration at a face velocity of 25 cm s^{-1} [3].

loading. It is interesting to note the difference in penetration between the cellulose filters and the glass fibre filters. The penetration figures shown were obtained by drawing air containing NaCl particles (0.6 to 1.7 μm diameter) through the filter and measuring the concentration of sodium downstream by means of flame photometry.

The purpose of the sampling will largely determine the type of filter used. For example, for analytical work where the filter has to be ashed before analysis, one must take account of the fact that both Whatman No. 1 and No. 4 have an ash content of 0.06%, whereas the ash content of No. 41 grade is 0.01%. Cellulose filters also contain trace impurities and this may, depending of course on the materials being examined, preclude their suitability for work where chemical analysis of the sample is to be carried out.

Although cellulose filters are still widely used in air pollution work, their limited efficiency relative to glass fibre and membrane filters makes them an inferior option for work where high accuracy is required.

Glass fibre filters

These filters offer less resistance to air flow than cellulose filters and, as may be seen in Fig. 4.3, the penetration of particles through the filter is much less than with cellulose filters, being less than 0.1% for the size of particles tested. Thus more of the particulate matter is collected on the surface of the filter than is the case with cellulose. Since the particles are more heavily concentrated on the surface this filter will appear blacker with a given concentration of material than would appear with a cellulose filter. In fact, if the light reflectance method is used to determine suspended matter collected, the sample on glass fibre would appear to be about 2.5 times more dense than an equal amount on cellulose.

One serious disadvantage of glass fibre filters is that they are rather fragile and extreme care should be taken when using them for gravimetric work to ensure that fibres do not break away in handling, particularly after being clamped into a sampling apparatus. In all cases, as with cellulose filters, the background contamination from the filter should be checked before sampling if subsequent sensitive analytical work is to be carried out.

One advantage of glass fibre filters is that they can withstand higher temperatures (up to 800° C) than cellulose (250 to 300° C) and vaporization of the carbonaceous content of the sample may be an important means of improving analytical resolution.

Because of their high efficiency and low flow resistance, glass fibre filters are usually the preferred medium for use with Hi-Vol samplers. Filters pre-stamped with identification numbers are available for this purpose. The major drawback with glass fibre is the high trace element background and the prevalence of artifact formation (see Chapter 1). Where reliable analysis of sulphate and nitrate is required, quartz fibre is the preferable filter substrate (Chapter 1).

Membrane and Nuclepore filters

Membrane and Nuclepore filters collect the particles predominantly on the surface and this makes them ideally suited for microscopic work, for which glass fibre and cellulose filters are unsuited.

Membrane filters are mostly prepared from one of several cellulose esters and the pore size of the filter can be controlled by the manufacturing process. These filters may be used to trap particles down to far less than 0.1 μm in diameter. One of the main drawbacks of membrane filters is that they have a high resistance to air flow and tend to be brittle if flexed. In use it is therefore advisable to support them by means of a porous plate or very fine stainless steel mesh.

Nuclepore filters are polycarbonate sheet which has been irradiated by neutrons and etched to give pores of constant dimensions. These bear certain similarities to membrane filters and the properties of both have been extensively investigated by Spurny and co-workers [4–6]. The mechanism of filtration of suspended particles is a product of several processes including impaction and diffusion precipitation and electrostatic attraction. Hence, efficiency is a complex function of many variables, including pore size, particle size, face velocity and filter loading. Some comparative efficiency data appear in Table 4.2 and further data on filter efficiency appear in Chapter 1. Although some workers have reported low filtration efficiencies at high face velocities [7], this appears to be by no means a general finding and no clear recommendation on this point can be made. In general, the smaller the pore size of the filter, the greater the efficiency but the lower the rate of flow readily achieved [6, 8]. Efficiency increases with filter loading and hence for long sampling periods a more effective collection of particulates will be achieved [5, 9]. Since particles passing the filter represent a loss in analytical accuracy, care should be exercised in selection of a filter. In general, membrane and Nuclepore filters are of higher efficiency than cellulose and glass fibre, and of the former the membrane filter is generally the better. For near-quantitative collection of particles under almost all conditions, the use of membrane filters of pore size $\leqslant 0.6$ μm is recommended, while those requiring more comprehensive data upon collection efficiencies are referred to the experimental studies of Spurny *et al.* [6] and Lui and Lee [8].

Table 4.2   Comparative efficiency of filters at 5 cm s^{-1} face velocity [5]

Analytical filter	Nominal pore size (μm)	Pressure drop (mm H$_2$O)	Efficiency for particles of 0.03 μm	Efficiency for particles of 0.3 μm
Nuclepore	0.50	701.0	0.987	0.993
	0.80	698.0	0.946	0.619
	1.00	127.0	0.868	0.522
	2.00	56.0	0.433	0.283
	5.0	35.0	0.184	0.144
	8.0	65.0	0.057	0.101
Membrane				
VUFS – Synthesia	0.30	903.0	0.998	0.991
VUFS – Synthesia	0.8	553.0	0.999	0.987
HA – Millipore	0.45	418.0	0.999	0.988
AA – Millipore	0.8	260.0	0.999	0.973
Acropore Gelman	0.8	51.0	0.993	0.970
OH – Millipore	1.5	607.0	0.872	0.511
SS – Millipore	3.0	97.0	0.999	0.912
AUFS – Synthesia	1.4	28.0	0.992	0.932
RUFS – Synthesia	2.4	27.0	0.945	0.792
PUFS – Synthesia	7.0	15.0	0.949	0.688
OS – Millipore	10.0	28.0	0.634	0.312
Fibre				
AGF – Gelman		688.0	0.999	0.990
PF – 41 – Whatman		48.0	0.589	0.224

Membrane filters can be made transparent for microscopic work by the addition of a few drops of immersion oil or other oils with a refractive index of 1.56. In addition, samples may be prepared for electron microscopy by taking a small piece of the exposed filter and floating it, sample uppermost, on a small quantity of acetone. The cellulose ester material will dissolve, leaving a thin film holding the dust. An uncoated electron microscope specimen grid may then be inserted carefully under the film by the use of forceps. When the film is lifted away from the container the acetone on the grid will evaporate, leaving the specimen ready for examination.

Membrane and Nuclepore filters have a very low ash content and low levels of background impurities and are thus ideally suited for some of the more sophisticated analytical work. In particular, Teflon membrane filters, although expensive, are highly favoured due to their low blank and chemical inertness. They are suited to the collection of sulphates and nitrates as they do not cause artifact formation [10].

4.2.2   Determination of total particulate pollutant concentrations

There are two main methods used for determining the concentrations of total particulate matter collected on filter papers. These are direct gravimetric determination of collected particles, and measurement of the soiling of a filter paper by the dimunition of light reflectance from it. In the USA smoke density is also measured by transmitted light and is referred to in terms of coefficient of haze (COH) units. Useful comparisons of techniques have been reported by Dalager [11] and Kretzschmar [12]. In the latter work, a good correlation was found between the results obtained by a gravimetric method, an integrating nephelometer (*vide infra*) and light reflectance, although the three methods do not give identical measurements of particulate concentrations.

4.2.2.1   *Light reflectance method*

In this method the darkness of the stained filters is assessed using a reflectometer. This instrument consists of a light source and a photo-sensitive element mounted together in a measuring head. In the model used in the UK the light beam passes through a circular hole in the centre of the photo-sensitive element. The measuring head fits snugly into a masking disc which restricts the illumination to an 0.5 inch diameter hole through which the stained filter is examined. Light reflected back from the filter paper falls on the sensitive surface of the photo-sensitive element and the current generated is measured by means of a milliammeter. The darker the stain the lower the intensity of the reflected light.

The results may be converted into concentrations of 'standard smoke' by use of calibration curves [13]. Alternatively the method is calibrated by running two sampling lines in parallel; one operating at the standard air flow of approximately 70 ft^3 (2 m^3) per day and a second line operating at an air flow sufficiently high to enable the amount of collected particulate matter to be weighed. Calibrations have been carried out in different regions where there was a predominance of

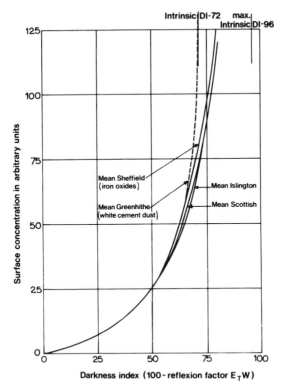

Fig. 4.4 British smoke calibration curves. EEL reflectometer, Whatman No. 1 filter paper, 1 inch diameter.

particulate emissions from sources, such as cement works and steel works, which would impart a different colour to the particulate matter. The calibration curves are shown in Fig. 4.4. The light reflectance method is suitable for most routine air pollution work but too much reliance should not be placed on its accuracy. For example, in areas where there is a high percentage of crystalline matter this could affect the amount of light reflected back to the photo-sensitive element. Some doubt has also been cast on the reliability of the calibration because of suspicions that owing to the bad design of the inlet orifice and the use of higher sampling rates, the calibration instrument could sample a particle size distribution different from that of the standard instrument.

In a recent evaluation of the smoke shade method, Edwards *et al.* conclude that reflectance, as measured by the British method, is controlled by the optical absorption coefficient of the aerosol. A good correlation between British smoke shade (BSS) using Whatman filters and an EEL reflectometer and sub-μm elemental carbon (EC) was reported.

$$EC(\mu g\ cm^{-2}) = (0.13 \pm 0.03)BSS(\mu g\ cm^{-2}) - (0.1 \pm 2.4)\ (r^2 = 0.68;\ n = 31)$$

Poorer correlations for the mass concentration of suspended particles were reported [14].

4.2.2.2   Gravimetric techniques

Because of doubts on the reliability of the light reflectance method for determining the amount of particulate matter sampled, most workers prefer to use a gravimetric technique as the standard method for the determination of particulate concentrations.

This technique is self explanatory in that a sample of particulate material is collected at a sampling rate sufficiently high for a reliable weighing of the particulates to be carried out. The higher sampling rate, however, brings in added complications in that a totally different size range of particles may be collected. When considering sampling systems, importance should be given to the design of the inlet orifice to ensure that the size fraction of the sampled particles is known. In some routine instruments air is drawn into the system through an inverted funnel. This may effectively prevent rain being drawn into the instrument but it is of too poor a design to enable an accurate particle size distribution to be reliably sampled.

When considering inlet orifice design, attention should be paid to work by Davies [15] on the correct design of dust sampling instruments. One instrument which would appear to fulfil the requirements is that designed by the Landesanstalt für Immissions und Bodennutzsschutz in Essen [16]. In this unit, air is drawn into the instrument through a sharp edged, parallel-sided orifice. One advantage of this instrument is that the filter paper is situated inside the orifice at the top, thereby eliminating the possible deposition of particles on the tubing. Despite a poor inlet design the Hi-Vol sampler (see Chapter 1) is widely used for the collection of large particulate samples for gravimetric analysis.

When examining instruments for long period sampling of particulates, care should be taken to ensure that there is an accurate method of measuring the air flow through the instrument. Normally, after a few hours operation, the dust loading on filters increases their resistance to flow, causing a subsequent drop in the air flow. In all cases, gas meters or other accurate flow monitoring devices should be incorporated. Pressure drop methods of estimating air flows can cause serious errors and should be avoided.

Before weighing the particulate matter collected on a filter, it should be equilibrated to constant weight in a controlled temperature and humidity environment. It should be noted that some filter materials, especially membrane filters, are themselves substantially sensitive to humidity variations.

Basic details of the US Hi-Vol sampler have been given in Chapter 1, which includes a discussion of the characteristics of the inlet as a function of particle size, wind velocity and orientation. In routine use, Hi-Vol samples are taken over 24 h or 48 h periods. Several variants of the basic sampler are available, some of which incorporate constant flow rate control. These controls rely upon a hot wire sensor located in the throat of the Hi-Vol, between the filter and the motor.

Dependent upon the signal from this sensor, the power to the motor is increased to compensate for increased flow resistance due to loading of the filter. In our experience, this compensation is effective only at rather low filter loadings. Hi-Vol samplers are generally supplied only with flow measurement by a rotameter attached to an orifice plate at the rear of the motor. These can be highly unreliable, and independent calibration using a hot wire device which is attached temporarily above the Hi-Vol filter is essential. These Hi-Vol flow calibrators are commercially available and take only a minute or so to give a reading of flow rate.

Prior to weighing, both the unexposed and exposed filters are normally equilibrated at 55% relative humidity to eliminate effects due to the hygroscopic nature of the filter and collected particles.

4.2.2.3   Other filter paper devices

Eight-port sampling instrument

Strictly speaking this is not merely a filter-using device but is rather a changeover device for converting the simple smoke and SO_2 instrument into a form suitable for semi-automatic use.

In practice, when using the simple instrument, difficulty may be found especially at week-ends in arranging to change filter papers and Drechsel bottles daily and this instrument has been designed to eliminate the difficulty. A diagrammatic sketch of the unit is shown in Fig. 4.5. The timing motor shown selects a number of predetermined periods (e.g. 8×24 h periods, or 8×3 h periods) when it will automatically change over sampling lines by switching on the valve changeover motor. This motor moves the upper slotted disc through

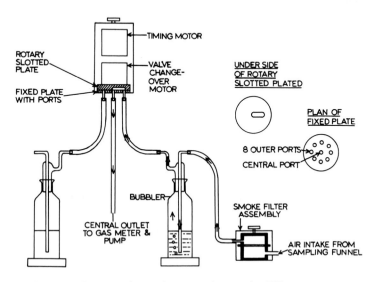

Fig. 4.5   Schematic diagram of a semi-automatic sampler [2].

45° (1/8 of a revolution) which connects the air intake in sequence to a number of standard daily instruments for predetermined periods. At the end of the sampling sequence, say once per week, the Drechsel bottles and filter papers are removed and analysed. This is probably one of the simplest automatic devices in use in air pollution analysis.

Automatic smoke instruments

Quite frequently in air pollution work there is a necessity for short period samples which cannot easily be obtained with simple instruments operated manually; the problem may be the study either of the diurnal variation of suspended particulate matter or how the variations in particulate concentrations near busy streets can be related to traffic flows. Such sampling can be achieved by the use of an automatic tape sampler. This instrument (see also Chapter 1) uses filter paper in the form of a reel. The filter strip is held in a 1 inch diameter clamp which can be released automatically by means of a solenoid operated by a timing motor. The timing motor period may be varied as desired but normal times of operation are for periods of 1, 3 or 24 h. Air is drawn through the filter clamp for a predetermined period after which the solenoid automatically raises the clamp. At the same time a second motor drives the take-up spool through one complete revolution thus winding up the filter reel and exposing fresh filter under the clamp. The solenoid is then de-energized, clamping the filter paper again. After exposure the strip of filter paper is removed for evaluation by light reflectance methods, or for chemical analysis of the collected material.

β-radiation monitors

Another method which has been used to monitor the concentration of particulate material in the atmosphere involves the use of the attenuation of intensity of β-radiation passing through solid material.

The instrument is basically an automatic smoke recorder in which ambient air is drawn through a high efficiency glass fibre filter. This filter is in the form of a reel which can be automatically moved forwards at intervals. A diagram of the basic system is shown in Fig. 4.6 showing the radioactive krypton-85 source mounted opposite two ionization chambers. The radiation is measured as ionization current in chamber 2 where it is partly absorbed by the chamber and the clean filter. In chamber 1 there is further attenuation caused by the dust on the filter. The two chambers have opposite polarity and a voltage drop proportional to the amount of dust is produced across the resistance connected to both chambers. The output, which is proportional to the particle mass on the filter, is amplified and recorded continuously on a chart or other type of recorder.

As already mentioned, the instrument samples continuously and the filter tape is moved forward automatically at predetermined intervals. As new filter tape appears before ionization chamber 1 this causes a voltage drop which is compensated by an automatic zero balance. The current from ionization chamber 2 is used as a reference current for the current from chamber 1 which continues to decrease as the dust concentration builds up on the tape. This eliminates variations which could be caused by variations in the thickness of the

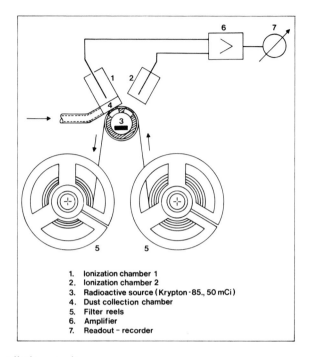

Fig. 4.6 β-radiation monitor.

filter. Some deviations can occur where there is a high proportion of heavier elements but allowance can be made for them.

Short-term portable sampling

A major constraint on air pollution work in the field occurs when sampling needs to be carried out in locations where mains power is either not available or not convenient for short periods. Lightweight portable, battery operated equipment [17] may be used for such work. Today there is a plentiful supply of small battery driven pumps but before they can be used in air pollution work they require meters for monitoring gas flows and in the field this is not always convenient. One unit which may be used is a small positive displacement suction pump, for example, a converted toy steam engine running in an oil bath and driven by a 6 V electric motor. The number of revolutions of the piston is monitored using a digital mechanical counter driven by the piston or the pump. Before use the units may be calibrated against wet gas meters.

4.2.2.4 *Piezoelectric mass monitors*

Another technique now available is the continuous microbalance. In this instrument air is drawn through a chamber containing a small electrostatic precipitator which precipitates particles from the air on to the surface of an oscillating crystal. This crystal is one of a matched pair and the precipitated particles cause its oscillatory frequency to change. This change is proportional to the weight of precipitated material. One drawback to this method, and also to the preceding β-radiation method, is that both are affected by moisture. In addition, the mass monitor has shown that weights of collected material can change if some of the material collected is volatile. This is not strictly a disadvantage of the method but rather gives some indication of the sensitivity of the method, because such slight changes would not be noticeable with simpler instruments. A detection limit of 1 μg is claimed. An example of the loss of volatile matter was the collection of a sample of cigarette smoke on the crystal followed by continued operation of the monitor on clean filtered air. After about 4 h of operation on filtered air the effective weight recorded by the crystal dropped by approximately 17%, probably caused jointly by the evaporation of moisture and some of the tarry matter. A similar test carried out after sampling normal atmospheric pollution produced no such drop in weight.

One disadvantage of the mass monitor is that periodically the crystal has to be cleaned, but this difficulty can be reduced by sampling intermittently over long periods. When using the more sophisticated techniques described in the last section provision may have to be made to record other parameters such as humidity if high accuracy is required.

4.2.3 Cascade impactors

The size distribution of particles collected on filters may be determined microscopically, but this is very laborious, and larger particles tend to obscure the smaller, making counting difficult. The cascade impactor, shown schematically in Fig. 4.7, separates particles into fractions on the basis of their aerodynamic impaction properties (see also Chapter 1).

There are several types of cascade impactors in use, probably the best known being the Andersen sampler [19, 20]. This instrument consists of as many as seven impaction stages, backed by an efficient filter. Each stage contains

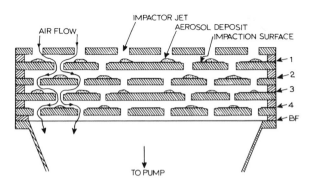

Fig. 4.7 Schematic representation of a cascade impactor [18].

accurately drilled holes (360 in each of the top 3 stages and 400 in each of the remaining four), and the stages are arranged so that beneath each hole there is a solid section of plate which is part of the next stage. The holes in each successive stage are smaller than those in the preceding, and as air is drawn through the instrument at a constant rate of 1 ft^3 min^{-1}, the effective velocity through each stage increases. Large particles of high inertia are aerodynamically impacted on the backing plate at the first stage, and progressively smaller particles are impacted at each stage as velocities increase. The theory of impaction processes is given by Butcher and Charlson [21], and although other cascade impactors are different in form and appearance (see Chapter 1), the basic principle is the same.

Because the aerodynamic impaction properties of particles are dependent upon several factors including density and shape and are also affected by the possibility of a particle being hollow, microscopic examination of particles collected on impactor plates will not show a high uniformity of physical size. If such examination is to be performed, sampling should not be extended over too long a period, or too large a particulate load will be collected, as particles appear only as small colonies under each hole in the preceding stage. If the sampled particles will later be examined by electron microscope techniques, it is possible to grind a circular groove in the impaction plates to fit either an electron microscope sample grid or a small palladium-covered copper disc to enable the collected sample to be transferred to such instruments without the tedious operation normally required for sample preparation. Because the impactor separates particles on the basis of their aerodynamic properties, rather than physical diameters, the cut-offs of the various stages are described in terms of 'aerodynamic diameters'. These are the diameters of spherical particles of unit specific gravity having the behaviour observed in the impactor. Typical particle size distributions measured in a cascade impactor are shown in Fig. 4.8; the particle size corresponding to the 50% point on the cumulative frequency axis is termed the mass median aerodynamic diameter (MMAD).

Another impactor in use is the high volume fractionating cascade impactor. This is a four stage instrument employing 12 inch diameter plates, each with 300 jets [22]. Impaction of particles is on to glass fibre filters or aluminium sheets which are themselves perforated, and this is a useful instrument for collection of larger samples. Details of the particle size fractions collected are given in Chapter 1. It suffers from several defects, however. It is based on the Hi-Vol sampler and no positive measurement of air flow is easily made; the design of air intake leaves room for improvement, and if glass fibre filters are used, extreme caution must be exercised to prevent fragments of filter breaking away.

Measurements of particle size distribution are extremely valuable, as they give an indication of the source of given particulate materials, and they allow determination of the proportion of particles in the respirable size range, these being the greatest hazard to health. Caution should, however, be exercised in the use of impactors and the interpretation of results, as distorted size distributions, attributed to particle bounce from dry impaction surfaces, have been reported [18]. Such errors may be avoided by the use of a virtual impactor [23] which divides particles into 0 to 2.5 μm (respirable) and 2.5 to 15 μm size ranges. In this

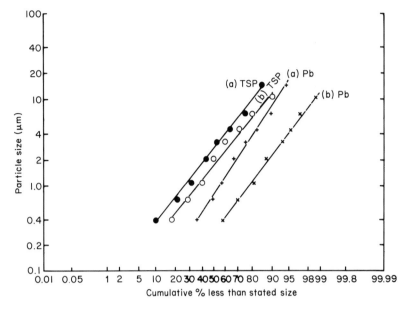

Fig. 4.8   Particle size distributions for lead and total suspended particulates (TSP) plotted as a cumulative frequency distribution: (a) an urban site; (b) a rural roadside site.

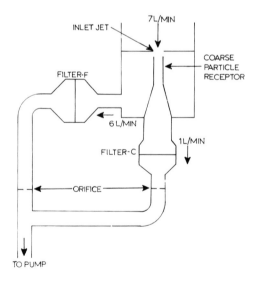

Fig. 4.9   Schematic view of a dichotomous sampler containing a single stage virtual impactor. Coarse particles are collected on filter C and fine particles on F [23].

instrument (Fig. 4.9), the large particles impact into a void, and both size ranges are collected, each on a separate filter. Further details are given in Chapter 1. This instrument is now widely used in US air sampling networks.

4.2.4 Light scattering techniques

4.2.4.1 *The integrating nephelometer*

Another method for analysing airborne particles is the light-scattering method. This method is becoming increasingly used because of the employment of automatic monitoring networks throughout the world and it is now the standard method for quantifying suspended particulate matter in Japan. One of the drawbacks of this system is that it suffers from interference by high humidity. A number of designs are available [24] but the basic principles are fairly similar. One, called an integrating nephelometer, is shown diagrammatically in Fig. 4.10 [25]. The theory of such instruments is described by Butcher and Charlson [21].

In this instrument (Fig. 4.10) the photomultiplier 'looks' down the centre of the main body of the instrument through holes in the discs forming the collimator and light trap and the cone of observation is defined by the first and fourth discs. The light source illuminates the sample air in the centre section of the pipe and any material causing light scatter will be seen by the photomultiplier.

The unit illustrated has provision for a rapid field calibration. The normal air supply is sealed off and clean purging air is introduced through orifices and fills the instrument with particle-free air with a sea-level Rayleigh scattering coefficient of 2.8×10^{-5} m^{-1} at approximately 460 nm. The upper scale calibration is provided by a white surface which is illuminated by light from the

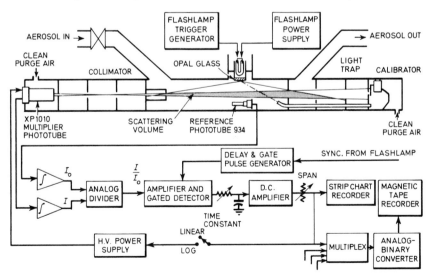

Fig. 4.10 Diagrammatic sketch of an integrating nephelometer [25].

flash lamp through a flexible light-pipe. A solenoid-actuated shutter uncovers a 1 mm hole in the end of the light trap through which the photomultiplier views this white surface. The signal thus introduced into the photomultiplier is about equal to that produced by the scattering of Freon-12; this is usually set at about half scale on the lowest range of the instrument which is 0 to 10^{-4} m^{-1}. These two points – i.e. clean air and a half scale signal – provide a check of calibration in the field in less than 1 min.

The papers by both Garland [24] and Ahlquist and Charlson [25] give details of methods used to eliminate the drift in sensitivity in the circuitry, caused mainly by the flashtube. The unit described by the latter workers has been designed for mobile work to measure, among other things, horizontal profiles of visibility in a city.

It should be borne in mind that the integrating nephelometer, which is commercially available, gives readings in terms of light scattering coefficients rather than gravimetric particle loadings. The relation of scattering to particle loading is, of course, particle size dependent, but for a typical atmospheric aerosol the mass loading of atmospheric particles may be inferred from the scattering coefficient with a quite small margin of error.

4.2.4.2   Aerosol particle counters

Other optical instruments of importance in the air pollution field are aerosol particle counters such as that manufactured by Royco Instruments. These instruments are of particular use in clean rooms etc. where limitations are laid down not only on total dust loading but also on size ranges. A diagrammatic sketch of the optical sensor of such an instrument is shown in Fig. 4.11. The particle sensor is a right-angle scatter, dark field optical system as can be seen. A sampling pump draws air through the optical viewing cell where it is intensely illuminated by a projecting lens system, a slit and a lamp. The sample air stream is viewed through a collection lens, a slit and a photomultiplier which are positioned at 90° to the main projection axis. Thus a small volume in the centre of the sensor head is optically defined by the lamp. When there is no particle passing through this volume, the photomultiplier sees a dark field. If a particle is present, light is scattered into the photomultiplier during the time that the particle is passing through the illuminated volume.

The amount of scattered light is a function of the particle diameter, and the photomultiplier produces an output pulse train where the height of each pulse is dependent upon the particle diameter. An electronic pulse height analyser is used to determine the number of particles that should be classified into each size range [21].

Because the projection lamp can vary in intensity and the photomultiplier gain can vary with temperature the instrument is provided with a built-in calibrator. When this is used a portion of the light from the lamp passes through the projection lens system, through the mechanical chopper, through a light pipe to the cell and photomultiplier. The calibrator simulates pulses of light from actual particles. The energy in the pulse train formed by the chopper is integrated and

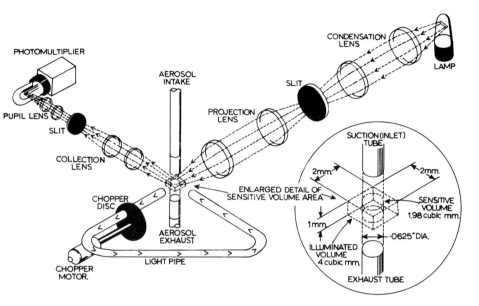

Fig. 4.11 The optical sensor of an aerosol particle counter.

measured with a front panel calibration meter. The amplification of the photomultiplier is adjusted with a front panel control until the calibration voltmeter indicates the proper values.

The diffusion battery, described in detail in Chapter 1, is a valuable means of separating sub-micrometer particles, a task for which impactors are not well suited.

The most usual application of a diffusion battery is in conjunction with a condensation nucleus counter (CNC). The CNC, which gives a numerical count of the total particle loading of a gas sample, is connected in sequence by an automatic valve to the ports of the diffusion battery. Thus different size fractions are sampled in turn, and by means of a computer program, the full particle size distribution is constructed. The disadvantage of the method is that no sample can be taken for subsequent chemical analysis.

Where samples are required for chemical analysis, it is unfortunately not possible to sample simultaneously from all ports of the diffusion battery. Indeed, air may be taken at a given time from only one port, and hence a sampling system combining a cascade impactor and diffusion battery, such as that described by Harrison and Pio [26] must be employed (see also Chapter 1).

4.2.5 The directional sampler

When field investigations of pollution levels are being carried out, one of the most common complaints to be investigated is of excessive pollution from a given direction affecting a particular area. With the simpler instruments described in this chapter it is very difficult to make such an assessment because of the extreme variability of meteorological conditions. The problem of comparing pollution levels can be solved by the use of an instrument in which sampling is controlled by wind direction; this makes it possible to give more meaningful pollution comparisons. The directional sampler performs such a function. A wind vane and anemometer, apart from giving a record of wind speed and direction, also control the operation of sampling equipment, and this is of course applicable to the sampling of gaseous pollutants as well as to suspended particulates.

The meteorological vane consists of a very low torque potentiometer of conducting plastic with a vane fixed to an extension to the potentiometer shaft and a stabilized voltage fed across the potentiometer. The anemometer is basically a small d.c. tachogenerator, calibrated in a wind tunnel, with anemometer cups fitted to the shaft. The outputs from both meteorological instruments are fed to a control box which contains three pairs of gate circuits. These circuits can be adjusted in such a way that the unit may be preset to switch on external instruments when the voltage outputs and consequently the angles of the vane are between certain preset values. In this way pollution may be measured automatically when the wind is blowing from given directions. Up to three different sectors may be selected.

When the wind speeds are low, however, directions tend to be extremely variable and directional sampling tends to be unsatisfactory. Under these conditions a limit switch is fitted to the anemometer output which eliminates directional sampling when the wind speed drops below a preset but variable value. If required, a socket may be provided for sampling to be carried out separately under these calm conditions.

In one survey, such an instrument was used to examine pollution entering an airport. Two such instruments were set up, one at the western end of the airport and one at the eastern end. The comparative values of smoke and SO_2 obtained at these two sites are shown in Table 4.3.

This gives some idea of the complication of air pollution work. In easterly winds, for example, the mean concentration of SO_2 entering the airport was 208 μg m^{-3} and the level of SO_2 leaving the airport at the western site was 91 μg m^{-3}. Since the airport itself was emitting little SO_2 a large percentage of this 91 μg m^{-3} must have come into the airport from the east. At the western site the mean level of SO_2 entering the airport was 152 μg m^{-3}, less than that which

Table 4.3 Pollution levels at airport (μg m^{-3})

Western site		Eastern site	
Smoke	Sulphur dioxide	Smoke	Sulphur dioxide
From the west		From the airport	
16	152	15	59
From the airport		From the east	
26	91	29	208

came in from the east. This would be expected as the airport is less heavily built up on the western side. Since little SO_2 is emitted at the airport, the results serve to illustrate the dilution of SO_2 in passing over the airport. In westerly winds concentrations decreased from 152 to 59 μg m^{-3}, while in easterly winds they dropped from 208 to 91 μg m^{-3}, in both cases a decrease of approximately 100 μg m^{-3}. The drop in smoke concentrations is less marked, suggesting some emission of particles at the airport.

4.3   Dustfall sampling

4.3.1   Introduction

A method for estimating dust deposition from stacks was given in an earlier section and it is important to know the pattern of deposition when attempting to measure it. The problems of such measurements are two-fold: firstly, if it is required to monitor general deposition in an area it is important to know how to avoid siting gauges at positions where the results could be seriously affected by one particular source; and secondly, if it is required to monitor deposition from a particular source the general deposition pattern must be known.

Deposit gauges to monitor the deposition of particulate matter are probably the earliest instruments used for air pollution measurement. Very little basic research has ever gone into the design of such gauges, however, and most of the designs leave much to be desired. For routine work they all fall into the same broad category, namely a bowl or cylinder with a horizontal, upward facing, collecting surface. Their major drawback is that the very presence of the gauge introduces disturbance into the falling pattern of particles, and it is conceivable that the results obtained may bear no relationship to true deposition figures in the area. In addition, it must be pointed out that the roughness of a particular area will affect the amount of deposition. Even in an ideal, open, grass-covered site the rate of deposition could change from day to day, all other things being equal, merely because of differences in the length of the grass. A concrete surface will also present a different collection efficiency to dust. Despite these drawbacks, deposit gauges still have their uses, but in recent years greater attention is being paid to collecting dust samples with the aim of identifying and isolating nuisance sources, rather than for the routine collection of data for statistical purposes.

When looking at dust deposition it must be remembered that when dust particles fall, only the very largest and densest fall at an angle even approaching that of the vertical. To illustrate this point further, some examples are given of angles of fall of particles of two different densities under wind speeds of 5 miles h^{-1} (2.2 m s^{-1}) (taken as average) and also 10 miles h^{-1} (4.5 m s^{-1}) (Table 4.4).

It can be seen that the heavier particles will fall close to the stacks, or sources, and if a gauge is sited to monitor general dustfall in an area well away from local sources, only the smaller particles will be collected in anything but higher wind speeds. Most of the particles which are being collected will be arriving at a very shallow angle to the horizontal, and under these conditions it is doubtful whether

Table 4.4   Angle of fall of particles

Particle size (μm)	Density	Angle of fall 2.2 m s⁻¹	(Degrees to horizontal) 4.5 m s⁻¹
1000	1	61	42
800	1	54	34
600	1	45	27
400	1	37	20
200	1	20	10
100	1	8	4
50	1	2	1
1000	2	75	61
800	2	70	54
600	2	64	45
400	2	56	37
200	2	35	20
100	2	15	8
50	2	4	2

a horizontal collecting surface is really ideal for the purpose. However, provided it is realized that deposits collected in horizontal deposit gauges represent only the particulate matter collected in the configuration offered by the gauge, and that such results may only represent a very approximate index of the dust loading in the neighbourhood, then it is acceptable to use them with reservation.

It should be recognized that the deposition velocity of atmospheric particles

$$v_g(\text{cm s}^{-1}) = \frac{\text{deposition flux } (\mu\text{g cm}^{-2}\text{ s}^{-1})}{\text{atmospheric concentration } (\mu\text{g cm}^{-3})}$$

is a complex function of particle size. It is not only large diameter particles which have a high deposition velocity (due to gravitational settling) but also very small particles ($<0.1\ \mu$m) due to their high Brownian diffusivity. Thus particles of a wide range of sizes will be collected.

4.3.2   Designs of national deposit gauges

There are, in various parts of the world, standard deposit gauges of different designs. As has already been mentioned, the very presence of the gauge affects the pattern of dust fall, an effect well illustrated by wind tunnel work by Pestel [27] who tested various types of gauges in a wind tunnel and using an artificial smoke tracer discovered the flow patterns generated round the gauges shown in Fig. 4.12.

It is not proposed to go into the detailed differences of all the different designs but to select three major types of design: (a) the British Standard gauge; (b) the French Standard gauge and (c) the Norwegian NILU Standard gauge, since they broadly represent the main design differences.

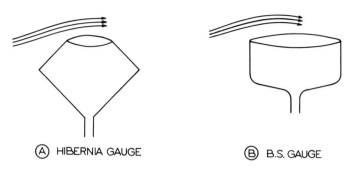

Ⓐ HIBERNIA GAUGE   Ⓑ B.S. GAUGE

Fig. 4.12 Flow visualization over gauges [27].

4.3.2.1 *The British Standard deposit gauge*

Details of the specifications of this gauge are given in BS 1747: 1951 [28] and a sketch is shown in Fig. 4.13. The gauge has been in use for many years but, because of doubts as to the value of the results obtained from it, its use is now becoming less common. It is now mainly used in areas where there are already in existence data collected over a number of years. The instrument consists of a glass or plastic collecting bowl into which dust falls. Dust washed down by rain falls into a large collecting bottle under the bowl. In periods of dry weather, however, the collected dust could remain in the top collecting bowl for long periods and there is a suspicion that in periods of high winds this dust could be blown out of the gauge because of the shallow nature of the collecting bowl. (The dimensions

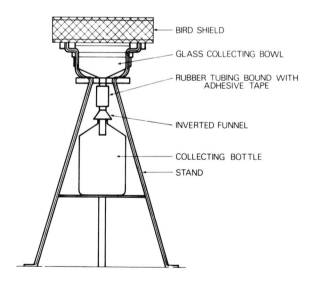

BIRD SHIELD

GLASS COLLECTING BOWL

RUBBER TUBING BOUND WITH ADHESIVE TAPE

INVERTED FUNNEL

COLLECTING BOTTLE

STAND

Fig. 4.13 British Standard deposit gauge. (After [28].)

are approximately 315 mm diameter; with a depth of 105 mm near the circumference, increasing to 180 mm near the centre. These dimensions give a ratio of 1.75:1 for diameter versus depth). As far as is known no checks have ever been carried out on this possible defect.

At the end of one month the collecting bowl is washed down carefully with a measured volume of water to wash any dust in the bowl into the collecting bottle below. The analysis of the samples is very simple indeed and is usually restricted to the determination of the total amount of water collected and its pH. This latter measurement is, however, of importance because recently there has been increased world-wide interest in the acidity of rainfall in projects dealing with the long-range transport of pollution. Routine analysis is restricted to the determination of the total weights of undissolved solids and the total weights of dissolved substances, such as chlorides, etc. These are carried out after extraneous matter such as twigs, leaves and dead insects have been removed from the samples. If the gauges are situated in regions, such as near steelworks, where there may be specific emissions, then the total weights of, say, iron collected should also be determined.

One main disadvantage of the standard deposit gauge is that, because of the extended exposure times, their ability to provide any directional resolution is limited. This is a serious disadvantage today because there is increased interest in using deposit gauges as 'troubleshooters' in identifying sources of a nuisance rather than for the routine collection of data over long periods.

4.3.2.2   *French Standard deposit gauge (Ref. NF, X43-006 (1972))*

This gauge, shown in Fig. 4.14a, is a development from the earlier British Standard deposit gauge and incorporates a modification added to prevent dust being blown out. The angles of the lower half of the collecting bowl are much steeper than the British gauge and this should result in less dust resting in the collecting bowl and being in danger of being blown out again in high winds. In order to reduce the danger even further, an inverted, truncated conical section has been added to the top and this should reduce this possible fault. It is not

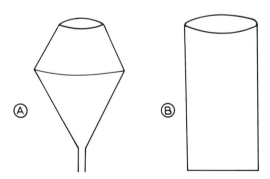

Fig. 4.14   Deposit gauges. (a) French Standard gauge; (b) NILU gauge.

known whether the French authorities have any data comparing their improved gauge with the original British unit but it should be less liable to loss of sample.

4.3.2.3   Norwegian NILU deposit gauge

This is the design which has been adopted in Norway and it is shown in Fig. 4.14b. This gauge has been shown as typical of the simpler, cylindrical types of deposit gauges, such as the Bergerhoff (Germany) or the USA, ASTM Dust Collector (Ref. D.1739-70). In all these designs the ratio of the diameter of the orifice to the depth is of the order 1:2 and this has been done in an attempt to reduce loss of sample by being blown out again. The top of the cylinder of the NILU gauge is shaped and bevelled at an angle of 45° downwards and outwards. In order to make doubly certain of reducing the loss of sample in high winds, it has also been suggested that enough water should be placed in the gauge at the start of the exposure period to ensure that there is liquid in the gauge throughout the exposure period, thereby preventing dust from blowing out. The amount will be determined by local climatic conditions. The other problems which arise with dust samples in liquids for long periods is the development of algae in the container and also the danger that the water could freeze in cold weather. Copper sulphate has been the material used to inhibit the growth of algae, but some workers favour a 5% initial concentration of 2-methoxyethanol which is not only an effective bactericide and algicide, but will also prevent frost damage. The 2-methoxyethanol may also be easily removed by evaporation (b.p. 124° C) prior to analysis.

Care should be taken in siting deposit gauges: there should be no object within 3 m of the gauge because such obstructions can affect the wind flow in the vicinity of the gauge. When gauges are sited in the vicinity of tall buildings, trees, etc., the angular height of such objects from the collecting surface should be less than 30°. Siting in the proximity of sources of dust should be avoided unless this is the primary reason for the tests. In order to avoid ground effects and to reduce interference from dust re-entrained into the air from the ground, collecting surfaces should be at a height of at least 1.5 m.

4.3.3   Short-term surveys

In order to obtain useful data on dust deposition it is necessary to carry out measurements for long periods using horizontal deposit gauges. Quite frequently, however, complaints are received of a dust nuisance which either comes from a particular direction or which occurs intermittently. If such determinations were made over periods of one month, the true effect of the nuisance would not be found because of variations in wind direction.

In the past, many methods have been suggested for carrying out short-term surveys using jam jars or Petri dishes because of the high cost of the standard gauge. Both of these containers are too small to be really effective and the Petri dish suffers from the added disadvantage that it is too shallow, so that the dust collected can easily be blown out again. Others have suggested the use of greased plates. In practice the greased plate is too effective a collector to give results comparable with actual values and in addition the grease can complicate subsequent examination of the sample. If organic solvents are used to dissolve away the grease, attack on the atmospheric deposits may occur, especially if they contain a high percentage of carbonaceous matter.

Cheap polythene containers, such as washing-up bowls, provide an extremely simple compromise. There may, however, be objections to these because of the wide collecting surface in relation to the depth (diameter 0.4 m; depth 0.15 m). There is, however, a large variety of polythene containers available and one of suitable dimensions can easily be found. If exposure is limited to periods of light winds the objection to the possibility of dust being blown out may not apply. Such polythene containers have been successfully tried in a number of field surveys for locating the source of a dust nuisance. The bowls are sited downwind of a suspected source for periods of one working day. Restricting the length of exposure to a period of steady wind direction increases the directional resolution of the sampling. If sufficient deposit is not collected in the space of one working day with bowls such as these, which have a diameter of approximately 0.4 m, it may be assumed that the dust problem is minimal. For such tests it is advisable to choose times when the meteorological conditions can be expected to be constant during exposure. If possible, sampling should not take place in wind speeds in excess of 15 miles h^{-1} (6.7 m s^{-1}), otherwise the dust may be blown out of the gauges. It is also advisable to avoid exposure during periods of rain because rain affects the deposition of solid material and can cause agglomeration, which makes subsequent identification difficult. After collection the dust should be carefully brushed out of the bowl and weighed prior to subsequent identification or analysis.

From such short period tests it is possible, knowing the time of exposure and the area of the collecting surface, to estimate the rate of deposition in mg m^{-2} day^{-1}, which is the standard unit used for quoting deposition results. Because of the short period of exposure, however, such values should be treated with caution. Although it may not be required in detail, it is useful to determine the approximate size range of the particles collected because, if the wind speed is known, approximate estimates can then be made of the likely distance over which the particles have travelled and this can assist the location of a source. This distance of travel will also depend on the height of emission, but once the material has been identified, all possible local sources can be checked to determine which one is likely to be the potential one after taking into account the different heights of emission and the likely distances over which the respective emissions will travel under the existing meteorological conditions.

Short-term deposition problems may be tackled in two ways. If the problem is simply to identify what appears to be a single source causing a nuisance, this can usually be solved by the use of one or two deposit bowls. If, on the other hand, the problem is to try to estimate the dust loading and major emission sources in a particular area, then a more extensive survey will be required. These will be discussed in turn.

4.3.3.1   *Single bowl surveys*

When complaints of dust nuisance are received in an area the first important point to be determined is the wind direction in which the nuisance occurs. When this is known one or two bowls can be sited in the affected area during these wind directions. The bowls should be sited at a height of about 1.5 m above the ground, and, where possible, there should not be an obstruction within 3 m. If there are tall buildings or trees in the neighbourhood, the angle between the line from the collecting surface of the bowl to such obstructions and the ground should be less than 30°.

The bowls should be placed in position early in the morning, when conditions are suitable, and exposed for a period of about 8 h. They should then be collected and the sample carefully brushed out and weighed. Subsequently, the sample may be analysed as required. If it is not possible to isolate the main offending source, it may be necessary to repeat the tests when the wind is blowing from other directions, changing the sites of the bowls such that they remain downwind of the suspected source.

4.3.3.2   *Larger surveys*

Where a major source of nuisance exists in an area, it will sometimes be useful to determine the effective dust loading in the area. For such work to be effective, it is important that dust from the source should be readily distinguishable microscopically or analytically, from general background dust in the area.

Probably the best way of describing this type of survey is to consider a typical project carried out in an industrial area where there was one major source of dust emission, namely a carbonizing plant. Ten sampling bowls were located in the area as shown in Fig. 4.15 which also gives the site numbers and the percentage amount of carbonized coal collected in each bowl. The method of operation was similar to that explained in the preceding section: when the meteorological conditions were suitable the bowls were set out in the morning and collected in the evening. After collection, the dust was weighed and, the surface area of the collecting bowls being known, the rate of deposition was estimated. The results of one typical test are shown in Table 4.5.

As would be expected the proportion of carbonized coal from the carbonizing plant decreased with distance from the plant, but increased again at sites 7 and 4, which were affected by a nearby coke storage area. Another interesting result which emerged from this one-day survey was the one from site 8, which showed a very high rate of deposition. Microscopic determination of the deposit indicated that the sample did not come from the carbonizing plant but from the nearby chimney of an oil fired installation, which was not only very low but was emitting large quantities of smoke: both these factors affected the deposit at this site. This one test illustrates the technique and shows what may be obtained from even one day's work. It must be realized, however, that to obtain reliable data more than one day's results would have to be obtained for them to be conclusive. In a case

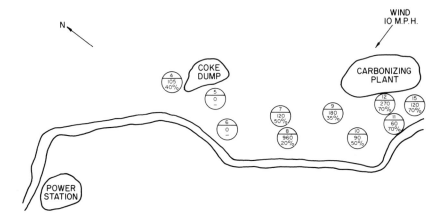

THE CIRCLES ARE CENTRED ON THE DEPOSIT BOWL SITES. THE NUMBERS ON THE CIRCLES REPRESENT –

TOP – IDENTIFICATION NUMBER OF THE SITE
MIDDLE – RATE OF DEPOSIT IN mg m^{-2} day^{-1}
BOTTOM – % COKE IN THE DEPOSIT

Fig. 4.15 Illustration of a short-term survey using a number of deposit bowls.

Table 4.5 Results of short-term survey

Site no.	Rate of deposition (mg m^{-2} day^{-1})	Percentage of carbonized coal
15	120	70
12	270	70
11	60	70
10	90	50
9	180	35
8	960	20[a]
7	120	50
6	0	—
4	105	40

[a] At this site 80% of the deposit consisted of cenospheres from an oil fired chimney.

such as this it is also advisable to obtain results from other directions, for example when the wind was blowing from the direction of the power station.

This illustration also serves to demonstrate that it is not advisable to make assumptions of expected pollution levels in an area, without being supported by actual measurements. This was demonstrated by the exceptionally heavy deposits found at site 8 in the last example.

4.3.4 British Standard directional deposit gauge

As explained in an earlier section the early British Standard deposit gauge has limitations, particularly when it is required to be used to locate a potential source of nuisance in an area. In order to improve the directional resolution for the determination of deposited matter for work in the vicinity of power stations, the Central Electricity Research Laboratory designed a directional deposit gauge [29]. A diagram of this gauge is shown in Fig. 4.16. The unit consists of four plastic tubes which are mounted as shown in the illustration with the vertical slots facing outwards at right angles to its neighbour. Dust trapped in each of the tubes is collected in a removable container at the base. After collection, a suspension of the dust is placed in a water-filled glass cell, and a measure of the dust loading (the

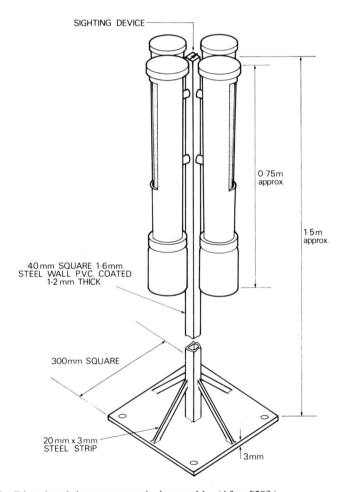

Fig. 4.16 Directional dust gauge typical assembly. (After [30].)

10-day percentage obscuration) estimated by the amount of obscuration of a beam of light passing through the cell. The unit is fully described in BS 1747: Part 5 [30]. The method advised in the British Standard which was designed to measure dinginess has limitations when applied to the identification of the source of a nuisance in a complex industrial area. Basically the BS method of analysing the deposit could only be applied where there is one source in an area; for example, a power station in open country. In such a case, a very large percentage of the total dust will come from the one source and that dust will have approximately reproducible composition and density. If, however, such a directional deposit gauge were installed in an industrial area, the density of the dusts collected can vary considerably (this will be illustrated in a later section on density separation) and this could affect the results obtained by the method of light obscuration mentioned earlier, because some of the dusts would tend to settle out much more rapidly than the less dense dusts. Secondly, if there is more than one source in the area, a simple obscuration test will not give useful data on the contribution from the different sources and some form of more sophisticated analysis will have to be carried out, such as microscopic examination.

When discussing the deposit gauges with horizontal collecting surfaces in an earlier section, it was mentioned that only very large particles fall out at angles approaching the vertical. Most airborne dust in the size range which will be found away from the immediate vicinity of emission sources will be falling at a fairly shallow angle to the horizontal and it could be argued that in such cases the vertical collecting surface of the directional deposit gauge should prove a more efficient collector than the horizontal collecting surfaces of the earlier gauges. However, the reverse should be the case with heavier particles.

When the directional deposit gauge was first tried in the field some comparisons were carried out between the two instruments by siting them close together to monitor dust deposition in certain types of areas:

(a) In an area containing a cement works
(b) In a typical industrial area
(c) Near a granite quarry where frequent blasting operations were carried out.

An approximate particle sizing of the deposits collected showed the following ranges:

(a) Cement works area   50– 100 μm
(b) Industrial area   200– 500 μm
(c) Granite quarry   100–1000 μm.

In these tests, in the three areas, the ratios of the weights collected in the directional deposit gauge to the weights of deposits collected in the older type of horizontal gauge were as in Table 4.6.

From the results in Table 4.6 there would appear to be some positive confirmation of the earlier statement that the smaller dust size fractions are collected preferentially by the directional deposit gauge.

When the directional deposit gauge is installed to monitor emissions from one major source the orifice should be sited in such a way that it is pointing directly to the source. However, in more congested areas with multiple sources, it is not

Table 4.6

Area	Ratio directional gauge to BS horizontal gauge	No. of monthly readings averaged
(a) Cement works	2.49:1	25
(b) Industrial area	0.65:1	12
(c) Granite quarry	0.70:1	5

possible to install the gauge ideally and the position arises where dust from particular sources is arriving at the site from a direction between the openings of two adjoining collectors in the gauge. In such cases the dust will collect in the two collectors and a large part of the directional resolution is lost. In practice the position appears to be more complex and dust tends to appear in collectors sited at angles pointing in the opposite direction. This is probably caused by turbulence generated by the configuration of the gauges. No evidence is available to date on this aspect but some preliminary tracer work in a wind tunnel appears to confirm this theory. Methods of analysing dusts collected in directional deposit gauges are discussed in a later section.

The points raised in this section have attempted to show that the pattern of dust deposition, particularly in a complex industrial area, can be very involved and it is worthwhile taking great care, not only in the choice of the correct instrument, but also in siting, and the analytical techniques which are used, to obtain the best results in this field of air pollution research. It is surprising just how little effective work has been done recently to improve our ability to understand dust deposition problems.

4.4 Physical techniques for classification of particulates

4.4.1 Density gradient separation

One of the simplest techniques in the analysis of dust samples is to make use of differences in density which occur throughout the whole range of atmospheric particles. The method is applicable to insoluble dusts above 10 μm in diameter and with densities up to 5 g cm^{-3}. In this technique dust samples are inserted into a tube containing two density gradient liquids mixed in such a way that the density of the liquid in the tube decreases almost linearly with height. When the tubes are centrifuged the dust separates out into its respective density bands and these bands may, if desired, be physically removed one by one and analysed separately.

4.4.1.1 Density gradient liquids

There are a number of suitable liquids which can be used for this work and some of them are shown in Table 4.7. In addition, Clerici solutions (50:50 thallium malonate and thallium formate in water) extend the range to the majority of materials likely to be encountered.

Table 4.7 Properties of density gradient liquids

Liquid	Density (g ml^{-1} at 20° C)	Vapour pressure (mm Hg at 20° C)
Bromoform	2.89	5
Tetrabromoethane	2.96	< 1
Di-iodomethane	3.32	1.25
Bromobenzene	1.52	5.5
Di-*n*-butyl phthalate	1.05	—
Acetone	0.79	285

Dibutyl phthalate has been found very suitable as a diluent as it has a low vapour pressure similar to that of tetrabromoethane and lower than that of di-iodomethane so that, when it is used as a diluent the resulting composite liquid is stable with respect to density over relatively long periods. When accurate control of the density of a composite liquid is not essential, or when an inherently stable liquid is not required, acetone may be conveniently substituted as a diluent.

The three heavy liquids are sensitive to light and heat and the breakdown products produce a red to dark red discoloration. The liquids should therefore be stored in the dark, in darkened containers.

4.4.1.2 *Recovery and cleaning of liquids*

When dibutyl phthalate is used as a diluent there is no simple method for the recovery of the heavy liquids, most of which are expensive. However, di-iodomethane may be recovered from dibutyl phthalate by placing the liquids in a suitable separating funnel and rapidly freezing the liquid using dry ice, or preferably liquid nitrogen ($-196°$ C) if available, or alternatively a mixture of ice, with brine, acetone or alcohol. After freezing, the solid is allowed to melt at room temperature; on melting the resulting liquid begins to separate out into its two component liquids. Repeated freezing and thawing (say 3 to 5 times) eventually leads to the separation of about 90% of the original di-iodomethane which forms a clear-cut layer in the bottom of the funnel. This process has the advantage of removing any free iodine in the di-iodomethane into the phthalate so that the pure di-iodomethane is recovered as a light straw coloured liquid. If acetone is used, this may be completely removed by streaming water through the mixture. The purified heavy liquid may then be separated off from the excess water and shaken with calcium chloride to remove the final traces of water.

Another way to clean the liquids is to use small quantities of Fuller's earth. The earth should be vigorously shaken or stirred with the liquid for a few minutes followed by filtration. This technique is simple to apply and in addition to decolorizing the liquid also removes traces of water and therefore may be used as an alternative method of drying a washed purified liquid.

4.4.1.3 *Preparation of the gradient*

One of the earlier methods of preparing a density gradient, described by Peters [31], was to take a number of glass tubes (12 by 1/4 inch) with one end sealed and, using a mixture of bromoform and bromobenzene, to make up the columns as follows: the bottom inch of the tube was filled with 100% bromoform, the next inch with 90% bromoform/10% bromobenzene, the next inch with an 80/20 mixture and so on until the last inch contained 100% bromobenzene. The columns were allowed to stand for at least 6 h to allow diffusion to occur between the density boundaries. The dust samples were then inserted and allowed to stand for about 24 h to allow the dust to settle out to its correct density level.

For air pollution work the long delay in preparation was considered too long for practical purposes, and there was some indication that, while there might be some mixing near the density boundaries, the gradient tended to be more stepwise than linear. In such a case where the dusts contained flat, or plate-like particles, these could settle out on a density boundary which was not representative of their true densities. The third, and most serious criticism of this technique was that, where dust contained carbonaceous matter, most of the liquids suitable for density gradient work would attack the deposits, dissolving out some of the tarry matter and ruining the identification.

One of the most simple methods of preparing a density gradient, although control of the actual gradient is less certain, is to take a test tube half-filled with the heavier liquid and then fill the upper half carefully with the lighter liquid. If the tube is held between the thumb and forefinger and is rhythmically shaken in a pendulum fashion the formation of the gradient can be followed by the behaviour of the striae in the liquid. This is a useful technique for rapid preliminary examination.

A technique used for routine work is that described by Muller and Burton [32] in which the entire operation of making up the gradient and centrifuging the sample could be done in 30 min, thus allowing much less time for the attack of the organic materials in the sample.

The unit used for the preparation of the columns consists of a small glass filter funnel with the stem inserted into a small separating funnel through an air-tight bung. This is the mixer unit and it is completed by the addition of an intermittently operated electromagnetic stirrer. This unit is energized twice per second to actuate a number of glass coated pins contained in the separating funnel. When a 50 ml centrifuge tube is being used the mixer is loaded by placing 25 ml of the denser liquid in the separating funnel and 25 ml of the lighter liquid in the filter funnel. The lower separating funnel stopcock is adjusted to give a rate of discharge of liquid such that the centrifuge tube is filled in about 5 to 10 min. At the same time the intermittent magnet is switched on. As the denser liquid begins to drip from the separating funnel it is replaced by an equal volume of the less dense liquid from the upper funnel. This small amount of less dense liquid will be mixed with the dense liquid by the agitation of the pins, thereby slightly reducing the density. In this way the density of the emerging liquid continuously decreases and, when the centrifuge tube is filled, the resultant density will be found to decrease almost linearly. It is not necessary to start with pure liquids. With pure di-iodomethane and acetone the density difference will be from 3.32 to 0.79 g ml^{-1}. However, both liquids can be diluted before preparing the gradient and thus expand any intermediate section of the gradient. For example, a gradient tube varying, say, from a density of 2.50 to 2.00 g ml^{-1} may be prepared by diluting the two liquids suitably before starting.

After the columns have been prepared, weighed samples of dust are added to the liquid. Prior to centrifuging, the columns are made up in pairs with the weights carefully adjusted and balanced beforehand to avoid damage to the centrifuge. For most work centrifuging at 3000 rev min^{-1} for 10–15 min is adequate to separate the dust into its respective density bands.

In the case of naturally occurring mineral ores it will be found that the size of the particles can affect the density separation. Figure 4.17, illustrative of work at Warren Spring Laboratory, demonstrates this point. The tube on the left contains mineral dust in the size fraction 10 to 14 mesh, the second 14 to 25; the third 25 to 32; the fourth 52 to 72 and the last 72 to 100. The mineral ore is separated out into its component fractions only when it is ground to around 100 mesh. When such liberation sizes are obtained the resolution of the density technique is very satisfactory. However, in air pollution work, as has already been mentioned, a great many atmospheric particles are hollow and this can complicate density resolution, as in the case of silica. Naturally occurring silica will settle out at a particular density level, but silica from a particular combustion process may be in the form of hollow spheres and these will settle out at other levels. Far from being a disadvantage, this can be very useful. A similar separation occurs with carbonaceous matter; if there is a coal-fired source in the vicinity, samples of dust will be found containing particles varying from almost completely unburnt coal through all the stages of carbonization to coke, resulting in a very broad band of carbonaceous matter.

Fig. 4.17 The effect of particle mesh size upon the resolution of density gradient fractionation.

When further detailed analysis is required, it is possible to centrifuge the sample in a co-axial centrifuge tube, shown in Fig. 4.18. When the sample has been centrifuged the splitter unit is placed over the ground glass joint in the top of the co-axial centrifuge tube. If pure di-iodomethane or mercury is carefully introduced into the side arm of the co-axial tube it will gradually force the entire density column up into the splitter unit. As each dust band passes through the two flat glass discs the splitter unit should be rotated through 180°, at which point the dust band suspended in the upper part of the splitter unit will drip through the stopcock into a collecting jar. The density liquid may then be removed by washing in acetone, and the dust simply evaporated to dryness for subsequent analysis. Each of the density bands may be collected separately if the process is repeated.

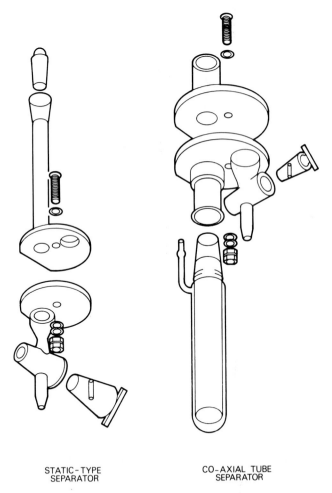

STATIC-TYPE
SEPARATOR

CO-AXIAL TUBE
SEPARATOR

Fig. 4.18   Co-axial centrifuge tube and splitter unit.

Some examples of this technique applied to air pollution work are now shown. Figure 4.19, illustrative of work at Warren Spring Laboratory, shows dust from two types of power stations; on the left, dust from an old-fashioned coal fired power station, and on the right, the deposit from a more modern pulverized coal power station. The broad band of carbonaceous matter may be seen in the left-hand sample.

Figure 4.20, also from work at Warren Spring Laboratory, shows three of the four samples taken from a directional deposit gauge exposed in an industrial area. In the left-hand tube which faced open country, the only visible band appears at a density of about 2.6 g cm^{-3} and this is composed of wind-blown silica. The centre tube shows a much stronger band at this density, together with traces of carbonaceous deposits just about half-way up the tube. The main

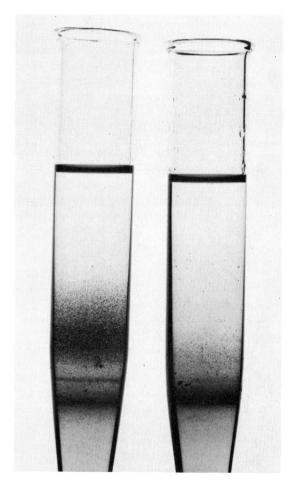

Fig. 4.19   Density gradient fractionation of dusts from a directional deposit gauge.

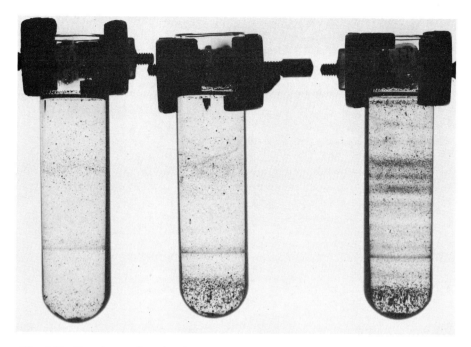

Fig. 4.20   Density gradient fractionation of dusts from a directional deposit gauge.

deposit near the base of the tube is silicon carbide (density 3.2 g cm^{-3}) and the deposit at the bottom of the tube, alumina (density 4.0 g cm^{-3}). Both of these deposits come from a nearby factory making grinding wheels. This sector of the directional deposit gauge was on the boundary of the factory. The right-hand tube, containing the sample taken directly in line with the factory, shows much heavier deposits of the three main types of dust shown in the centre tube. In addition there are traces of cenospheres from an oil fired chimney near the top of the tube. These spheres, although of carbonaceous matter, are hollow and therefore tend to float if undamaged.

4.4.2   Dispersion staining

A large percentage of atmospheric particulates are carbonaceous and, therefore, opaque; however there is a sufficient abundance of transparent particulates in the atmosphere to justify the use of this technique [33, 34] which is an optical method of imparting a colour to transparent substances to simplify identification. It is of considerable value, for example, in determining the percentage of quartz, or other dangerous dusts, in air samples. The technique is based on the difference in the refractive index of transparent solid particles and the liquid medium in which they are immersed. This difference in refractive index causes refraction to take

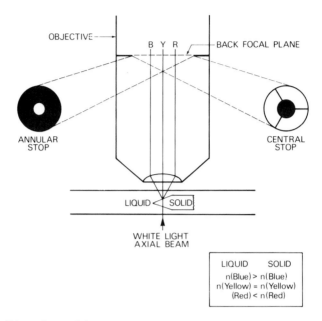

Fig. 4.21 Dispersion staining.

place at the crystal boundaries; the greater the difference, the greater the refraction. A diagrammatic sketch, Fig. 4.21, shows the principle of the technique. In this sketch the refracted blue and red radiation are shown on the left and right respectively. After being focused by the objective lens the rays pass to the back focal plane where either an annular or central stop is placed as shown in the illustration. Each of the stops gives coloured particle boundaries. The annular stop shows a colour containing those wavelengths near that at which both particle and liquid show a match in refractive index. The central stop shows colours complementary to those shown by the annular stop, i.e. light which has been highly refracted by the particle in that particular liquid medium.

With this technique, which uses white light, good centralized axial illumination is essential. The condenser lens should be accurately centred with respect to the objective and stopped down to give optimum colour development.

In the work reported in the literature [33, 34] a large number of atmospheric particles were examined and dispersion curves drawn for them. A typical series of such curves, extracted from these papers, are shown in Fig. 4.22. If this figure is used to explain the technique, let it be assumed that the unknown material is lead nitrate. If crystals of this compound are immersed in a dispersion liquid of refractive index 1.777, i.e. where the two materials have a matching refractive index, the material will appear to be greenish in colour, and this will be found to have a wavelength of 486 nm. If, on the other hand, a liquid with a refractive index of 1.783 is used, the colour of the crystal will appear red, at 620 nm.

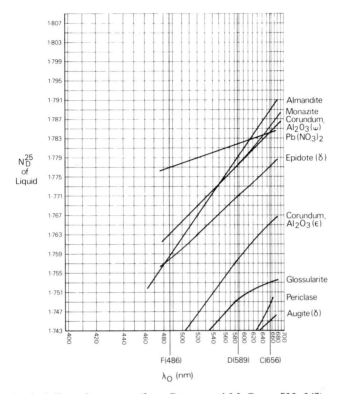

Fig. 4.22   Typical dispersion curves (from Brown and McCrone [33, 34]).

There are a number of uses for this technique in the air pollution field; for example, the determination of the percentage of quartz in a sample. In such an application a dust sample containing quartz dust is sized microscopically using standard illumination. If the preparation is then examined using a dispersion staining objective, the quartz particles may be made to appear coloured and it is then a matter of carrying out a particle size distribution analysis of the coloured particles to obtain the necessary details of the quartz content.

The technique has also been used in conjunction with samples from the directional deposit gauge. For example, in the illustration shown earlier in the vicinity of a factory making grinding wheels the contribution from the factory consisted of mainly alumina and silicon carbide, the amounts of which can be determined by dispersion staining.

4.4.3   Microscopic techniques

The normal technique for the identification of atmospheric dusts is by their characteristic shape, structure and colour. This is a technique which can be learned only by experience, and some necessary background information is given in this section. Although the composition of dusts will change from area to area, it will usually be found that in urban areas a large percentage of dusts come from combustion sources. There are, of course, innumerable industrial processes emitting specific dusts, but it will initially be found preferable to concentrate on the identification of combustion products.

The composition of atmospheric dusts will be found to vary considerably at any one site under different wind directions. Because of such changes it is, therefore, important that satisfactory sampling be carried out and not to base dust identifications on dust swept up from flat surfaces such as window sills.

If samples are collected from chimney stacks then the proper sampling technique detailed in the relevant British Standard should be followed [35]. However, if it is desired to perform a preliminary examination of dusts from a particular source and provided care is used, this may be carried out using grits collected in chimney arresters. Such samples will tend to contain larger particles than will be found in the atmosphere. It is possible to build a small but effective dust collector, but it must be emphasized that this simple unit is for the purpose of collecting a preliminary sample and it should not be assumed that the dust collected is representative. Briefly the method is to draw a sample of dust from a chimney into a small deposition chamber by means of a suction pump. A filter paper should be inserted in the line to the pump to protect it from contamination by the dust. If the tube from the chimney is approximately 1/4 inch diameter the deposition chamber should have a diameter of about 6 inch. The chamber should have a removable base so that a Petri dish may be inserted. As air is drawn from the chimney into the chamber the reduction in gas velocity will cause particles to be deposited out and they will be collected in the Petri dish. The length of sampling will depend on the dust loading in the chimney and the size range of the particles.

After collection the dust samples may be removed from the Petri dish and mounted on microscope slides for examination or to be used as reference samples. Initially it may be found difficult to pick up the individual particles to transfer them to slides but this operation can be simplified by the use of a very sharply pointed needle which can be prepared from a short length of tungsten wire heated to red heat and then plunged into a dish containing sodium nitrite. During the subsequent reaction the tungsten will be oxidized and it is possible to burn the metal away to a very fine point. If this point is drawn across the fingers after it is cooled, a fine film of skin grease will be deposited on the needle which will be adequate to cause adhesion of individual particles.

4.4.3.1   *Mounting samples*

In the past it was common practice to mount samples in liquids such as Canada Balsam. Unfortunately, with atmospheric particles containing carbonaceous matter, it will be found that the mounting liquid will attack the particles and if the mounted sample is retained for any length of time the preparation will turn dark brown. However, a simple but effective method can be used to mount samples for reference purposes, but the method is not suitable for more sophisticated optical work such as dispersion staining. The method involves placing a drop of Durafix diluted in amyl acetate (30% Durafix to 70% amyl acetate by volume) on a microscope slide in the centre of the cover-slip ring. The droplet will spread out inside the ring and when the liquid becomes 'tacky', the dust sample should be carefully sprinkled over the slide. With experience it is possible to gauge the correct time to add the dust. If the operation is successful, the particulates will be held on top of a very thin film of Durafix after the liquid has evaporated. If the particles are placed on the slide too early they will become embedded in the liquid and the dried film of Durafix over the particles will make subsequent microscopic examination almost impossible.

If more sophisticated optical work is planned, some of the more modern thermoplastic mounting media may be used. A number of different grades are available, but for air pollution work it is suggested that a grade be used which is solid at room temperatures. For mounting, the medium is heated until it melts. After mounting, the preparation is allowed to cool and solidify. These thermoplastic media have a light yellowish tint but this does not affect identification.

4.4.3.2 *Identification of dusts and reference library*

There are a number of aids for identifying atmospheric particles such as two atlases of atmospheric dusts; one by the CEGB [36] and one by McCrone [37]. However, many beginners will find it difficult to identify these photomicrographs with actual dust samples under the microscope. Difficulties arise from the limited depth of focus of the microscope which tends to complicate identification for beginners, and from the difficulty of accurate colour reproduction. These problems may be circumvented by making up a reference library of dusts from known sources for comparison with samples for identification. The work involved in building up this library gives the beginner experience in recognizing the different types of dust and also the component parts of dust samples from particular sources. Initially the library should concentrate on the more common dusts in the area, and continue with less common samples as experience is gained.

4.4.3.3 *Description of dusts from different combustion and industrial sources*

Mineral matter in British coals can be divided into three main classes, i.e. shales and clays, sulphide minerals and carbonate minerals. In most coals the mineral matter will fuse at temperatures between 1200 and 1600° C under oxidizing conditions, but may fuse at lower temperatures under reducing conditions. Generally speaking, the flame or fuel-bed temperatures of small appliances will be below the fusion temperature. In larger installations, where the ash is heated beyond its fusion temperature while 'airborne', as in a pulverized fuel plant, it will tend to become spherical, whereas if it reaches the same temperature when resting on the fuel bed, it is unlikely to become spherical but will tend rather to be rounded. It is, therefore, possible to determine approximate fuel bed conditions from the physical characteristics of the ash particles.

Domestic fires

Typical fly ash from domestic sources consists of small unfused particles, mostly colourless or white but with some pale yellow material. In addition, there may be small black and red iron oxide particles. Generally the dust emitted from domestic sources will not be distinguishable from the constituents in typical dust samples because of the small quantity involved compared with other sources. Owing to the low flue gas flow in domestic chimneys, these particles emitted rarely exceed 10 μm.

Shell type boilers

This type of small, hand fired boiler is still extensively used in the UK in industry for steam raising. Although the fuel bed temperatures are in the range at which many coal ashes fuse, only a small percentage of the ash becomes completely spherical; the bulk of the particles are merely rounded. Coal or coked coal particles are also to be seen in deposits from such boilers. Deposits from this type of boiler are usually in the range 10–100 μm.

Stoker fired boilers

The carbonaceous matter from such boilers consists of irregularly shaped coke or partially coked coal particles. Most of the ash consists of spherical particles with a dull olive green or brown coloration; these are usually referred to as 'smoky' spheres. These 'smoky' spheres may be regarded as clearly indicative of stoker fired boilers. The coloration of the spheres is not strong, and the percentage of spheres in dust samples from this source tends to vary. Particulates from small economic boilers contain mainly semi-fused mineral matter, coke particles and only a few spheres. The dust is similar in appearance to that from the large hand fired boilers. On the other hand, dusts from the larger boilers contain a large percentage of spheres which are mainly colourless, pale yellow or orange, together with the 'smoky' spheres. Puffy particles of partially carbonized coal are also found and these are generally larger than the ash particles with which they are associated. When ash from this type of boiler is compared with ash from a pulverized fuel boiler, it will be found to contain more coloured material.

Pulverized fuel boilers

Fuel burned in these boilers reaches maximum temperatures of 1500–1600° C. The fuel is burned in suspension, and, because of this, most of the mineral matter emitted emerges as fully spherical, colourless spheres. In addition, black magnetite spheres may be found. The magnetite is formed by the decomposition of the iron sulphide minerals in the coal. Upon heating, these minerals are oxidized, releasing SO_2 and leaving a black oxide of iron (Fe_3O_4) residue. As this occurs in suspension, the magnetite fuses into spheres. In pulverized fuel boilers, coloured spheres are rare and there is also usually a small amount of combustible matter, usually angular coke fragments or puffy coke coal.

Oil fired boilers

Since fuel oils have a low mineral content, the fly ash content from industrial boilers should contain little ash. However, the small amount of mineral matter there is, is usually more highly coloured than that of coal and consists of green, yellow, orange, red and black particles about 1–5 μm in diameter. The green coloration of the ash, which is specific to oil firing, is due to nickel and vanadium compounds. This is not the best way of identifying a particular dust as coming from an oil fired source. The main deposit from such boilers consists of cenospheres with little ash and it is this combination which will point to the oil fired boiler as the culprit. The cenospheres vary in appearance, between a fine lace-like structure, similar to those from a strongly swelling coal, to a black matt surface without any apparent structure. In the case of oil fired boilers which are badly run, samples of shiny, dark green spheres around 50 μm in diameter may also be found. These are almost completely unburnt oil.

The carbonization industry

Owing to conversion to natural gas, dust from this industry is not very common today. Coke particles produced by carbonization of a swelling coal are harder and denser than those found in normal fly ash and, on occasions, it is only by careful examination that the porous nature of the coke can be seen.

Coke ovens

Dust emissions from this process comprise two different types: particles of unburnt coal from the oven charging process, and swollen, rounded particles of partially carbonized coal. During discharging of the ovens and the subsequent quenching of the coke, angular fragments of coke and small ash particles appear. Dust emission during quenching is small.

Horizontal retorts

This method of coke production is now obsolete. Dust emission during the charging process consists of particles of puffy, swollen coal. A typical dust from a horizontal retort would contain small particles of coal at various stages of carbonization. Angular fragments of coke and occasional coke cenospheres can also be found.

Intermittent vertical retorts

This operation is similar to that for coke ovens and the dust produced is similar in appearance to that from coke ovens and horizontal retorts.

Continuous vertical retorts

This is a less dusty operation than other methods of carbonizing coal, because the coal passes through a seal before reaching the hot zone of the retort. The only dust produced by this technique is that arising from the coal and coke handling operations.

The cement industry

Cement is manufactured by heating limestone or chalk with a small percentage of clay. In the wet process the mixture is fed as a slurry into the top of a long inclined rotary kiln. During its passage down the kiln the mixture is heated to 1450° C. The sintered product is cooled and ground with a small amount of gypsum. Most of the dust produced in a cement works is removed by electrostatic precipitators. The dust consists of fine, colourless or pale-yellow particles with a small proportion of typical pulverized fuel fly ash. The dust tends to be hygroscopic and it helps to heat it to about 110° C for 1 h prior to examination. If this is inadequate and further segregation is required, this may be achieved by preparation of a slurry in acetone. Water should not be used for this purpose, because some of the dust is water soluble.

The ceramic industry

Raw materials are heated in kilns which can be fired by a number of methods. The general features of the emissions depend on the kiln temperature. In small bottle kilns, where the temperature of the fuel bed is around 1150° C, the major part of the dust consists of amorphous soot with small, rounded or semi-fused ash particles. Kilns fitted with mechanical stokers have higher fuel-bed temperatures and the ash emission contains semi-fused particles, but no spherical particles. However, the amount of dust emitted from large kilns is small, unless smoke is being emitted. Even so, the dust will not be emitted in quantities large enough for identification.

The other type of emission from this industry is caused by grinding and handling of raw materials. This emission, however, tends to be very localized and should not cause any nuisance outside the works area, except in high winds.

Iron and steel industry

Blast furnaces

These are used to produce basic iron, and the charge, consisting of iron ore, metallurgical coke and limestone is fed into the top of the furnace where the temperature reaches about 1500° C in the hottest zone. Most of the particulate emissions from this process are caused by 'hanging', i.e. where there is a sudden drop of part of the furnace charge, which causes the rapid ejection of large quantities of gas heavily laden with dust. The dust emitted has a size range up to 30 μm and consists of black, sharp-edged matt-surfaced iron particles, with similar amounts of coke, as well as pale-yellow and colourless limestone and slag. The iron particles, being denser than either coke or slag particles, settle out more quickly than the other constituents.

Sintering plant

This plant is usually associated with blast furnaces, and makes some use of dust collected by dedusting the blast furnace, which can contain up to 25% iron oxide. The dust is mixed with fine iron ore and coke breeze and burned to produce a sintered mass. The dust emission tends to be similar to that produced by blast furnaces.

Open hearth furnaces

This process converts pig iron or scrap iron into steel. The furnaces may be fired either by producer gas or heavy fuel oil, and the temperatures can reach 1800° C. The dust emitted consists mainly of iron oxides and minerals such as limestone which is used as a flux to form slag. The dust is yellowish brown in colour and is composed of black magnetite spheres and yellow, orange and colourless material from the flux and slag. The size range of the particles from this process is below 5 μm, and it is unlikely that dust from this process will settle readily.

Electric arc furnaces

These furnaces operate at high temperatures to produce alloy steels, and they create a fume of iron oxide with particles up to 5 μm. The bulk of the emission, however, is in the form of ferric oxide fume with a size range below 1 μm, which does not have any distinguishable features.

Deseaming mills

In the deseaming mills the surface layer of scale is removed from the ingots by burning off with oxyacetylene torches. The emissions consist of black, shiny spheres of magnetite and also yellow fume. The magnetite spheres are usually between 10 and 20 μm but they can be as large as 100 μm in diameter. Owing to their high density, the larger spheres will not be expected to travel far, and only the smaller spheres are likely to be found at any distance from the mills.

Cupolae

These installations are used for melting cast iron or convertor iron, and the emission varies from a coarse grit of 100 μm to a fine fume of 0.5 μm. The coarse grit consists of angular, matt-surfaced, black iron fragments with slightly reddened, oxidized surfaces in places. Smaller quantities of yellow and orange translucent 'slag' material and coke are also found. Despite the high temperatures, the particles show no sign of rounding or sphere formation.

The sections dealt with here represent some of the major sources of dust which are likely to be encountered. However, in some areas, these dusts may be in the minority. In such a case it will be necessary to compile a list to cover such materials.

4.4.3.4   Dust identification table

An attempt has been made to summarize the different types of dust by arranging them as shown in Fig. 4.23. This figure has been arranged so that the dust may be identified by a process of successive elimination. The presence or absence of particular physical characteristics in dusts lead by stages to the final conclusion. Although dusts will be found which are not dealt with here, the illustration will assist in the identification of a large percentage of the dusts encountered.

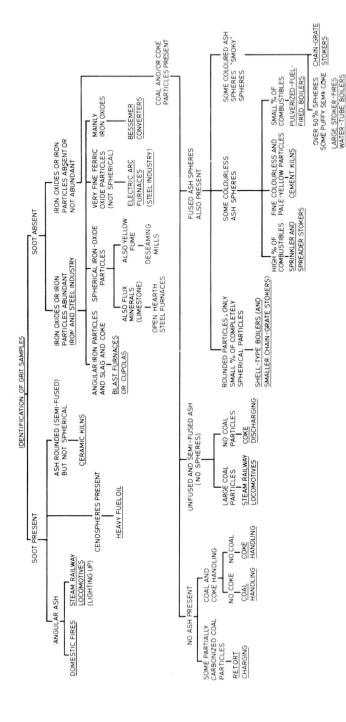

Fig. 4.23 Table for identification of dust samples.

4.4.4   Determination of asbestos

Recently, great concern has arisen over the hazards of exposure to asbestos dust. Three conditions, all of which may take many years to develop, have been identified in workers occupationally exposed. These are fibrosis of the lung, known as asbestosis; cancer of the lung consequent upon fibrosis and mesothelioma, a rare malignant tumour of the pleura or peritoneum [38]. Two forms of asbestos have been used predominantly: crocidolite (blue asbestos) and chrysotile (white asbestos), but the former is not now in use due to the greater hazard associated with it, but will still exist in older installations.

Sampling asbestos dust is normally performed using membrane filters as these have a high collection efficiency and are suitable for the subsequent analytical procedures. Because of the chemical complexity of asbestos, no simple analysis is available, and optical techniques have been recommended for determination of asbestos in the occupational environment [39]. Detection is based upon the characteristic fibrous nature of asbestos particles which may be readily recognized and counted. Frequently, asbestos dust concentrations are expressed in terms of fibres per unit volume of air, although more conventional units of mass per unit volume are also in use. After collection on membrane filters, the fibres may be counted by light [39], or electron microscopy [40, 41], following dissolution or low temperature ashing of the filter material for the latter technique.

In the industrial atmosphere, such counting techniques are applicable, as is X-ray diffraction. The concentrations of asbestos encountered in ambient urban atmospheres are far lower, however, and the concentrations of other minerals are frequently higher. Samples collected on membrane filters have been successfully analysed by electron microscopy [40, 41]. An X-ray diffraction technique for urban samples, after collection by electrostatic precipitation, proved too insensitive for the detection of chrysotile asbestos, however, even in the vicinity of an asbestos factory [42]. Another method used successfully in analysis of ambient air utilizes the i.r. absorption of chrysotile for quantification [43].

4.4.4.1   *Membrane filter method* [39]

This method is fully discussed in a leaflet issued by the Asbestosis Research Council and is designed specifically for measuring concentrations encountered occupationally in factories, workshops and constructional sites.

The method is restricted to a count of fibres of a length greater than 5 μm and having a length/breadth ratio of at least 3:1. The sample is collected on a membrane filter (0.8 μm pore size) fixed in an open filter holder. The length of sampling will be determined by the expected concentrations and some recommendations are given in Table 4.8.

When the sample has been taken the dust should be fixed immediately. There are several ways of doing this and one recommended method is to drip a solution of poly(methyl methacrylate) (Perspex) in chloroform on to the filter, while clean air is being drawn through it. On return from field sampling sufficient triacetin

Table 4.8

Anticipated asbestos dust concentration (fibres ml^{-1})	Sample volume (ml)
<2	10 000–20 000
2– 4	5 000–10 000
4–12	2 000– 5 000
>12	Pro rata

(glycerol triacetate) should be placed on a microscope slide to give a circle the same diameter as the filter and to serve as a base for it. The filter should be carefully removed from the sampling head using forceps and placed dust-side uppermost on top of the triacetin. After about 3 min, place a clean cover-slip of the same diameter over the filter and press gently to remove air bubbles which may be trapped underneath.

The dust should be examined under transmitted light with phase contrast at a magnification of × 500. With an even distribution of fibres, counting can proceed by counting the fibres in random fields; the number of fields needed for a total count of 200 fibres should be noted. If the deposit is sparse or unevenly distributed then 100 fields are observed. It is suggested that a multi-stage method of choosing fields be used, i.e. 10 filter grids are selected scattered over the sample area and within each grid 10 random fields are counted, giving 100 fields in all.

The formula for estimating the fibre counts (i.e. fibres longer than 5 μm) is:

$$\text{Concentration} = \frac{D^2}{d^2} \times \frac{N}{n} \times \frac{1}{V} \text{ fibres ml}^{-1}$$

where: D=diameter of entire dust deposit; d=diameter of field of view; N=number of fibres counted; n=number of fields examined; V=volume of sample (ml).

4.4.4.2 Infrared technique for ambient atmospheres

Concentrations of asbestos likely to be found in ambient air are very much lower than may be encountered in industry as already discussed. This rules out a counting technique because of the large number of fields which have to be examined, even to find one fibre. In addition, sampling carried out in the past has shown most asbestos fibres in ambient air to be in the region of 0.3 μm in length, which eliminates optical microscopy.

A method has been reported which makes use of the strong infrared absorption band at 2.72 μm possessed by chrysotile, the most commonly used of the asbestos minerals [43]. Atmospheric samples are collected on membrane filters and ashed at 450° C for about 1 h, during which process the material of the filter and any carbonaceous matter present in the sample is oxidized or volatilized. Under these conditions decomposition of the chrysotile is barely detectable, and no correction to the calibration is necessary. The residue is incorporated in a potassium bromide disc and the amount of chrysotile present determined from the i.r. spectrum.

One possible defect of the method is that serpentine minerals also exhibit strong absorption bands at the same wavelength and it may be necessary to detect whether these minerals are present in the samples to avoid interference.

4.4.4.3   *Transmission electron microscope methods for ambient atmospheres*

In non-occupational environments, very low levels of asbestos fibres are found. In remote areas there are around 40–100 electron microscope-visible fibres per cubic metre and in urban areas around 0–2400 fibres [44]. At these low levels, the transmission electron microscope (TEM) provides the best means of analysis. However, many of the mineral fibres found in ambient air are not asbestos [45], and positive confirmation of the identity of the mineral by selected area electron diffraction, or energy dispersive X-ray analysis, both of which are available facilities on a TEM, is necessary.

Burdett and Rood [46] report a method for transference of collected fibres from a membrane filter directly on to an evaporated carbon film suitable for TEM analysis. Steen and co-workers [47] have presented a detailed appraisal of TEM methods and provide lengthy recommendations of sampling and analytical procedures applicable to asbestos in ambient air. Preliminary investigations are recommended prior to selection of the most appropriate [47].

4.4.5   Determination of particle size distribution

In any estimation of the particle size distribution of a sample of atmospheric particulates it is advisable to take into account some measure of density as well as of size, particularly if the sample is taken from a general urban or industrial area where a wide range of densities can be expected to be found. Depending on the overall distribution of sizes it may be necessary to carry out an initial separation of the particles in an instrument such as the cascade impactor to avoid having large and small particles on the one slide, where the larger particles could obscure the small particles.

It is not possible to lay down hard and fast rules on sampling for a particle size distribution because the length of sampling will depend not only on the concentration of particulates in the area. If a cascade impactor is used it may be necessary to take the sample in two parts: (a) a shorter sampling period for the smaller size fraction and (b) a much longer period for the upper fraction of the dust, where the number of particles is likely to be less in comparison and thus a particle count would be less accurate. Whatever sampling system is used it should be ensured that the sample taken is meaningful and time should not be wasted, for example, analysing the size distribution of particles which have collected on surfaces like window sills under unspecified conditions, because samples taken from such surfaces can yield misleading results.

4.4.5.1 *Sieve techniques*

With coarser materials a size distribution can be carried out using a number of standard sieves. The range of sizes of sieves varies from country to country, but the sieves available in the UK to British Standard specifications have the mesh sizes shown in Table 4.9.

Table 4.9 British Standard sieve sizes

BS Sieve No.	Mesh size (μm)
16	1000
18	840
22	710
25	590
30	500
36	420
44	350
52	297
60	250
72	210
85	177
100	149
120	125
150	105
170	88
200	76
240	62
300	53

For smaller sizes US standard sieves are available which extend the range down to:

270	53
325	44
400	37

Size distribution by sieves is normally confined to the size determinations of larger dust and grit from industrial chimneys to check the efficiency of control techniques for dust abatement.

4.4.5.2 *Microscope techniques*

For a full description of particle size distribution work by the use of the microscope the reader is referred to the work of Chamot and Mason [48]. This section will be confined to description of the usual method used in air pollution work.

Because of the irregular shaped particles normally found in air pollution work, one of the first problems which arises when carrying out determinations of size distribution is how the dimensions or diameter of such particles are to be expressed. The most frequently used method is known as 'Martin's' diameter. This is the horizontal, or west-to-east dimension of each particle which divides the projected area of the particle into two equal halves and this is illustrated diagrammatically in Fig. 4.24. This dimension is obtained by moving a sample on a microscope stage past the lens system which contains a calibrated scale in such a way that as each particle is bisected by the scale the dimensions are estimated.

Fig. 4.24   Illustration of Martin's diameter.

This may appear to be a very approximate method of measuring the size of irregularly shaped particles, but provided a large number of particles is counted, it will be found to be sufficiently accurate. Another method is to use the circles on one of the special air pollution graticules discussed later and to estimate the size most closely approximating to the irregularly shaped particle being viewed. Depending on the irregularities in particle shapes the total count can vary from 200 to 2000.

Micrometers and graticules

Before sizing can be carried out it is necessary to calibrate any micrometers or graticules in use at each range of magnifications. This is done using a stage micrometer, which in its simplest form is an etched scale, 1 mm in length and divided into 100 equal divisions, i.e. each division is 10 μm in length. The eyepiece graticule may also be a simple etched scale which is calibrated against the stage micrometer. However, for air pollution work a number of special graticules have been designed to simplify these size determinations. These are shown in Fig. 4.25a–c. The Patterson graticule consists of nine rectangles and a series of numbered globes and circles of progressive diameters. In use, the graticule should be moved over a given particle until the circle which most closely represents the irregular dimension is found.

The Porton graticule has a wide range of applications and is of later design than the Patterson. The right-hand side of the rectangle is divided vertically as shown. The diameters of the globes and circles increase in a geometric progression of two.

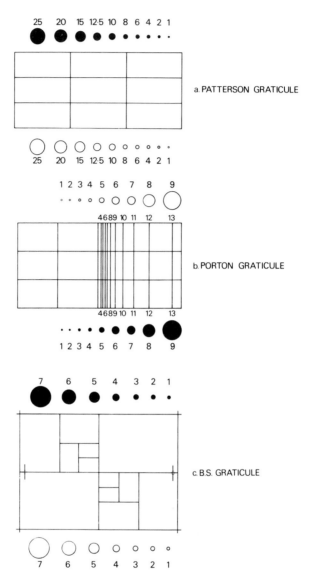

Fig. 4.25 Graticules in common use.

The third most commonly used graticule is the BS graticule [49] which is also illustrated in Fig. 4.25c. The main rectangle of the grid is subdivided in such a way that the rectangles formed give two each of 1/4, 1/8, 1/16 and 1/32 and four of 1/64 of the total area of the grid. In addition, the diameters of the seven different sized circles are in the ratio:

Circle no.	Numerical value of diameter
1	1.00
2	1.41
3	2.00
4	2.83
5	4.00
6	5.66
7	8.00

Counting is carried out by moving the slide preparation containing the particles past the eyepiece graticule to enable the respective particle diameters to be measured. The movement of the microscope stage may be done either mechanically or by setting up the preparation off-centre on a rotating stage and then rotating it. Where possible a combination of eyepiece and objective magnification should be chosen to give about 10 or 12 size classes. During the count, tallies should be kept of the numbers of particles in the different size classes and Table 4.10 gives a typical table of results of a count.

Table 4.10 Particle size count

Eyepiece graticule size class	Total no. of particles per class
1	8
2	70
3	130
4	105
5	90
6	72
7	42
8	21
9	12
10	2

The next stage in the preliminary calculations is to take the product of the number of particles in each size class (n), and (a) the diameter class, (b) the square of the diameter class, (c) the cube of the diameter class and (d) the fourth power of the diameter class as shown in Table 4.11.

If the calibration of the eyepiece graticule is taken as 1.8 μm per division, then the following average values may be determined from the above calculations.

Mean particle diameter $= \Sigma nd/\Sigma n = 2430/552 \times 1.8 \ \mu m = 7.9 \ \mu m$
Area mean diameter $\quad = \Sigma nd^3/\Sigma nd^2 = 73 \ 506/12 \ 554 \times 1.8 \ \mu m = 10.5 \ \mu m$
Mass mean diameter $\quad = \Sigma nd^4/\Sigma nd^3 = 473 \ 690/73 \ 506 \times 1.8 \ \mu m = 11.6 \ \mu m$

Other calculations, such as cumulative percents by number, surface or mass,

Table 4.11   Preliminary calculations of size distribution

Eyepiece size Class (d)	n	nd	nd^2	nd^3	nd^4
1	8	8	8	8	8
2	70	140	280	560	1 120
3	130	390	1 170	3 510	10 530
4	105	420	1 680	6 720	26 880
5	90	450	2 250	11 250	56 250
6	72	432	2 592	15 552	93 312
7	42	294	2 058	14 406	100 842
8	21	168	1 344	10 752	86 016
9	12	108	972	8 748	78 732
10	2	20	200	2 000	20 000
Total	552	2 432	12 554	73 506	473 690

etc., may be made from the figures shown in Table 4.11. A discussion of the meaning of different expressions of particle diameter is given by Ledbetter [50].

As was mentioned earlier in this section, some correction should be made where there are obvious differences in density of the particles. In such cases the data shown in Table 4.10 should be duplicated for each density class. If this is not done the figure obtained for the mass mean diameter in particular will be seriously in error.

References

1. Vandegrift, A. E., Shannon, L. J., Sallee, E. E. *et al.* (1971) *J. Air Pollut. Control Assoc.*, **21**, 321–8.
2. Warren Spring Laboratory (1966) *National Survey of Smoke and Sulphur Dioxide Instruction Manual*, Ministry of Technology, London.
3. Whatman Leaflet AP-69, W. & R. Balston Ltd, Maidstone, Kent, England.
4. Spurny, K. and Lodge, J. P. (1968) *Staub-Reinhalt. Luft*, **28**, 1–10.
5. Spurny, K. R., Lodge, J. P., Frank, E. R. and Sheesley, D. C. (1969) *Environ. Sci. Technol.*, **3**, 453–64.
6. Spurny, K. R., Lodge, J. P., Frank, E. R. and Sheesley, D. C. (1969) *Environ. Sci. Technol.*, **3**, 464–8.
7. Seeley, J. L. and Skogerboe, R. K. (1974) *Anal. Chem.*, **46**, 415–21.
8. Lui, B. Y. H. and Lee, K. W. (1976) *Environ. Sci. Technol.*, **10**, 345–50.
9. Biles, B. and Ellison, J. Mc. K. (1975) *Atmos. Environ.*, **9**, 1030–2.
10. Harrison, R. M. (1984) *CRC Crit. Rev. Anal. Chem.*, **15**, 1–64.
11. Dalager, S. (1975) *Atmos. Environ.*, **9**, 687–91.
12. Kretzschmar, J. G. (1975) *Atmos. Environ.*, **9**, 931–4.
13. British Standards Institution, Determination of Concentration of Suspended Matter, BS 1747: Part 2, 1969.

14. Edwards, J. D., Ogren, J. A., Weiss, R. E. and Charlson, R. J. (1983) *Atmos. Environ.*, **17**, 2337–41.
15. Davies, C. N. (1954) *Dust is Dangerous*, Faber and Faber, London.
16. Herpetz, E. (1969) *Staub-Reinhalt. Luft*, **29**, 12–18.
17. Barrett, C. F. and Parker, J. (1963) *Int. J. Air Water Pollut.*, **7**, 995–8.
18. Dzubay, T. G., Hines, L. E. and Stevens, R. K. (1976) *Atmos. Environ.*, **10**, 229–34.
19. Anderson, A. A. (1958) *J. Bacteriol.*, **76**, 471–84.
20. May, K. R. (1964) *Appl. Bact.*, **12**, 37–43.
21. Butcher, S. S. and Charlson, R. J. (1972) *An Introduction to Air Chemistry*, Academic Press, London.
22. Burton, R. M., Howard, J. N., Penley, R. L. *et al.* (1973) *J. Air Pollut. Control Assoc.*, **23**, 277–81.
23. Dzubay, T. G. and Stevens, R. K. (1975) *Environ. Sci. Technol.*, **9**, 663–8.
24. Garland, J. A. (1972) Proceedings, *Conference at Inst. Mech. Eng.*
25. Ahlquist, N. C. and Charlson, R. J. (1968) *Environ. Sci. Technol.*, **2**, 363–6.
26. Harrison, R. M. and Pio, C. A. (1981) *J. Air Pollut. Control Assoc.*, **31**, 784–7.
27. Pestel, E. (1963) *Strömungstechnische Untersuchungen von Staubniederschlagmess-geräten*, Westdeutscher Verlag, Köln, Nr 1183.
28. British Standards Institution, Deposit Gauges, BS 1747: Part 1, 1969.
29. Lucas, D. H. and Moore, D. J. (1964) *Int. J. Air Water Pollut.*, **8**, 441–53.
30. British Standards Institution, Directional Dust Gauges, BS 1747: Part 5, 1972.
31. Peters, D. W. A. (1962) *The Examination of Soil with Particular Reference to the Density Gradient Test*, British Academy of Forensic Sciences Teaching Symposium No. **1**, 27–36, Sweet and Maxwell, London.
32. Muller, L. D. and Burton, C. J. (1965) *The Heavy Liquid Density Gradient and its Application on Ore Dressing Mineralogy*, 8th Commonwealth Mining and Metallurgical Congress, Melbourne.
33. Brown, K. M. and McCrone, W. C. (1963) *Microscope and Crystal Front*, **13**, 311–22.
34. Brown, K. M. and McCrone, W. C. (1963) *Microscope and Crystal Front*, **14**, 39–54.
35. British Standards Institution, Simplified Methods for the Measurement of Grit and Dust Emissions from Chimneys, BS 3405: 1961.
36. Hamilton, E. M. and Jarvis, W. D. (1962) *Central Electricity Board Monograph No. RD/P/21*, London.
37. McCrone, W. C. (1967) *The Particle Atlas*, Ann Arbor Science Publ., Ann Arbor.
38. Bruckman, L. and Rubino, R. A. (1975) *J. Air Pollut. Control Assoc.*, **25**, 1207–15.
39. Asbestosis Research Council (1971) Tech. Note No. 1: *The Measurement of Airborne Asbestos by the Membrane Filter Method*.
40. Holt, P. F. and Young, D. K. (1973) *Atmos. Environ.*, **7**, 481–3.
41. Nicholson, W. J., Rohl, A. N. and Ferrard, E. F. (1970) *Proc. Second International Clean Air Congress*, 136–9, Washington, DC.
42. Rickards, A. L. and Badami, D. C. (1971) *Nature*, **234**, 93–4.
43. Gadsden, J. A., Parker, J. and Smith, W. L. (1970) *Atmos. Environ.*, **4**, 667–70.
44. Murchio, J. C., Cooper, W. C. and De Leon, A. (1973) Cited in *Asbestos: an Information Resource*, Department of Health and Welfare, Washington, DC, Publ. No. (NIH) 79-1681, 1978, p. 56.
45. Spurny, K. R., Stober, W., Opiela, H. and Weiss, G. (1979) *Sci. Tot. Environ.*, **11**, 1–40.
46. Burdett, G. J. and Rood, A. P. (1983) *Environ. Sci. Technol.*, **17**, 643–8.
47. Steen, D., Guillemin, N. P., Buffat, P. and Litzistorf, G. (1983) *Atmos. Environ.*, **17**, 2285–97.

48. Chamot, E. M. and Mason, C. W. (1959) *Handbook of Chemical Microscopy, Vol I*, 3rd edn, Wiley, New York.
49. British Standards Institution, Eyepiece and Screen Graticules for the Sizing of Particles, BS 3625: 1963.
50. Ledbetter, J. O. (1972) *Air Pollution, Part A: Analysis*, Dekker, New York.

5 Metal analysis

R. M. HARRISON

5.1 Introduction

Metal pollutants are emitted into the atmosphere from numerous sources including combustion of fossil fuels (including leaded petrol), metal smelters and alloy refineries, cement manufacturing plants and municipal incinerators [1]. Metals and metallic compounds exist in the atmosphere in three distinct physical forms: solid particulate matter, liquid droplets (mists) and vapours. The size range of airborne particulate matter is broad [1], and particles represent by far the most common form of metallic air pollution. Table 5.1 indicates the levels of metals in suspended particulate matter as measured by the US National Air

Table 5.1 Average and maximum concentration of airborne pollutants measured at urban stations by the US National air sampling network, 1964–65 [2]

Pollutant	Concentration (μg m^{-3})[a] Arithmetic average	Maximum
Total suspended particulate matter	105	1254
Antimony (Sb)	0.001	0.160
Arsenic (As)	0.02	
Beryllium (Be)	<0.0005	0.010
Bismuth (Bi)	<0.0005	0.064
Cadmium (Cd)	0.002	0.420
Chromium (Cr)	0.015	0.330
Cobalt (Co)	0.0005	0.060
Copper (Cu)	0.09	10.00
Iron (Fe)	1.58	22.00
Lead (Pb)	0.79	8.60
Manganese (Mn)	0.10	9.98
Molybdenum (Mo)	<0.005	0.78
Nickel (Ni)	0.034	0.460
Tin (Sn)	0.02	0.50
Titanium (Ti)	0.04	1.10
Vanadium (V)	0.050	2.200
Zinc (Zn)	0.67	58.00

[a] Bi-weekly 24 h samples.

Sampling Network [2]. Figures such as these give an indication of levels likely to be encountered during ambient air monitoring and may thus be used to estimate the volume of air which needs to be sampled to allow collection of a quantity of a metal which is compatible with a proposed analytical technique.

5.2   Analysis of particulate matter

5.2.1   General sampling considerations

Airborne particles may be collected by impingers, electrostatic precipitators or filters. Filters are currently much favoured because of their ease of use and high efficiency of collection for small particles. Organic membrane filters are suitable for the collection of airborne particles as small as 0.03 μm in diameter, although efficiency is reduced with increasing pore size of the filter and with increasing particle velocity at the face of the filter and thus sampling rate [3–5]. Glass fibre filters, although not having such a high collection efficiency [6–8], allow very fast passage of air and are used with high volume (Hi-Vol) samplers. Several other filter materials are also available. Detailed consideration of filter efficiencies is given in Chapters 1 and 4.

An important consideration in selecting a filter medium, whatever the subsequent method of analysis, is the background metal content of the filter material. Table 5.2 shows levels of metallic impurities in various filter materials [9–11]. Dams *et al.* [10] evaluated a number of filter materials and impaction surfaces for their impurity content, flow properties, retention properties and tensile strength (where appropriate) and their general suitability for neutron activation analysis of aerosols. It was concluded that Whatman No. 41 cellulose was the preferred filter medium for neutron activation work, and Vogg and Haertel [12] have similarly recommended the use of cellulose paper filters because of the low levels and homogeneous distribution of the metallic impurities in them. The disadvantage, however, lies in their limited collection efficiency (Chapters 1 and 4). Zoller and Gordon [11], also using neutron activation analysis, report the levels of impurities in Millipore (cellulose ester membrane) and Delbag (polystyrene) filters. Continuous washing of some cellulose paper filters may be used to remove water-soluble impurities, without adverse effect upon the filtration properties [13]. Silver membranes are of little use in trace element analysis, but are useful when subsequent X-ray diffraction analysis is to be carried out.

Sampling of particulate matter from gas streams such as stack gases should be isokinetic, but for sampling relatively stagnant ambient air this is not essential, and generally not feasible as wind velocity and direction are seldom constant. Passage of an air sample through a bent tube prior to collection of particles may lead to deposition and loss of particles by impaction at the bend. Also, in a bent tube different flow patterns for smaller and larger particles result from differing particle momenta and this can lead to uneven deposition on to a filter which will introduce error if only a segment of the filter is analysed. Since airborne particles

Table 5.2 Metallic impurities in filter materials (ng cm^{-2})

Metal	Polystyrene (Delbag) [10]	Cellulose ester (Millipore 0.45 µm) [10]	Cellulose paper (Whatman No. 41) [10]	Glass fibre [9]	Organic membrane [9]	Silver membrane [9]
Ag	<2		2			
Al	20	10	12			
As				80		
Ba	<500	<100	<100			
Be				40	0.3	200
Bi					<1	
Ca	300	250	140			
Cd					5	
Co	0.2	<1	0.1		0.02	
Cr	2	14	3	80	2	60
Cu	320	40	<4	20	6	20
Fe	85	<300	40	4000	30	300
Hg	1	<1	0.5			
Mg	<1500	<200	<80			
Mn	2	2	0.5	400	10	30
Mo					0.1	
Ni	<25	<50	<10	<80	1	100
Pb				800	8	200
Sb	1	3	0.15	30	0.1	
Sn				50	1	
Ti	70	5	10	800	2000	200
V	<0.6	0.09	<0.03	30	0.1	
Zn	515	20	<25	160 000	2	10

are subject to gravitational settling, a probe pointing upwards will collect more particles than one which points downwards, but the effect is only significant when sampling relatively large particles (greater than 30 µm diameter). Particles may be lost in a sampler by electrostatic attraction to surfaces within the sampler, particularly glass and plastics. Ideally, therefore, particles should reach the collector as soon as possible after entering the sampling apparatus. Additionally, all sampling equipment should be rigorously tested for leaks before use if quantitative sampling is required.

Although some methods of analysis of particulate matter for metals require no pretreatment of the collected sample prior to analysis, for most analytical methods it is necessary to destroy organic matter associated with the sample and, as a dissolved sample is normally required, to render the sample soluble. Wet ashing with acids is frequently a suitable procedure [14, 15], as is dry ashing in a muffle furnace followed by dissolution in strong acids. Losses, assumed to be a result of volatilization, may occur with some metals under normal dry ashing conditions (550° C for 1 h) and Thompson and co-workers [16, 17] report a low temperature ashing procedure, for glass fibre and membrane filters, which uses an oxygen plasma and leads to much reduced losses (Table 5.3). Instruments for low temperature ashing are now commercially available. The presence of other interfering substances in a sample of atmospheric particulate matter was found to have little effect upon metal recovery when using the low temperature ashing procedure [16, 17].

Table 5.3   Effect of ignition method on metal recovery [16, 17]

Metal	Recovery (%)	
	Low temp. ashing	*Muffle furnace at 550° C*
Ba	97	99
Cd	92	53
Co	96	97
Cr	112	100
Cu	98	92
Mo	98	116
Mn	99	107
Ni	97	99
Pb	101	46
Sb	99	46
Sn	95	87
Ti	95	92
Zn	96	39

Kometani *et al.* [14] advance the theory that apparent losses of metals during dry ashing result from the formation of insoluble metal silicates. They show that ashing of paper filters at 500° C, with or without prior wetting with sulphuric acid (H_2SO_4), gives a recovery of metals from atmospheric particulates comparable to that achieved by wet ashing with nitric-perchloric acids or low temperature ashing. The use of glass fibre filters, or of silica, glass or porcelain-glazed crucibles leads to a low recovery of metals.

Once prepared, both sample solutions and standard solutions of metals may not keep satisfactorily. Acid solutions may leach lead from some types of glass, soda glass in particular, and from some plastics (e.g. PVC). Additionally, both glass and plastics may adsorb a number of metal ions. Struempler [18] showed adsorption of Ag, Pb, Cd, Zn and Ni on to borosilicate glass surfaces, which was prevented, for all but Ni, by acidification to pH2. Polyethylene containers adsorbed Ag, Pb and Ni, but not Cd nor Zn, and Ag adsorption could be prevented by acidification to pH2. Cadmium is lost from aqueous solution during storage in glass containers, but only at alkaline pH values [19]. Dilute aqueous solutions of mercury have poor keeping properties, even in acidic solution [20, 21], but may be kept satisfactorily by addition of 5% v/v $HNO_3 + 0.01\%$ dichromate. Silver is also adsorbed by borosilicate glass, with slightly reduced adsorption at acid pH values [22], and storage is assisted by the addition of sodium thiosulphate [23]. Although acidification may assist the preservation of dilute solutions of metals, lengthy storage cannot be guaranteed. It is, therefore, advisable to prepare fresh dilute standards daily, and to investigate the keeping properties of sample solutions if they are not to be analysed immediately upon preparation.

Polypropylene containers have been shown to be serious sources of Cd and Zn contamination [18]. Lubricants in ground-glass joints and stopcocks may also lead to contamination and, for trace analysis, PTFE stopcocks should be used wherever possible.

The sampling of mists from the atmosphere is a problem occasionally encountered. Silverman and Ege [24] report that liquid impingers, electrostatic precipitators, filter papers and cotton plugs moistened with glycerol have all been used for collection of chromic acid mists from air. A comparative evaluation of these techniques showed a thick (0.026 inch), absorbent filter paper to be the best collection medium with an efficiency under the conditions used of 99.9% [24].

5.2.2 Analytical methods involving no pretreatment of the sample

Analytical methods for metals and applications of these to the analysis of air pollutants will be described. For a full explanation of the theory and practice of a given technique, the reader is referred to more comprehensive texts on the subject. Although an exhaustive review of the literature regarding each technique is not attempted, examples of the applications of techniques are given, and full experimental details of at least one procedure are described.

5.2.2.1 *X-ray emission analysis*

A metal target bombarded with accelerated electrons emits X-rays. The emitted spectrum is in two parts: a continuous spectrum with a definite short wavelength limit (white radiation), and a number of sharp high intensity lines characteristic of the target element. Bombarding electrons knock atomic electrons from their orbitals, and when other electrons 'jump' from outer orbitals to fill the vacant orbital, they lose energy by emission of radiation in the X-ray wavelength region. These effects arise from inner orbitals, and valency-level electrons have little effect on the energetics of the process. Hence, the chemical state of the emitting atom has virtually no effect upon the emitted wavelength. If, instead of using electrons, X-rays are used to bombard the metal, absorption occurs with a secondary emission of X-rays (fluorescence), characteristic of the irradiated metal. In this case, the continuous emission spectrum is absent, but the secondary emission is less intense than the primary emission caused by electron bombardment. In addition to the secondary X-rays emitted, scatter of the incident primary radiation also occurs, as well as scattering involving a change in wavelength

(Compton scattering) and diffraction by crystalline substances, which obeys the Bragg equation.

Using appropriate instrumentation, the intensity of the secondary emissions may, by comparison with a suitable standard, be used to give a quantitative measure of a metal. In a mixture of elements the secondary emission of one element may be absorbed by another and re-emitted at a different wavelength. Hence complex inter-element effects arise in which the emissions of some elements are enhanced while others are reduced, and calculation or comparison with appropriate standards is necessary for estimation of the elemental composition of the sample. Macdonald has given an excellent account of both practical and theoretical considerations in X-ray analysis [25].

Gilfrich *et al.* [26] have compared several X-ray fluorescence techniques for the analysis of airborne particulate matter collected on filters. They indicate that X-ray analysis is rapid, needing no sample pretreatment, non-destructive and applicable to elements irrespective of their position in the periodic table, including all elements beyond atomic number 11 (sodium). Numerous elements may be analysed simultaneously and results are comparable to those obtained using neutron activation analysis [27] and atomic absorption analysis [15, 26, 28, 29]. In a comparison of wavelength dispersive and energy dispersive techniques, multi-channel wavelength dispersive instruments were found to be the most satisfactory [26]. Typical detection limits for X-ray fluorescence analysis of particulate material collected on filter papers, using a 100 s or 10 min analytical period, appear in Table 5.4.

Because of the small values for the total quantities of samples collected by air filtration (typically up to 1 mg cm^{-2}) the metals are in a 'dilute' form and the inter-element effects which are encountered in bulk samples are not normally significant. Size effects may, however, be significant for larger particles, in general those greater than 5 μm in diameter [26]. For known particle size distributions the effect is calculable, but for normal air pollution samples this is not possible. For optimum precision, samples collected by fractionating impactors are necessary.

Two calibration procedures are most commonly used. Dilute standard solutions of the metals being analysed may be spotted on to filter discs and allowed to dry leaving a thin film. Calibration curves produced from such standards are linear up to a few hundred μg per cm^2 [26, 28, 30]. Alternatively, insoluble metal salts chemically produced in fine particle form by precipitation from aqueous solution, and deposited on filter discs by filtration of the aqueous suspension, may be used [15, 29]. Besides weighed lead metal samples vaporized on to aluminium foil, comparison with wet chemical procedures has also been used for calibration [31]. Pradzynski and Rhodes [32] describe the development of standards based upon both drying of solutions on glass fibre filters, and for non-volatile trace elements by spiking 10 μm quartz powder with the element of interest and deposition from air suspension. Billiet *et al.* [33] report a novel procedure in which a solution of various elements and a radioactive tracer (^{24}Na) is mixed with the water-soluble polymer methylcellulose. An even thin film is spread on a glass plate and dried; the trace element concentrations are

Table 5.4   Detection limits for X-ray fluorescence analysis [26, 28]

	Detection limit (ng cm^{-2})		
Metal	*Energy dispersive; radioisotope sources* [28]	*Wavelength dispersive; X-ray tube excitation (Rh Tube)* [26]	*Energy dispersive; X-ray tube excitation (W Tube-Ni foil)* [26]
Al		85	
As	100		
Ca	60	29	140
Co	80		
Cr	240		
Cu	50	49	
Fe	80	30	120
K		18	220
Mn	120		
Mo	30		
Ni	60		
Pb	110	260	110
Se		150	81
Sr	40		
Ti	50		
V	30	33	90
Zr	30		
Zn	30	51	110

Gilfrich *et al.* [26] used a 100 s count, and Rhodes *et al.* [28] used a 10 min count.

determined by measurement of the tracer activity. The method is of high precision and accuracy and applicable to a wide range of trace elements. Breiter *et al.* [34] have reviewed the many methods available for preparation of thin film X-ray fluorescence (XRF) standards. Data from X-ray fluorescence analyses may be speedily processed by the use of on-line computing facilities [28, 35].

One problem inherent in the use of cellulose filters is that substantial penetration of particles into the body of the filter may occur and a consequent correction for X-ray absorption by the filter material is necessary [36]. Glass fibre is a more suitable filter substrate in this respect. Absorption corrections may be overcome by folding the filter with loaded side inwards, which enhances sensitivity and makes absorption corrections simpler and more accurate [37].

X-ray techniques for the examination of individual particles are available. The electron probe microanalyser allows the bombardment of particles with an electron beam, typically about 1 μm in diameter. X-ray emission is then analysed for characteristic emission wavelengths, allowing a quantitative analysis of the composition of the particle under the electron beam. Pure elements are used as standards. For samples of greater diameter than the electron beam, the technique may be used to give information on the distribution of an element over the sample, the electron beam being made to scan over an area of sample; monitoring scattered electrons or emitted X-rays allows a 'picture' of the sample to be built up. Combination with an optical microscope is also possible.

In order to avoid electric charge build-up or overheating under the electron beam, a thin coating of a conducting substance on the sample is frequently required. Ter Haar and Bayard collected airborne particulates on Millipore filters which were subsequently dissolved in acetone [38]. The centrifuged particles were mounted in a film of nitrocellulose on Be coated with a 300 Å layer of carbon. Use of the electron probe microanalyser allowed determination of the elemental content of particles of different sizes. Particulate matter emitted from vehicle exhausts was analysed in a similar fashion. The use of acetone as a solvent carries some disadvantages as many metal salts will have an appreciable solubility in this medium. Particulate matter deposited on the bark of trees growing near busy roads has also been examined [39]. Sections of dried bark were mounted on a carbon block and covered with a thin carbon coating prior to electron probe analysis.

Electron microscopy and microanalysis (EMMA), which combines high resolution electron microscopy with X-ray microanalysis, allows size determination and semi-quantitative elemental analysis of particles as small as 0.05 μm. Its application to the analysis of urban airborne particulate material has been described [40].

X-ray diffraction measurements allow the identification of homogeneous crystalline particles, and even allow discrimination between different crystalline forms of a compound. Techniques such as EMMA and X-ray diffraction are useful as speciation methods rather than routine analytical procedures and are discussed in this context in Chapter 11.

In laboratories where the equipment is available, particle (proton) induced X-ray emission (PIXE) is favoured as a means of analysis of airborne particles [41]. In a comparative study of PIXE and energy-dispersive XRF of air particulates, Ahlberg and Adams [42] report that using the criterion of equal numbers of counts in the spectra, PIXE gives about three times lower detection limit for air filters and thirty times lower for a cascade impactor sample.

Experimental procedure for X-ray fluorescence analysis of Pb, Zn, Co, Cu, Ni, Fe, Mn *and* Cr [15].

Air sampling

Samples (24 h) of approximately 1500 m^3 are drawn through 8 × 10 inch sheets of Schleicher and Schuell (S & S) No. 589 Green Ribbon filter paper in high-volume samplers. Discs (25 mm) are cut using a special die, and these are held firmly in the spectrometer by a clamping ring to an aluminium mask having an 11/16 inch diameter opening. The area exposed to the X-ray beam is 0.371 inch^2.

Preparation of standards

Standard solutions (1 mg ml^{-1}) of the metals are prepared using chloride salts, if available, and sufficient acid to prevent hydrolysis. Transfer hydrochloric acid

(HCl) (5 ml) and de-ionized water (50 ml) to a 100 ml volumetric flask and add appropriate amounts of the standard metal solutions. Solutions containing nitrates are first combined in a small conical flask, treated with perchloric acid (HClO₄) (1 ml) (Care needed), evaporated to white fumes to decompose all nitrates and, after cooling, quantitatively transferred to the volumetric flask. Fill the flask to the mark with de-ionized water. Samples collected from urban air using the above sampling procedure dictated that 1 ml of the multimetal solution should contain 50 μg of Pb and Fe, 10 μg of Zn and Cu, and 2 μg of Ni, Co, Mn and Cr. Of necessity, these amounts are adjusted to the analytical problem at hand. Transfer an aliquot (1.0 ml) of the multimetal standard solution to a 50 ml beaker, add de-ionized water (10 ml), HCl (2 to 3 drops) and metacresol purple indicator (0.1%; 1 drop). Neutralize the solution by dropwise addition of NH₄OH (1 + 1) and add one drop in excess. Add a clear solution of 2% sodium diethyldithiocarbamate (5 ml), swirl the mixture and allow to stand for 5 min. (The carbamate solution should be stored in a plastic bottle and refrigerated when not in use.) Filter the liquid slowly through a 25 mm 0.8 μm Millipore disc and wash with de-ionized water (about 2 ml). Dry the disc thoroughly at room temperature, or more rapidly by placing it on a glass surface at 50 to 60° C using a lightweight plastic ring on top to prevent curling of the paper.

Instrumental

Luke *et al.* [15] used a General Electric XRD-6S air or helium path instrument equipped with a four-crystal changer, bulk sample holder, dual (Tungsten/Chromium) target X-ray tube, and dual (scintillation and flow proportional) counter tube detector system with solid-state electronics. The analytical line employed for X-ray counting is the K_{α} doublet for all elements except lead, for which L_{α_1} line is used. The tungsten target of the EA75 X-ray tube is operated at 50 kVP and 75 mA. Other parameters include a lithium fluoride (200) analysing crystal, 10-mil Soller slit and helium path. Take a 10 s count at the goniometer setting for the X-ray line of each element to be determined on a clean Millipore disc, a blank S & S paper disc, the multimetal standard disc, and all S & S paper samples to be analysed. For each analytical line, subtract the total intensity measured on the Millipore blank from that obtained on the multi-element standard. Similarly, subtract the intensity on the S & S blank from those found on the pollution samples. The number of μg of each metal per inch² of sample may then be calculated from the formula

$$\mu\text{g in}^{-2} = 1/0.371 \times \frac{\text{Net counts sample}}{\text{Net counts standard}} \times \mu\text{g of metal on standard disc}$$

5.2.2.2 *Radioactivation methods*

In radioactivation analysis, the sample is irradiated with neutrons or charged particles without pretreatment. Interaction of the sample with the bombarding particles causes the formation of different isotopes of the elements in the sample, or conversion to isotopes of neighbouring elements. Many of these isotopes produced are radioactive, and measurement of their activity and comparison with standards allows quantitative analysis of the elements in the sample. Gibbons and Lambie have reviewed the theory and practice of radioactivation analysis [43]. Advantages of the technique are speed, simplicity, specificity and non-destruction of the sample [44]; the major disadvantage is the need for a high intensity source of activating particles.

The process most commonly used in activation analysis is irradiation with a high thermal neutron flux, generally produced by a nuclear reactor. Measurement of the characteristic γ-ray emissions of the isotopes produced is then used for analysis of the sample. Several neutron activation techniques for the analysis of airborne particulate matter collected on filters have been described [10–12, 45–47]. Sample filters are each sealed in a polyethylene vial and irradiated for 5 min with neutrons. γ-ray counts after 3 and 15 min allow the determination of 13 elements. A further irradiation at a higher neutron flux for 2 to 5 h, followed by γ-ray counts after 20 to 30 h and 20 to 30 days of cooling allow the determination of a further 20 elements [45]. Standards are prepared by deposition of solutions containing a mixture of the elements to be analysed upon blank filters. After drying, the standards are sealed in polyethylene and irradiated in a manner identical to that used for the samples [27, 45].

Neutron activation analysis of airborne particulate matter has been criticized because, although numerous elements can be analysed, many cannot because of inadequate sensitivity, interferences or high blanks. Few low molecular weight elements can be analysed, the analysis of important elements such as Pb and Cd is not reported, and several other important metals may only be analysed after a 20 to 30 day decay period [48]. Hence, the application of neutron activation analysis is subject to severe limitations. The sensitivity of the technique for a given element is dependent upon the overall composition of the sample, owing to interference effects, but Dams *et al.* [45] quote typical detection limits, shown in Table 5.5. For those metals determined by neutron activation analysis, limits of detection compare quite favourably with those for emission spectrographic and conventional atomic absorption methods [11, 45]. Instrumental neutron activation analysis (INAA) has been applied to the analysis of 2500 m^3 air samples collected on glass fibre filters in the US National Air Sampling Network [49], allowing determination of 26 elements. Quantitative analysis of 32 elements in the US National Bureau of Standards urban air particulate reference standard has also been achieved using INAA [50].

The scope and applications of charged particle activation analysis appear limited [51], although fairly sensitive methods for lead analysis have been developed [52, 53].

Experimental procedure for neutron activation analysis [45]

Air sampling

Air is drawn through 25 mm or 47 mm diameter polystyrene filters at up to 12 l min^{-1} cm^{-2} for 24 h using a high vacuum pump.

Table 5.5 Sensitivity of neutron activation analysis for determination of trace elements in aerosols [45]

Element	Decay time before counting	Detection limit (μg)	Minimum concn. in urban air – 24 h sample (μg m^{-3})
Al	3 (min)	0.04	0.008
Ca	3	1.0	0.2
Ti	3	0.2	0.04
V	3	0.001	0.002
Cu	3	0.1	0.02
Na	15	0.2	0.04
Mg	15	3.0	0.6
Mn	15	0.003	0.000 6
In	15	0.000 2	0.000 04
K	20–30 (h)	0.075	0.007 5
Cu	20–30	0.05	0.005
Zn	20–30	0.2	0.02
As	20–30	0.04	0.004
Ga	20–30	0.01	0.001
Sb	20–30	0.03	0.003
La	20–30	0.002	0.000 2
Sm	20–30	0.000 05	0.000 005
Eu	20–30	0.000 1	0.000 01
W	20–30	0.005	0.000 5
Au	20–30	0.001	0.000 1
Sc	20–30 (day)	0.003	0.000 004
Cr	20–30	0.02	0.000 25
Fe	20–30	1.5	0.02
Co	20–30	0.002	0.000 025
Ni	20–30	1.5	0.02
Zn	20–30	0.1	0.001
Se	20–30	0.01	0.000 1
Ag	20–30	0.1	0.001
Sb	20–30	0.08	0.001
Ce	20–30	0.02	0.000 25
Hg	20–30	0.01	0.000 1
Th	20–30	0.003	0.000 04

Areas of filter of 0.8, 0.8, 1.6 and 13 cm^2 used for counts at 3, 15 min, 20–30 h and 20–30 day respectively.

Activation and analysis

(a) *Short-lived isotopes.* For the analysis of elements giving rise to short-lived isotopes, each sample is packaged in a polyethylene vial, and placed in a rabbit which carries it through a pneumatic tube to a position near the core of the reactor, where it is irradiated for 5 min at a flux of 2×10^{12} neutrons cm^{-2} s^{-1}. At the end of this period, the sample is returned to the laboratory where it is manually transferred to a counting vial and carried to the counting room. 3 min after irradiation, commence a count of 400 s live-time duration, and follow by a count of 1000 s live-time starting 15 min after irradiation. These and subsequent counts are performed on a 30 cm³ Ge(Li) detector coupled to a 4096 channel analyser. The detector is housed in an iron shield and operated, in an air-conditioned room, at a gain of 1 keV per channel. The observed resolution (reported by Dams *et al.* [45]) was 2.5 keV full-width at half maximum (FWHM) for the ^{60}Co 1332 keV photopeak and a peak to Compton ratio of 18/1. Table 5.6 includes the isotopes determined by the first 2 counts.

All spectra are recorded on 7-track magnetic tape for further analysis. Conversion of counting rates under the various peaks to concentrations is

Table 5.6   Nuclear properties and measurement of isotopes [45]

Element	Isotope	Half-life	t (irradiate)	t (cool)	t (count) (s)	Gamma-rays used (keV)
Al	^{28}Al	2.31 (min)	5 (min)	3 (min)	400	1778.9
Ca	^{49}Ca	8.8	5	3	400	3083.0
Ti	^{51}Ti	5.79	5	3	400	320.0
V	^{52}V	3.76	5	3	400	1434.4
Cu	^{66}Cu	5.1	5	3	400	1039.0
Na	^{24}Na	15 (h)	5	15	1000	1368.4; 2753.6
Mg	^{27}Mg	9.45 (min)	5	15	1000	1014.1
Mn	^{56}Mn	2.58 (h)	5	15	1000	846.9; 1810.7
In	^{116m}In	54 (min)	5	15	1000	417.0; 1097.1
K	^{42}K	12.52 (h)	2–5 (h)	20–00 (h)	2000	1524.7
Cu	^{64}Cu	12.5	2–5	20–30	2000	511.0
Zn	^{69m}Zn	13.8	2–5	20–30	2000	438.7
As	^{76}As	26.3	2–5	20–30	2000	657.0; 1215.8
Ga	^{72}Ga	14.3	2–5	20–30	2000	630.1; 834.1; 1860.4·
Sb	^{122}Sb	2.75 (day)	2–5	20–30	2000	564.0; 692.5
La	^{140}La	40.3 (h)	2–5	20–30	2000	486.8; 1595.4
Sm	^{153}Sm	47.1	2–5	20–30	2000	103.2
Eu	^{152m}Eu	9.35	2–5	20–30	2000	121.8; 963.5
W	^{187}W	24.0	2–5	20–30	2000	479.3; 685.7
Au	^{198}Au	2.70 (day)	2–5	20–30	2000	411.8
Sc	^{46}Sc	83.9	2–5	20–30 (day)	4000	889.4; 1120.3
Cr	^{51}Cr	27.8	2–5	20–30	4000	320.0
Fe	^{59}Fe	45.1	2–5	20–30	4000	1098.6; 1291.5
Co	^{60}Co	5.2 (year)	2–5	20–30	4000	1173.1; 1332.4
Ni	^{58}Co	71.3 (day)	2–5	20–30	4000	810.3
Zn	^{65}Zn	245	2–5	20–30	4000	1115.4
Se	^{75}Se	121	2–5	20–30	4000	136.0; 264.6
Ag	^{110m}Ag	253	2–5	20–30	4000	937.2; 1384.0
Sb	^{124}Sb	60.9	2–5	20–30	4000	602.6; 1690.7
Ce	^{141}Ce	32.5	2–5	20–30	4000	145.4
Hg	^{203}Hg	46.9	2–5	20–30	4000	279.1
Th	^{233}Pa	27.0	2–5	20–30	4000	311.8

accomplished by subjecting a few standard solutions containing well-known mixtures of the same elements to the same irradiation and counting sequence. In order to avoid possible errors due to coincidence summing or to broadening of peaks at high counting rates, sample sizes are generally adjusted to make counting rates of sample and standard of comparable magnitude.

Small corrections for variations of both neutron flux and rabbit placement from irradiation to irradiation are accomplished by co-irradiation of a titanium foil flux monitor with each sample. It is counted for 20 s at 13 min after the end of the irradiation, between the 2 sample counts. If the analysis rate does not exceed 1 sample in 40 min, the same flux monitor may be used repeatedly with less than 1% of the original 5.8 min ^{51}Ti remaining in the next count. Net counting rates of the sample spectrum are normalized to an arbitrary titanium activity, equivalent to a reference neutron flux.

(b) *Long-lived isotopes*. The same sample or another portion of the same air filter is then irradiated in the reactor core at a higher flux (1.5×10^{13} neutrons cm^{-2} s^{-1}) for 2–5 h. Each is individually heat-sealed in a polyethylene tube and irradiated with 8 others, plus a standard mixture of elements in a polyethylene bottle, 4 cm in diameter, lowered into the reactor pool. Cooling of the samples during irradiation is achieved by allowing the pool water to circulate through several holes cut in the container bottle. Standards are prepared by depositing 100 μl each of 2 well-balanced mixtures of the appropriate elements on to a highly pure substrate (ashless filter paper) and allowing to dry, then sealing inside polyethylene tubes.

After irradiation, the samples and standards are transferred to clean containers and counted once for 2000 s live-time after 20 to 30 h of cooling and then for 4000 s live-time after 20 to 30 days of cooling. Table 5.6 includes the elements determined from these counts. Errors resulting from thermal neutron flux gradients over the bottle dimension were less than 5%, provided the samples were confined to a single horizontal layer of vertically oriented tubes at the bottom of the bottle and the bottle was rotated through 180° half-way through the irradiation time. Fast neutron flux gradients were about twice as large as thermal gradients, but the only fast neutron reaction used is in the determination of nickel.

Interferences

Some prominent photopeaks are not used because of interferences by neighbouring peaks of other isotopes. In other cases corrections are necessary. The ^{75}Se (279.6 keV) interferes with the monoenergetic ^{203}Hg (279.1 keV), but a correction based on the spectrum of pure ^{75}Se can be applied since the interference in air pollution samples is usually less than 20% of the ^{203}Hg activity. The measurement of ^{64}Cu (511.0 keV) is complicated by interference caused by external pair production of high energy γ-rays. In typical samples, 15 h ^{24}Na is the most important source of γ-rays after a decay period of 20 h, and a correction, usually less than 10%, may be applied to the apparent ^{64}Cu activity. Interferences by threshold reactions have been calculated and checked experimentally, and in typical aerosol samples the only reaction affecting a calculated concentration by greater than 2% is ^{27}Al(n, p) ^{27}Mg. Once the aluminium concentration is known, the appropriate correction can be applied to the magnesium concentration.

5.2.3   Methods involving pretreatment of the samples

5.2.3.1   *Emission spectrography*

At ambient temperatures, in the absence of external stimuli, almost all atoms exist in the electronic ground state. External stimuli, however, may be used to excite atoms or molecules electronically, such that the resultant excited species return to the ground state by emission of energy in the form of light. In emission spectrography, a sample is deposited upon a graphite electrode which is used as one of two electrodes between which a d.c. arc is passed. The energy liberated in the arc causes volatilization of the sample, which is excited and emits light at discrete wavelengths (lines) characteristic of the elements present. Dispersion of the light according to wavelength by a prism or grating allows separation of the emission lines which are recorded on a photographic plate, the intensity of light recorded by the plate from the various lines being a measure of the quantity of the different elements present in the sample.

Although the d.c. arc is still widely used for sample volatilization, numerous refinements are available which may improve sensitivity [54]. Standards of composition similar to that of the sample are used to calibrate the instrument. Because of matrix effects, such standards should resemble the sample, both chemically and physically, as closely as possible. All possible variables which may affect the analysis should be controlled as closely as possible. Such factors include the uniformity of the graphite electrodes and photographic plates and chemicals, and the temperature and humidity of the room where the analysis is performed.

Emission spectrography has not been widely applied to the analysis of metallic air pollutants. Most notably, however, it has been used for many years by the US National Air Sampling Network [2, 6, 55–58]. Samples must be ashed and taken up in acid prior to analysis [6]. Although the method allows simultaneous determination of many metals it is reported to be only semi-quantitative despite the application of considerable analytical skill, to lack sensitivity for a number of metals and to be limited by blanks in the filter substrate [48]. The sensitivity of the method in the determination of a number of metals is shown in Table 5.7. Extrapolation to other elements may be made by using the data of Zoller and Gordon [11] and Mitteldorf [54] for relative sensitivities.

The very high sensitivity of emission spectrography for analysis of Be has led to recommendation of the method for the determination of this metal in air [59]. Normal particle collection techniques are used, and the Be may be determined, after pretreatment of the sample, using graphite rod electrodes and a d.c. arc [60], or a rotating graphite electrode and an a.c. current arc [61]. Using the latter technique, as little as 0.002 μg Be ml^{-1} may be analysed, corresponding to 0.001 μg m^{-3} in a 20 to 30 min air sample. Alternatively, if a Whatman No. 40

Table 5.7 Limits of detection for emission spectrographic analysis [58]

Metal	Analytical line (Å)	Minimum concn. detectable in urban air (μg m^{-3})[a]	Minimum concn. detectable in non-urban air (μg m^{-3})
Be	2348.6	0.0008	0.000 16
Bi	3067.7	0.0011	0.000 2
Cd	2288.0	0.011	0.004
Co	3453.5	0.0064	0.002
Cr	2677.2	0.0064	0.002
Cu	2824.4; 3274.0	0.01	0.001 5
Fe	2457.6	0.084	0.006
Mn	2933.1	0.011	0.006 0
Mo	3170.3	0.0028	0.000 5
Ni	3003.6	0.0064	0.001 6
Pb	2663.2	0.04	0.01
Sb	2877.9	0.040	0.006
Sn	2840.0	0.006	0.001 8
Ti	3199.9	0.0024	0.000 48
V	3183.4	0.0032	0.000 48
Zn	3345.0	0.24	0.08

[a] Using 26% of a 24 h high-volume sample (2000 m^3 total).

filter paper is used, it may be rolled and pressed into the crater of a graphite electrode. After charring thermally and addition of barium chloride as a carrier, a d.c. arc is used to determine Be with a high sensitivity [62].

A technique for continuous monitoring of Be in air involves drawing the polluted air through a spark between copper electrodes [63]. Photo-electric monitoring of the 3130.4 Å Be line and display on a chart recorder allows continuous recording of Be concentrations. The instrument has a range up to 20 μg m^{-3} of Be with a detection limit of 0.5 μg m^{-3}.

An ingenious procedure which uses a graphite cup electrode as a filter for atmospheric particulate matter has recently been reported [64]. Particulate material collected by the graphite cup is volatilized by a d.c. arc in a specially modified source of an emission spectrograph. Very high filtration efficiency was found, and the use of an indium internal standard allowed quantitative analysis of collected metals with a standard deviation in the 10 to 20% standard deviation range for the elements studied. Table 5.8 shows the reported analytical wavelengths and detection limits for a sampling period of 30 min. The workers did, however, fail to acknowledge or investigate the possible adsorption of organometallic compounds, such as lead alkyls, or of elemental mercury by the graphite cup. Hence, in the case of certain metals, there is uncertainty regarding the analysis when the metal is present in the air in other than particulate forms.

Table 5.8 Detection limits for combined sampling – analysis method [55]

Element	Wavelength (Å)	Absolute detection limit (ng)	Detection limit for air (30 min sample) (μg m^{-3})
Al	3082.15	1	0.03[a]
Be	3130.41	0.1	0.003
Co	3453.50	1	0.03
Cr	2843.25	0.5	0.015
Hg	2536.51	0.5	0.015
Mg	2852.12	0.5	0.015[a]
Mn	2576.10	0.5	0.015
Mo	3798.25	1	0.03
Ni	3414.76	0.5	0.015
Pb	2833.06	3	0.09
Ti	3349.03	1	0.03
V	3183.98	2	0.06
W	4008.75	5	0.15
Zn	3345.02	3	0.10

[a] The calculated detection limits for aluminium and magnesium in air are not normally attainable as a result of high levels of these elements as impurities in the graphite cup electrodes.

Experimental procedure for emission spectrographic analysis [6]

Air sampling

Samples of about 2000 m^3 of air are drawn through 8 × 10 inch flash fired glass fibre filters over 24 h using a high-volume sampler.

Standardization

Prepare a stock solution of the following metals, with the concentrations indicated: Be, Bi, Cd, Co, 1 μg/0.05 ml; Sb, Cr, 2 μg/0.05 ml; Mo, Ti, 4 μg/0.05 ml; Ni, V, Sn, 8 μg/0.05 ml; Mn, 20 μg/0.05 ml; Cu, Fe, Pb, 100 μg/0.05 ml; Zn, 200 μg/0.05 ml. Salts should be better than or at least analytical grade. Make working standard solutions by diluting the stock solution to give concentrations 1/2, 1/4, 1/8 ... 1/1024 of the original One stronger standard is made by doubling the quantity on the electrode.

Add a HNO$_3$ extract of the glass filter (63 inch^2 of filter = 40 ml acid extract) directly to the electrodes in order to duplicate the conditions obtained in the particulate sample solution. This compensates for the acid soluble material present in the glass filter, some of which could interfere in the metal analysis, and eliminates the need for making separate corrections for the metals present in the filter.

Three replicate standard plates are made as follows. Warm electrodes of high purity graphite (V crater with centre post) for 10 min in an oven at about 80° C. Briefly immerse the crater portion into a hot 20% solution of paraffin in redistilled benzene, then oven dry for a few minutes. Then fill the crater 1/3 full with a lithium chloride–graphite mixture (1 part lithium chloride and 2.5 parts graphite), transfer 0.05 ml of the glass filter extract (equal to 1/800 of the filter) to the electrode, and add one drop of methanol to distribute the solution evenly. Partially dry the electrodes in an oven, then pipette 0.05 ml of the working standard (0.10 ml for the stronger standard) and add one drop of methanol. Dry the electrode at 80° C, then heat at 105° C for 1 h.

Arc the loaded electrodes as the anode for 30 s at about 7 A d.c. (NASN use a Baird 3-meter-grating spectrogaph [6]). Prior to arcing set the current at 8.25 A closed circuit and do not adjust further during the arcing period. Maintain gap at 3 mm, slit 6 mm × 25 μm. Sharpen pointed counter electrodes with a pencil sharpener.

A single-step sector, passing 1/8 of the uninterrupted beam, is interposed between the slit and the lens. Eastman 103-0 plates are used, and the first order of the region 2170 Å to 3590 Å recorded. Process the plates by standard procedures (3 min D-19, 20° C water rinse; 10 min in Kodak acid fix and 25 min wash). Measure line intensity with an NSL Spec. Reader.

Analysis lines for Cd, Be, Fe, Pb, Cr, Cu, Sn, Sb, Mn, Ni, Bi, Mo, V, Ti, Zn and Co shown in Table 5.7 are used. Plot calibration curves for each metal as a ratio of % transmittance of line plus background to % transmittance of adjacent background versus μg of metal on the electrode on log–log graph paper. Data from the curves can be consolidated in chart form.

Analysis

In order to permit more complete removal of inorganic materials than is possible by simple acid extraction, first muffle samples (26% of the original) at 500° C for 1 h to burn off organic matter. Then make two extractions with 40 ml portions of 1:1 redistilled HNO_3 at slightly below the boiling point for 1 h. After filtering the resulting solutions through Whatman No. 42 filter paper, evaporate them to 3 to 4 ml and make up to 10.4 ml (0.05 ml of this solution is equivalent to 1/800 of the sample).

Load electrodes (prepared as described above, except the addition of glass filter extract is omitted) with 0.05 ml of the sample solution. Arcing conditions, plate processing and densitometry are the same as standard plate conditions. By use of the numerical value of the line plus background to background ratio, amounts of each metal on the electrode are obtained by reference to the chart.

$$\mu\text{g of metal on electrode} \times 800/\text{m}^3 \text{ of air sampled} = \mu\text{g of metal m}^{-3} \text{ of air}$$

5.2.3.2 Ring oven methods

The ring oven is a relatively cheap and simple piece of apparatus [65]. A clean piece of filter paper is fixed horizontally on the oven surface, and a sample is introduced at its centre. Addition of solvent at the centre of the paper causes the sample to be washed outwards in a circle by capillary movements (Fig. 5.1). At a certain distance from the centre, the sample solution reaches a ring, heated to

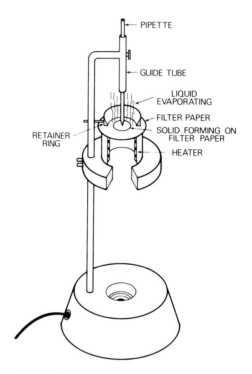

PIPETTE

GUIDE TUBE

LIQUID
EVAPORATING

FILTER PAPER

RETAINER
RING

SOLID FORMING ON
FILTER PAPER

HEATER

Fig. 5.1   Essential features of the ring oven [67].

about 15° C above the boiling point of the solvent used for washings. The solution is consequently dried, leaving a ring of sample. The filter paper may then be cut into segments, each of which is analysed for a given metal or radical. Analysis is by the addition of highly sensitive specific or selective organic colorimetric or fluorimetric reagents. Selectivity may be enhanced by the use of masking agents or other processes. Once the test colour has been developed, visual comparison with rings produced using standard samples allows accurate (90 to 95% in the μg range) quantitative determination of the unknown.

West and co-workers [66, 67] have described the application of ring oven techniques to air pollution analysis. Samples may be collected by filtration on a range of materials, or by impaction or electrostatic precipitation. Dissolution of the sample is then the normal procedure, followed by analysis of an aliquot using the ring oven. Alternatively, samples collected by filtration may be transferred to the ring oven by cutting a triangular portion, about 0.5 in per side from the filter and using it as the sample. This section of filter is fixed at the centre of the ring oven, beneath the normal clean filter paper, and washing out of the sample is performed in the usual manner. In the case of samples collected with an automatic tape sampler, if the sample spot is smaller than the heated ring of the ring oven, the section of tape containing the appropriate spot may be transferred directly to the centre of the oven and analysed as described. West reports that small air samples (1 m^3) collected using an automatic tape sampler allow as little as 2 to 3 μg of a constituent metal of the air sample to be determined with good accuracy [66].

A one-dimensional concentration technique inspired by the ring oven procedure has been used to determine lead at the 10 to 100 ng level collected on air filters, by a colorimetric reaction with sodium rhodizonate [68].

Table 5.9 shows the detection limits of the ring oven method for a number of metals under ideal conditions [66, 67, 69]. Interference effects of other ions may affect the results obtained as many reagents are not entirely specific, and this fact must always be taken into account when performing analysis by this technique.

Table 5.9 Detection limits for ring oven procedure [66, 67, 69]

Metal	Detection limit (μg)^a	Working range (μg)
Al	0.01	0.03– 0.5
Be	0.01	0.01– 0.2
Cd	0.075	
Co	0.02	0.04– 0.5
Cr	0.15	0.3 – 1.0
Cu	0.04	0.1 – 0.5
Fe	0.01	0.01– 0.5
Mn	2.0 (malonic acid [67]) 0.075 (benzidine [66])	2.0 –10.0
Ni	0.08	0.10– 1.0
Pb	0.015	
Sb	0.08	0.1 – 1.0
Se	0.08	0.1 – 0.5
V	0.01	0.01– 3.0
Zn	0.04 (*o*-mercaptothenalaniline [67]) 0.075 (ammonium mercury thiocyanate [66])	0.05– 1.0

^{a} In the total ring.

Experimental procedure for ring oven analysis [66]

Air sampling

Samples are collected by filtration on any filter medium with a low metal blank.

Sample preparation and concentration

Cut out a triangular portion, averaging 0.5 in per side, from the filter. Place the triangle centrally upon a circle of filter paper (e.g. Whatman No. 40) and fix it in place. Membrane filters, made of cellulose acetate, may be fixed by slight moistening of the paper filter with acetone, and pressing on to the membrane filter triangle when the acetone is almost dry. The use of too much acetone may cause clogging of the pores of the filter paper, or occlusion of collected particulate matter by cellulose acetate, and is to be avoided. The filter paper with triangle in place is put in position across the surface of the ring oven, the triangle on the underside. The addition of 8 to 10 portions (5 μl) of 0.1 N HCl to the centre of the filter paper causes transfer and concentration of the sample in a ring on the filter paper.

If paper filters are used for air sampling, a triangular portion is used as above, but is fixed to the receiving filter paper with tiny dots of glue, applied by a fine glass needle. The triangle is placed above the receiving filter paper, and centred directly beneath the guide tube on the oven. Samples collected on glass fibre filters may be treated in a similar manner.

Analysis

The concentrated sample, in the form of a ring, is divided by cutting the receiving filter paper into segments, each of which is analysed for a particular metal.

(a) Al. Spray the sector with a saturated solution of morin in methanol, dry and bathe in 2 N HCl. Observe under u.v. light, while still moist. A yellow–green fluorescent line indicates Al.

(b) Cd. Spray the sector with a solution of cadion in ethanol, containing some NaOH. A red–pink line denotes Cd.

(c) Cr. Oxidize the chromium (III) to chromate with a 1:1 mixture of 10% H_2O_2 and NH_3 solution. (Fill a fine glass capillary with oxidizing mixture and run the capillary rapidly along the circular line on the section; thus only a very narrow moistened line is obtained.) Dry the sector in a drying oven and spray with a freshly prepared 1% solution of diphenylcarbazide in ethanol. Finally dip into 2 N H_2SO_4. A violet line appears if Cr is present.

(d) Co. Fume the sector over NH_3 solution, apply 5% disodium hydrogen phosphate to mask iron, and spray with a 1% solution of 1-nitroso-2-naphthol in acetone. A red–brown line appears if Co is present.

(e) Cu. Fume the sector over NH_3 solution and spray with a 1% solution of dithio-oxamide in ethanol. An olive–green to black line indicates Cu.

(f) Fe. Fume the sector in HCl, and if necessary over bromine water, and finally spray with a 1% potassium ferrocyanide solution. A blue line denotes Fe.

(g) Pb. Treat the sector with freshly prepared 0.2% aqueous solution of sodium rhodizonate and fume over HCl while the yellow colour of the reagent disappears. A violet to red line indicates Pb.

(h) Mn. Treat the sector with 0.05 N KOH solution and spray with a 0.05% solution of benzidine (Care: Carcinogenic Reagent) in dilute acetic acid. A blue line indicates Mn.

(i) Ni. Fume the sector over ammonia solution and spray with a 1% solution of dimethylglyoxime in ethanol. A red line denotes Ni. If the ring zone contains iron, it is advisable to bathe the developed sector in a tartrate solution.

(j) **Zn.** Moisten the ring zone on the sector with ammonium mercury thiocyanate solution (3.3 g of ammonium thiocyanate + 3 g of mercury (II) chloride are dissolved in 5 ml of water without warming). Bathe and agitate the sector in a 0.02% cobalt solution. A blue line denotes Zn. Blue spots, which might eventually appear on other parts of the paper, should be ignored.

Calibration

A standard solution is prepared so as to have a concentration of about 0.01 mg of metal ml^{-1}. Standard rings are prepared by adding 10, 20, 40, 60, 80 and 100 μl of the standard to each of 6 separate filter discs using a 10 μl pipette. The added metal solutions are washed with 5×10 μl portions of 0.1 N HCl to transfer the metal to the ring zone. Appropriate reagents are then used to develop a standard series of colours. Unknowns are determined by comparison with the standards, greater precision being achieved by running the sample 3 times, each time using a different sample size.

5.2.3.3 *Polarography*

A technique which has long been applied to the analysis of metals, polarography is unlikely to be a preferred technique for air pollution analysis. Although it is less sensitive than many methods currently available for the analysis of metals, the application of the technique to the analysis of airborne particulate matter has been described [67, 70].

The technique is electrochemical and the polarographic cell in which the sample solution is placed contains 2 electrodes. A pool of mercury in the bottom of the cell serves as a large non-polarizable reference electrode, while a stream of small mercury drops (typically 2 to 6 s per drop), known as a dropping mercury electrode, serves as a polarizable active electrode with a continuously renewed surface [71]. A steadily increasing voltage is applied to the cell and the current passing through the cell is measured. A slow rise in current with applied potential is found, with much faster rises occurring at characteristic potentials corresponding to the electrolysis of specific chemical species. A relatively high concentration of a 'supporting electrolyte' is included in the sample solution and an appropriate choice of this electrolyte can cause separation of the electrolysis steps of the different metal ions in the solution, allowing quantitative determination of 5 or 6 substances from a given polarogram (Fig. 5.2). West [67] reports that although sensitivities as high as 0.1 μg ml^{-1} may be attained for some metals, in general a more realistic figure is 1 μg ml^{-1}. Sample volumes of around 1 ml may be used. Some improvement in sensitivity has resulted from refinements applied to conventional polarographic techniques [71, 72].

A major problem in polarographic analysis is the determination of a constituent of low concentration in the presence of large amounts of more easily electrolysable substances, which, in the case of metal analysis, may include organic materials. In general, two techniques are available to overcome this problem. Firstly, selective extraction of the minor constituent is a valuable

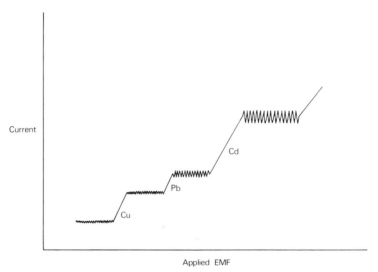

Fig. 5.2   A polarogram [67].

method, and this may allow a pre-concentration if extraction is into an organic solvent [73]. Secondly, complexing of the interfering substances to produce less readily electrolysable forms may prove successful.

Application to air pollution analysis has been described by Levine [70]. Samples collected by impingement or electrostatic precipitation were wet ashed, and then dissolved in an appropriate supporting electrolyte. Polarographic analysis of the sample and comparison with a series of standards prepared in a similar manner to the sample, and hence in a comparable matrix, enabled the determination of Pb, ZnO fumes, Cd fumes, chromic acid fumes or mists and Mn fumes in air to be carried out.

Experimental procedure for polarographic analysis of Pb and Cd [70]

Air sampling

The air is sampled with a modified Greenberg–Smith impinger or an electrostatic precipitator, and in an industrial environment about 1 m^3 of air is sampled.

Analysis for Pb

Transfer the collected sample to a casserole, add 6 N HCl (5 ml), 6 N HNO_3 (1 ml) and evaporate to dryness on a steam bath. Take up the residue in exactly 5 ml of 20% citric acid solution. Pour the contents of the casserole into the electrolytic cell, expel oxygen and polarograph at 1/5 the galvanometer sensitivity (0.015 μA mm^{-1}) in an applied voltage range of -0.4 to -0.8 V. Measure the step height and read the concentration on the calibration curve.

Calibration curve for Pb

Prepare a standard solution of lead nitrate containing 50 μg lead ml^{-1}, and add the following quantities of Pb to casseroles containing water (75 ml), 6 N HCl (5 ml) and 6 N HNO$_3$ (1 ml): 250, 200, 150, 100 and 50 μg. Evaporate to dryness on the steam bath. Take up in exactly 5 ml of 20% citric acid, pour into the electrolytic cell and polarograph as above. A graph plotted of step heights against concentration in μg ml^{-1} is linear.

Analysis of Cd

Transfer the sample from the impinger or electrostatic precipitator into a casserole, add 6 N HCl (5 ml), and evaporate to dryness on a steam bath. Take up the residue with exactly 5 ml of 20% citric acid solution containing 3 drops of 0.1% methyl red solution per 100 ml. Pour into the cell and polarograph at 1/20 the galvanometer sensitivity (0.06 μA mm^{-1}) and at a voltage range to include the empirical half-wave potential of about -0.6 V. Measure the step height and read the concentration on the calibration curve.

Calibration curve for Cd

Prepare a standard solution of a Cd salt having a concentration of 25 μg Cd ml^{-1}. Using this solution, add the following quantities of Cd to casseroles containing water (75 ml) and 6 N HCl (5 ml): 250, 200, 150, 100, 50 and 25 μg. Evaporate the solutions to dryness, take up the residues in 5 ml of 20% citric acid solution containing 3 drops of 0.1% methyl red solution per 100 ml, and polarograph as above. A graph is plotted of step height against concentration in μg ml^{-1}.

5.2.3.4   *Anodic stripping voltammetry*

This technique is closely related to polarography, and samples are prepared in a similar manner. Application of a negative voltage is used to cause electrodeposition of metal ions from the sample solution on to a Hg or solid electrode. Subsequent linear variation of the electrode potential in an anodic direction causes rapid dissolution with a corresponding sharp current peak. Quantitative deposition and stripping of an ion allows calculation of the concentration of that ion from the integrated stripping current. Alternatively, a reproducible fraction of the ion is electrodeposited and subsequently stripped. The stripping current is measured and the concentration of the ion determined by comparison with standards. Since the deposition causes a very substantial concentration of the ions, the sensitivity is increased substantially over that of conventional polarography, at best by several orders of magnitude [74].

Anodic stripping voltammetry has found application in water analysis [74] and has been comparatively assessed with other instrumental techniques for the analysis of metals in gasoline and other matrices [75]. Air samples collected over 24 h on glass fibre filters have been analysed for Pb, Cd, Cu and bismuth by anodic stripping voltammetry [76], and Colovos *et al.* [77] used the technique to determine Pb, Cd, Cu and Zn. In the latter study samples were collected on Millipore filters and ashed in a low temperature asher after addition of 0.2 ml 0.1 M potassium sulphate, and prior to analysis were digested with hydrofluoric acid in a PTFE bomb.

MacLeod and Lee [78] have described the analysis by anodic stripping voltammetry of metals collected in a 2 h sample by an AISI tape sampler. After ashing, the spot tape sample was dissolved in acid and diluted prior to analysis. Airborne Cd (7 to 350 ng), Pb (80 ng to 2.4 μg) and Cu (6 ng to 1 μg) were determined and replicate determinations with standard solutions show a maximum relative standard deviation of less than 12% due to instrumental variability or operation error. It was suggested that the method could be easily extended to measure other ASV responsive metals including Zn, Bi, Ag, Tl and Sb. With a view to further enhancing sensitivity, Ryan and Siemer [79] analysed samples of Pb, Cu and Cd collected on a porous graphite filter using a voltammetric cell of volume only 0.1–0.3 ml.

Experimental procedure for ASV analysis of Pb, Cd *and* Cu [78]

Air sampling

Samples are collected using AISI tape samplers, in which ambient air is drawn through a circular portion 1 inch in diameter of a continuous strip of Whatman No. 4 filter paper 2 inch wide. A sampling interval of 2 h is used, after which the tape is automatically advanced.

Reagents

Buffer: 0.1 M NaCl–sodium acetate solution made with ACS reagent grade chemicals. The buffer is cleaned and stored in an Environmental Science Associates (ESA) reagent cleaning system 2014P at a reduction potential maintained at -1400 mV.

Water is de-ionized or double distilled in an all-glass still before use and redistilled lead-free grade nitric and perchloric acids are used to prepare 1:1 (v/v) perchloric–nitric acid solution. All glassware is precleaned by soaking for 48 h in a 10% by volume solution of perchloric acid in water.

Procedure

Cut a circle 0.75 inch in diameter (representing 56.24% of the total exposed area) from the centre of each AISI tape sampler spot using a nickel-burnished steel cork-borer. Place each cut-out in a borosilicate glass sample boat and ash for 30 min in a low-temperature asher (Tracerlab LTA 600) at 200 W with an oxygen flow of 80 cm^3 min^{-1}. After 30 min, only a light residue is left. Carefully add 1:1 perchloric–nitric acid solution (0.1 ml) to the boat and transfer the dissolved sample quantitatively with water to a 10 ml volumetric flask. This sample is stable for 2 weeks.

All instrumental parameters are held constant for the analysis. Keep plating times and stirring times to within 5 s and keep the stripping rate constant throughout pH is controlled by the addition of identical amounts of buffer and acid solution each time. The mercury coating is regulated by recoating the electrodes overnight with a dilute mercury solution; slight changes are compensated by running standards on each cell several times a day.

Pipette a 5 ml aliquot of buffer into each cell and place the cells on the cell holders. At equal intervals (30 s) switch the cells to the plating mode and turn on the nitrogen stream. 15 s before the end of each plating time turn off the nitrogen stream. With the cell turned to 'strip' turn the sweep rate from 'reset' to 'on' and strip the cell. Do this to each cell in turn. Using an ESA 4-cell multiple anodic stripping unit, MASA 2014, with 4-cell extension unit PMI 1014S the instrumental settings are as follows: hold the unit with a positive sweep rate of 60 mV s^{-1}, auto sweep hold 'on' at 1070 mV, the initial potential at -1100 mV and the plating potential of the cells at -1100 mV. Set the current range at 0.5 mA full scale with a recorder sweep of 10 s inch^{-1}.

Add a 1.0 ml aliquot of sample to the buffer in each cell and repeat the plating and stripping sequences. At the end of the stripping sequence, each of the cells is again stripped to ensure that all the metals are removed from the electrode.

Run standards by addition to the sample in the cell, giving a recorder trace of buffer plus sample plus standard, and thus matrix effects are taken into account.

Peak heights of the recorder trace are proportional to the quantity of each metal present in the solution, allowing calculation of the concentration of metal present.

Unused paper-tape is similarly analysed to determine the metal blank.

5.2.3.5 *Spark source mass spectrometry*

Mass spectrometry is a highly sensitive analytical technique which has, until recently, found little application to inorganic substances. The advent of the spark source, however, has allowed the volatilization and ionization of relatively involatile inorganic substances. The sample is coated on electrodes across which a spark is induced by application of an electrical potential. Three basic types of source are used [80]. In the RF spark source an a.c. potential in the form of pulses of a few microseconds duration is applied to the electrodes causing sparking. The second type of source, the vacuum vibrator, involves 2 electrodes with a low potential difference (10 to 30 V) between them. Mechanical vibration of one electrode causes it to make contact with the other, once in each vibration, and upon parting again a spark is generated across the electrodes. Thirdly, in the direct current hot arc source, a condenser is used to build up a charge between stationary electrodes. A high voltage trigger causes spark formation and discharge of the condenser, after which a new cycle starts.

Conventional mass spectrographic or spectrometric focusing techniques focus the ions formed in the spark on the basis of their mass/charge ratio. Detection may be on a photographic plate [80], or by scanning with an electrostatic and electromagnetic analyser followed by detection by an electron multiplier [81].

Brown and Vossen [81, 82] have described the application of this technique to air pollution analysis. A 10.8 m^3 air sample collected on a nitrocellulose or cellulose acetate filter was pretreated by low temperature ignition of the nitrocellulose or dissolution of the cellulose acetate in acetone after addition of graphite. After removal of the filter material and addition of a 5 μg silver standard, the sample was ground and pressed as a tip on pure graphite electrodes. All elements except hydrogen and helium may be determined using the technique and in an urban air sample of 10.8 m^3, elements may be determined with an estimated precision of plus or minus 30% s.d. A detection limit of 10 pg is attainable [82].

The technique has also been used to determine metals in dustfall [81], in fly ash and in other matrices and a comparison of these results with those from other instrumental analytical techniques has been reported [75].

Experimental procedure for spark source mass spectrometry [82]

Air sampling

Air is drawn through a nitrocellulose filter pad for 9.5 h at 19 litre min^{-1}.

Sample preparation

Place the filter pad in a clean silica boat with 0.1 g of high purity graphite (Graphite type USP supplied by Ultra Carbon Corporation, Michigan, USA). Add high purity ethanol (1 ml) as a wetting agent and 100 μl of an aqueous solution of silver nitrate containing 5 μg of Ag as internal standard. Exercising extreme caution, ignite the alcohol, causing slow combustion of the nitrocellulose, and finally to ensure complete ashing place the boat in an oven at 450° C for 1 h.

Transfer the resultant mixture to a desiccator to cool, and then grind thoroughly in a clean agate vial with an agate ball pestle. After grinding, compress portions of the sample on to a support substrate of pure carbon. Encapsulation of the sample rods (3/8 inch length; 3/32 inch diameter) in polyethylene during the pressing operation prevents contamination.

Procedure

Mount the sample electrodes in the source sample clamps so that the sample tips form the analysis gap in front of the first slit. A vibrator is used in place of the micromanipulators for manual adjustment of the electrode gap. Record 5 scans at a suitable multiplier gain. Using an AEI MS702 instrument, a multiplier gain of 10^5 allowed a limit of detection of 0.004 μg m^{-3} for Pb. A scan time of 9 min allows examination of all the elements in the periodic table, apart from hydrogen and helium. Spark excitation parameters were as follows: RF voltage, 30 kV; spark pulse repetition rate, 1000 pulses s^{-1}; spark pulse length, 100 μs. The ion extraction voltage was 20 kV, and cryo-absorption pumping was found to be advisable.

Calculation of results

Measure the height of a peak of an isotope of each element and the peak obtained from the m/e 107 isotope for Ag, the internal standard. For an Ag internal standard of 5 μg, and an air sample of 10.8 m³, element concentrations are calculated from the following expression

$$5 \times \frac{A_i}{A_s} \times \frac{I_s}{I_i} \times \frac{\text{at wt}_i}{\text{at wt}_s} \div 10.8 = \text{concn. in } \mu\text{g m}^{-3}$$

where A_i = peak height impurity isotope; A_s = peak height $^{107}\text{Ag}^+$; I_s = isotope abundance of ^{107}Ag; I_i = isotope abundance of impurity isotope; at wt$_i$ = atomic weight impurity element; at wt$_s$ = atomic weight of silver. Determination of the metal content of a blank filter is advisable.

5.2.3.6   *Spectrophotometry and fluorometry*

Spectrophotometric methods, other than atomic absorption, have long been used in air pollution analysis, but because of their generally inferior sensitivity, they are steadily being replaced by more modern techniques. Airborne particulate matter is collected by a normal procedure, ashed and a solution prepared in the usual manner. By addition of a reagent, a complex compound of the metal is formed, which absorbs light in the visible or u.v. region. Spectrophotometric measurement of the light absorption in a narrow band of wavelengths, and comparison with standards, allows calculation of the concentration of the metal complex. Microgram quantities of metals may be determined, but since many reagents form light-absorbing compounds with a range of metals, selectivity must be enhanced by fixing of oxidation states, the use of masking agents and control of pH. Colorimetric, spectrophotometric and related methods are described in detail by Cheng [83].

Spectrometers and spectrophotometers vary considerably in design and sophistication, but Fig. 5.3 shows in block form the basic design of a simple single beam spectrometer. Light from the source is passed through a cell containing the

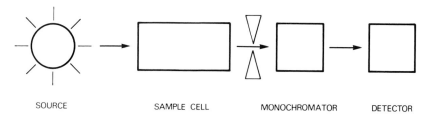

SOURCE   SAMPLE CELL   MONOCHROMATOR   DETECTOR

Fig. 5.3   Essential features of a simple single beam spectrometer.

sample solution. The beam then passes to a monochromator which selects a narrow band of wavelengths, normally at the absorption maximum of the metal compound, and transmits light in that band to the detector, a photocell which measures the intensity of the beam. The light absorption of a cell containing the reagents, but not added sample, is also measured and used to zero the instrument. The transmittance of the sample, T, is defined as the ratio of the radiant power transmitted by a sample to the radiant power incident upon the sample [84], and the absorbance, $A = \log_{10} 1/T$.

Normally, either transmittance or absorbance is measured by the instrument and this may be used to estimate the concentration of metal in the sample by use of a calibration curve. Beer's law applies to most measurements and this may be stated as $a = A/bc$ where a, the absorptivity, is a constant for the given light absorbing substance; c is the concentration of the substance in solution and b is the sample length, or internal length of the sample cell. The molar absorptivity, ε, also known as the molar extinction coefficient, is a commonly quoted figure and is a constant for a given compound

$$\varepsilon = a \times \text{Molecular weight}$$

An important extension of Beer's law is the 'law of additivity' which states that absorption of one molecular species will be unaffected by the presence of other species, absorbing or otherwise. Hence in a mixture of absorbing substances, the absorbances are additive.

Literature relating to spectrophotometric reagents and determination of metals is legion. A valuable list of reagents, molar absorptivities, specificities and experimental methods has been compiled by IUPAC [85]. The application of these methods to the analysis of air pollutants has been described in detail for many metals by West [67], the American Public Health Association [86] and by Jacobs [87], and hence will not be reviewed in detail here.

Probably the most frequently described determination by a spectrophotometric method is that of airborne lead with dithizone [67, 86–91], reported to have a threshold sensitivity of 0.2 μg Pb and a precision on any given result of plus or minus 0.1 μg Pb [88, 89]. Dithizone may, by adjustment of conditions, also be used for determination of other metals such as Hg, Ag, Bi, Sb, Cd and Zn [67, 92]. Spectrophotometric methods have also been recommended for the determination *inter alia* of As [93], Be [94, 95] and Ni [96] in air. Detection limits normally lie within the region of 0.1 to 1 μg of the metal [67].

A method reported for determination of chromic acid mists in air involves direct impregnation of the paper used for air filtration with the reagent, and glycerol as a humectant. The Cr is determined by comparison of the colour of the filter with standards after completion of air sampling [24, 97].

Absorption of light in the visible or u.v. region by a molecule involves an electronic excitation. One mechanism by which the molecule may lose energy is first by loss of vibrational energy to return to the ground vibrational state of the excited electronic state, and subsequent emission of light energy, which allows a return to the ground electronic state. The emitted light is known as fluorescence, and is of longer wavelength than the absorbed exciting radiation [83]. Fluorescence measurements, using a fluorometer or spectrofluorometer, usually involve irradiation of the sample by monochromatic light in the u.v. region. Measurement of fluorescent intensity at the emission maximum, via a second monochromator, allows determination of the concentration of the fluorescent compound, upon comparison with standard samples. Be may be determined as the morin (2', 4', 3, 5, 7-pentahydroxyflavone) complex by fluorescent measurement [58, 76, 80] and high sensitivity may be attained, allowing measurement of as little as 0.01 μg of Be. Selenium (4^+) may also be determined by fluorescence after reaction with 2, 3-diaminonaphthalene [86].

Experimental procedure for determination of Pb *with dithizone* [91]

Air sampling

Approximately 150 to 200 m^3 of air are drawn at 20 litre min^{-1} through a 47 mm glass fibre filter.

Reagents

Water is distilled or de-ionized before use. Nitric acid–perchloric acid solution is prepared by mixing conc. HNO_3 (300 ml) with 72% perchloric acid (200 ml). HNO_3 (1 + 4) is prepared by dilution of conc. HNO_3 (200 ml) with water to 1 litre. Dithizone solution is made by dissolution of diphenyl-thiocarbazone (40 mg) in chloroform (1 litre). It is stored at room temperature in the absence of direct light. Disodium EDTA solution is prepared by dissolution of disodium ethylene-diaminetetracetate (5 g) in water (500 ml). Buffer solution is prepared by solution of dibasic ammonium citrate (400 g), hydroxylamine hydrochloride (10 g) and potassium cyanide (40 g) in water (1 litre), and subsequent mixing with conc. (sp. gr. 0.90) ammonium hydroxide (2 litres).

Procedure

Thallium, stannous tin and trivalent indium can interfere if present in quantities greater than 20, 100 and 200 μg respectively, when a modified procedure must be employed. In the absence of such interferences proceed as below.

Remove the filter from the filter holder and place it, exposed side down, in a 150 ml beaker and add nitric–perchloric acid mixture (10 ml). Digest on a hot-plate to fumes of perchloric acid (use a fume hood, and exercise extreme care), until all of the dark carbonaceous material has oxidized. Add HNO_3 (1 + 4) (20 ml), mix, crush the filter with a glass rod and allow to cool. Filter the sample through a thin, rapid-filtering paper hardened to great wet strength, directly into a lead-free 100 ml volumetric flask. Rinse the glass fibre with water (3 × 20 ml), make up to 100 ml, stopper and shake the volumetric flask to mix well.

Pipette a suitable aliquot (normally 10 ml) of sample to a 200 ml modified absorption cell and add HNO_3 (1 + 4) [20 ml], water (25 ml), buffer (50 ml), mix and cool to room temperature. The modified absorption cell consists of a normal 10 mm path length spectrometer cell with a 200 ml stoppered glass bulb fused to the top (Fig. 5.4). Add dithizone solution (10 ml) and shake vigorously for 30 s. Insert the cell in the spectrophotometer and measure the absorbance of the lower layer at 510 nm, using air as reference. Add disodium EDTA solution (5 ml), shake vigorously for 90 s and measure the absorbance of the lower layer from 1 to 5 min after the two layers separate. The difference between the two absorbance readings represents the quantity of Pb present in the aliquot. If the initial absorbance reading is greater than 2.0, add additional dithizone solution (10 ml

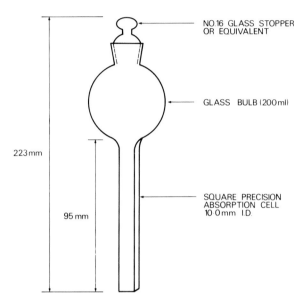

NO.16 GLASS STOPPER
OR EQUIVALENT

GLASS BULB (200 ml)

223 mm

SQUARE PRECISION
ABSORPTION CELL
10.0 mm I.D.

95 mm

Fig. 5.4 Modified absorption cell [91].

portions) to the sample to dilute the lead dithizonate colour, shake vigorously for 30 s and repeat the absorbance measurements before and after adding EDTA solution.

Calibration

A lead standard is prepared by solution of lead nitrate (0.1599 g; analytical grade) in water (about 200 ml). After addition of conc. HNO_3 (10 ml) the solution is diluted to 1 litre with water. Further dilution of the above solution (20 ml) to 1 litre provides the necessary lead standard (2 μg lead ml^{-1}). Add HNO_3 (1 + 4) [20 ml], water (25 ml) and buffer solution (50 ml) to a 200 ml modified absorption cell, mix and cool to room temperature. Add dithizone solution (10 ml) and shake vigorously for 30 s. Insert the cell into the spectrophotometer and measure the absorbance of the lower layer as 510 nm using air as reference. Add standard lead solution (2 μg ml^{-1}; 10 ml), shake the mixture vigorously for 30 s and measure the absorbance due to the added lead. Add further standard lead solution (10 ml) and repeat the above procedure.

Calculation

Calculate the calibration factor

$$F = X/(Y - Z)$$

where $X = \mu$g lead in calibration sample; $Y =$ absorbance after adding Pb; $Z =$ absorbance before adding Pb.

Calculate the particulate Pb

$$C_p(\mu g\ m^{-3}) = (A - B) - (C - D) \times F \times (35.3/H) \times (G/10) \times (100/E)$$

where A = sample absorbance before EDTA treatment; B = sample absorbance after EDTA treatment; C = blank absorbance before EDTA treatment; D = blank absorbance after EDTA treatment; E = volume (ml) of aliquot removed from 100 ml volumetric flask; G = volume of dithizone solution (ml); H = volume of air sampled (ft^3).

The blank values are obtained by analysis of an unexposed filter by the above procedure.

Experimental procedure for fluorimetric determination of Be [86]

Sample collection

Ambient air is sampled for 24 h through a suitable filter.

Sample preparation

(Cellulose or membrane filter sample.) Transfer the filter to a Vycor dish and wet with 1:5 H_2SO_4. Heat on a hot-plate until charring occurs, then add a few drops of conc. HNO_3 and evaporate to dryness. Fire with a Meeker burner until carbon-free (a muffle furnace may be used). Allow the dish to cool, wet the ash with mixed acid (50% conc. HNO_3–50% conc. H_2SO_4) and evaporate to fumes on a hot-plate or burner. Repeat if not carbon free. Cool, add a few drops of 1:1 H_2SO_4 and heat to strong sulphur trioxide fumes. Remove from hot-plate, cool and add enough water to dissolve salts. Transfer to a 50 ml volumetric flask and make up to volume. (Samples containing beryl ore or high fired beryllium oxide which is difficult to put in solution should be decomposed using a potassium fluoride, sodium pyrosulphate fusion [98].)

Analysis

Pipette sample solution (1 ml) into a 15 ml graduated centrifuge tube. Add 0.05 M aluminium nitrate solution (1 ml), 25% w/v ammonium chloride solution (1 ml) and phenol red (1 drop), and then enough 3 N NH_3 solution to make the solution alkaline. Allow to stand for 10 min, and then centrifuge at 2000 rev min^{-1} for 5 min, rotate the tube 180° in the centrifuge and centrifuge for 5 min further. Pour off and discard the supernatant liquor. Dissolve the precipitate in 4 N NaOH (1 ml), add 25% w/v potassium cyanide (1 drop) and make up to 10 ml with water. Centrifuge for 10 min at 2000 rev min^{-1} and pour the liquid into a Coleman cuvette (19 mm diameter × 100 mm long). Measure the fluorescence in a fluorimeter set at 100% fluorescence with a solution of quinine (4 g litre^{-1}) in 0.1 N H_2SO_4. The fluorimeter should be equipped with a filter transmitting approximately 436 nm between lamp and sample and a filter between sample and photocell which transmits a maximum fluorescent emission of 550 nm. Add morin reagent (1 ml) to the sample, mix and read the fluorescence immediately. The morin reading, minus the sample reading without morin, equals the net instrument reading. The morin solution is prepared by dissolution of morin (20 mg) in ethylene glycol (5 ml) and dilution to 100 ml with water. Subsequent dilution of 12 ml of this stock solution to 200 ml with water gives the working solution. The stock solution is stable for 1 month if stored in a dark bottle in the refrigerator, and the working solution, which must stand at least 8 h prior to use, may be stored in a dark bottle in the refrigerator.

Calibration

The equivalent of 0.1 g of pure beryllium powder (the available powder is not pure) is dissolved in 1:1 H_2SO_4 (15 ml). After cooling it is diluted to 1 litre with water. This stock solution (100 μg beryllium ml^{-1}) is then diluted ten-fold (10 μg Be ml^{-1}), one hundred-fold (1 μg Be ml^{-1}) and 2000-fold (0.05 μg Be ml^{-1}). The latter solutions are prepared daily before use. Add 2, 5 and 10 ml portions of Be standard solution (0.05 μg ml^{-1}) and a 1 ml portion of (1 μg Be ml^{-1}) to filters similar to those used for sample collection. Run the impregnated filters through the above procedures.

Calculation

$$\text{Total Be in sample} = \frac{\text{ml sample}}{\text{ml aliquot}} \times \frac{\text{net sample I.R.}}{\text{net standard I.R.}} \times \mu\text{g Be in standard}$$

I.R. = instrument reading; ml sample = volume of dilution of total sample; ml aliquot = portion of sample used for analysis; net sample I.R. = I.R. sample with morin minus I.R. sample without morin; net standard I.R. = I.R. with standard minus I.R. of standard (without morin).

$$\mu\text{g Be m}^{-3} \text{ of ambient air} = \frac{\text{Total of } \mu\text{g Be}}{\text{Volume of air sampled (m}^3)}$$

The procedure is recommended for analysis of 0.01 to 1.0 μg of Be in 10 ml of solution, but greater sensitivity can be achieved. Samples can be stored indefinitely without loss of Be.

5.2.3.7   Atomic spectroscopy

An atomic absorption spectrometer bears much similarity to a u.v./visible spectrometer. In place of a cell of sample, however, a flame fills the sample space (Fig. 5.5). Aspiration of the sample solution into the flame causes chemical reduction of metal ions to ground state atoms, the light absorption of which is measured. Atomic absorption spectra are line spectra, unlike molecular spectra which are in the form of broad bands. The atomic absorptions are discrete lines of narrow bandwidth at wavelengths which are characteristic of the given element.

The light source, a hollow cathode lamp in which the cathode is coated with the metal to be determined, emits light at the discrete absorption wavelengths. Introduction of sample into the flame causes absorption of some of the incident light and after passage through the monochromator, which selects one absorption line, the reduction in beam intensity is measured by the detector. In

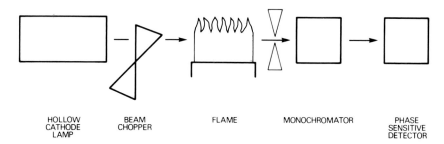

HOLLOW   BEAM   FLAME   MONOCHROMATOR   PHASE
CATHODE   CHOPPER   SENSITIVE
LAMP   DETECTOR

Fig. 5.5   A simple atomic absorption spectrometer.

double beam spectrophotometers, a second reference light beam from the same hollow cathode lamp by-passes the flame and is received at the detector, in alternate pulses to the sample beam and the ratio of beam intensities, after passage through the monochromator, is measured.

The wavelength range of atomic absorption measurements is limited by light absorption by the flame, most usually air–acetylene, at shorter wavelengths, and by the atmosphere at longer wavelengths. This working range, in the u.v. and visible region, includes all metals and semi-metals but excludes most non-metallic elements. It is reported that 65 elements may be determined by atomic absorption methods, although sensitivities vary substantially [86, 99, 100].

The absorption laws applying to spectrophotometric measurements apply to atomic absorption, and most commercial instruments read out in absorbance units. Beer's law of the linear relationship between absorbance and sample concentration applies to most measurements, but deviations occur, particularly at higher concentrations and it is normally advisable to calibrate over the full range of measurements. Additionally, since instrumental parameters are liable to change, it is normal practice to calibrate the instrument at least daily.

Two techniques related to atomic absorption are also currently in use. Some elements, notably the alkali metals and alkaline earths, are electronically excited by the flame, to an extent sufficient for the effect to be analytically useful. The return to the ground electronic state is by emission of light at discrete characteristic wavelengths, and measurement of emission intensities in the absence of a hollow cathode lamp allows determination of metal concentration by comparison with standards. This technique is known as emission spectroscopy or flame photometry. The other related technique, atomic fluorescence spectroscopy, does involve use of a light source. Absorption of light by the atoms in the flame occurs and results in electronic excitation. Return to the ground state may be by fluorescent light emission, and the measurement of fluorescent intensity allows determination of metal concentrations. Atomic fluorescence may be more sensitive than conventional atomic absorption spectroscopy [101].

Detection limits for some metals by conventional atomic absorption techniques are shown in Table 5.10. Double beam instruments normally allow a limit of detection several times lower than that attainable with a single beam. In

Table 5.10 Atomic absorption detection limits

Metal	Analytical line (nm)	Flame[a] (μg ml^{-1})	Sampling boat (ng)	Graphite atomizer (pg)
Ag	328.1	0.002	0.2	0.1
Al	309.3	0.02[b]		2
As	193.7	0.05[c]	20	10
Be	234.9	0.002[b]		3
Ca	422.7	<0.0005		
Cd	228.8	0.002	0.1	0.1
Co	240.7	0.01		5
Cr	357.9	0.003		10
Cu	324.7	0.001		2
Fe	248.3	0.005		3
Hg	253.6	0.25	20	
K	766.5	<0.002		
Mg	285.2	<0.0001		
Mn	279.5	0.002		0.2
Mo	313.3	0.02[b]		3
Na	589.0; 589.6	<0.0002		
Ni	232.0	0.002		10
Pb	283.3	0.01	1	2
Sb	217.6	0.04		20
Se	196.0	0.05[c]	10	50
Sn	224.6	0.01[c]		100
Sr	460.7	0.002		5
V	318.3; 318.4; 318.5	0.04[b]		100
Zn	213.9	<0.001	0.03	0.05

[a] Double beam instrument; [b] Nitrous oxide–acetylene flame; [c] Argon–hydrogen-entrained air flame. Data cited for Perkin-Elmer spectrophotometer and heater graphite atomizer.

estimating absolute detection limits it must be borne in mind that normally several ml of samples are required to obtain one reading. The introduction of the 'Sampling Boat' technique, in which an aliquot of sample (10 μl–1 ml) is dried in a tantalum boat and then inserted into the flame causing rapid volatilization of the metal [102], and the advent of flameless atomizers has allowed a substantial improvement in detection limits (Table 5.10). Flameless atomizers involve use of a graphite tube, cup, rod or ribbon whose temperature is controlled and may be rapidly varied [103, 104]. Hence an aqueous sample (about 20 μl) may be dried at 100° C, ashed at 550° C and atomized by raising the temperature to 2000° C, the graphite assisting chemical reduction of the metal ions. A sharp peak, corresponding to atomization, is recorded.

Although spectral interferences, involving overlap of absorption lines, are not a problem in atomic absorption measurement, other types of interference may occur. Chemical and ionization interferences occur for some metals, but these are well documented and may be overcome [99]. Matrix effects are common in air pollution measurements by atomic absorption, and must be taken into account. Physical properties of the matrix (e.g. viscosity) affect the rate of sample aspiration into the flame when using conventional techniques. Additionally, other materials present in the matrix affect the rate of atomization of the metal, and hence affect the absorption signal. It is not possible when dealing with air pollution samples to make standard solutions containing identical matrix materials to those in the sample, and hence use of the method of standard additions [105] is advisable, and its use in air pollution measurements has been described [106–108]. Three or more aliquots of sample are taken, and the first diluted to a known volume with solvent. The other aliquots are diluted with solvent to the same volume after addition of suitable quantities of standard solution of the metal to be determined. The two, or more, additions are different, and are calculated to approximately double or treble the concentrations of metal in the diluted solutions. A graph of measured absorbance against standard addition may then be plotted after A.A. measurement (Fig. 5.6). Upon extrapolation to zero absorbance, the intercept on the concentration axis corresponds to the concentration of the metal in the diluted solution.

Extraction of the metal using a complexing agent and organic solvent may be beneficial in that interference may be avoided and samples concentrated [73, 109]. Conventional flame techniques are typically two to four times more sensitive when using organic solvents.

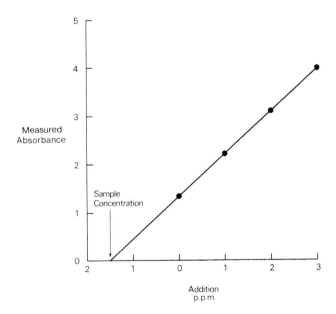

Fig. 5.6 The method of additions.

Atomic absorption methods for determination of metals in airborne particulate matter have been reviewed by Hwang [9]. Pb in air has been the subject of intensive study, using both conventional [16, 17, 106–108, 110–115] and flameless [116–119] methods. Techniques have also been described for analysis of Cd [120, 121], Fe [107, 110], V [122] and ranges of metals [16, 17, 73, 107, 111–113, 123, 124] in airborne particulate matter.

Flameless atomization procedures are so sensitive that air samples as small as 200 ml may be used to measure lead [119] at levels as low as 0.1 μg m^{-3}. Filter and reagent blanks, however, tend to become the limiting factor in such measurements. The blanks may be avoided by continuous introduction of sample into the instrument, simply by pumping or drawing polluted air through the flame or flameless atomizer. Such methods have been devised for continuous monitoring of metals in air. These include determination of Pb at levels from 0.16 μg m^{-3} of air [125, 126], Cd from 0.005 μg m^{-3} of air [127], Ag from 3 μg m^{-3} of air [128], and Pb in vehicle exhaust gases [129] and smelter fume [130]. Robinson and Wolcott [131] have described some of the difficulties inherent in the calibration of such techniques. Techniques using a graphite cup from an atomic absorption flameless atomizer as an air filter have a very low blank, and are hence also capable of very high sensitivity [118, 132].

Before analysis of particulate matter collected on an air filter, extraction is necessary. Recommended procedures vary tremendously in the strength of the extraction reagent used, and each should be judged on its merits for the task in hand. While strong oxidizing acids are necessary for some samples [133], dilute acid and ultrasonic treatment may be adequate for extraction of lead and cadmium from unashed filter samples [116] and is accepted by USEPA for lead analysis [134]. In the case of aluminium, hydrofluoric acid treatment is essential to liberate the metal from silicate matrices [135], and indeed for any sample where a significant proportion of the analyte is expected to reside within silicate lattices, HF treatment is required.

Flameless atomic absorption methods are particularly liable to interference and the standard additions method is normally routine. In addition, matrix modification may be needed to overcome severe interferences; for example the addition of magnesium is valuable in the analysis of As in marine air particulates [136]. Suitable methods for low volume air sampling and flameless AAS analysis have been reported for Cd, Cu, Fe, Pb and Zn [133]; Al, Ca, Cd, Co, Cr, Fe, Mg, Mn, Ni, Pb and Zn [117] and for Cd, Pb and Mn in size-fractionated particles [137]. Automation of flameless AAS analysis of Pb in air particulates has been reported [138]. In a comparative study, a variety of wet ashing and AAS analytical procedures for Cd and Pb showed good precision and accuracy [139].

Two cold vapour atomic absorption methods have been described for the analysis of mercury in air particulates. In the first, mercury is collected on quartz fibre filters, released by pyrolysis and collected on a gold-coated sand absorber prior to thermal desorption into the cold vapour AAS [140]. In the other report, 5% $KMnO_4$ and concentrated HF are added to an air filter in a sealed polyethylene bottle. When dissolution of the sample is complete, analysis is by conventional cold vapour AAS and recoveries of Hg exceed 90% [141].

Experimental procedure for atomic absorption analysis of As, Ba, Cd, Cr, Cu, Fe, Mn, Pb, Tl, V, *and* Zn [16, 17]

Air sampling

High-volume air samples (24 h) are taken through 8×10 in glass fibre filters (7×9 in exposed surface) at 50 to 60 ft^3 min^{-1} (about 2200 m^3 total).

Sample preparation

Using a plastic template as a cutting guide, cut a strip from the filter for analysis. The amount of filter taken depends upon the type of sample: for urban samples a 7×1 in strip is cut, for non-urban samples a 7×2 in strip, and for industrial samples the size may be reduced to an extent consistent with the expected levels of airborne metals. Ash the strip in a low-temperature asher at approximately 150° C for 1 h at 250 W and 1 mm chamber pressure with an oxygen flow of 50 ml min^{-1}. Alternatively, the filter strip may be ashed in a muffle furnace at 500 to 550° C, but losses of individual metals of up to 55% may occur. Place the ashed filter in a glass thimble which is placed in an extraction tube (Fig. 5.7). A 125 ml Erlenmeyer flask with a 24/40 female joint is charged with constant boiling (about 19%) HCl (8 ml) and 40% HNO$_3$ (32 ml). The flask is attached to the extraction tube, and the tube fitted with an Ahlin condenser. Reflux the acid over the sample for 3 h, during which sample and extraction thimble should remain at the temperature of the boiling acid.

Remove the extraction tube and condenser from the Erlenmeyer flask and fit the flask with a thermometer adapter which serves as a spray retainer. Concentrate the extracted liquid to 1 to 2 ml on a hot-plate and allow to cool and stand overnight. Quantitatively transfer the concentrated material to a graduated 15 ml centrifuge tube with 3 washings of 5 to 10 drops of diluted acid. Dilute urban samples to 4.4 ml per 1 in strip (40 ml per 63 in^2 of filter) and non-urban samples to 3.0 ml per 2 in strip (13.3 ml per 64 in^2 of filter). Following dilution, centrifuge the sample at 2000 rev min^{-1} for 30 min and then decant the supernatant liquor into polypropylene tubes which are capped and stored prior to analysis. 1 ml from each solution is diluted to 10 ml for analysis by atomic absorption.

Analysis

Determine the metal concentration of the solution by comparison of the absorbance of the sample solution to the absorbance of the standard metal solutions. General instrumental operating parameters for a Perkin-Elmer Model 303 are shown in Table 5.11. Working standards are prepared daily from stock solutions of 500 to 1000 μg ml^{-1} of each metal. Working standard concentrations will vary according to the sensitivity and detection limits of each metal. Results are calculated by multiplying the number of μg of each metal per ml of sample extract by the appropriate dilution factor and dividing by the number of m^3 of air represented by the sample.

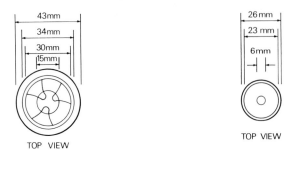

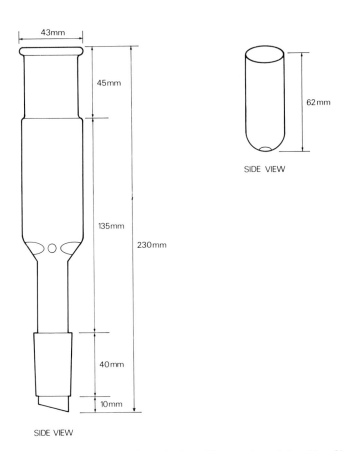

Fig. 5.7   Sample holder and extraction tube for acid extraction of glass fibre filter papers [16, 17].

Table 5.11 Operating and analytical parameters [16, 17]

Element	Wavelength (nm)	Range ($\mu g\ ml^{-1}$)	Minimal detect. conc. ($\mu g\ m^{-3}$)[a]
As	193.7	1.00– 50	0.02
Ba[b]	553.6	1.00–100	0.02
Ca	422.7	0.01– 8	0.0002
Cd	228.8	0.10– 20	0.0002
Cr	357.9	0.10– 20	0.002
Cu	324.7	0.05– 20	0.001
Fe	372.0	0.50– 40	0.01
Mn	279.8	0.05– 20	0.001
Ni	232.0	0.20– 20	0.004
Pb	217.0	0.10– 40	0.002
Tl	276.8	0.50– 50	0.01
V	318.4	0.50– 50	0.01
Zn	213.8	0.01– 2	0.0002

[a] Based on a 2000 m^3 air sample.
[b] Nitrous oxide–acetylene burner required.

Note
(i) The ten-fold dilution prior to analysis is necessary to reduce the dissolved solids content to less than 0.5%, and thereby eliminate matrix effects.
(ii) Silica extracted from glass-fibre filters and from the particulate matter itself can also cause significant interference with Fe, Ca, Mn and Zn. This interference may be overcome by allowing the acid extracts to stand for 12 to 24 h followed by centrifugation. The acid extract is then separated from the precipitated silica by decantation into a polypropylene tube.
(iii) All glassware is soaked in 20% HNO_3 for 2 to 6 h and washed with distilled water prior to use. HNO_3 and HCl acids are distilled in all-glass apparatus to remove metal contaminants.
(iv) Blank glass fibre filters are analysed to determine the metal background. The use of membrane filters permits the analysis of the extremely low levels of elements such as Fe, Ba and Zn that are found in high concentrations in glass fibre filters.

Experimental procedure for atomic absorption analysis of Al, Be, Bi, Ca, Co, Cr, Cd, Cu, Fe, K, Li, Mg, Mn, Na, Ni, Pb, Rb, Si, Sr, Ti, V *and* Zn [142]

Air sampling
Air is sampled through polystyrene filters (microsorban Type 99/97, Delbag-Luftfilter GMBH, Berlin, Germany) (25 × 20 cm sheet) for 24 h using a high-volume sampler (about 2000 m^3 sample).

Sample preparation

The filter is folded, placed in a 20 ml beaker, covered with a watch glass and ashed overnight (12 to 16 h) at 400 to 425° C in a muffle furnace. After ashing, transfer the residue to the PTFE cup of an acid digestion bomb (Parr Instrument Co., Moline, Ill.). Scrape the residue out of the beaker using a PTFE-coated spatula and add 65% HNO_3 (2 ml) and 30% HCl (0.5 ml) to the beaker. Warm the beaker on a hot-plate (60° C) and cool before pouring the solution into the PTFE cup. Add 65% HNO_3 (2 ml) to the beaker and warm on the hot-plate (60° C) for 5 min. Cool the beaker and pour the acid solution into the PTFE cup. The procedure is repeated with 65% HNO_3 (2 ml) and 30% HCl (1 ml) and finally 40% HF (3 ml) is added directly to the PTFE cup (Extreme caution). Place the cover on the PTFE cup and seal the cup in the acid digestion bomb. Heat the bomb in an oven at 125° C for 6 h, and after digestion cool in a freezer to −5° C for 2 h.

Open the bomb, remove the PTFE cup and allow to reach room temperature. If necessary, the volume of solution is adjusted to a known volume in polypropylene labware (normally the final volume of solution on opening the bomb is 10.0 ml plus or minus 0.2 ml, in which case this step may be omitted). An aliquot (2 ml) is taken and placed in a polypropylene bottle (0.5 oz) containing 10 000 ppm potassium solution (1 ml) and de-ionized water (7 ml). This solution, designated Aliquot A (1.8 M hydrofluoric acid; 1.9 M HNO_3; 0.4 M HCl; and 1000 ppm K approximately) is used to measure Be, Cs, Li, Rb, Ti and V.

Aliquot A (1 ml) is added to another polypropylene bottle (0.5 oz) containing 5000 ppm potassium solution (1 ml) and de-ionized water (3 ml). The resultant solution, designated Aliquot B (0.4 M hydrofluoric acid; 0.4 M HNO_3; 0.1 M HCl; and 1000 ppm K) is used to measure Si.

Evaporate the sample solution left in the PTFE cup to dryness on a hot-plate (about 70° C). Dissolve the residue in 10 M HNO_3 (10 ml) and evaporate to dryness. Redissolve the residue in 10 M HNO_3 (10 ml) and evaporate to half its original volume. Quantitatively transfer the sample to a 25 ml volumetric flask containing 6250 ppm lanthanum in 8 M HNO_3 (4 ml) and 10 000 ppm Cs (2.5 ml). Wash the PTFE cup several times with de-ionized water, and add the wash to the volumetric flask. Dilute the sample to 25 ml with de-ionized water. This solution (1.5 M HNO_3; 1000 ppm Cs and 1000 ppm La), designated Aliquot C is used to determine Bi, Cd, Co, Cr, Cu, Mn, Ni, Pb and Sr.

Add Aliquot C (0.5 ml) to a 50 ml volumetric flask containing 6250 ppm La in 10 M HNO_3 (8 ml), and 10 000 ppm Cs (5 ml). Dilute the sample to 50 ml with de-ionized water and the solution (1.5 M HNO_3, 1000 ppm La, 1000 ppm Cs), designated Aliquot D is used to determine Ca, Fe, Mg, Na and K. Add Aliquot C (1 ml) to a (10 ml) volumetric flask containing 6250 ppm La in 10 M HNO_3 (1.6 ml) and 10 000 ppm Cs (1 ml). Dilute to 10 ml, and the solution (1.5 M HNO_3; 1000 ppm La; and 1000 ppm Cs, designated Aliquot E is used to measure Al and Zn.

All standards used in the analysis are prepared so as to have as nearly as possible the same matrix as the samples. Interferences may be checked by use of the standard additions method. Standard stock solutions (1000 ppm) of the metals are prepared from reagent grade chemicals, and diluted working standards prepared fresh daily. High purity HCl, HNO_3 and hydrofluoric acid are used and lanthanum nitrate hexahydrate and caesium nitrate solutions are prepared from the purest grades available.

The elements Cs, Li and Rb are best determined by atomic emission and Al, Be, Ca, Li, Mg, Si, Sr, Ti and V require a nitrous oxide–acetylene flame for atomic absorption measurement. All other metals are determined by conventional A.A. techniques. Figure 5.8 shows the analytical scheme and Table 5.12 the practical detection limits for the procedure.

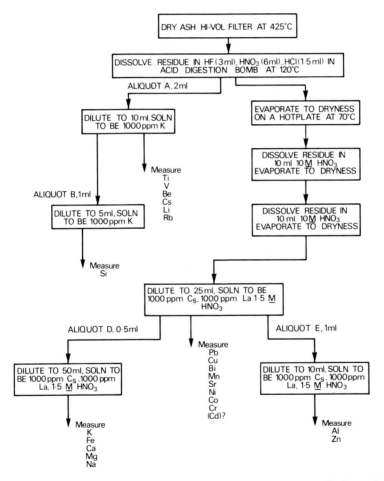

Fig. 5.8   Flow chart for the atomic absorption analysis of Hi-Vol atmospheric particulate samples collected on polystyrene filters [142].

Table 5.12 Practical detection limits for atomic absorption analysis of samples collected on polystyrene filters [142]

Element	$\mu g\ m^{-3}$ in air sample[a]
Al	0.4
Be	0.000 04
Bi	0.002
Ca	0.03
Cd	0.0003
Co	0.000 8
Cr	0.000 3
Cs	0.000 04
Cu	0.003
Fe	0.07
K	0.04
Li	0.000 3
Mg	0.01
Mn	0.000 3
Na	0.03
Ni	0.002
Pb	0.007
Rb	0.000 5
Si	0.8
Sr	0.003
Ti	0.07
V	0.006
Zn	0.03

[a] The amount of metal required to provide a signal which is equal to twice the s.d. of the blank signal for the same element when 2000 m^3 of air is sampled.

Procedure for analysis of Na, K, Ca *and* Mg *in low-volume air samples*

Air is sampled at a rate of 4–6 litres min^{-1} through a 47 mm diameter 0.45 μm membrane filter (e.g. Millipore type HAWP 04700). The filter is placed in an acid-washed polyethylene beaker and 2 N Analar nitric acid (10 ml) added. The beaker is warmed to 40–50° C, or placed in an ultrasonic bath for 30 min. The extract is carefully decanted into a 25 ml volumetric flask, and the filter washed with a further 10 ml aliquot of 2 N nitric acid and finally the volumetric flask is made up to volume with 2 N nitric acid.

The filter extract solution is analysed for Na and K by flame photometry after calibration of the instrument with standards in the range 0–8 $\mu g\ ml^{-1}$ Na and 0–3 $\mu g\ ml^{-1}$ K made up in 2 N nitric acid. Analysis for Ca and Mg is by flame atomic absorption using manufacturer's recommended instrumental parameters and prior calibration in the range 0–8 $\mu g\ ml^{-1}$ Ca and 0–2 $\mu g\ ml^{-1}$ Mg, all standards prepared in 2 N nitric acid. Unexposed filters are extracted in parallel with the exposed to give filter and reagents blanks which are subtracted before calculation of airborne concentrations.

The method is applicable also to the analysis of other metals (e.g. Pb) soluble in 2 N nitric acid under these conditions.

Experimental procedure for determination of Pb *by flameless atomic absorption* [116]

Air sampling

Air is sampled through Whatman No. 41 cellulose filter paper using either a low vacuum high-volume sampler (20×25 cm filter holder; initial rate 70 $m^3 h^{-1}$) or a high vacuum pump (10 cm diameter filter holder; 20 $m^3 h^{-1}$) for 24 h, equivalent to 1200 m^3 and 400 m^3 of air respectively. For Pb determination a 10 cm^2 piece is cut from a 20×25 cm filter and 1/8 segment from the 10 cm diameter filter. Alternative types of filter paper give a higher collection efficiency (Chapters 1 and 4).

Sample preparation

Transfer the filter sector into 0.1 M HNO_3 (about 50 ml) and expose it to ultrasonic vibration (Branson ultrasonic cleaner, consisting of an ultrasonic generator Model LG-150 and an ultrasonic tank-type transducer LTH-60) for 5 min. Decant the solution and repeat ultrasonic treatment in 0.1 M HNO_3 (30 ml) for 5 min. Wash the beaker several times with 0.1 M HNO_3, combine all extracts and washings and dilute to 100 ml in a volumetric flask. Treat a blank filter in the same way. Use only glassware pretreated with HNO_3 for 24 h.

Determine Pb by injecting 10 to 50 μl of the extract with an Eppendorf micropipette into the graphite cell (Perkin-Elmer HGA-70) (Programme 5 – drying at 100° C and charring at 490° C; atomization voltage, 9 V, 2400° C) fitted to a Perkin-Elmer Model 303 double beam A.A. spectrophotometer equipped with a deuterium background corrector. Measure the absorbance due to lead at the 283.3 resonance line, running both sample and blank at least twice. Subtract the blank absorbance value from that of the sample filter and determine the concentration from a calibration curve.

Calibration

Prepare a standard solution of Pb (1000 ppm) from lead nitrate (1.599 g) dissolved in 0.1 M HNO_3 and diluted with 0.1 M HNO_3 to 1 litre. Dilute the stock solution with 0.1 M HNO_3 to concentrations from 20 to 250 μg Pb $litre^{-1}$. The dilutions are prepared weekly. Inject aliquots of standards, of the same fixed volume as that used for the sample, into the graphite furnace. Plot the absorbance against the quantity of lead present in the furnace, a graph which has been found to be linear up to 5 ng lead. 1% absorption was found for 0.1 ng Pb, and the method may be used to determine 0.01 μg m^{-3} of Pb, as described.

Application to Cd [121]

Air samples are collected and prepared exactly as for Pb, with the first ultrasonic treatment being lengthened to 10 min. In this instance, the instrumental parameters are Programme 4 and atomization voltage of 9 V: drying at 100° C; charring at 330° C and atomization at 2400° C. Absorbance peak heights are measured at the 228.8 nm Cd line. If one half of a 10 cm diameter filter is taken, the dilution is made to 50 ml and 50 μl is injected, Cd levels as low as 0.2 ng m^{-3} may be determined with a s.d. of 10%.

5.2.3.8   *Other analytical methods*

Titrimetric procedures have long been applied in air pollution analysis. Because of their generally low sensitivity and selectivity they are of limited application, but may be of value in the industrial hygiene field where levels of airborne metals are generally higher than in ambient air in other locations and a limited range of metals is present at an appreciable level. Thus, for instance, Fe (100 μg to 400 mg) may be determined by titration with ceric sulphate or potassium dichromate [86], and Jacobs [87] describes the determination of a range of metals by direct or indirect titration. An iodine microtitration method using an amperometric titrator for the determination of airborne arsenic has also been recommended [93].

Other procedures applicable to the analysis of metals may also be applied to air pollution analysis. Be, collected on glass fibre filters may be converted to the bis-trifluoroacetylacetone complex and determined by gas chromatography. The method has a limit of detection of 4×10^{-14} g Be and readily determines levels of Be in urban air (about 3×10^{-4} μg m^{-3}) from a portion of a filter used to sample 2200 m^3 of air over a 24 h period [143].

Other more specialized observations and techniques have included electron spin resonance (e.s.r.) spectra of ferromagnetic particles of airborne dust collected on filters. A clear relationship was found between the e.s.r. signal and the amount of Fe on the filter, and this allowed calculation of airborne Fe in a 100 m^3 air sample [144].

Anodic deposition of Pb, followed by isotope ratio mass spectrometry is reported to be applicable to the analysis of 10 ng to 10 μg of Pb and has been used for determination of Pb collected on filters [145]. The method is particularly valuable where isotope ratio data are required.

Experimental procedure for gas chromatographic determination of Be [143]

Air sampling and sample preparation

Air samples (about 2200 m^3) are collected on glass fibre filters in a manner identical to that of Thompson *et al.* [16, 17], as described in detail under experimental procedures for atomic absorption analysis. Ashing and extraction of the filter are performed in an identical manner. After dilution to 4.4 ml with distilled water and centrifugation at 2000 rev min^{-1} for 30 min, the supernatant is transferred to a polypropylene tube, ready for analysis.

Reagents

0.164 M trifluoroacetylacetone [H(tfa)] is prepared by diluting freshly distilled H(tfa) (2 ml) in high purity benzene (100 ml). Store the solution in a silanized 100 ml borosilicate glass volumetric flask. H(tfa), both neat and in benzene solution is subject to slow decomposition on standing. Ethylenediaminetetraacetic acid (EDTA) – buffer solution is prepared by weighing disodium EDTA monohydrate (5.15 g), sodium acetate trihydrate (85 g) and glacial acetic acid (6.25 ml) and dissolving the reagents in de-ionized water (500 ml). Standard Be(tfa)$_2$ solutions are prepared by dissolving Be(tfa)$_2$, purified by sublimation, in high purity benzene. Further dilutions are made to a resultant concentration of 1.0×10^{-11} g beryllium μl^{-1}. Prepare this solution every 2 weeks.

Standard solutions are stored in borosilicate glass volumetric flasks, cleaned in Chromorge cleaning solution and silanized to reduce any active sites. Silanization is achieved by filling the flasks with 20% hexamethyldisilazane (HMDS) in benzene and standing overnight. The reaction vessels (5 ml culture tubes) are treated similarly.

Analysis

1 ml of the *aqua regia* filter digest is pipetted into a 5 ml borosilicate glass culture tube fitted with a screw cap. EDTA – sodium acetate buffer solution (1 ml) is added to the culture tube. Then add 3 N NaOH solution to a final pH of 5.5 to 6.0; normally from 0.8 to 1.3 ml is required. The tube, containing a small PTFE-coated magnetic stirrer is capped and inserted into an oil bath maintained at 93° C on a Corning stirring hot-plate. Heat the sample and stir vigorously for 10 min. Cool the tube and add 0.164 M H(tfa) in benzene (1 ml). Cap the tube and place above the stirring mechanism of the hot-plate and stir for another 15 min at ambient temperatures.

Stand the tube until aqueous and organic layers separate. Transfer the organic layer to a 2-dram vial with a medicine dropper and add 0.1 N NaOH solution (2 ml). Shake the mixture quickly by hand for 5 s to remove excess H(tfa) from the organic layer. Separate the phases immediately by withdrawing the organic layer with a medicine dropper and transferring it to another 2-dram vial which is then capped. This step must be performed rapidly and reproducibly or intolerably large amounts of Be will be lost and both precision and accuracy will suffer.

Repetitively inject $5 \times 1 \mu l$ aliquots into the GC. The mean of peak heights from the Be(tfa)$_2$ peaks in the unknown solution are compared with the average from 5 analyses of a standard solution of Be(tfa)$_2$. Over the concentration range used the response is essentially linear so peak height ratios can be used for calculation of Be found.

Instrumental parameters

Ross and Sievers [143] used a Hewlett-Packard Model 402 high efficiency GC equipped with a ^{3}H source electron-capture detector. Column: 2 m × 3 mm internal diameter borosilicate glass column packed with 2.8% W-98 silicone (Union Carbide) on Diataport S (Hewlett-Packard). Carrier gas: methane, 10%; argon, 90%; 54.5 ml min^{-1}. Column temp: 110° C; detector temp: 200° C; on-column injection (no additional heat at site of injection).

The blank from reagents and filter was found to be 4% of the Be in typical urban air samples. The analytical method is sensitive to as little as 4×10^{-14} g Be, corresponding to less than 10^{-6} μg Be m^{-3} of air.

5.3   Gases and vapours

5.3.1   General sampling considerations

Few metallic elements and compounds are of significant volatility at ambient temperatures. Those that are present in air in the vapour phase, however, provide a totally different sampling problem from that of the particulate forms, although analytical techniques may be similar. Some volatile metal compounds may be determined by continuous monitoring at certain concentrations (e.g. nickel carbonyl, tetraethyl lead), while others require a pre-concentration stage. Pre-concentration may be by adsorption tube, liquid scrubber, thermal decomposition to an involatile form, or freeze-out. No technique can be assumed to give a quantitative collection, and in all cases collection efficiency must be established by calibration. The effect of temperature upon such procedures should also be investigated. It must also be recognized that the efficiency of recovery of collected compounds from adsorbents and freeze-out traps may be less than quantitative.

Choice of unsuitable materials for the construction of apparatus and sample lines can lead to considerable error. Some types of tubing, including glass, may strongly absorb gaseous compounds. Hence, initially low results may be obtained, while subsequent desorption into a gas stream containing low levels of the compound may cause false positive readings. Adsorption on surfaces and tubing may also catalyse reactions with other adsorbed compounds and lead to error. Thin tubing may be significantly permeable to gaseous pollutants, and both outward diffusion of sample and inward diffusion of the surrounding atmosphere may occur. Greases and oils used in seals may strongly absorb gaseous materials; PTFE, which is self-lubricating, is recommended for making such seals. Surface effects and diffusion effects will be most pronounced if samples are stored, and where possible rapid analysis of grab samples is desirable. Mechanical leakage may be a large source of error, and corrections for temperature and pressure are important when calculating pollutant levels.

Particulate matter in an air sample may foul rate meters and total volume meters, as well as blocking orifices in impinger-type collectors. It is frequently necessary to use a pre-filter to avoid such problems, although the possibility of adsorption of gaseous pollutants upon particulate matter must be borne in mind. In some cases, such as the collection of lead alkyls with iodine monochloride solution, efficient prior filtration of lead particulate matter is essential if analysis of particulate Pb with the volatile lead alkyls is to be avoided.

5.3.2.1 *Metal carbonyls*

Metal carbonyls are formed in industrial processes, both by intent and as undesired by-products of catalyst decomposition. Hence, analytical procedures for Ni and Fe carbonyls have been devised for use in the context of industrial hygiene. These compounds also occur in town gas, and Densham *et al.* [146] have reviewed the analytical methods available for their determination. Jacobs [87] describes a method for determination of iron and nickel carbonyls in air which involves thermal decomposition to the metal in a silica or glass tube, and subsequent analysis of the deposited metal. This method, however, is reported not to be quantitative at the thermal decomposition stage [146].

Numerous liquid and solid adsorbents have been used for collection of metal carbonyls from air [147]. An acidified solution of chloramine-B in alcohol has been used [148], as well as alcoholic iodine [149, 150]. These methods involve a subsequent spectrophotometric determination of the metal as a complex, and this limits their sensitivity. Analysis by atomic absorption spectrometry might be expected to allow a substantial improvement in sensitivity. Filtration of the air sample initially is necessary to ensure elimination of particulate metal.

Brief *et al.* [151] describe the collection of nickel carbonyl in 10–15 ml of 3% HCl in a midget impinger. Reaction with α-furildioxime in the presence of chloroform forms a yellow complex which is determined spectrophotometrically. Collection efficiency is greater than 90%, but Cr, Fe and V interfere with the analysis. Sampling for 20 to 30 min at 0.1 ft^3 min^{-1} gives a detection limit of 0.0008 ppm of nickel carbonyl. For collection of iron pentacarbonyl Brief *et al.* [152] use 3% HCl mixed with an I/KI solution immediately prior to use. Air is sampled at 2–3 litre min^{-1} through 10 ml of this solution in a fritted bubbler, after prior removal of particulate matter by filtration. Collection was found to be virtually quantitative. After reduction to the ferrous state, Fe is determined spectrophotometrically as its 1,10-phenanthroline complex, and interferences may be overcome by extraction of Fe, as ferric chloride, into isopropyl ether. In a 50 litre air sample 0.009 ppm of iron carbonyl may be detected.

Densham *et al.* [146], using a sintered bubbler containing iodine monochloride in glacial acetic acid, were able to collect both iron and nickel carbonyls. After evaporation of the reagent and addition of the complexing agent, determination of the metal by visual colorimetric means is carried out. Sensitivities were 0.006 ppm for nickel carbonyl and 0.01 to 0.03 ppm for iron carbonyl.

A continuous procedure for measurement of metal carbonyls in town gas is also described by Densham *et al.* [146]. Polluted town gas was used to supply the combustion gas stream of an atomic absorption instrument. In order to obtain blank readings, the gas stream was purified by prior passage through activated carbon, and the absorbance measured. The detection limits were 0.002 ppm of nickel carbonyl in the gas, and 0.01 ppm of iron carbonyl. Subsequent improvements in atomic absorption instrumentation should allow a substantial lowering of these detection limits.

A continuous monitor for metal carbonyls in air, unable to discriminate between Ni and Fe, has been described by McCarley *et al.* [147]. Polluted air impinges upon a hot borosilicate glass plate causing thermal decomposition of metal carbonyls and deposition of the metal. An ingenious optical system produces a collimated beam of plane polarized light which is incident upon the surface of the glass at the Brewsterian angle. In this arrangement, clean glass reflects no light, and reflectance is due solely to deposited metal. Concentrations of 0.05 to 4 ppm of nickel carbonyl in air may be continuously determined.

It is also reported that a direct field instrument which is commercially available is suitable for continuous monitoring of nickel carbonyl in the 10 to 1500 ppb range [149]. Kincaid *et al.* [150] give a brief description of a continuous monitoring instrument in which nickel carbonyl is reacted with Cl_2 or Br_2 in the gas phase forming a smoke of nickel halide. The light scattering caused by the smoke is measured, and the intensity of scattered light is related to the concentration of nickel carbonyl in the contaminated air. Levels of a fraction of 1 ppm may be determined.

A commercially available i.r. monitor with a 40 m path length absorption cell is capable of determining nickel carbonyl with a detection limit of 0.2 ppb, and by automatically measuring absorbance at three different wavelengths avoids interference by considerable excess concentrations of CO.

Experimental procedure for determination of nickel carbonyl [151]

Air sampling

Air or process gas is sampled by bubbling through 10 to 15 ml of 3% HCl in a midget impinger for 30 to 60 min at 3 litres min^{-1}. A filter assembly precedes the bubbler to ensure that no nickel solids enter the bubbler.

Analysis

Transfer the liquid to a 60 ml separatory flask and add in sequence:

(a) 2 drops of phenolphthalein
(b) 4 drops of ammonium hydroxide (about 20 to 30% aqueous solution)
(c) 20% aqueous NaOH to the phenolphthalein endpoint, plus 3 drops excess
(d) 3.0 ml α-furildioxime solution (1% α-furildioxime in 1:1 alcohol:water)
(e) 5 ml chloroform.

After shaking for at least 1 min, allow the chloroform layer to separate from the aqueous phase and draw the chloroform layer into a test tube. The colour may be compared with pre-established standards to determine μg of Ni in the chloroform. For greater accuracy the final colour is compared against reagent blanks in 1 cm cells in a spectrophotometer set at 435 nm. The volume of air sampled and the weight of Ni collected are then used to calculate the Ni concentration in the gas.

Sensitivity

The method is visually sensitive to at least 0.002 ppm, and using a spectrophotometer a sensitivity of 0.0008 ppm has been reported for 20 to 30 min sampling periods. No serious interferences are known, although Cu gives a brown precipitate which can be removed by washing the chloroform extract twice with 10% ammonium hydroxide. Cr can be tolerated up to about 0.1% and Fe and V up to about 0.5%.

Iron carbonyl [152]

Air sampling

The absorption solution is made up in 2 parts and mixed just prior to use. One part consists of 3% v/v HCl and the other I_2–KI solution. The latter is prepared by dissolving I_2 (4 g) and KI (10 g) in water, filtering the solution through glass wool, diluting the filtrate to 100 ml and storing it in a brown bottle. When required the absorption solution is prepared by mixing 1 ml of the I_2–KI solution to 10 ml of the HCl solution.

Prior to use all glassware used in sampling and analysis must be rinsed with 10% HCl and then with distilled water. The air sample is bubbled through 10 ml of the absorption solution in a fritted bubbler, designed for this volume of solution, at a rate of 2 to 3 litres min^{-1} and about 50 litres of air are sampled. A high efficiency filter precedes the bubbler to collect particulate iron before it reaches the sampler.

Analysis

Transfer the sample to a 50 ml volumetric flask, using a minimum of wash water. Then, add in order:

(a) 1% w/v sodium sulphite solution dropwise until the I_2 is just decolorized (this solution is prepared fresh daily)
(b) 20% w/v hydroxylamine hydrochloride (1 ml) reduces Fe^{3+} to Fe^{2+}
(c) 30% w/v sodium acetate solution – the volume added is that found necessary to adjust the pH of a reagent blank to 5.0
(d) 1% w/v 1,10-phenanthroline solution (2 ml).

Make up the liquid volume to 50 ml with water. (If the 1,10-phenanthroline solution changes colour, discard it and prepare a fresh solution.) Allow at least 10 min for colour development and determine the absorbance at 508 nm spectrophotometrically.

Standardization

Electrolytic iron wire is cleaned with fine sandpaper and dissolved in 15% v/v H_2SO_4 to make a stock solution of 100 mg litre^{-1}, which is diluted to create standards of 1 to 10 μg ml^{-1}. The method is sensitive to 1 μg Fe, and will detect 0.009 ppm of iron carbonyl in a 50 litre air sample. Interferences include Cr, Cu, Ni, Co, Zn, Cd and Hg. If the presence of these metals is suspected, make up the sample solution to 7 to 8 N in HCl, and extract the ferric chloride into isopropyl ether. After re-extraction of the iron into water, reduce with hydroxylamine, and form the colour complex in acetate-buffered solution as described previously. A new calibration curve should be developed.

5.3.2.2   *Hg and its compounds*

Hg may be found in air in several distinct chemical and physical forms. Elemental mercury is a common air pollutant, present as the vapour, or as an aerosol of liquid droplets. Organically bound mercury can exist in air as monoalkyl mercury compounds such as methylmercury chloride [CH_3HgCl] or dialkyl mercury compounds such as dimethylmercury [$(CH_3)_2Hg$], both present predominantly in the vapour phase. Inorganic mercury compounds are relatively involatile and are present in air as particles, although some, notably mercuric chloride, may be present substantially in the vapour phase. Methods for the cold vapour AAS analysis of particulate Hg in air are cited in Section 5.2.3.7.

The more sophisticated analytical procedures discriminate between the various forms of Hg in air; an important facet in view of the differing toxicology associated with the different chemical forms. It must be emphasized, however, that many sampling and analytical procedures either measure only one form of the metal or measure the sum of the concentrations of all forms, and this may be acceptable only in industrial situations where adequate knowledge of the predominant form of airborne mercury is available.

Impingers and scrubbers have been used for sampling Hg in air. Jacobs [87] describes a method utilizing I_2/KI for collection of Hg from air, followed by a colorimetric determination. Panek used acidified 6% potassium permanganate for collection, following with a colorimetric determination with dithizone [153]. Linch *et al.* [154], however, found poor collection of diethylmercury by acid permanganate and very inefficient collection of dimethyl- and diethylmercury by I/KI. Crystalline iodine was found to retain little diethylmercury, but collected elemental Hg. They did, however, find that 0.1 N iodine monochloride in 0.5 M HCl gave highly efficient recovery of alkyl, dialkyl, inorganic and elemental mercury including particulate forms. Determination of collected Hg by a dithizone procedure was performed. This method has, however, been criticized for high and irreproducible blanks, instability of collected Hg compounds and interference by high levels of SO_2.

Elemental Hg in the air may be directly determined by continuous atomic absorption measurement at 253.7 nm [87]. Commercially available instruments are fairly sensitive but respond to various organic vapours also present in air. Although sensitivity to molecular absorption by organic compounds is not as high as for atomic absorption by mercury, under some circumstances, errors may arise. A typical instrument will measure mercury levels of 5 μg to 1 mg m^{-3} of air.

Hg vapour in air may be collected by amalgamation on Ag wool [155] or gauze [156, 157]. Subsequent release by thermal de-amalgamation and elution into an atomic absorption or u.v. spectrophotometer allows determination of the Hg without interference from organic air pollutants. For 24 h sampling times ambient Hg levels from 15 ng m^{-3} to 10 μg m^{-3} may be determined using Ag wool collection [155] or from 5 ng m^{-3} to 100 μg m^{-3} using sampling times of 2 h or less and an Ag gauze collector [156, 157]. Scaringelli *et al.* [158] describe collection of all forms of Hg by adsorption on charcoal, backed by a fibre filter to ensure efficient collection of particulate matter. Subsequent heating of the sampler and nitrogen elution frees the collected compounds which are pyrolysed to form elemental Hg, and collected on Ag wool. Heating of the silver wool causes de-amalgamation of the mercury which is determined by atomic absorption at 253.7 nm. Alternatively, using purified activated charcoal [159], direct analysis of the charcoal by neutron activation may be used to determine mercury concentrations down to 0.5 ng m^{-3}. A porous graphite tube plated with gold has been shown to be an efficient collector of particulate and elemental Hg [160] Subsequent insertion of the tube in the carbon-rod atomizer of an atomic absorption instrument allows determination of the collected Hg.

Conventional flame atomic absorption of Hg is of poor sensitivity. A flameless cold vapour technique has been developed, however, involving reduction of Hg compounds in solution using stannous chloride or hydrazine hydrate and introduction of the resultant elemental mercury as vapour into a cell in the sample beam of an atomic absorption instrument [161, 162]. Alternatively, the Hg may be electrolytically amalgamated on to Cu wire and released into a cell in the spectrophotometer light beam by heating the wire [163]. The detection limit is 0.2 ng of Hg. Methods involving the chemical reduction of Hg compounds and the flushing of the element from solution may also introduce volatile organic materials into the light path, thus causing erroneous results. This may be overcome by prior ashing of the sample [156] or the use of a background correction device [162], commonly available as an accessory on double beam atomic absorption spectrophotometers. After suitable chemical treatment, such analytical methods may be applied to particulate mercury collected by filtration or impingement, or samples of mercury and its compounds collected by absorption in a reagent. The method is of high sensitivity and is specific. A valuable critical report of sampling and analytical techniques for mercury in stationary sources has been compiled by Driscoll [164].

Elegant techniques for separation collection and analysis of different forms of Hg have been described. Henriques *et al.* [165] separated particulate Hg initially by filtration through a Millipore filter. A gold filter, coated with a gold–silicon alloy collected elemental Hg vapour, a scrubber containing acid permanganate solution trapped methylmercury (and other readily oxidized mercury compounds) and, last in the train, a pure gold filter collected dimethylmercury and other gold-soluble mercury compounds. Analysis was by vapour atomic absorption.

Braman and Johnson [166] described a technique suitable for determination of several forms of Hg at levels down to 0.1 to 1 ng m^{-3}. A glass wool filter removed particulate matter above 0.3 μm from an air sample. This was followed by adsorption tubes containing respectively siliconized Chromosorb W treated with hydrogen chloride (collects mercuric chloride type compounds and particulate matter which pass through the filter); Chromosorb W treated with caustic soda (methylmercury chloride type compounds); silvered glass beads (elemental mercury) and finally gold coated glass beads (dimethylmercury). High collection efficiencies were found. Hg was eluted from the adsorption tubes by heating with helium carrier gas, and passage through a d.c. discharge chamber caused excitation of the Hg and emission at 257.3 nm was measured. The detection limit of the analytical procedure was about 0.01 ng.

Partial information on speciation is provided by the method of Trujillo and Campbell [167], in which particulate Hg is collected on a pre-filter, organomercury compounds on Carbosieve B and elemental mercury vapour on silvered Chromosorb P. The mercury is analysed by thermal desorption, with a working range of 0.3 ng–2.5 μg and a precision of 5% at the lower end of the range.

Scheide and co-workers [168] have reviewed the question of calibration of mercury analysers. This may be via a continuous generation of Hg vapour, or injection of a known volume from a static standard atmosphere.

Experimental procedure for determination of elemental Hg [155]

Air sampling

Hg in ambient air is sampled by drawing air at a known rate for a fixed period of time through cleaned collectors, each containing 1 to 2 g of Ag wool and connected in series. For field sampling purposes a flow rate of 100 ml min^{-1} for 24 h is anticipated. Flow rate, collection time and number of collectors is varied to suit the expected Hg level. After use the collector is disengaged from the pump and capped to prevent contamination. Ag wool in an 8 mm internal diameter tube has been shown to retain 3 to 4 μg of Hg per gram of Ag wool before breakthrough.

Collectors are 100 mm long × 5 mm internal diameter borosilicate glass tube equipped with ball joints on the ends and packed snugly with 1 to 2 g of cleaned Ag wool. The Ag wool is Fisher micro-analysis grade and is initially cleaned by placing it in a furnace at 800° C for 2 h. All collectors are permanently wrapped with exactly 100 cm of 22 gauge Nichrome heating wire.

Analysis

After calibration of the analytical system (Fig. 5.9) using a similar collector, the collector to be analysed is clamped into the system by ball-joint clamps, the carrier gas flow is adjusted to 200 ml min^{-1} and the collector is heated for 30 s at 24 V producing a temperature of 400° C (after one 30 s heating period the collector is Hg free, and ready for future use). The gas train, constructed of 0.25 in Tygon tubing and Kel-F coated valves, carries the Hg vapour to a 20 cm, 3.3 cm, internal diameter, fused silica absorption cell, heated to 90° C with heating tape to prevent condensation of Hg, placed in the light beam of an atomic absorption spectrophotometer set up for mercury measurement at 253.65 nm (Long *et al.* [155] used a Perkin-Elmer 403 instrument). Very short lengths of Tygon tubing are used for connections owing to Hg absorption problems, which are far more acute with rubber or PTFE. Air entering the analytical train is pre-filtered through Ag wool and activated charcoal. The atomic absorption signal is recorded on a pen-recorder and integrated by an Automatic Digital Integrator.

Calibration

The injection of known amounts of air saturated with Hg vapour on to a collector

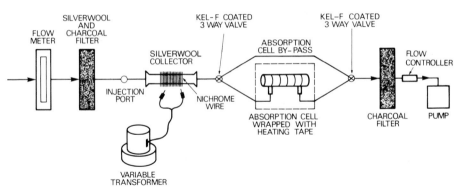

Fig. 5.9 Analytical system for mercury analysis [155].

in the analytical train at a carrier flow of 200 ml min^{-1} is used to standardize the detection system. The Hg is subsequently released by heating, and analysed in the usual manner (direct injection gives lower results). Three 1 litre borosilicate glass bottles containing sufficient Hg to cover the bottoms and equipped with serum caps are used as standardization reservoirs. They are maintained at 20.0 plus or minus 0.1° C in a constant temperature bath. At 20° C and 1 atm pressure 1.0 cm^3 of air contains 13.19 ng Hg. By using gas-tight syringes of volume 0.02 to 100 cm^3, 0.26 to 1319 ng can be introduced into the analytical system through the injection port.

Interferences

SO_2 and NO_2 do not interfere significantly. H_2S gives a negative interference of about 6% at levels of 13 to 650 μg m^{-3} (significantly above normal ambient levels of H_2S). Dimethylmercury is not collected significantly (less than 1%). Very high levels (20 μg injections) of chlorine attack the Ag wool, but a basic scrubber before the collector will eliminate this problem.

Sensitivity

A detection limit of 0.3 ng is found and calibration curves are reproducible to within 11% at 0.5 ng, and to within 3%, relative s.D., beyond 6 ng of Hg.

5.3.2.3 *Volatile Pb compounds*

A small proportion of lead in urban air is in the form of vapour phase lead alkyls. It appears that, while lead is relatively stable in dialkyl, trialkyl and tetra-alkyl forms, the only forms of any real significance in air are inorganic lead and tetra-alkyl lead [169, 170]. The latter may take the form of tetramethyl lead (TML), tetraethyl lead (TEL), or the three mixed methylethyl derivatives $(CH_3(C_2H_5)_3Pb; (CH_3)_2(C_2H_5)_2Pb; (CH_3)_3C_2H_5Pb)$.

Total tetra-alkyl lead may be determined by collection in iodine monochloride and subsequent analysis by extraction into methylisobutyl ketone as the ammonium pyrrolidinecarbodithioate [171] or dithizone [172] complex and atomic absorption determination. This method, however, is sensitive to all lead in the solution and relies upon both a highly efficient separation of particulate inorganic lead by pre-filtration of the air stream, and a low lead blank. Such requirements are not always met and results by this method may be open to considerable doubt [169]. An elegant modification to this method utilizes the fact that tetra-alkyl lead collected in iodine monochloride is converted quantitatively to dialkyl lead [173]. After addition of EDTA to complex inorganic lead, the dialkyl lead is extracted with dithizone in CCl_4, from which it may be back-extracted into a dilute HNO_3/H_2O_2 solution and determined by flameless atomic absorption. The method as originally proposed by Hancock and Slater [173] was designed for a 1 h sampling period, with a detection limit of 0.04 μg (Pb) m^{-3}. Further development of the method has enabled its use for sampling periods of 6 h [174] or 24 and 48 h, the latter providing a detection limit of 0.25 ng (Pb) m^{-3}, sufficient for determination of tetra-alkyl lead in maritime air [175].

The separate determination of individual tetra-alkyl lead compounds in air requires pre-concentration and gas chromatographic separation. Pre-concentration may be upon gas chromatographic packing material at $-78°$ C [176], or upon glass beads at $-130°$ C [177]. Subsequent rapid desorption into a flame atomic absorption instrument was used by Harrison et al. [176] for analysis of total tetra-alkyl lead in air at concentrations down to 0.01 μg (Pb) m^{-3} in a 30 min air sample. Analysis may however be by gas chromatography–mass spectrometry [178], although Radziuk et al. [179] found that the characteristic lead isotope ratio could not be found in ions with m/e values corresponding to alkyl lead fragments. They concluded that considerable interferences arise from other high molecular weight compounds in ambient air, and reported this technique as being unsatisfactory. An alternative and specific method of detection subsequent to gas chromatographic separation is provided by flameless atomic absorption [177], or by a microwave plasma using the 405.78 nm lead emission line [180]. The detection limits of these techniques are shown in Table 5.13.

A continuous atomic absorption procedure designed for measurement of airborne lead in industrial plants manufacturing lead alkyl compounds has been described [181]. The polluted air is used as the supply for a specially designed burner, and measurement of atomic absorption allows determination of organic Pb in air at levels down to 1 μg m^{-3}, although particulate lead is not discriminated.

Experimental 24–48 h method for the determination of total volatile alkyl lead and particulate lead in air

(a) *Tetra-alkyl lead*

Reagents

All reagents were Analar grade, unless otherwise stated. The use of 'water' means de-ionized-distilled water.

Table 5.13 Methods for determination of tetra-alkyl lead in air

Method	Approximate detection limit[a]
Activated carbon/AAS	5 μg Pb[b]
Iodine monochloride/flameless AAS	2 ng Pb
Adsorption tube:	
flame AAS	0.2 ng Pb
G.C. – flameless AAS	0.04 ng Pb
G.C. – microwave plasma	0.006 ng Pb

[a] Detection limits refer to TML.
[b] Limited by lead blank in activated carbon.

Iodine monochloride 1 M. To 111 g of potassium iodide dissolved in 400 ml of water, add 445 ml of conc. HCl (S.G. 1.18). Slowly add 75 g of potassium iodate and stir until dissolved (several hours). Cool and make up to 1000 ml. Store in a glass stoppered bottle.

Iodine monochloride 0.1 M. Dilute the stock solution with water.

Buffer. Dissolve 16 g of citric acid monohydrate and 80 g of hydrated sodium sulphite in 600–700 ml of water. Add 320 ml of conc. ammonia solution (S.G. 0.88). Make up to 2000 ml with water. Store in a glass container.

EDTA 0.1 M. Dissolve 37 g of disodium dihydrogen EDTA in water and make up to 1000 ml. Store in a polythene bottle.

Dithizone in carbon tetrachloride. Dissolve 5 mg of dithizone in 250 ml of Aristar carbon tetrachloride (B.D.H. Ltd). Store in a darkened glass container in a fridge.

Acid solution. Add 20 ml of Aristar hydrogen peroxide and 20 ml of Aristar nitric acid to 2000 ml of water.

Sampling

Add 50–60 ml iodine monochloride to a darkened 125 ml gas bubbler, modified by extending the central tube in the fashion of a midget impinger to within 3–4 mm of the base of the bubbler, and by narrowing the outlet of the tube to a diameter of *c.* 1 mm. Draw pre-filtered air (see below and Fig. 5.10) at 1–3.5 litres min^{-1} through the bubbler. Measure the flow rate before the filter at the beginning and end of each sample period. Use a second bubbler filled with water, followed by a U-tube containing activated charcoal, to remove iodine monochloride vapour in order to protect the pump (Charles Austen, Model M361).

Analysis

To store before analysis, transfer the iodine monochloride to a 100 ml graduated cylinder and note the approximate volume. Pour into a darkened 100 ml bottle. Rinse bubbler with *c.* 5 ml water, add to the bottle via the graduated cylinder. To analyse, transfer to a Pyrex separating funnel with Teflon stopcock together with a further 5 ml washing of the bottle.

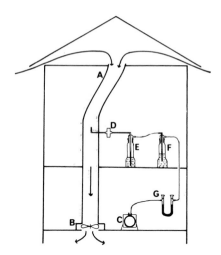

Fig. 5.10   The air sampling system for particulate and organic lead in air. A = stainless steel duct; B = fan; C = pump; D = stainless steel filter holder; E = modified bubbler with iodine monochloride; F = bubbler with water; G = activated carbon trap.

For direct analysis, transfer the iodine monochloride to a 100 ml graduated cylinder and note the approximate volume. Pour directly into the separating funnel with bubbler washings.

Rapidly add buffer to the separating funnel (from an Oxford Pipettor, Model 10/30) in the measured proportion of 20 ml of buffer per 15 ml of iodine monochloride. Swirl to reduce any free iodine formed initially. Then add 5 drops of 0.1 M EDTA solution and swirl to mix.

Add 5 ml of dithizone in carbon tetrachloride to the separating funnel and shake vigorously for 30 s, venting after 5–10 s. Do not allow any iodine monochloride/buffer solution to enter the bore of the stopcock. Allow to stand until the lower organic layer is clear, then run off into a Teflon test tube (this provides a better separation than a Pyrex test tube but is not essential), taking care to leave a little of the organic layer in the separating funnel. Repeat with 5 ml of carbon tetrachloride.

To the dithizone solution in the test tube add 2 ml of the acid solution (from an Oxford Dispenser) and shake for 30 s. Allow to settle for several minutes, then take 20 μl aliquots (with a Gilson Pipetman) from the top acid layer for lead analysis by flameless atomic absorption spectrometry (see below).

(b) *Particulate lead*

Sampling

Pre-filter the air entering the bubblers using a 0.45 μm membrane filter (Millipore HAWP 04700) held in a stainless-steel filter holder with a Teflon gasket. Duct the air to the filter holder and thence to the bubblers through 6 mm o.d. Teflon tube, connected with 1/4 in Swagelok connectors (stainless steel and Teflon were used to avoid any adsorption problems with the TAL vapours). Measure the flow by attaching a calibrated flow meter (Flow Bits Ltd, Basingstoke) via the Swagelok connector to the filter holder while drawing air through the system. Regulate the air flow with a valve (Milli Mite 1300 series, Hoke International Ltd).

Analysis

The particulate lead was analysed by an adaptation of the method of Janssens and Dams [116], similar to the current US Environmental Protection Agency recommended method [134]. Place the filters in 100 ml polypropylene beakers (Azlon) and add 10 ml of 10% Aristar nitric acid. (The volume added was actually adjusted between 5 and 70 ml according to the lead loading of the filter, judged by experience.) Place the beakers in an ultrasonic bath for *c*. 15 min. Determine the lead concentration by injection of 20 μl aliquots of the acid solution directly into a heated graphite atomizer.

(c) *Lead analysis*

Perform the lead analysis by flameless AAS (Perkin-Elmer HGA 76 with a Perkin-Elmer 305 in this work) using the 283.3 nm lead absorbing line, with the following operating conditions: (1) drying – ramp rate 2 to 105° C, hold 5 s; (2) ashing – ramp rate 2 to 550° C, hold 5 s; (3) atomization – ramp rate 1 to 2100° C, hold 8 s (rate $1 = 430°$ C s^{-1}, rate $2 = 12.5°$ C s^{-1}). Run the purge gas under mini flow condition. Calibrate against inorganic lead standards made up in the acid solution for tetra-alkyl lead and 10% nitric acid for particulate lead, over the range 0–100 ng (Pb) ml^{-1}. The calibration is almost linear over this range.

Non-atomic absorption may be checked on the adjacent 280.3 nm non-absorbing line. This was found to be non-existent for both the TAL and particulate lead samples.

Run reagent blanks for the TAL analysis and determine the filter blank for the particulate lead analysis.

(d) *Sample collection*

Sample collection was generally performed using a specially constructed sampling box (Fig. 5.10). Air was drawn under the roof (coated inside with inert FEP Teflon film) and down a central stainless-steel tube using an extractor fan. Air was then drawn from this air stream through the sample train. At some sites, ambient air was sampled by drawing air down a short length of Teflon tube protruding from a window by 30–50 cm.

All sampling was performed in duplicate and each sample analysed separately.

References

1. Lee, R. E. and von Lehmden, D. J. (1973) *J. Air Pollut. Control Assoc.*, **23**(10), 853–7.
2. US Dept of Health, Education and Welfare (1966) Air Quality Data from the National Air Sampling Networks and Contributing State and Local Networks (1964–65), Public Health Service, Division of Air Quality, Cincinnati, Ohio.

3. Spurny, K. R., Lodge, J. P., Frank, E. R. and Sheesley, C. D. (1969) *Environ. Sci. Technol.*, **3**(5), 453–64.
4. Spurny, K. and Lodge, J. P. (1968) *Staub-Reinhalt. Luft.*, **28**(5), 1–10.
5. Spurny, K. and Pich, J. (1965) *Coll. Czech. Chem. Commun.*, **30**, 2276–86.
6. US Dept of Health, Education and Welfare (1962) Air Pollution Measurements of the National Air Sampling Network – Analyses of suspended Particulates (1957–61), Public Health Service Publication No. 978, US Government Printing Office, Washington, DC.
7. Cohen, A. L. (1973) *Environ. Sci. Technol.*, **7**(1), 60–1.
8. Pate, J. B. and Tabor, E. C. (1962) *Am. Ind. Hyg. Assoc. J.*, **23**, 145–50.
9. Hwang, J. Y. (1972) *Anal. Chem.*, **44**, 20A–7A.
10. Dams, R., Rahn, K. A. and Winchester, J. W. (1972) *Environ. Sci. Technol.*, **6**(5), 441–8.
11. Zoller, W. H. and Gordon, G. E. (1970) *Anal. Chem.*, **42**(2), 257–65.
12. Vogg, H. and Haertel, R. (1973) *Fresenius' Z. Anal. Chem.*, **267**(4), 257–60.
13. Gandrud, B. W. and Lazrus, A. L. (1972) *Environ. Sci. Technol.*, **6**(5), 455–8.
14. Kometani, T. Y., Bove, J. L., Nathanson, B. *et al.* (1972) *Environ. Sci. Technol.*, **6**(7), 617–20.
15. Luke, C. L., Kometani, T. Y., Kessler, J. E. *et al.* (1972) *Environ. Sci. Technol.*, **6**(13), 1105–9.
16. Thompson, R. J., Morgan, G. B. and Purdue, L. J. (1969) *Anal. Instr.*, **7**, 9–17.
17. Thompson, R. J., Morgan, G. B. and Purdue, L. J. (1970) *At. Absorpt. Newsl.*, **9**(3), 53–7.
18. Struempler, A. W. (1973) *Anal. Chem.*, **45**(13), 2251–4.
19. King, W. G., Rodriguez, J. M. and Wai, C. M. (1974) *Anal. Chem.*, **46**(6), 771–3.
20. Feldman, C. (1974) *Anal. Chem.*, **46**(1), 99–102.
21. Newton, D. W. and Ellis, R. (1974) *J. Environ. Qual.*, **3**(1), 20–3.
22. Dyck, W. (1968) *Anal. Chem.*, **40**(2), 454–5.
23. West, F. K., West, P. W. and Iddings, F. A. (1966) *Anal. Chem.*, **38**(11), 1566–70.
24. Silverman, L. and Ege, J. F. (1947) *J. Ind. Hyg. Toxicol.*, **29**(2), 136–9.
25. Macdonald, G. L. (1971) *Comprehensive Analytical Chemistry*, Vol. IIC (eds C. L. Wilson and D. W. Wilson), Elsevier, Amsterdam.
26. Gilfrich, J. V., Burkhalter, P. G. and Birks, L. S. (1973) *Anal. Chem.*, **45**(12), 2002–9.
27. Hammerle, R. H., Marsh, R. H., Rengan, K. *et al.* (1973) *Anal. Chem.*, **45**(11), 1939–40.
28. Rhodes, J. R., Pradzynski, A. H., Hunter, C. B. *et al.* (1972) *Environ. Sci. Technol.*, **6**(10), 922–7.
29. Hwang, J. Y. (1970) *Talanta*, **17**, 118–21.
30. Beitz, L., Haase, J. and Weichert, N. (1974) *International Symposium – Environment and Health*, Paris, CEC-WHO-EPA, Luxembourg.
31. Bowman, H. R., Conway, J. G. and Asaro, F. (1972) *Environ. Sci. Technol.*, **6**(6), 558–60.
32. Pradzynski, A. H. and Rhodes, J. R. (1976) *ASTM Spec. Publ. 598*, 320.
33. Billiet, J., Dams, R. and Hoste, J. (1980) *X-Ray Spectrom.*, **9**, 206.
34. Breiter, D. N., Pella, P. A. and Heinrich, K. F. J. (1977) *Nat. Bur. Stand. (US), Spec. Publ. 464*, 527.
35. Epler, R. J. (1974) *Environ. Sci. Technol.*, **8**(1), 28–30.
36. O'Connor, B. H., Kerrigan, G. C., Thomas, W. W. and Gasseng, R. (1975) *X-Ray Spectrom.*, **4**, 190.

37. Van Grieken, R. E. and Adams, F. C. (1976) *X-Ray Spectrom.*, **5,** 61.
38. Ter Haar, G. L. and Bayard, M. A. (1971) *Nature*, **232,** 553–4.
39. Heichel, G. H. and Hankin, L. (1972) *Environ. Sci. Technol.*, **6**(13), 1121–2.
40. Yakowitz, H., Jacobs, M. H. and Hunneyball, P. D. (1972) *Micron*, **3,** 498–505.
41. Walter, R. L., Willis, R. D., Gutknecht, W. F. and Joyce, J. M. (1974) *Anal. Chem.*, **46,** 843.
42. Ahlberg, M. S. and Adams, F. C. (1978) *X-Ray Spectrom.*, **7,** 73.
43. Gibbons, D. and Lambie, D. A. (1971) *Comprehensive Analytical Chemistry*, Vol. IIC (eds C. L. Wilson and D. W. Wilson), Elsevier, Amsterdam.
44. Iddings, F. A. (1969) *Environ. Sci. Technol.*, **3**(2), 132–40.
45. Dams, R., Robbins, J. A., Rahn, K. A. and Winchester, J. W. (1970) *Anal. Chem.*, **42**(8), 861–7.
46. Gordon, C. M. and Larson, R. E. (1964) *NRL Quarterly on Nuclear Science and Technology*, Naval Research Laboratory, Washington, DC, pp. 17–22.
47. Brar, S. S., Nelson, D. M., Kline, J. R. *et al.* (1970) *J. Geophys. Res.*, **75**(15), 2939–45.
48. Altshuller, A. P. (1972) *Analytical Chemistry: Key to Progress on National Problems* (eds W. W. Meinke and J. K. Taylor), *Nat. Bur. Stand. (US), Spec. Publ.* 351.
49. Lambert, J. P. F. and Wiltshire, F. W. (1979) *Anal. Chem.*, **51,** 1346.
50. Greenberg, R. R. (1979) *Anal. Chem.*, **51,** 2004.
51. Swindle, D. L. and Schweikert, E. A. (1973) *Anal. Chem.*, **45**(12), 211–15.
52. Riddle, D. C. and Schweikert, E. A. (1974) *Anal. Chem.*, **46**(3), 395–8.
53. Parsa, B. and Markowitz, S. S. (1974) *Anal. Chem.*, **46**(2), 186–9.
54. Mitteldorf, A. J. (1965) *Trace Analysis – Physical Methods* (ed. G. H. Morrison), Interscience, New York.
55. Homan, R. E. and Morgan, G. B. (1968) 19th Pittsburgh Conference on Analytical Chemistry and Applied Spectroscopy, American Chemical Society, Washington, DC.
56. Morgan, G. B., Ozolins, G. and Tabor, E. C. (1970) *Science*, **170,** 289–96.
57. Tabor, E. C. and Warren, W. V. (1958) *AMA Arch. Ind. Health*, **17,** 145–51.
58. National Air Pollution Control Administration (1968) Air Quality Data from the National Air Surveillance Networks and Contributing State and Local Networks, 1966 edn, APTD68–9, 167P, Washington, DC.
59. American Industrial Hygiene Association (1969) Analytical Guide: Beryllium, *Am. Ind. Hyg. Assoc. J.*, 103–5.
60. Cholak, J. and Hubbard, D. M. (1948) *Anal. Chem.*, **20**(1), 73–6.
61. Smith, R. G., Boyle, A. J., Fredrick, W. G. and Zak, B. (1952) *Anal. Chem.*, **24**(2), 406–9.
62. Fitzgerald, J. J. (1957) *AMA Arch. Ind. Health*, **15,** 68–73.
63. Churchill, W. I. and Gillieson, A. H. C. P. (1952) *Spectrochim. Acta*, **5,** 238–50.
64. Seeley, J. L. and Skogerboe, R. K. (1974) *Anal. Chem.*, **46**(3), 415–21.
65. Weisz, H. (1970) *Microanalysis by the Ring Oven Technique*, 2nd edn, Pergamon Press, Oxford.
66. West, P. W., Weisz, H., Gaeke, G. C. and Lyles, G. (1960) *Anal. Chem.*, **32**(8), 943–6.
67. West, P. W. (1968) *Air Pollution*, 2nd edn, Vol. II (ed. A. C. Stern), Academic Press, New York, pp. 147–85.
68. Ronneau, C. J.-M., Jacob, N. M. and Apers, D. J. (1973) *Anal. Chem.*, **45**(12), 2152.
69. West, P. W. and Thabet, S. K. (1967) *Anal. Chim. Acta*, **37,** 246–52.
70. Levine, L. (1945) *J. Ind. Hyg. Toxicol.*, **27**(6), 171–7.

71. Taylor, J. K., Maienthal, E. J. and Marinenko, G. (1965) *Trace Analysis – Physical Methods* (ed. G. H. Morrison), Interscience, New York.
72. Ferrett, D. J., Milner, G. W. C., Shalgosky, H. I. and Slee, L. J. (1956) *Analyst*, **81,** 506–12.
73. Sachdev, S. L. and West, P. W. (1970) *Environ. Sci. Technol.*, **4**(9), 749–51.
74. Mancy, K. H. (1972) *Analytical Chemistry: Key to Progress on National Problems* (eds W. W. Meinke and J. K. Taylor), *Nat. Bur. Stand. (US), Spec. Publ.* 351.
75. von Lehmden, D. J., Jungers, R. H. and Lee, R. E. (1974) *Anal. Chem.*, **46**(2), 239–45.
76. Harrison, P. R. and Winchester, J. W. (1971) *Atmos. Environ.*, **5**(10), 863–80.
77. Colovos, G., Wilson, G. S. and Moyers, J. (1973) *Anal. Chim. Acta*, **64,** 457–64.
78. MacLeod, K. E. and Lee, R. E. (1973) *Anal. Chem.*, **45**(14), 2380–3.
79. Ryan, M. D. and Siemer, D. D. (1977) *Environmental Analysis*, Academic Press, London, p. 105.
80. Roboz, J. (1965) *Trace Analysis – Physical Methods* (ed. G. H. Morrison), Interscience, New York.
81. Brown, R. and Vossen, P. G. T. (1971) *Int. Symp. Ident. Meas. Environ. Pollut. (Proc)*, (ed. B. Westley), pp. 427–31.
82. Brown, R. and Vossen, P. G. T. (1970) *Anal Chem.*, **42**(14), 1820–2.
83. Cheng, K. L. (1965) *Trace Analysis – Physical Methods* (ed. G. H. Morrison), Interscience, New York.
84. Spectrometry Nomenclature (1973) *Anal. Chem.*, **45**(14), 2449.
85. IUPAC (1963) *Tables of Spectrophotometric Absorption Data of Compounds used for the Colorimetric Determination of Elements*, Butterworth, London.
86. American Public Health Association (1977) *Methods of Air Sampling and Analysis*, 2nd edn, Washington, DC.
87. Jacobs, M. B. (1967) *The Analytical Toxicology of Industrial Inorganic Poisons*, Interscience, New York.
88. Robinson, E. and Ludwig, F. L. (1967) *J. Air Pollut. Control Assoc.*, **17**(10), 664–9.
89. Cholak, J. (1964) *Arch. Environ. Health*, **8**(2), 222–31.
90. American Industrial Hygiene Association (1969) Community Air Quality Guides: Lead. *Am. Ind. Hyg. Assoc. J.*, 95–7; Analytical Guides: Lead, 102–3.
91. American Society for Testing and Materials, ASTM Method D3112–72T, ASTM, Philadelphia, Pa.
92. Saltzman, B. E. (1953) *Anal. Chem.*, **25**(3), 493–6.
93. American Industrial Hygiene Association (1964) Hygienic Guide Series: Arsenic and Its Compounds, *Am. Ind. Hyg. Assoc. J.*, 610–13.
94. Hiser, R. A., Donaldson, H. M. and Schwenzfeier, C. W. (1961) *Am. Ind. Hyg. Assoc. J.*, 280–5.
95. McCloskey, J. (1967) *Microchem. J.*, **12,** 32–45.
96. American Industrial Hygiene Association (1966) Hygienic Guide Series: Nickel, *Am. Ind. Hyg. Assoc. J.*, 202–5.
97. Ege, J. F. and Silverman, L. (1947) *Anal. Chem.*, **19**(9), 693–4.
98. Sill, C. W. (1961) *Anal. Chem.* **33,** 1684–6.
99. Kahn, H. L. (1968) *Adv. Chem. Ser.* No. 73, American Chemical Society, pp. 183–229.
100. Weberling, R. P. and Cosgrove, J. F. (1965) *Trace Analysis – Physical Methods* (ed. G. H. Morrison), Interscience, New York.
101. Browner, R. F., Dagnall, R. M. and West, T. S. (1970) *Anal. Chim. Acta*, **50,** 375–81.
102. Kahn, H. L., Peterson, G. E. and Schallis, J. E. (1968) *At. Absorpt. Newsl.*, **7**(2), 35–9.

103. Woodriff, R. (1969) *Trace Subst. Environ. Health – 3*, Proc. 3rd Univ. Mo. Ann. Conf., pp. 297–303.
104. Manning, D. C. and Fernandez, F. (1970) *At. Absorpt. Newsl.*, **9**(3), 65–70.
105. Beukelman, T. E. and Lord, S. S. (1960) *Appl. Spectros.*, **14**(1), 12–17.
106. Burnham, C. D., Moore, C. E., Kanabrocki, E. and Hattori, D. M. (1969) *Environ. Sci. Technol.*, **3**(5), 472–5.
107. Jackson, G. B. and Myrick, H. N. (1971) *Internat. Lab.*, 41–7.
108. Hwang, J. Y. (1971) *Can. Spectrosc.*, **16**, 43–5; 53.
109. Koirtyohann, S. R. and Wen, J. W. (1973) *Anal. Chem.*, **45**(12), 1986–9.
110. Lundgren, D. A. (1970) *J. Air Pollut. Control Assoc.*, **20**(9), 603–8.
111. Lee, R. E., Patterson, R. K. and Wagman, J. (1968) *Environ. Sci. Technol.*, **2**(4), 288–90.
112. Morgan, G. B. and Homan, R. E. (1967) 18th Pittsburgh Conference on Analytical Chemistry and Applied Spectroscopy, American Chemical Society, Washington, DC.
113. Kneip, T. J., Eisenbud, M., Strehlow, C. D. and Freudenthal, P. C. (1970) *J. Air Pollut. Control Assoc.*, **20**(3), 144–9.
114. Burnham, C. D., Moore, C. E., Kowalski, T. and Krasniewski, J. (1970) *Appl. Spectrosc.*, **24**(4), 411–14.
115. Purdue, L. J., Enrione, R. E., Thompson, R. J. and Bonfield, B. A. (1973) *Anal. Chem.*, **45**(3), 527–30.
116. Janssens, M. and Dams, R. (1973) *Anal. Chim. Acta*, **65**, 41–7.
117. Begnoche, B. C. and Risby, T. H. (1975) *Anal. Chem.*, **47**, 1041–5.
118. Woodriff, R. and Lech, J. F. (1972) *Anal. Chem.*, **44**(7), 1323–5.
119. Matousek, J. P. and Brodie, K. G. (1973) *Anal. Chem.*, **45**(9), 1606–9.
120. Zdrojewski, A., Quickert, N. and Dubois, L. (1973) *Int. J. Environ. Anal. Chem.* **2**, 331–41.
121. Janssens, M. and Dams, R. (1974) *Anal. Chim. Acta*, **70**, 25–33.
122. Sachdev, S. L., Robinson, J. W. and West, P. W. (1967) *Anal. Chim. Acta*, **37**, 12–19.
123. Sachdev, S. L., Robinson, J. W. and West, P. W. (1967) *Anal. Chim. Acta*, **38**, 499–506.
124. Hwang, J. Y. and Feldman, F. J. (1970) *Appl. Spectrosc.*, **24**(3), 371–4.
125. Loftin, H. P., Christian, C. M. and Robinson, J. W. (1970) *Spectrosc. Lett.*, **3**(7), 161–74.
126. Robinson, J. W. (1972) Proc. International Symposium: Environmental Health Aspects of Lead, pp. 1099–105, Amsterdam, CEC–EPA, Luxembourg.
127. Robinson, J. W., Wolcott, D. K., Slevin, P. J. and Hindman, G. D. (1973) *Anal. Chim. Acta*, **66**, 13–21.
128. Edwards, H. W. (1969) *Anal. Chem.*, **41**(10), 1172–5.
129. Clayton, P. and Wallin, S. C. (1973) CCMS/CPPSD Conference: Ann Arbor, Michigan.
130. White, R. A. (1967) *J. Sci. Instr.*, **44**, 678–80.
131. Robinson, J. W. and Wolcott, D. K. (1973) *Anal. Chim. Acta*, **66**, 333–42.
132. Siemer, D., Lech, J. F. and Woodriff, R. (1973) *Spectrochim. Acta*, **28B**, 469–71.
133. Geladi, P. and Adams, F. (1979) *Anal. Chim. Acta*, **65**, 41.
134. Environmental Protection Agency (1978) *Federal Register*, **43**, 46246.
135. Pilate, A., Geladi, P. and Adams, F. (1977) *Talanta*, **25**, 512.
136. Walsh, P. R., Fasching, J. L. and Duce, R. A. (1976) *Anal. Chem.*, **48**, 1014.
137. Peden, M. E. (1977) *Nat. Bur. Stand. (US) Spec. Publ. 464*, 367.

138. Pickford, C. J. and Rossi, G. (1978) *Analyst*, **103,** 341.
139. Eller, P. M. and Haartz, J. C. (1977) *Am. Ind. Hyg. Assoc. J.*, **38,** 116.
140. Dumarey, R., Heindryckx, R. and Dams, R. (1980) *Anal. Chim. Acta*, **116,** 111.
141. Kermoshchuk, J. O. and Warner, J. O. (1981) *Proc. Conf. Heavy Metals in the Environment*, CEP Consultants, Edinburgh, 603.
142. Ranweiler, L. E. and Moyers, J. L. (1974) *Environ. Sci. Technol.*, **8**(2), 152–6.
143. Ross, W. D. and Sievers, R. E. (1972) *Environ. Sci. Technol.*, **6**(2), 155–8.
144. Strackee, L. (1968) *Nature*, **218,** 497–8.
145. Barnes, I. L., Murphy, T. J., Gramlich, J. W. and Shields, W. R. (1973) *Anal. Chem.*, **45**(11), 1881–4.
146. Densham, A. B., Beale, P. A. A. and Palmer, R. (1963) *J. Appl. Chem.*, **13,** 576–80.
147. McCarley, J. E., Saltzman, R. S. and Osborn, R. H. (1956) *Anal. Chem.*, **28**(5), 880–2.
148. Belyakov, A. A. (1960) *Zavodskaya Laboratoriya*, **26,** 158–9.
149. American Industrial Hygiene Association (1968) Hygienic Guide Series: Nickel Carbonyl, *Am. Ind. Hyg. Assoc. J.*, 304–7.
150. Kincaid, J. F., Stanley, E. L., Beckworth, C. H. and Sunderman, F. W. (1956) *Am. J. Clin. Pathol.*, **26,** 107–19.
151. Brief, R. S., Venable, F. S. and Ajemian, R. S. (1965) *Am. Ind. Hyg. Assoc. J.*, **26,** 72–6.
152. Brief, R. S., Ajemian, R. S. and Confer, R. G. (1967) *Am. Ind. Hyg. Assoc. J.*, **28,** 21–30.
153. Panek, J. (1973) *Cesk. Hyg.*, **18**(5), 244–9.
154. Linch, A. L., Stalzer, R. F. and Lefferts, D. T. (1968) *Am. Ind. Hyg. Assoc. J.*, **29,** 79–86.
155. Long, S. J., Scott, D. R. and Thompson, R. J. (1973) *Anal. Chem.*, **45**(13), 2227–33.
156. Corte, G., Dubois, L. and Monkman, J. L. (1973) *Sci. Tot. Environ.*, **2**(1), 89–96.
157. Corte, G. L., Thomas, R. S., Dubois, L. and Monkman, J. L. (1973) *Sci. Tot. Environ.*, **2,** 251–8.
158. Scaringelli, F. P., Puzak, J. C., Bennett, B. I. and Denny, R. L. (1974) *Analyt. Chem.*, **46**(2), 278–83.
159. Van der Sloot, H. A. and Das, H. A. (1974) *Anal. Chim. Acta*, **70,** 439.
160. Siemer, D., Lech, J. and Woodriff, R. (1974) *Appl. Spectrosc.*, **28**(1), 68–71.
161. Hatch, W. R. and Ott, W. L. (1968) *Anal. Chem.*, **40**(14), 2085–7.
162. Hwang, J. Y., Ullucci, P. A. and Malenfant, A. L. (1971) *Can. Spectrosc.*, **16,** 100–6.
163. Brandenberger, H. and Bader, H. (1968) *At. Absorpt. Newsl.*, **7**(3), 53–4.
164. Driscoll, J. N. (1974) *Health Lab. Sci.*, **11,** 348–53.
165. Henriques, A., Isberg, J. and Kjellgren, D. (1973) *Chemica Scripta*, **4,** 139–42.
166. Braman, R. S. and Johnson, D. L. (1974) *Environ. Sci. Technol.*, **8,** 996–1003.
167. Trujillo, P. E. and Campbell, E. E. (1975) *Anal. Chem.*, **47,** 1629.
168. Scheide, E. P., Hughes, E. E. and Taylor, J. K. (1979) *Am. Ind. Hyg. Assoc. J.*, **40,** 180.
169. Harrison, R. M. and Perry, R. (1977) *Atmos. Environ.*, **11,** 847.
170. Harrison, R. M. and Laxen, D. P. H. (1978) *Environ. Sci. Technol.*, **12,** 1384.
171. Purdue, L. J., Enrione, R. E., Thompson, R. J. and Bonfield, B. A. (1973) *Anal. Chem.*, **45,** 527.
172. Colwill, D. M. and Hickman, A. J. (1973) *TRRL Report LR545*, Dept of Environment, London.
173. Hancock, S. and Slater, A. (1975) *Analyst*, **100,** 422–9.
174. De Jonghe, W. and Adams, F. (1979) *Anal. Chim. Acta*, **108,** 21.
175. Birch, J., Harrison, R. M. and Laxen, D. P. H. (1980) *Sci. Tot. Environ.*, **14,** 31–42.

176. Harrison, R. M., Perry, R. and Slater, D. H. (1974) *Atmos. Environ.*, **8**, 1187–94.
177. De Jonghe, W. R. A., Chakraborti, D. and Adams, F. C. (1980) *Anal. Chem.*, **52**, 1974.
178. Laveskog, A. (1970) *Proc. Second Int. Clean Air Congr.*, Air Pollution Control Association, Washington, DC, p. 549.
179. Radzuik, B., Thomassen, Y., Van Loon, J. C. and Chau, Y. K. (1979) *Anal. Chim. Acta*, **105**, 255.
180. Reamer, D. C., Zoller, W. H. and O'Haver, T. C. (1978) *Anal. Chem.*, **50**, 1449.
181. Thilliez, G. (1967) *Anal. Chem.*, **39**, 427–32.

6 Nitrogen and sulphur compounds*

R. M. HARRISON

6.1 Introduction

The gaseous compounds of sulphur and nitrogen which are of interest in atmospheric pollution studies fall into three main chemical groups – oxides, hydrides and organic compounds of sulphur and nitrogen.

Of the oxides of sulphur only SO_2 and SO_3 are important air pollutants. SO_2 is a major pollutant causing widespread concern. The main source is the combustion of fossil fuels, when most of the sulphur present in the fuel is oxidized to SO_2. Other major sources include the metallurgical, cement, petroleum refining and miscellaneous chemical process industries. Motor vehicles are a relatively minor source of SO_2 since refined motor fuel normally has a low sulphur content.

There are a number of well documented deleterious effects of atmospheric SO_2, such as damage to vegetation of all kinds, deterioration of textiles and corrosion of metals and building materials [1]. Also sulphate aerosols, produced as a result of oxidation of atmospheric SO_2, contribute significantly to the aerosol burden of the atmosphere, giving rise to loss of visibility and acidic precipitation. Health hazards of SO_2 are less easily defined. The gas is toxic at high concentrations ($TLV = 5$ ppm) but there is no clear evidence that any injurious effects on the health of city dwellers are directly attributable to SO_2 itself. It is believed, however, that SO_2 in combination with other air pollutants, e.g. smoke, can be injurious to health.

SO_3 is also produced during the combustion of fossil fuels but to a much lesser extent than SO_2. Chemical installations, such as those manufacturing H_2SO_4, may also constitute sources of SO_3. In view of the extreme reactivity of SO_3, emissions are closely controlled to prevent damage to plant, personnel etc. On contact with water vapour in the atmosphere SO_3 is rapidly converted into sulphuric acid aerosol. The presence of free SO_3 in the atmosphere has never been demonstrated and is, in fact, very unlikely. Analysis for SO_3 is therefore of concern only at the source.

The oxides of nitrogen which are of major concern in atmospheric pollution studies are nitric oxide (NO) and nitrogen dioxide (NO_2). The higher oxides of nitrogen dinitrogen trioxide (N_2O_3) and dinitrogen tetroxide (N_2O_4) exist in

---

* A revision of the chapter by R. A. Cox, *Handbook of Air Pollution Analysis*, first edition.

equilibrium with NO and NO_2 but at atmospheric concentrations of the latter the N_2O_3 and N_2O_4 components are negligible. Similar remarks apply to the other higher oxides nitrogen trioxide (NO_3) and dinitrogen pentoxide (N_2O_5) which are believed to be important intermediates in the photochemical smog-forming reactions. The major source of nitrogen oxides is combustion when fixation of atmospheric nitrogen occurs at the high flame temperature. The oxides are emitted mainly as NO which is normally rapidly oxidized to NO_2 by atmospheric O_3 and free radicals (see Chapter 3). Motor vehicle exhaust contributes a sizeable fraction of the total emissions of oxides of nitrogen. As well as stationary combustion sources, the manufacture of nitric acid and nitrate fertilizer are sources of NO and NO_2.

As with sulphur oxides, nitrogen oxides may have many deleterious effects [2]. NO is non-toxic but NO_2 is a powerful lung irritant. Adverse human health effects have been observed as a result of long term exposure to concentrations as low as 0.1 ppm NO_2. Concentrations greater than 100 ppm are lethal to most species. NO_2 assists corrosion of metals, deterioration of textiles and can damage vegetation. The oxides of nitrogen are also precursors in the formation of photochemical smog and their oxidation product, nitric acid, contributes to the aerosol burden in the atmosphere. The chemical reactivity of the oxides of nitrogen in the atmosphere is a major factor warranting their control.

The gaseous hydrides of sulphur and nitrogen, namely hydrogen sulphide (H_2S) and ammonia (NH_3), are pollutants of secondary importance but may present considerable problems in specific locations. The major sources of H_2S include pulp and paper manufactures and refining and coking operations; ammonia is emitted during fertilizer manufacture and sewage treatment. The toxic, odorous and corrosive properties of H_2S are well known; also, H_2S in the atmosphere, is rapidly oxidized to SO_2 with its associated effects. Deleterious effects of ammonia are mainly associated with its role in the formation of atmospheric particulate matter.

Emissions of organic sulphur and nitrogen compounds are, in volume terms, comparatively minor. They are, however, of great importance from the point of view of odour nuisance, a field that is receiving a growing amount of attention. The main sources are industries which process natural products such as wood pulp, paper and animal offal, as well as miscellaneous chemical processes. Few odorous compounds have been definitely identified but mercaptans and other organosulphur compounds have been detected as pollutants from paper manufacture. The organic nitrogen containing esters, e.g. peroxyacetylnitrate (PAN), are characteristic products of photochemical smog and will be considered in Chapter 7. Miscellaneous compounds such as HCN, amines, carbon disulphide are pollutants associated with certain chemical process industries but are not normally encountered in ambient air.

It should be noted that many of the gases classified above as pollutants are also emitted from natural sources, e.g. H_2S from swamps, NH_3 from animal urine and amines from chicken dung. Interest in the behaviour and status of trace gases in the natural atmosphere has further stimulated research into sensitive analytical techniques for the reliable measurement of low concentrations of gases in air.

The optimum analytical technique to be employed in a particular monitoring exercise will depend on the concentration range which is likely to be encountered and the time-variation of the pollutant concentration. These factors will largely depend on where the measurements are to be made. Locations for air pollution measurements may be broadly divided into five categories, i.e. source, source vicinity, urban and industrial regions, rural regions and remote regions. The range of concentrations of several gaseous nitrogen and sulphur pollutants which may be expected in these regions are given in Table 6.1. Concentrations in the vicinity of a particular source will depend greatly on the source strength and wind

Table 6.1 Concentration ranges of pollutant gases in different locations

Gas	Concentration (ppb)				
	Source	*Source vicinity*	*Urban*	*Rural*	*Remote*
SO_2	2×10^6	10^3	20–500	5 –50	1
NO_x	10^6	10^3	20–500	5 –50	3
H_2S	–	> 50	1–10	0.1– 1	0.1
NH_3	–	10^3	2–25	2 –25	5

direction and large short-term variations in concentration may be observed. At locations increasingly removed from the sources the short-term variations become progressively less. In order to obtain meaningful information on peak levels when the concentrations are varying rapidly (e.g. kerbside measurements), real-time continuous measurements are desirable. On the other hand for most measurements of a 'background' type, e.g. determination of general pollutant level in a particular region, time resolution of less than one hour is not normally necessary.

The range of concentration of pollutants in the atmosphere is roughly 4 to 5 orders of magnitude. At the present time satisfactory analytical techniques are available for measurement of the common gaseous S and N pollutants at the 0.1 to 1.0 ppm level. Improvement of existing methods and new instrumental techniques now allows reasonably reliable continuous measurement down to 0.01 ppm. The analytical problems involved in the measurement of trace gas concentrations at the ppb level and below are, however, considerable and for most gases have yet to be satisfactorily resolved.

In this chapter analytical methods, both manual and instrumental, for the oxides and hydrides of S and N are discussed. The general methods for the analysis of hydrocarbons given in Chapter 8, are also applicable to organic compounds containing N or S. Some specific methods for organosulphur and nitrogen compounds are discussed in the present chapter. The practical details given in Section 6.3 refer primarily to methods for ambient air measurement but some of the instrumental techniques are suitable for source analysis.

6.2   Basic analytical techniques

The methods for the analysis of gaseous sulphur and nitrogen pollutants may be classified as chemical and physical. The chemical methods, which have been developed from techniques used in the chemical process industries, involve trapping the gas in a suitable medium followed by chemical or electrochemical analysis of the trapped material. Physical methods involve direct measurement of a physical or optical property either of the pollutant itself or following its interaction with another compound. The measurement may be preceded by chromatographic separation.

Since the atmosphere contains a wide variety of trace gases, interference by other gases is a major factor to be considered in the analysis of a given pollutant. Since many chemical methods rely on properties such as the acidity, oxidizing or reducing capabilities of the gases to be analysed, they are subject particularly to interference by other gases. Particulate material in the air may interfere chemically or, especially in the case of continuous instrumental methods, by physical contamination. Interfering substances may be removed either selectively or by discrimination either during sampling or at the analytical stage.

The errors in the analysis of gases in the atmosphere are expressed in terms of the degree of accuracy and precision. Inaccuracy arises from interferences, variable and indeterminate collection efficiency, calibration errors, etc., and is usually the main source of analytical error. For most methods the analytical precision is a minor source of error, provided the procedures are carried out carefully. Frequently an unreasonable degree of accuracy and precision in measurement is specified. There is no need to obtain a concentration to the nearest 0.01 unit if differences between effects are not noticeable to the nearest unit. In ambient air, pollutant concentrations vary greatly in time and place and an overall analytical error of 10% will generally be quite satisfactory for the interpretation of field data. In obtaining cause-and-effect relationships in laboratory experiments a greater degree of accuracy may be required, but higher accuracy and precision is easier to achieve in controlled laboratory conditions. It is also unnecessary to obtain exactitude in sampling which is greater than the precision of the analytical procedures and vice versa.

6.2.1   Sampling techniques

The methods used for the sampling of gaseous atmospheric pollutants are surveyed in Chapter 1. For the chemical analysis of gaseous S and N compounds, sample concentration is usually required. This is achieved by passing the air sample through a suitable trapping medium, e.g. an absorbing solution or an impregnated filter paper. The latter is convenient for automatic sequential sampling utilizing a filter paper tape. The choice of sampling parameters, e.g. flow rate, absorbent volume, time, etc., will depend on the expected concentration range of the pollutant, the time period over which the average concentration is required, the collection efficiency of the medium and the limitations of the analytical method.

Techniques requiring sample concentration cannot be used for continuous real-time measurement of pollutant concentrations. Physical and electro-chemical instrumental methods offer more scope for this type of measurement since the sensitivity of modern instruments is sufficient to dispense with sample concentration, at least for measurements in urban air. Air is simply drawn into the instrument continuously and the amount of pollutant gas entering the sensing device is monitored. Alternatively, physical sensing techniques may be applied to discrete samples, as for example in gas-chromatographic analysis. By making automatic sequential analysis, to give a series of spot measurements, a reasonable approach to continuous monitoring can be made, provided the analysis time is short. Gas chromatographic analysis may also be applied to samples collected at a remote location and transported to the laboratory. However, this approach is not recommended for S and N gases in view of possible losses on containment and during transport.

In the design of samplers for gases, it is important to ensure that there is no significant loss or modification of the pollutant during transport from the free atmosphere to the absorption medium or sensing device. Most of the gaseous S and N compounds are chemically reactive and tend to be absorbed on containing materials, particularly in moist environments. Sample probes should be as short, clean and dry as possible and should be of a suitably inert material, e.g. borosilicate glass, polytetrafluoroethylene (PTFE) or other fluorocarbon polymer. Most metals and the more common types of plastic tubing, e.g. polythene and PVC, should be avoided. Couplings and valves should, as far as possible, be avoided on sampling lines, but if used should be of high quality stainless steel or inert plastic.

Interfering particulate material is most conveniently removed at the sampling stage by filtration of the gas stream. The filter should be of a suitably inert material to minimize absorption of the gas of interest. Similarly, the filter holder should be constructed so that the gas stream is only exposed to an inert surface such as glass or PTFE. In order to minimize adsorption, it may be necessary to heat the filter, particularly when sampling at high relative humidity (see Section 6.3.1.1 for SO_2 analysis).

Selective removal of interfering gaseous substances at the sampling stage has also been widely utilized, particularly in monitoring instruments based on chemical methods of analysis. This is effected by exposure of the sample gas to an absorbent (either a liquid or solid) which removes the interfering component(s) but allows the pollutant of interest to pass on to the collection or sensing unit. Pretreatment may also be used to convert chemically the pollutant to a compound which is more suitable for analysis (e.g. oxidation of NO to NO_2). The efficiency of any sample pretreatment must be carefully assessed to ensure that it is quantitative under the conditions operating. In view of the possibility of enhanced absorption losses or inadvertent modification of the sample, pretreat-ment of the sample gas should only be applied if significant interference is expected which cannot be removed at a later stage of the analysis.

6.2.2   Analytical methods – chemical

The chemical methods for the analysis of gaseous sulphur and nitrogen pollutants may be broadly classified as acidimetric, colorimetric and coulometric techniques.

6.2.2.1   *Acidimetric methods*

Acidimetric techniques have been widely used for the routine analysis of SO_2 and involve determination of the free acid (H^+ ion) produced following the absorption of SO_2 in an oxidizing solution (e.g. dilute H_2O_2) where it is converted to sulphuric acid (H_2SO_4). The free acid may be determined by titration or electrically, by conductivity or pH measurements. Automatic measurement of the change in conductivity of a solution exposed to SO_2 provides the basis for continuous acidimetric analysis of the gas. Any other gas which is absorbed rapidly by aqueous solution to yield strong acid will give a response but normally SO_2 is the most abundant atmospheric constituent which behaves in this way. A more serious interference comes from alkaline substances (e.g. NH_3) which neutralize the acid. Acid and alkaline particulate matter can also interfere and the acidimetric method for the determination of SO_2 has been widely criticized on account of these interferences.

6.2.2.2   *Colorimetric methods*

A wide variety of colorimetric methods have been applied to N and S gases. The technique involves interaction in solution of the gas or its hydrolysis or oxidation products with a colour-forming reagent, followed by spectrophotometric measurement of the colour. The optical absorbance is proportional to the concentration of the component of interest.

Variations of the basic colorimetric method include turbidimetric measurement of a colloidal suspension (e.g. barium sulphate from the sulphate ion in solution), measurement of optical density of a colour produced on impregnated filter paper (e.g. lead acetate stain method for H_2S) and spectrofluorimetric determination of ions in solution.

Colorimetric methods can be highly specific and sensitive. The colour reagent may be incorporated in the absorbing solution allowing the colour to develop as the sample is taken. Reactions of this type have been utilized in automatic colorimetric analysers in which pumped sample gas and reagent(s) are continuously mixed and the exposed reagent passes to a flow colorimeter where the absorbance is measured (Fig. 6.1). It is necessary to have controlled flows of both reagent and sample gas for reproducible results. Reproducibility and efficiency of absorption of the gas in the reagent are important. A double beam colorimeter system has advantages for stability, compensating against lamp emission, voltage and reagent optical density. When instruments are designed for air pollution work, reagents for seven or eight days are carried in storage bottles.

An alternative approach to colorimetric analysis involves absorption in a

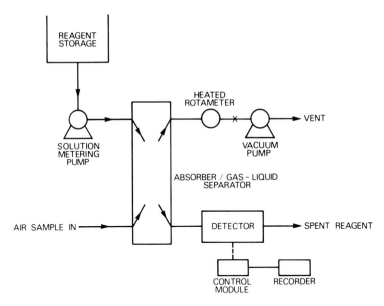

Fig. 6.1   Block diagram of continuous colorimetric analyser.

medium which traps the pollutant gas in a stable, non-volatile form (e.g. SO_2 as sulphate ion, NO_2 as nitrite ion). The samples may then be stored for analysis at a later date. Filter paper tape impregnated with a suitable absorbent provides a useful medium for the collection of large numbers of samples for colorimetric analysis of SO_2 and NH_3. The filter samples may be stored dry in sealed bags and extracted into solution when the analysis is to be carried out. For small-scale monitoring exercises, simple liquid bubbler absorbers are more convenient.

Of the chemical techniques, colorimetric methods are the most versatile. Manual procedures giving sensitivity to approximately 10 ppb are available for most of the gaseous N and S pollutants. The colour response can be conveniently calibrated using standard solutions of the corresponding ions and any interfering substances can often be eliminated during sample preparation. The labour involved in the analysis of a large number of individual samples has been reduced by the advent of automatic chemical analysers. These instruments perform automatically the 'wet chemical' operations normally carried out manually in the laboratory. The heart of the instrument is a multichannel proportioning-pump which dispenses sample solution and reagents into a continuous-flow system. Processes such as mixing, heating, etc. are performed automatically on the sample stream and finally optical density is measured and the results displayed on a chart recorder. Because of the accurately standardized automatic procedure, precision is frequently better than that obtained in the best manual methods and a very high rate of sample analysis is possible. Technicon, a familiar company in this field, has a large range of analysing systems covering many pollutants.

6.2.2.3   *Coulometric methods*

Coulometric analysis involves measurement of the electrical current produced when strongly oxidizing or reducing pollutant gases react with potassium iodide or bromide solution in an electrochemical cell. Two general types of coulometric analysers have been employed; one involves the principle of coulometric internal electrolysis (galvanic action – Hersch Cell [3]) and the other is described as an amperometric coulomer. The latter principle was used in early systems for coulometric analysis of O_3 [4] e.g. 'Mast' ozone meter, see Chapter 7 and requires reagent replacement and an applied external voltage. Galvanic cells which utilize a cyclic oxidation – reduction process and require no applied potential, are usually used for the coulometric analysis of N and S gases. In the Beckmann NO_2 analyser (Fig. 6.2), the sample gas is drawn through the detector cell containing buffered KI electrolyte which is circulated past the electrodes. The cell contains a Pt cathode and a C anode with a galvanic potential difference between them. When NO_2 enters the cell it reacts with the iodide ion in the following reaction

$$NO_2 + 2I^- + 2H^+ \rightarrow I_2 + NO + H_2O$$

The iodine produced is reduced at the cathode in the electrochemical reaction

$$I_2 + 2e^- \rightarrow 2I^-$$

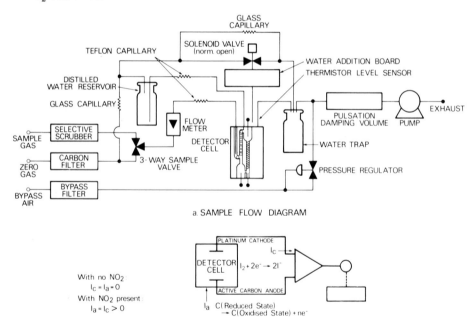

a. SAMPLE FLOW DIAGRAM

b. ELECTROCHEMICAL REACTIONS OF THE CELL

Fig. 6.2   Coulometric analyser for NO_2 (Beckman).

and a corresponding oxidation reaction occurs at the anode

C (reduced state)$\rightarrow$C (oxidized state)$+$ne$^-$

Thus the iodine liberated allows the passage of a current in the anode–cathode circuit which is proportional to the amount of NO_2 entering the cell. With no oxidizing agent entering the cell the residual current is zero.

A more complex indirect galvanic cell is used for the coulometric analysis of SO_2 and other reducing gases. The sample gas is introduced into the anode side of the detector cell which contains iodide solution having electrogenerated iodine present. The iodine is reduced by SO_2 to iodide in the reaction:

$$SO_2 + I_2 + 2H_2O \rightarrow SO_4^{2-} + 2I^- + 2H^+$$

resulting in a lowering of the amount of iodine to be reduced at the cathode. A reference cell in a bridge circuit measures the difference, i_d, between the constant electrogenerating current (anode output) and the cathodic output current, i_d being proportional to the amount of SO_2 entering the cell.

If the reaction of the oxidizing or reducing gas in the electrolyte occurs rapidly with 100% efficiency the electrical current is directly related by Faraday's law to the amount of gas entering. Thus for a given controlled sample flow rate, coulometric analysis can give an absolute measurement of concentration without the requirement of calibration against standard mixtures or solutions. The main problem with coulometric analysis of a given component arises from interference by other oxidizing and reducing components which may be present in the sample. For example, in the analysis of NO_2 a positive interference is given by O_3, peroxides, peroxyacylnitrates, Cl_2, etc. and a negative interference by SO_2, H_2S and other sulphides. The success of coulometric instruments for reliable ambient air measurements relies heavily on the design of suitably selective sample scrubbers to remove interfering substances. Several commercial manufacturers claim reasonably selective measurement of NO, NO_2, SO_2, H_2S in the 0.01 to 1 ppm range. Although once very popular, most coulometric instruments have now been supplanted by analysers based upon other measurement principles.

6.2.2.4   *Miscellaneous chemical methods*

An instrumental technique using specific ion electrodes has been introduced (Bran and Lubbe Ltd) in which gas sample and liquid reagent are pumped into an absorption or reaction chamber and reacted reagent passes at intervals into the measuring chamber where an ion-selective electrode measures the concentration of the selected ion. The main application is in the analysis of hydrogen fluoride (HF) but the system can be used for NH_3, hydrogen cyanide (HCN) and H_2S.

Another new electrochemical technique, which has been applied to the measurement of higher concentrations of pollutants, e.g. in stack gases, is the selective redox cell. The selectivity is based on variation of the electrode potential of the cell. For an oxidation reaction, only gases with oxidation potentials below the electrode potential will be oxidized and, for reduction, only gases with reduction potentials above the electrode potential will be reduced. SO_2 can be detected in the presence of NO_2 by oxidation and NO_2 in the presence of SO_2 by reduction. Analysers based on this system are simple, low in cost and have sealed replaceable and interchangeable cells (e.g. Dynasciences Sensors, Envirometrics Inc., Faristors). Because response times are slow and sensitivity moderate, source measurements are likely to remain their main area of use.

6.2.3   Physical methods

Most of the physical methods currently used for the analysis of gaseous N and S pollutants involve optical measurements of some kind. The optical techniques include chemiluminescence, fluorescence and absorption spectroscopy. Gas chromatographic methods have recently been developed for the S gases but for gaseous N pollutants, application of this powerful technique has not yet been successful.

6.2.3.1   *Chemiluminescence*

The phenomenon of chemiluminescence occurs when part of the energy of an exothermic chemical reaction is released as light. The factors affecting light emission from chemiluminescent gas reactions are exemplified in the reaction between NO and O_3, which is now widely used in the analysis of nitrogen oxides [5]. The fast reaction between NO and O_3 in the gas phase produces excited NO_2 molecules which lose their energy either by light emission or by quenching collisions with other molecules present:

$$NO + O_3 \rightarrow NO_2 + O_2$$

$$NO_2 \underset{M \; \rightarrow NO_2 \; (M = N_2, \; O_2, \; H_2O, \; etc.)}{\overset{\nearrow NO_2 + h\nu \; \text{chemiluminescence}}{}}$$

The light emission (I) is given by an equation of the form:

$$I = \text{constant} \times [NO][O_3]/[M]$$

where $[M]$ is proportional to the total pressure. For a fixed pressure, I is dependent only on the NO and O_3 concentrations. If either of these components is held constant (which is effectively achieved by arranging for one gas to be in large excess), the light emission is proportional to the concentration of the other gas. The emission from a given chemiluminescent reaction has a characteristic spectral composition and the spectral region of interest may be selected using optical filters and/or a suitable choice of photomultiplier response characteristics. Thus high selectivity can be achieved for the analysis of a particular gas.

Figure 6.3 shows the basic layout for a chemiluminescence analyser for NO. The sample gas, containing NO, and a stream of ozonated air, is mixed in a reaction chamber which is positioned adjacent to the end-window of a photomultiplier tube. The pressure and flow of gas through the reaction chamber are maintained constant by a vacuum pump in conjunction with critical orifice capillaries and pressure regulators on the inlet lines. Typical operating pressures

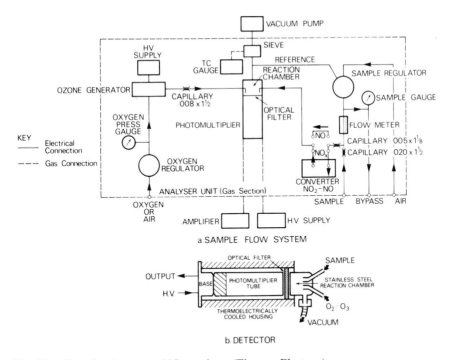

Fig. 6.3 Chemiluminescence NO_x analyser (Thermo-Electron).

are 0.01 to 0.05 atm. In operation, the system gives a continuous signal proportional to the amount of NO entering the reaction cell and thence the NO concentration in the sample gas is measured. The system is calibrated using standard NO gas mixtures. The response is linear over a wide range and a sensitivity down to less than 1 ppb can be obtained. When NO_2 is to be measured the sample gas is passed through a converter consisting of a heated stainless steel tube in which NO_2 is decomposed to NO. The total $NO + NO_2(NO_x)$ is subsequently measured and the NO_2 obtained by difference. In some instruments the NO_2 is determined by sequential measurement of NO_x and NO, while others contain two reaction chambers which analyse NO_x and NO respectively on a continuous basis, thus providing continuous NO_2 readings.

Another type of chemiluminescence detector, which has proved particularly useful in the analysis of gaseous S compounds, is the flame-chemiluminescence or flame photometric detector (FPD) [6]. When S containing compounds are burned in a fuel-rich hydrogen flame an intense chemiluminescence in the 300 to 425 nm spectral region results from the radiative recombination of atomic sulphur:

$$S + S + M \rightarrow S_2 + M$$
$$S_2 \rightarrow S_2 + h\nu$$

If the emission is monitored using a narrow band filter (394 ± 5 nm) a specificity ratio of approximately 20 000:1 for sulphur compounds compared with other components giving chemiluminescence in the flame (e.g. hydrocarbons) is achieved. The response is roughly proportional to the square of the sulphur concentration in the sample gas but a linear output can be obtained by logarithmic amplification of the photocurrent. The minimum detectable concentration on current instruments is approximately 1 ppb. Analysers, based on FPD, function as total sulphur monitors unless chromatographic separation of the components is carried out. In most air pollution work SO_2 is the predominant sulphur component and total sulphur monitoring is adequate. Around refineries and other sources producing H_2S and mercaptans some discrimination must be made either by GC or selective absorption from the sample gas. Some manufacturers provide selective adsorption tubes which remove interferent gases.

6.2.3.2   Fluorescence

Fluorescent emission occurs when molecules absorb radiant energy at one wavelength and re-emit part of that energy at another wavelength. In gases, fluorescence is a low pressure phenomenon, the emission being normally quenched to undetectable levels at pressures near atmospheric. However, given a suitably intense monochromatic light source, the method can in principle provide the basis for selective analysis of trace gases (e.g. SO_2) which have a strong absorption band from which fluorescence emission occurs [7]. Instruments are available for the specific measurement of SO_2 in the 0.5 to 1000 ppm and 0.5–1000 ppb ranges utilizing fluorescence emission. These employ a pulsed u.v. source which irradiates the sample gas flowing continuously through the optical cell. The fluorescence emission is detected by a photomultiplier viewing at 90° to the excitation beam. Optical filters are used to select narrow bandwidths for the exciting and emitted radiation. The disadvantage of the method for ambient air analysis is the rather lengthy response time (*c.* 2 min) in relation to the flame photometric analyser (*c.* 25 s).

6.2.3.3   Absorption spectroscopy

Several instruments and techniques for the measurement of gaseous N and S compounds have been based on the absorption of i.r. and u.v. radiation. The absorption spectra are specific fingerprints for compounds absorbing in those regions and the information contained in the absorption spectrum of a sample gas is adequate to give a specific measurement of each of the absorbing compounds in the sample. The problem arises in extracting the information from the spectrum. Simple methods such as non-dispersive i.r. and u.v. analyses are non-specific and have only moderate sensitivity. Non-dispersive i.r. analysis is, however, widely used for source measurement of carbon containing compounds and to a lesser extent for NO. Dispersive i.r. instruments (i.e. spectrometers) are specific but even when long pathlengths are used the sensitivity is barely adequate for monitoring ambient air. Furthermore, i.r. spectrometers are complicated, delicate and expensive instruments and are not readily utilized for continuous operation. Two new methods for reducing spectral data to a simple quantitative output have been recently developed and incorporated into commercially available instruments. These are correlation spectrometry and derivative spectrometry, both of which are primarily used in the u.v. region for the determination of SO_2 and NO_2.

In correlation spectrometry (Fig. 6.4) the incoming light signal (i.e. the light being sampled) is dispersed by a grating spectrometer. Instead of the normal exit slit there is a correlation mask which is a photographic replica of the spectrum of the compound of interest, with slits corresponding to the main absorption peaks.

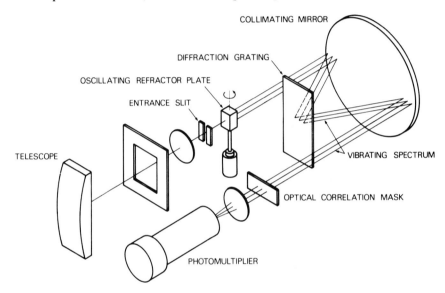

Fig. 6.4 Correlation spectrometry.

When the spectrum is vibrated across this mask, a beat signal is obtained when the incident light bears the absorption pattern of the compound being monitored. Although absorption of other gases may overlap on some peaks, no other gas will correlate over the whole spectrum and so the signal from the desired component is much enhanced. The Barringer Research Correlation spectrometer [8] has mainly been used as a remote sensing instrument for NO_2 and SO_2 using daylight as a source. The total amount of pollutant in the optical path is measured and expressed in units of ppm meters. Using a high intensity xenon arc-lamp as a source, measurements can be made over a fixed pathlength to give average concentrations in a given location.

The second derivative spectrometer (Spectrometrics of Florida Ltd) incorporates an oscillating inlet slit in an u.v. grating spectrometer [9]. The varying angle of incidence on the grating produces a signal with amplitude proportional to the second derivative of the absorption spectrum. A substantial increase in the sensitivity and selectivity over direct absorption spectrometry results, giving minimum detectable levels in the ppb range for SO_2, NO and NO_2. Ambient air is continuously aspirated through the absorption cell and each component measured in turn on a 3 to 7 min cycle.

At the present time there is much research activity in the field of air pollution monitoring by advanced spectroscopic techniques [10]. In particular the advent of lasers has provided stimulus to the remote optical sensing field. In the future these techniques will almost certainly supersede existing methods for many monitoring applications. Current techniques for remote sensing of pollutants are discussed in Chapter 10.

6.2.3.4   *Gas chromatography*

Gas chromatography is one of the most versatile and selective methods for gas analysis. The basic principles of the method as applied to the analysis of atmospheric pollutants are described in Chapter 8. The main difficulty in the application of this method to the analysis of S and N pollutants arises from their high reactivity and consequent absorption loss and irreproducible transfer through the chromatographic system. For the measurement of S compounds, however, the pioneering work of Stevens *et al.* [11] has led to considerable advances in technique. By the use of PTFE components and specially developed column materials, together with selective flame photometric detection, quantitative analysis of gaseous S pollutants in ambient air by gas chromatography is now possible.

There have been numerous attempts to separate and analyse NO and NO_2 at ppm levels by gas chromatography [12]. Most workers have experienced serious loss or modification of these gases on the column packings. The best results have been obtained with porous polymer column packings, e.g. Porapak. Even if these problems could be overcome the problem of a suitable detector remains. Rare gas ionization detectors or electron capture detectors seem to offer the best prospects.

6.2.3.5   *Other physical methods*

Although mass spectrometric techniques have been widely used for gas analysis in the laboratory, application to the routine measurement of low concentrations of S and N gases in ambient air has not proved practical. Without prior separation and sample concentration, sensitivity is too low and the mass spectra are impossibly complicated for quantitative analysis.

A technique which is currently being developed for trace gas analysis is photo-ionization mass spectrometry [13]. This involves ionization using high energy u.v. radiation from a rare gas (krypton or argon) resonance lamp followed by mass spectrometric detection. As only those molecules which have ionization potentials less than the u.v. excitation energy will be ionized, a measure of selectivity is thereby gained. This may prove a useful method for the determination of NO which has a relatively low ionization potential compared with other low molecular weight gases present in the atmosphere.

6.3   Experimental section

In this section the methods of analysis of the individual gaseous S and N pollutants are surveyed and practical details of several recommended wet chemical methods are given. The practical details include procedures for sampling, analysis and calibration. Instrumental methods are also discussed but it is not possible to evaluate critically the performance of individual instruments from each manufacturer. Instead a brief description of the operation, specification and calibration of selected instrument types is given. Finally, at the end of this section practical details of general methods for the preparation of standard gas mixtures are given.

6.3.1   Analysis of SO_2

Since SO_2 is such a widespread pollutant many methods have been devised for the analysis of this gas at levels found in the atmosphere [14]. Among the principles employed are acidimetry, colorimetry, electrochemistry, flame photometry, emission and absorption spectroscopy. The main factors determining the optimum technique are the type of environment to be sampled, the availability of resources, both financial and manpower, and the operational requirements, i.e. continuous, periodic or 'spot' recordings. Many of the above principles have been adapted for automatic operation, but it should be pointed out that the most severe limitation on all commercial instruments and on many manual methods for SO_2 analysis is the lack of sensitivity. Generally the lowest realistic detection limit of instrumental methods is about 0.01 ppm. Since the SO_2 concentration in most non-urban environments is <0.02 ppm an improvement in sensitivity of at least a factor of 10 is at present required to satisfy the needs of environmental researchers.

6.3.1.1   *Chemical methods*

The recommended procedures for manual determination of SO_2 are the West–Gaeke colorimetric technique [15] and the hydrogen peroxide method [16]. There are a number of variations of the latter method inasmuch as the H_2SO_4 which is formed when SO_2 is trapped in aqueous H_2O_2 can be determined either titrimetrically or colorimetrically as SO_4^{2-}. The latter is to be preferred since it is more specific, and using the procedure given below SO_2 concentrations down to 0.001 ppm can be measured. Determination as SO_4^{2-} provides a useful method for the measurement of SO_2 collected on impregnated filter tape [17]. Analysis of SO_2 based on its redox reactions with halogens in solution is widely used in commercial electrochemical instruments. Manual iodometric techniques have also been described [18]. The main problem with the redox system is interference from other oxidizing and reducing substances and, although manufacturers of commercial devices claim to have reduced the interference to acceptable levels, the additional procedures for the removal of interfering substances make this method unattractive for manual analysis.

West–Gaeke (colorimetric) method (ISC method no. 42401-01-69T [19])

Principle

SO_2 is absorbed by aspirating a measured volume of air through a solution of potassium tetrachloromercurate (TCM). The stable non-volatile dichlorosulphitomercurate ion is formed in this procedure. Addition of solutions of purified, acid-bleached pararosaniline and formaldehyde leads to the formation of intensely coloured pararosaniline methyl sulphonic acid. The pH of the final solution is adjusted to 1.6 ± 0.1 by the addition of a prescribed amount of 3 M phosphoric acid to the pararosaniline reagent since the extinction coefficient of the product depends on pH ($\varepsilon = 47.7 \times 10^3$ litres mol^{-1} cm^{-1} at $\lambda_{max} = 548$ nm and pH 1.6).

The method is applicable for SO_2 concentrations of 0.01 to 5 ppm, the lower limit of detection being 0.3 μl SO_2 per 10 ml TCM which corresponds to 0.01 ppm SO_2 in 30 litres air. Absorption efficiency of TCM falls off at concentrations of SO_2 below this and therefore lower detection limits cannot be achieved by sampling larger volumes unless the absorption efficiency is determined separately using, for example, radioactive sulphur dioxide ($^{35}SO_2$) [20] or a standard SO_2 gas mixture.

The principal interfering compounds are oxides of nitrogen, O_3 and heavy metals. The effects of these are minimized in the experimental procedure by:

(a) The addition of a solution of sulphamic acid which destroys any nitrite ion formed by the absorption of oxides of nitrogen in the TCM solution [21]

(b) Allowing any dissolved O_3 to decay by delaying analysis for 20 min after sample collection [22] and

(c) Addition of ethylenediamine-tetra-acetic acid disodium salt (EDTA) to the TCM solution to complex heavy metals that can interfere by oxidation of the SO_2 before it can react with the TCM [22].

Apparatus

The SO_2 may be sampled either in midget or standard fritted bubblers, a midget impinger or standard impinger. The sample probe should be of borosilicate glass, stainless steel or PTFE and, if a prefilter is used, it should be heated when sampling at a relative humidity (r.h.) $> 70\%$ (see Section 6.3.1.1). A pump with a capacity of up to 2.5 litres min^{-1} (for midget samplers) or up to 15 litres min^{-1} for standard samplers is required for aspiration. The gas volumes are measured with a calibrated rotameter or a wet or dry gas meter. Alternatively a high volume pump and a critical orifice flow-meter may be employed.

Reagents

Analytical grade chemicals should be used. The pararosaniline dye should have an assay of greater than 95%.

(a) Absorbing reagent – 0.04 M potassium tetrachloromercurate (TCM) K_2HgCl_4: dissolve 10.86 g mercuric chloride (Poison), 5.96 g of potassium chloride, 0.066 g of EDTA (disodium salt) in water and make up to 1 litre. The pH should not be less than 5.2 or SO_2 absorption efficiency may be impaired. This solution is stable for six months.

(b) Sulphamic acid: dissolve 0.6 g of sulphamic acid in 100 ml of water (stable for a few days if protected from atmospheric oxidation).

(c) Buffer solution (for assay procedure): 100 ml of 0.1 M sodium acetate–acetic acid (pH = 4.69).

(d) Phosphoric acid (H_3PO_4): 3 M H_3PO_4 – dilute 205 ml H_3PO_4 (85%) to 1 litre.

(e) 0.2% Pararosaniline stock solution. The pararosaniline dye needed to prepare this reagent should yield a TCM-reagent blank of not more than 0.17 absorbance units (A.U.) at 22° C, should give a calibration curve with standard sulphite solutions of slope 0.746 ± 0.04 A.U. $(\mu g\ ml^{-1})^{-1}$ for 1 cm cells, and must have an absorbance maximum at 540 nm when assayed in a buffered solution of 0.1 M sodium acetate–acetic acid. To make the stock solution take 0.200 g pararosaniline dye and dissolve in 100 ml of 1 M HCl in 100 ml glass stoppered graduated cylinder. If the pararosaniline does not meet the requirements it may be purified by repeated solvent extraction with 1-butanol. The assay of pararosaniline in the stock solution is carried out as follows: dilute 1 ml of stock solution to 100 ml in a volumetric flask with distilled water. To a 5 ml aliquot in a 50 ml flask, add 5 ml of 1 M sodium acetate–acetic acid buffer and dilute to 50 ml with distilled water. After 1 h determine the absorbance at 540 nm with a spectrophotometer using 1 cm cells. The assay is given by

$$\% \text{ pararosaniline} = \frac{\text{Absorbance} \times 21.3}{\text{grams taken}}$$

(f) Pararosaniline reagent. To 20 ml of stock solution in a 250 ml flask add an additional 0.2 ml of stock solution for each 1% less than 100% assay in the stock, followed by 25 ml 3 M phosphoric acid (H_3PO_4). Dilute to volume with distilled water. The reagent is stable for at least nine months.

(g) Formaldehyde, 0.2%. Dilute 5 ml of 40% formaldehyde to 1 litre with water. Prepare daily.

(h) Standard sulphite solution. Dissolve 0.400 g sodium sulphite (or 0.300 g sodium metabisulphite) in 500 ml boiled and cooled distilled water. Sulphite solutions are unstable and must be freshly standardized before use. This is achieved by adding excess iodine and back titrating with sodium thiosulphate which has been standardized against potassium iodate or dichromate (primary standard).

(i) Dilute sulphite solution. Pipette accurately 2 ml of freshly standardized sulphite solution and make up to 100 ml with 0.04 M TCM. If stored at 5° C this solution is stable for one month.

Procedure

Place a measured quantity of absorbing reagent (10 to 20 ml for a midget impinger; 75 to 100 ml in a standard absorber) and connect up the sampling probe and metering system. Flow rates for midget impingers should be between 0.5 to 2.5 litres min^{-1} or up to 15 litres min^{-1} with large absorbers. Within these ranges the sampling efficiency should be $>98\%$. Sample for sufficient time to give between 0.5 and 3.0 μg SO_2 per ml absorbing solution. Shield the solution from direct sunlight during sampling and storage. Keep cool during storage, and if a precipitate forms, remove it by centrifugation. For analysis of a 10 ml sample transfer it quantitatively to a 25 ml volumetric flask with approximately 5 ml of distilled water for rinsing. For high concentrations or larger volumes, aliquots may be taken at this point. Leave the sample for 20 min to allow any O_3 present to decay. Meanwhile prepare a reagent blank using 10 ml of exposed reagent in a 25 ml flask. To each flask add 1 ml of 0.6% sulphamic acid and allow 10 min for the destruction of any nitrite ion from oxides of nitrogen. Accurately pipette 2 ml of formaldehyde (0.2%), then 5 ml of the pararosaniline reagent. Make up to 25 ml and determine the absorbance of both solutions against distilled water after 30 min.

Calibration

Accurately pipette graduated amounts of the dilute sulphite solution (e.g. 0, 1, 2, 3, 4, 5 ml) into 25 ml flasks and make up to approximately 10 ml with 0.04 M TCM. Add the remaining reagents as described in the procedure and measure the absorbances. The total absorbances, plotted as a function of μg SO_2 (total), should give a linear plot which intercepts to within 0.02 absorbance units of the blank. The calibration factor B is the reciprocal of the slope of the line. The concentration of SO_2 in the air sample is then given by:

$$SO_2 \text{ (ppm/v)} = \frac{(A - A_0)\, 0.382B}{V}$$

where A, A_0 are the sample and reagent blank absorbances respectively, 0.382 is the volume in μl of 1 μg SO_2 at 760 Torr and 25° C and V is the sample volume in litres (corrected to 760 Torr and 25° C).

The calibration may be alternatively carried out by sampling from a standard source of SO_2 in air obtained, for example, using a permeation tube [23, 24] (see Section 6.3.8.2). This procedure has the added advantage that any losses resulting from adsorption in the sampling probe or inefficiency of the absorber can be taken into account.

Hydrogen peroxide (H_2O_2)/sulphate method (using automatic colorimetric analysis)

Principle

The air sample containing SO_2 is sucked through a heater to raise its temperature by 10° C before passing through an absolute prefilter to remove particulate material and into a simple bubbler containing 1 volume H_2O_2 where the SO_2 is rapidly oxidized to involatile H_2SO_4. By heating the air losses of SO_2 by adsorption are reduced to a negligible level even when sampling air of up to 96% r.h. The sulphate is determined by the method of Persson [25] which was developed for use with a Technicon Autoanalyser. The decrease in the light absorbance of a barium–thoranol complex, when barium is removed from it by sulphate ions, is measured at 520 nm. The reactions which are carried out in a weakly acid solution (perchloric acid) of aqueous iso-propanol may be represented as:

$$Ba^{2+} + thoranol \rightleftharpoons Ba\text{–}thoranol\ complex$$

$$Ba\text{–}thoranol\ complex + SO_4^{2-} \rightarrow BaSO_4 + thoranol$$

where thoranol is 1-(orthoarsenophenylazo)-2-naphthol-3,6-disulphonic acid sodium salt (known also as Thoron, Thorin, naphtharson, APANS). The method is very susceptible to foreign ions, either anions which complex barium, or cations, particularly polyvalent cations. Ammonium ions do not interfere, however. Dilute H_2O_2 solutions do not interfere with the method and it is possible to analyse the bubbler samples without pretreatment. The sulphate content of 100 vol. H_2O_2 used to prepare absorbent solutions may be significant. Therefore, standards and autoanalyser wash solutions are prepared from the same batch as used in making up absorbent solutions.

A limit of detection for SO_4^{2-} in solution of 0.1 μg ml^{-1}, which in 40 ml of absorbent corresponds to 2.7 μg SO_2, is attainable. Analytical precision at this level is of the order of 10% increasing to 0.4% at 3 μg ml^{-1} sulphate ion. Flow rates of up to 30 litres min^{-1} through the absorber can be used and give a limit of detection of the order of 0.001 ppm SO_2 from a 30 min sample.

Apparatus

The sampling unit consists of a heated Pyrex inlet tube, a filter holder and a standard impinger assembled as shown in Fig. 6.5. The prefilter can be either Whatman 41 or a Microsorban polystyrene absolute filter (50 mm diameter). The heater consists of a standard 100 Ω resistor, wire, wound on a hollow ceramic former which fits closely round the sample tube, and is supplied with 30 to 40 V a.c. In order to prevent condensation the filter holder and bubbler entry tube are enclosed in a box of expanded polystyrene during sampling. The system is aspirated with a pump capable of delivering > 30 litres min^{-1}, through a conventional metering system.

A Technicon autoanalyser is used for the analysis of the sulphate ions. The analyser should preferably be equipped with silicone rubber pump tubes since the more usual 'solvaflex' tubes are attacked by the isopropanol solutions and have a normal operation time of only two days before replacement is necessary. A diagram of the analyser flow system is shown in Fig. 6.6.

Reagents

Analytical grade chemicals and water should be used.

(a) Absorbing solution, 1 vol. H_2O_2. Dilute 10 ml of 100 vol. H_2O_2 (30%) to 1 litre with distilled water. Store in a polythene container in the dark. (Prepare weekly.)

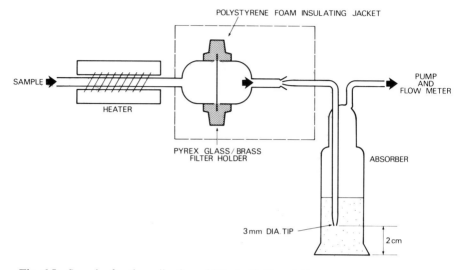

Fig. 6.5 Sampler for the collection of SO_2 in H_2O_2 solution.

(b) Stock barium perchlorate $[Ba(ClO_4)_2]$ 0.1 M perchloric acid. Weigh 0.90 g $Ba(ClO_4)_2$ and dissolve in 1 litre of 0.1 M perchloric acid $HClO_4$, prepared by dilution of 8.6 ml of $HClO_4$ (70% sp. gr. 1.66) with distilled water.

(c) Working barium perchlorate/$HClO_4$. Dilute 10 ml of stock solution to 1 litre with iso-propanol.

(d) Thoranol. Weigh out 0.20 g of Thoranol and dissolve in 1 litre distilled water.

(e) Standard potassium sulphate (K_2SO_4). Weigh out accurately 0.183 g of K_2SO_4 and make up to 1 litre in a volumetric flask using 1 vol. H_2O_2 solution prepared with the same 100 vol. H_2O_2 as above. This solution contains 100 μg ml^{-1} SO_4^{2-}.

Procedure

Place a 40 ml aliquot of absorbent solution in the impinger and assemble the sampling train with a fresh filter. Switch on the heater and allow it to warm up for a few minutes before sampling. When sampling dry atmospheres ($<60\%$ r.h.) heating is not necessary. Sample at the chosen flow rate (1 to 30 litres min^{-1}) for sufficient time to obtain between 1 and 10 μg ml^{-1} SO_4^{2-}. If large volumes are sampled, particularly at low r.h. evaporative losses must be determined by weighing the collection vessel at the beginning and end of sampling. The SO_4^{2-} samples are quite stable and may be stored for several weeks if well stoppered.

For analysis of a series of samples a small amount (~ 5 ml) of each sample is placed in sample cups spaced in alternate positions on the analyser carousel. The

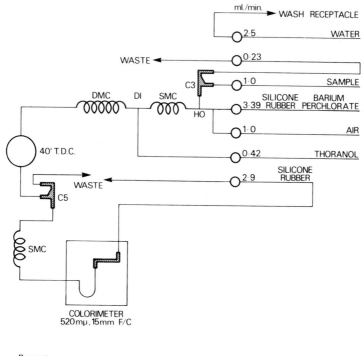

Fig. 6.6 Technicon autoanalyser manifold for the determination of SO_4^{2-} in the range 0 to 10 ppm.

cups in the intermediate positions are filled with unexposed reagent. Standards containing known amounts of SO_4^{2-} in the range 0 to 12 $\mu g\ ml^{-1}$ are prepared by the appropriate volumetric dilution of the standard K_2SO_4 solution. These are also placed in alternate positions on the carousel with unexposed absorbing reagent in between. The samples are then run on the analyser and the peak height corresponding to the depletion in absorbance of each solution at 520 nm measured from the recorder trace. A calibration plot is prepared from the standards and the sulphate content of the samples (in $\mu g\ ml^{-1}$) read off from the graph. The plot should be linear up to approximately 12 $\mu g\ ml^{-1}$ SO_4^{2-} but is usually curved at higher concentrations. Separate standards should be run with each batch of samples. The SO_2 concentration in the air sample is given by

$$SO_2\ (ppm/v) = \frac{\mu g\ ml^{-1}\ SO_4^{2-} \times 0.255 \times V_a}{V}$$

where V_a is the volume of absorbing solution corrected for evaporation loss, V is the volume of air (litres at 760 Torr and 25° C) and 0.255 is the volume of SO_2 (in μl at 760 Torr and 25° C) corresponding to 1 μg SO_4^{2-}.

Variations

The above analytical procedure using an autoanalyser is to be recommended when high sensitivity and analysis of a large number of samples is required. However, the H_2SO_4 formed when SO_2 is collected in dilute H_2O_2 may be determined by a number of different manual methods, e.g. the barium sulphate turbidimetric method, [26] or the barium chloranilate colorimetric method [27].

The Thoron method using the autoanalyser may also be used for measurement of SO_2 collected on filter paper tapes. The tapes (Whatman 41) are impregnated with 25% potassium carbonate in a 10% glycerol/water mixture and dried under an i.r. lamp. The SO_2 is collected as sulphite which is oxidized to SO_4. A collection efficiency of greater than 95% can be achieved at face velocities of up to 70 cm s^{-1}. The filters are extracted at a 70° C in a 1 vol. H_2O_2 solution which is then cooled and passed through a cation exchange resin (Zeocarb 225) to remove potassium ions which interfere. The sample solutions are then analysed as above.

Impregnated filter methods based upon the use of KOH [28] and tetrachloromercurate [29] impregnants are also available. The former reagent gives very high collection efficiencies at face velocities up to 90 cm s^{-1} and relative humidities in the range 30–100% [29].

Instrumental chemical methods for analysis of SO_2

Electrical conductivity analysers

A variety of continuous automatic conductivity analysers for SO_2 are marketed. The absorbing liquid is usually aqueous H_2O_2 which flows through the absorption cell and the change in conductivity resulting from H_2SO_4 formation is measured. The method is non-selective and is therefore not recommended for precise measurements of SO_2 in the atmosphere, but can be useful for measurement of SO_2 in industrial environments. Sensitivity ranges of between 0 to 0.2 ppm and between 0 to 20 ppm SO_2 are available.

Coulometric analysers

The reducing action of SO_2 on free halogen/potassium halide solutions (I or Br) is utilized for coulometric SO_2 analysers. The available commercial instruments differ slightly in mode of operation and sensitivity. The Philips PW 9700 (Fig. 6.7) claims the lowest detection limit of 4 ppb SO_2 and is designed for three months unattended operation. Interference from H_2S, O_3 and Cl_2 is reduced to <1% on a mol for mol basis by a silver wire scrubber. The main interferents are NO_2 (<5%) and mercaptans (100%). The instrument incorporates an automatic valve for sample, calibration and zero-check selection. In the zero position the sample air flows through activated charcoal to remove SO_2, giving a 'zero' signal from the coulometric cell. In the 'calibrate' position a known amount of SO_2 is introduced to the air stream from a built-in permeation tube. The response time

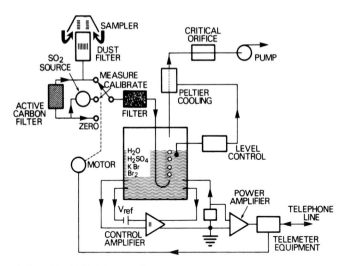

Fig. 6.7   Philips PW 9700 SO$_2$ monitor.

(95% of final value) of coulometric analysers for SO$_2$ is of the order of 3 to 5 min. Sensitivity ranges of between 0 to 0.2 and between 0 to 200 ppm are available.

6.3.1.2   *Physical analysis of SO$_2$*

The recommended procedure for physical analysis of SO$_2$ in the atmosphere is gas chromatographic separation followed by flame photometric detection (FPD). A 'plug' sample is analysed, but with an automatic gas sampling valve sequential samples may be taken at intervals down to 3 min depending on analysis time. An alternative arrangement, which is used in most commercial FPD analysers, is to dispense with chromatographic separation and continuously pass sample air into the FPD detector, thereby obtaining measurement of 'total S'. A line filter can be incorporated to remove particulate S.

Gas chromatographic method for SO$_2$ *analysis*

   Principle

An air sample is aspirated into the sample loop of a gas sampling valve. The valve is then switched to inject the sample on to a chromatographic column where SO$_2$ is separated from other gaseous S compounds. The eluent passes into a FPD detector and the detector signal amplified and recorded.

   Detector response is roughly proportional to the square of the gaseous S$_2$ concentration for concentrations up to approximately 1 ppm [30, 31]. The range of electrical response used is approximately four orders of magnitude giving a minimum detectable concentration of 5 to 10 ppb. The actual detection limit will also depend on the retention time of SO_2 on the column and baseline stability after injection.

The major factor affecting reproducibility is the potential adsorption problem. The use of PTFE flow components is imperative and contact of the sample gas with metal should be avoided in all instances. With a suitably designed system, a reproducibility as high as $\pm 1.5\%$ can be achieved at concentrations of the order of 0.05 ppm. There are no known interferences with this method for the analysis of SO_2 in air. The response may be affected by high concentrations of hydrocarbons in special environments.

Apparatus

Three major components are required

  (i) A sample injection system operated manually or automatically activated with a timer
 (ii) A chromatographic column for the separation of S compounds
(iii) A flame photometric detector with associated amplifier and recorder.

In addition 1/8 in PTFE tubing and couplings for the sample gas lines and columns and a sample pump are required. Figure 6.8 shows a flow diagram of the system.

Suitable six-port gas sampling valves (GSV) are the Chromatromix models R6031SV (manual) or R6031SVA (PTFE-bodied, automatic). In addition a

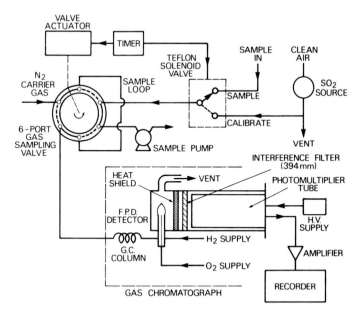

Fig. 6.8   Automatic gas chromatograph–flame photometric detector for SO_2 and other compounds.

PTFE three-way solenoid valve is required if calibration samples are to be taken automatically. An industrial cam timer is programmed to activate the automatic GSV. Automatic GSVs are normally pneumatically operated at 40 to 60 psig. The sample loop should be of PTFE and 10 ml in volume. The sample pump should be capable of drawing 1 litre min^{-1} through the sample loop.

Three different types of column have been successfully employed for the separation and analysis of ppb concentrations of SO_2 in the atmosphere. All columns are packed in 1/8 in PTFE tubing and the following stationary phases can be used:

(a) 36 ft length packed with 40 to 60 mesh PTFE powder flow coated with approximately 10% polyphenyl ether and 0.5% H_3PO_4. Practical details are given by Stevens *et al.* [11]. After conditioning at 140° C for 6 h this column will separate H_2S, SO_2, methyl mercaptan (CH_3SH), ethyl mercaptan (C_2H_5SH) and dimethyl sulphide (CH_3SCH_3). At a flow rate of 100 ml min^{-1} and at a temperature of 50° C, the retention time of SO_2 is approximately 2 min.

(b) 1.4 m length packed with graphitized carbon–black modified by treatment with 0.5% H_3PO_4 and 0.3% Dexsil (Bruner *et al.* [32]). At a flow of 125 ml min^{-1} and 60° C a separation and analysis of H_2S, SO_2 and CH_3SH can be obtained in 1 min. The column should be conditioned at 100° C for 24 h before use.

(c) Deactivated silica gel columns: Hartmann [31] has described a column for the analysis of ambient SO_2 (and H_2S) consisting of a 3 inch length packed with 100/120 mesh commercial deactivated silica gel (Deactigel) which was further deactivated by washing in turn with concentrated HCl, H_2O and acetone (procedure of Thornsberry [33]). At a flow of 80 ml min^{-1} and at 50° C the retention time of SO_2 was approximately 2 min. Deactivated silica gel columns for the analysis of ppb concentrations of sulphur compounds are commercially available, e.g. Supelco 'Chromasil 310'. Similarly columns (a) and (b) are available from some chromatographic suppliers.

Any commercial chromatograph equipped with a flame photometric detector can be used to house the column. The column should be connected directly into the FPD burner base where the fuel gas (H_2) and make-up O_2 is added. (N_2 may be used as make-up gas if air is used as carrier gas.) The other end of the column is connected to the GSV by as short a length of PTFE tubing as possible.

Reagents

The only materials required are the column packings, carrier gas (N_2 or air) H_2 and O_2 in cylinders fitted with good quality pressure regulators. For calibration, a gas containing an accurately known concentration of SO_2 in the range 0.01 to 1 ppm is required; this is most conveniently obtained from permeation tubes (see Section 6.3.8.2).

Procedure

After setting up the sampling and chromatographic equipment, set the gas flows to the required rates. The optimum gas flows depend somewhat on burner design; normally about 100 ml min^{-1}. H$_2$ is used with a total carrier plus make-up of the same order (100 ml min^{-1}). Ignite the FPD burner and establish a satisfactory baseline on a high sensitivity. Before carrying out air sampling, some time should be spent establishing optimum operating conditions and response characteristics using the SO$_2$ calibration source. In particular, since the response relationship of the FPD is of the form

$$R = [S]^n$$

(R = response, e.g. peak area; $[S]$ = S gas concentration), the value of n, which depends on the type of S gas and the operating conditions, must be determined.

Using a calibration gas of constant SO$_2$ concentration and repeated injection of samples, optimize conditions of gas flow and temperature to give maximum response consistent with resolution of SO$_2$ and minimum baseline noise. Inject several samples for each test, as 3 to 4 injections may be required to give a constant signal. The response is calibrated by injecting SO$_2$ at different concentrations within the operating range and recording the peak area. Plot the SO$_2$ concentration versus response (in A) on log–log paper and determine n from the slope. The plot should be approximately linear in the range 0.01 to 1.0 ppm and n should lie in the range 1.7 to 2.1. Some FPD detectors are equipped with linearizer amplifiers in which case the response is linearly related to the SO$_2$ concentration.

The equipment can now be used for ambient air sampling. For automatic operation the timing sequence will depend on the analysis time and the frequency of sampling required. At least 60 s should be allowed to purge the sample loop with the air sample, at a flow rate of a few hundred ml min^{-1}. After injection the loop is purged with carrier gas until the next sample is required, and during this period the automatic GSV control system should be de-energized.

Calibration samples may be run at any desired interval if a continuous SO$_2$ source is incorporated into the system. Timing of the three-way PTFE solenoid valve should be appropriately synchronized with the GSV. For accurate continuous monitoring it may be desirable to run alternate air-calibration samples so that the column is regularly conditioned with SO$_2$ even if the ambient level is very low. In this way, possible absorption losses are minimized.

Commercial instruments for SO$_2$ monitoring

Ambient air – total S

Instruments supplied by Bendix (Model 8300), Meloy Laboratories (Model SA 285, etc.) and other manufacturers monitor total S by continuous injection of an air sample into an FPD. These systems have similar specifications, i.e. log/linear amplification with six linear switched ranges 10^{-9} to 10^{-4} A FSD corresponding to a working concentration range of 0.01 to 1.0 ppm SO$_2$. Automatic zero and calibration modes are provided for continuous unattended operation, but a source of SO$_2$ must be supplied separately. Pulsed gas phase fluorescence analysers for SO_2 are also available (e.g. Thermo Electron) with a detection limit of 0.5 ppb and a response time of 2 min.

Commercial chromatographic sulphur gas analysers, operating basically on the procedures discussed above, are also available (e.g. Tracor Model 270 HA). These instruments can be used for the specific measurement of SO_2 and also H_2S, CH_3SH, carbon oxysulphide (COS), carbon disulphide (CS_2) with a minimum detection limit of 5 to 10 ppb.

Source monitoring for SO_2

Flame photometric detectors can be used for source monitoring (i.e. SO_2 concentrations in the range 10 to 10 000 ppm) if an air dilution system is employed (e.g. Meloy Laboratories Model FSA190 flue gas analyser). However, a more promising technique for this application is u.v. fluorescence analysis of SO_2. Two commercially available instruments utilizing this technique are available (Thermo Electron Model 40, Celesco Model 5000). Complete selectivity for SO_2 is claimed.

6.3.2 Analysis of SO_3

SO_3 in stack gases can be measured by conversion in solution to SO_4^{2-}. Two major problems arise; firstly, the difficulty of quantitative sampling from the hot, moist environment of the stack and, secondly, SO_2 is usually present in the flue gases at higher concentrations than SO_3, and can interfere.

An instrumental method has been developed [34] (EEL 147 SO_3 monitor) in which the gas is sampled continuously through an air-heated probe kept between 200 to 500° C so that the SO_3 does not condense nor the SO_2 oxidize. After filtration the gas is extracted with 4:1 isopropanol/H_2O solution which converts the SO_3 to H_2SO_4. Some SO_2 dissolves in solution, the remainder being swept away in a gas/liquid separator. The liquid is transferred by air injection into the reaction bed and dissolved SO_2 is stripped by the air. The solution passes through a bed of barium chloranilate crystals where the H_2SO_4 reacts to form barium sulphate and the soluble, highly coloured acid chloranilate ion. The colour which is proportional to SO_3 concentration is measured spectrophotometrically. The range of detection is 0.1 to 250 ppm SO_3. The main interference comes from SO_2, a small amount of which may be oxidized in solution to give H_2SO_4. Manual methods based on collection in aqueous isopropanol have also been reported [35].

An interesting gas chromatographic method for the determination of SO_3 in the presence of SO_2 has been reported [36]. The gas sample containing SO_3 is passed continuously through a bed of oxalic acid crystals where SO_3 reacts to form CO and CO_2, whilst SO_2 does not react. The CO in the effluent is then analysed by conventional gas chromatography using a katharometer detector. Any CO in the original sample will, however, interfere.

In the ambient atmosphere gaseous SO_3 does not exist in the free state but forms H_2SO_4 aerosol.

6.3.3 Analysis of H₂S

6.3.3.1 *Chemical methods*

A widely used colorimetric method for the analysis of H_2S in ambient air is based upon absorption in an alkaline cadmium hydroxide suspension, followed by conversion of the precipitated sulphide to methylene blue with *N, N*-dimethyl-*p*-phenylenediamine and ferric chloride with spectrophotometric determination [37]. Low recoveries have been encountered when applied to low concentrations, the major cause being photodecomposition of the sulphide during sampling and storage [38]. Significant improvement was obtained by adding 1% STRactan 10 (arabinogalactan) to the absorbent. The sensitivity is sufficient for determination of H_2S down to 1 ppb.

A variety of methods for analysis of H_2S based on collection with filters impregnated with heavy metal salts ($Pb(^{2+})$, $Hg(^{2+})$ and $Ag(^{+})$) have been reported [39]. The formation of sulphides results in a stain which is measured either by its optical density (compared to a blank section of filter) or by reflectance. Lead acetate has been widely used as the substrate in semi-continuous monitors utilizing filter paper tape but serious interference results from bleaching of the stain by O_3, SO_2, NO_2 and light, and the sensitivity is humidity dependent. These shortcomings, which are especially important at the concentrations normally encountered in air pollution work (0 to 40 ppb), led Pare [40] and later Hochheiser and Elfers [41] to develop the mercuric chloride ($HgCl_2$) paper tape method. The tapes (Whatman No. 4) are impregnated with a mixture of $HgCl_2$, urea and glycerol. After sample collection the tapes are developed by exposure to NH_3 vapour for up to 12 h and the optical density of the reaction spots measured. The effective concentration range for 2 h samples collected at 5 litres min^{-1} is 0.5 to 15 ppb H_2S with minimal interference from O_3, NO_2 and SO_2.

A manual method based on lead acetate filters has been developed by Okita *et al.* [42]. Impregnated cellulose membrane filters are dissolved after sampling in a mixed organic solvent and the resultant brown suspension is measured with a spectrophotometer. SO_2 and O_3 interference can be overcome by prefilters and NO_2 does not interfere below 0.2 ppm. The minimum detectable H_2S concentration is 2 ppb; no response was obtained by Okita *et al.* for lower concentrations, regardless of sampling time.

Natusch *et al.* [43] have described a method for measuring trace levels of atmospheric H_2S with a lower detection limit of 5×10^{-6} ppm. H_2S is collected on a silver nitrate ($AgNO_3$) impregnated filter paper. The silver sulphide (Ag_2S) is then dissolved in sodium cyanide ($NaCN$) solution and analysed fluorimetrically using very dilute fluorescein mercuric acetate. The reduction of fluorescence intensity, which is proportional to the sulphide concentration, was measured on a Perkin-Elmer Model 203 spectrofluorimeter. Fluorimetric analysis has also been applied to H_2S trapped in aqueous alkaline solution [44]. While the sensitivity of this latter method is adequate for background determinations, the collected sulphide ion is unstable and this necessitates that analysis follow soon after sampling. The main drawback of fluorimetric analysis is the high cost of fluorimeters.

Microcoulometric titration by bromine in an electrochemical cell is the basis of a method for continuously monitoring sulphur compounds including H_2S, primarily in the process industries [45]. Sensitivity for H_2S is in the 5 to 30 ppb range. Selective filters may be used to monitor ambient air for SO_2, H_2S, mercaptans, alkyl sulphides and disulphides.

Manual method for H_2S *analysis* (methylene blue method)
(ISC method 42402-01-70T) [19]

Principle

H_2S is collected by aspirating a measured volume of air through an alkaline suspension of cadmium hydroxide $Cd(OH)_2$. The sulphide is precipitated as cadmium sulphide (CdS) to prevent air oxidation of the sulphide which occurs rapidly in aqueous alkaline solution. STRactan 10 is added to minimize photodecomposition of the precipitated CdS. The collected sulphide is determined by spectrophotometric measurement of the methylene blue produced by reaction with a strongly acid solution of N, N-dimethyl-p-phenylenediamine and ferric chloride.

The methylene blue reaction is highly specific for low concentrations of sulphide. Strong reducing agents (e.g. SO_2) inhibit colour development and NO_2 and O_3 can give a slight negative interference. The minimum detectable amount of sulphide is 0.008 μg ml^{-1} corresponding to an H_2S concentration of about 1 ppb in a 2 h sample collected at the maximum recommended sampling rate of 1.5 litre min^{-1}. The method is applicable to concentrations of H_2S up to 100 ppb using the procedure below and higher concentrations can be measured with larger collection volumes and shorter sampling times. At low concentrations (<10 ppb) collection efficiency is variable and is affected by the type of scrubber, the bubble pattern and H_2S concentration.

Apparatus

A midget impinger is used to contain the absorbent and is aspirated with a pump having a minimum capacity of 2 litres min^{-1}. A rotameter or gas meter can be used to measure sample volume.

Reagents

(Use analytical grade chemicals and keep solutions under refrigeration.)

(a) Amine-H_2SO_4 (stock): To 30 ml of distilled water add 50 ml of concentrated H_2SO_4 (r.d. 1.84). After cooling add 12 g of N, N-dimethyl-p-phenylenediamine dihydrochloride and mix until complete solution.

(b) Amine-H_2SO_4 (working): Dilute 25 ml of the above stock to 1 litre with 1:1 H_2SO_4.

(c) Ferric chloride solution: Dissolve 100 g ferric chloride hexahydrate $(FeCl_3 . 6H_2O)$ in H_2O and make up to 100 ml.

(d) Ammonium phosphate solution: Dissolve 400 g of diammonium phosphate $(NH_4)_2HPO_4$ in H_2O and dilute to 1 litre.

(e) Absorbing solution: Dissolve 4.3 g of cadmium sulphate octahydrate $(3CdSO_4 \cdot 8H_2O)$ and 0.3 g of NaOH in separate portions of water, mix, add 10 g of STRactan 10 (arabinogalactan) and dilute to 1 litre. The solution should be freshly prepared (it is only stable for 3 to 5 days) and shaken vigorously before each aliquot of absorbing reagent is taken.

(f) Standard sulphide solutions: Aqueous sulphide solutions are unstable, being subject to air oxidation. Stock solutions must be made up with freshly boiled and cooled distilled water and, for accurate work, should be standardized for each calibration routine. During solution preparation and handling, oxidation can be minimized by flushing receptacles with O_2-free N_2. A stock solution containing approximately 400 μg sulphide ion ml^{-1} is made by dissolving in 1 litre 0.1 M NaOH, either gaseous H_2S (300 ml from a gas syringe through a septum) or sodium sulphide monohydrate crystals (approximately 3 g, weighed after washing with distilled water and drying quickly on filter paper). Standardize the solution with standard iodine and thiosulphate solutions. Dilute standard sulphide solutions (4 μg sulphide ion ml^{-1}) are prepared by diluting 10 ml of the freshly standardized stock to 1 litre with boiled distilled water.

Procedure

Aspirate the air sample through 10 ml of the absorbing solution in a midget impinger at 1.5 litre min^{-1} for a selected period of up to 2 h. Excessive foaming may be controlled by the addition of 5 ml ethanol just prior to sampling.

For analysis add 1.5 ml of the amine working solution to the absorbing solution in the impinger. Add 1 drop of $FeCl_3$ solution (if SO_2 is likely to exceed 10 μg ml^{-1} in the sample solution add 2 to 6 drops) and transfer the solution to a 25 ml volumetric flask. Add 1 drop of ammonium phosphate solution (or more if necessary) to discharge the yellow colour of the ferric ion and make up to volume with distilled water. Allow to stand for 30 min (50 min if extra $FeCl_3$ was added) for the colour to develop. The colour is measured at 670 nm against a reagent blank prepared with unexposed absorbing solution.

Calibration

Calibration may be carried out using standard sulphide solutions or by sampling gas mixtures containing known concentrations of H_2S from permeation tubes. The latter method is to be preferred for concentrations < 10 ppb since the variable collection efficiency may then be taken into account in the standardization.

(a) Aqueous sulphide method: Place 10 ml of the absorbing solution in each of a series of 25 ml flasks and add the diluted standard sulphide solution, equivalent to 1, 2, 3, 4 and 5 μg H_2S to the flasks. Add 1.5 ml amine working solution, mix and add 1 drop of $FeCl_3$ solution to each flask. Make up to volume and allow 30 min before determining the absorbance at 670 nm against a sulphide free blank. Prepare a standard plot of absorbance versus μg H_2S ml^{-1}.

(b) Permeation tube method: Use a permeation tube which emits approximately $0.1~\mu l~H_2S~min^{-1}$ at 25° C and 1 atm. Permeation tubes containing H_2S are calibrated under a stream of dry N_2 to prevent deposition of sulphur on the tube walls. A total dilution gas flow of at least 20 litres min^{-1} is required to provide concentrations at the lower end of the working range. Samples are taken from the standard gas mixture using a fixed sample volume and a plot of H_2S concentration against absorbance in the final solution is prepared.

Calculation

H_2S concentrations are calculated from the standardization plots after allowance for volumetric factors. Gas volume should be corrected to 25° C and 760 Torr pressure.

Stability of samples

Although CdS is reasonably stable towards oxidation, the analysis should be completed within 24 h of sampling. The absorbing solution should be shielded from light during both sampling and storage. Black paint or aluminium foil wrapping on the impinger will help prevent photodecomposition of the sulphide.

Instrumental chemical methods

An automated H_2S analyser based on the methylene blue colorimetric method is marketed by Technicon Corporation. The detection limit on a 0 to 100 ppb scale is claimed to be 2 ppb. Instruments for semicontinuous measurements of H_2S in industrial atmospheres based on the lead acetate paper tape method are also available commercially (e.g. Fleming Instruments Type 523, Maihack Mono-colour H_2S analyser). The minimum detectable concentration on these instruments is approximately 50 ppb H_2S.

6.3.3.2 *Physical methods*

Stevens et al. [11] and Bruner et al. [31] have reported the application of a gas chromatographic, flame-photometric detector (GC–FPD) system for the auto-mated GC measurement of H_2S (together with SO_2, methyl mercaptan and dimethyl sulphide – see Section 6.3.1.2). The limit of detection by this method is approximately 2 ppb for H_2S. If greater sensitivity is required, pre-concentration must be carried out prior to gas chromatography. Braman et al. [46] describe the use of gold-coated glass beads to pre-concentrate H_2S, giving an ultimate detection limit of 0.1 ppt for a 100 litre sample, although some interferences were found. Using cryogenic trapping with liquid nitrogen, Sandalls and Penkett [47] were able to analyse H_2S and other gaseous sulphur compounds at concentrations down to 0.02 ppb.

Infrared analysis of sulphides is non-discriminating, all sulphides absorbing at similar wavelengths. At present sensitivity is inadequate for ambient monitoring using i.r. Remote sensing of ambient H_2S using laser techniques is likely to be developed in the future [10].

Gas chromatographic methods for ambient H_2S *analysis*

Practical details of a gas chromatographic sulphur analyser, which is suitable for measurements of H_2S are given in Section 6.3.1. Pecsar and Hartmann [48] obtained better resolution of H_2S on a PTFE column (prepared by the method of Stevens *et al.* [11]) by using air as carrier gas. This enabled a shorter (12 ft) column to be used. Calibration of the GC–FPD system is carried out using standard mixtures of H_2S obtained from either permeation tubes [30] or an exponential dilution flask [31].

6.3.4   Analysis of organic S compounds

Interest in the organo-sulphur compounds is largely concerned with the measurement and control of effluents from the pulp and paper industries, and also with the gas industry where they are used as artificial odorants for natural gas. The compounds of interest are primarily the lower molecular weight mercaptans, dimethyl sulphide and dimethyl disulphide.

6.3.4.1   *Chemical methods*

As the organic sulphides are strong reducing agents, coulometry offers, in principle, a sensitive method of detection. The main problem is to distinguish between the various organic S compounds and also SO_2 and H_2S which in a coulometric detector behave similarly. Adams *et al.* [45] studied the retention efficiencies of a large number of impregnated membrane filters for five different S gases, H_2S, SO_2, methyl mercaptan (CH_3SH), dimethyl sulphide (DMS) and dimethyl disulphide (DMDS). A bromine coulometric microtitration cell (i.e. galvanic cell with electrogenerated bromine) was used to measure the sulphur gases not trapped by the experimental filters. A series of filters was developed which enabled the above five gases to be determined from measurements on sequential samples. Air was sampled through appropriate filter combinations to provide a continuous, stepwise analysis. Sodium bicarbonate (5%) removed SO_2 but left over 90% of the other compounds unaffected. H_2S was separated from the other compounds by using a zinc chloride–boric acid membrane. Silver membrane filters retained H_2S and CH_3SH but left 95% of the other three unchanged. Mercuric nitrate–tartaric acid retained DMS and CH_3SH. Filters impregnated with $AgNO_3$ retained all compounds except SO_2. Minimum detectability was SO_2, 25 ppb; H_2S, 10 ppb; CH_3SH, 15 ppb; DMS, 25 ppb; and DMDS, 5 ppb. An instrument using preselective filtration was used in a field study of sulphur gas concentrations in the vicinity of a Kraft paper mill [38].

For the selective determination of low molecular weight mercaptans in air the colorimetric method using *N, N*-dimethyl-*p*-phenylenediamine reagent is recommended [49]. This provides a sensitive and reproducible manual method for concentrations down to 2 ppb. A colorimetric method has also been developed for the measurement of carbon disulphide in industrial atmospheres [50]. This involves collection in an ethanolic solution of diethylamine-copper acetate contained in a fritted bubbler, the colour developing in the absorbing solution. Maximum sensitivity is in the ppm range.

Colorimetric determination of mercaptans in air
(ISC method 43901–01–70T) [19]

Principle

Mercaptans are collected by aspirating a measured volume of air through aqueous mercuric acetate–acetic acid solution. A red complex is produced when mercaptans react with N, N-dimethyl-p-phenylenediamine (DMPDA) and $FeCl_3$ in strongly acid solution and this is measured spectrophotometrically.

In addition to the low molecular weight mercaptans, H_2S and dimethyl sulphide can also be determined using DMPDA. These compounds commonly co-exist with mercaptans in industrial emissions and are potential interfering substances. Appropriate selection of sampling and analytical procedures minimizes this interference. Thus H_2S may cause a turbidity in the absorbing solution which must be filtered before continuing the analysis. In the analytical procedure 100 μg H_2S may give an absorption at 500 nm equivalent to up to 2 μg CH_3SH. Interference from dimethyl sulphide is negligible since this compound is not trapped in aqueous mercuric acetate. Tests have shown that SO_2 (300 ppm) and NO_2 (6 ppm) do not interfere.

The minimum detectable amount of CH_3SH is 0.04 μg ml^{-1} in the final liquid volume of 25 ml. In a 200 litre air sample this corresponds to 5 μg m^{-3} or 2.5 ppb CH_3SH. Precision is within 2.6% for the C_1 to C_6 mercaptans. The procedure given is suitable for the range 2 to 100 ppb but volumes can be modified to accommodate higher concentrations.

Apparatus

Sample absorption is carried out in a midget impinger fitted with a coarse frit. An air pump with a flow-meter or gas meter capable of aspirating and measuring a flow of 2 litres min^{-1} is required.

Reagents

Analytical grade reagents should be used and kept under refrigeration when not in use:

(a) Amine-HCl (stock) solution: Dissolve 5.0 g DMPDA hydrochloride salt in 1 litre concentrated HCl. If kept cool and dark, the reagent is stable for six months.

(b) Reissner solution: Dissolve 67.6 g $FeCl_3$. $6H_2O$ in distilled water, dilute to 500 ml and mix with 500 ml of an aqueous solution containing 72 ml freshly boiled, concentrated HNO_3 (r.d. 1.42).

(c) Colour developing solution (prepared freshly): Mix 3 vol. of DMPDA with 1 vol. of Reissner solution.

(d) Absorbing solution: Dissolve 50 g mercuric acetate (Poison) in 400 ml distilled water and add 25 ml glacial acetic acid. Dilute to 1 litre. The mercuric acetate must be free of mercurous salts to prevent precipitation of mercurous chloride during colour development.

(e) Standard lead mercaptide (stock): Weigh 156.6 mg of crystalline lead mercaptide and make up to 100 ml with absorbing solution. This solution contains the equivalent of 500 μg CH_3SH ml^{-1}.

Procedure

Aspirate the air sample through 15 ml of the absorbing solution in a midget impinger at 1.0 to 1.5 litres min^{-1} for a period of up to 2 h. Note the air volume sampled. Quantitatively transfer the sample to a 25 ml volumetric flask and dilute to approximately 22 ml with washings and distilled H_2O. Add 2 ml of colour reagent, make up to volume and mix well. Prepare a blank using 15 ml unexposed reagent and 2 ml colour reagent, diluted to 25 ml. After 30 min measure the colour at 500 nm against the blank.

Calibration and calculation

Transfer aliquots of the diluted standard mercaptide solution into a series of 25 ml flasks, dilute each with 15 ml absorbing solution and develop the colour as for the samples. Determine the absorbance of each sample and prepare a plot of absorbance against μg ml^{-1} CH_3SH. Alternatively, a standard mixture of CH_3SH in air may be prepared using a permeation tube. The permeation rate and dilution flow should be arranged to give concentrations in the range 1 to 25 ppb. By using the standard sampling procedure at different concentrations a plot of absorbance versus ppb CH_3SH can be prepared. This calibration includes any corrections for collection efficiency of the absorbing solution.

The concentration of mercaptans expressed in terms of CH_3SH is given by

$$\text{ppb } CH_3SH = \frac{A \times 0.510 \times B}{V}$$

where A is the absorbance of the solution measured against the reagent blank, 0.510 is the volume (μl) of 1 μg CH_3SH at 25° C and 760 Torr, B is the calibration factor (μg per A.U.) and V is the air sample volume in m^3 at 25° C and 760 Torr.

6.3.4.2   *Physical methods*

Undoubtedly the most useful and sensitive method for the determination of organo-sulphur compounds is GC analysis. Flame ionization detection can be used if the sulphur compounds are selectively sampled to separate sulphides from hydrocarbons, e.g. using impregnated filters [51]. The optimum system is the flame photometric detector with its high selectivity and sensitivity for S compounds. A variety of columns and techniques have been reported in the literature for the GC analysis of organic S compounds [52] but for ambient air analysis the techniques developed by Stevens *et al.* [11] and Bruner *et al.* [31] for SO_2, H_2S and low molecular weight organo-sulphur compounds are recommended. Practical details of a suitable GC–FPD system for measurement of H_2S, SO_2, CH_3SH and DMS in air are given in Section 6.3.1.2. It should be noted that the silica-gel based columns for H_2S and SO_2 analysis are unsuitable for the analysis of mercaptans and sulphides since the latter are too strongly retained. However, CS_2 and carbon oxysulphide (COS) may be analysed using Chromosil 310 (Supelco Ltd) column packing.

A simple, low cost GC–FPD system for measurement of organo-sulphur compounds is marketed by United Analysts Ltd (LRS Odour Chromatograph). Designed principally as a portable instrument for 'spot sample' measurement of organic S compounds in natural gas, the system can also be applied to air analysis. The minimum detectable concentration is 5 ppb DMS in a 10 ml sample. The measurement of organo-sulphur compounds could be automated by the use of an automatic gas sampling valve. The cryogenic pre-concentration procedure of Sandalls and Penkett [47] is applicable to DMS giving a detection limit of 0.02 ppb.

6.3.5   Analysis of oxides of nitrogen – NO and NO_2

The methods of analysis of NO and NO_2 are closely related in so far as all chemical methods for NO involve its oxidation to NO_2 and the widely adopted chemiluminescence method provides a useful method for NO_2 following its conversion to NO. A review by Allen [12] gives a comprehensive survey of methods for the analysis of these gases.

6.3.5.1   *Chemical methods*

Both colorimetric and coulometric analyses are used for the measurement of atmospheric NO_2. Colorimetric methods involve the determination of nitrite ion formed by hydrolysis of NO_2 in aqueous solutions. The determination of nitrite is based on the Griess–Isolvay reaction in which a pink coloured azo-dye complex is formed between sulphanilic acid, nitrite ion and α-napthylamine in acid medium. Various modifications of this method have been proposed but the one that is most widely employed is that introduced by Saltzmann in which the colour is developed in the absorbing solution [53]. The chief disadvantage of the latter method is that samples cannot be stored for more than a day. Although samples collected in alkali absorbent are stable for longer periods, the many methods utilizing this technique (e.g. Jacobs–Hochheiser method [54]) have been shown to give unreliable results due to poor collection efficiency and uncertain stoichiometry for the conversion of NO_2 to nitrite ion. A recently reported technique [55] based on absorption of NO_2 in ethanolamine, has been shown to be suitable for long-period atmospheric sampling. Nash [56] has reported efficient absorption of NO_2 in alkaline solutions after the addition of a small amount of guaicol (*o*-methoxy-phenol).

The main uncertainty involved in the methods based on the Griess–Isolvay reaction arises from the stoichiometry of conversion of NO_2 to nitrite ion in solution – the nitrite equivalence or 'Saltzman factor', the value of which has been the subject of some controversy (see [12] for a detailed summary). From the various reports it can be concluded that the nitrite equivalence factor appears to be influenced by many parameters including reagent composition, purity, temperature, the NO_2 concentration in the gas sample, the sample flow rate and bubbler design.

Saltzman [53] determined a value of nitrite ion $\equiv 0.72$ NO_2 (gas) for his original procedure and has since obtained further confirmation for this equivalence [57]. Scaringelli *et al.* [58] obtained a value of 0.764 ± 0.005 from several hundred analyses with the original reagent using permeation tubes to prepare gravimetrically standardized NO_2 mixtures. Provided the standard procedure is rigidly followed a value of 0.72 is recommended when using the Saltzman method but if any modification is made the factor should be determined using a standard NO_2 source.

Addition of 0.1% sodium arsenite to aqueous sodium hydroxide absorbing solution in a modification of the Jacobs–Hochheiser method has been shown to give a constant high NO_2 collection efficiency of 82% [59, 60]. There are no significant interferences at normal urban pollutant levels, and the method can give accurate, precise measurements of ambient NO_2 [61] with a reported detection limit of 8 μg m^{-3}. Several air monitoring networks have adopted the method.

Mulik and co-workers [62], using an empirical approach, developed an absorbing reagent containing triethanolamine, guaicol (*o*-methoxyphenol) and sodium metabisulphite. The collection efficiency was 93% in the range 20–750 μg m^{-3} NO_2 without apparent interference. Independent testing of the method showed good collection efficiency but considerable imprecision [63]. The method is useful for 24 h sampling; analysis may be delayed up to 7 days.

Coulometric analysis of NO_2 utilizes the oxidizing action of NO_2 on potassium iodide solutions. Efficient removal of other oxidizing and reducing pollutants is necessary for meaningful NO_2 measurements.

A number of systems have been devised for the oxidation of NO to NO_2 for analysis of atmospheric NO. The recommended system involves passage of the sample through a tube containing chromic oxide supported on an inert inorganic material. Other oxidizing systems that have been employed are solid manganese dioxide/potassium hydrogen sulphate ($MnO_2/KHSO_4$) [64], acid potassium permanganate bubblers [65], sodium dichromate impregnated filter papers [66] and oxidation in the gas phase by O_3. The former methods can suffer from deterioration of the reagents giving rise to low and variable oxidation efficiency; O_3 oxidation systems require careful design to ensure that excess O_3 does not lead to oxidation of NO_2. Any excess O_3 must be removed in a scrubber to avoid interference with the subsequent analysis for NO_2. The NO concentration may be obtained either by difference from simultaneous measurements of total $NO + NO_2$ in the air sample or directly by removal of the ambient NO_2 prior to NO oxidation.

Arsenite method for analysis of NO_2

Principle

Nitrogen dioxide is collected by bubbling air through a sodium hydroxide–sodium arsenite solution to form a stable solution of sodium nitrite. The nitrite ion produced during sampling is reacted with phosphoric acid, sulphanilamide and

N-1-(naphthyl)ethylenediamine dihydrochloride to form an azo dye and then determined colorimetrically. The method is applicable to collection of 24-h samples in the field and subsequent analysis in the laboratory.

The range of the analysis is 0.04 to 2.0 μg NO_2^- ml^{-1}. Beer's law is obeyed through this range (0 to 1.0 absorbance units). With 50 ml absorbing reagent and a sampling rate of 200 cm^3 min^{-1} for 24 h, the range of the method is 20 to 750 μg m^{-3} (0.01 to 0.4 ppm) nitrogen dioxide. A concentration of 0.04 μg NO_2^- ml^{-1} will produce an absorbance of approximately 0.02 with 1 cm cells.

Nitric oxide is a positive interferent. The presence of NO can increase the NO_2 response to 5 to 15% of the NO_2 sampled. The interference of sulphur dioxide is eliminated by converting it to sulphate ion with hydrogen peroxide before analysis.

The relative standard deviations for sampling NO_2 concentrations of 79, 105 and 329 μg m^{-3} are 3, 4 and 2% respectively.

Collected samples are stable for at least 6 weeks.

Apparatus

(a) Probe. Teflon, polypropylene or glass tube with a polypropylene or glass funnel at the end.

(b) Absorption tube. Polypropylene tubes 164 $\times$ 32 mm, equipped with polypropylene two-port closures. Rubber stoppers cause high and varying blank values and should not be used. A glass-tube restricted orifice is used to disperse the gas. The tube, approximately 8 mm o.d.–6 mm i.d., should be 152 mm long with the end drawn out to 0.6–1.0 mm i.d. The tube should be positioned so as to allow a clearance of 6 mm from the bottom of the absorber.

(c) Moisture trap. Polypropylene tube equipped with two-port closure. The entrance port of the closure is fitted with tubing that extends to the bottom of the trap. The unit is loosely packed with glass wool to prevent moisture entrainment.

(d) Membrane filter. Of 0.8 to 2.0 μm porosity.

(e) Flow control device. Any device capable of maintaining a constant flow through the sampling solution between 180–220 cm^3 min^{-1}. A typical flow control device is a 27 gauge hypodermic needle, 3/8 in long. (Most 27 gauge needles will give flow rates in this range.) The device used should be protected from particulate matter. A membrane filter is suggested. Change filter after collecting 10 samples.

(f) Air pump. Capable of maintaining a pressure differential of at least 0.6–0.7 of an atmosphere across the flow control device. This value includes the minimum useful differential, 0.53 atmospheres, plus a safety factor to allow for variations in atmospheric pressure.

(g) Calibration equipment. Flow-meter for measuring airflows up to 275 cm^3 min^{-1} within $\pm$2%, stopwatch, and a precision wet test meter (1 litre/revolution).

Reagents

Sampling:

(a) Sodium hydroxide. ACS reagent grade.

(b) Sodium arsenite. ACS reagent grade.

(c) Absorbing reagent. Dissolve 4.0 g sodium hydroxide in distilled water, add 1.0 g of sodium arsenite and dilute to 1000 ml with distilled water.

Analysis:

(a) Sulphanilamide. Melting point, 165–167° C.

(b) N-(1-naphthyl)-ethylenediamine dihydrochloride (NEDA). Best grade available.

(c) Hydrogen peroxide. ACS reagent grade, 30%.

(d) Sodium nitrite. Assay of 97% $NaNO_2$ or greater.

(e) Phosphoric acid. ACS reagent grade, 85%.

(f) Sulphanilamide solution. Dissolve 20 g sulphanilamide in 700 ml distilled water. Add, with mixing, 50 ml concentrated phosphoric acid and dilute to 1000 ml. This solution is stable for one month, if refrigerated.

(g) NEDA solution. Dissolve 0.5 g of NEDA in 500 ml of distilled water. This solution is stable for one month, if refrigerated and protected from light.

(h) Hydrogen peroxide solution. Dilute 0.2 ml of 30% hydrogen peroxide to 250 ml with distilled water. This solution may be used for one month, if protected from light and refrigerated.

(i) Standard nitrite solution. Dissolve sufficient desiccated sodium nitrite and dilute with distilled water to 1000 ml so that a solution containing 1000 μg NO_2^- ml^{-1} is obtained. The amount of $NaNO_2$ to use is calculated as follows:

$$G = \frac{1.500}{A} \times 100$$

G=amount of $NaNO_2$, grams; 1.500=gravimetric factor in converting NO_2 into $NaNO_2$; A=assay, per cent.

Procedure

Sampling:

Assemble the sampling apparatus in the sequence = probe, adsorption tube, moisture trap filter, flow control device, pump. Components upstream from the absorption tube may be connected, where required, with Teflon or polypropylene tubing; glass tubing with dry ball joints; or glass tubing with butt-to-butt joints with Tygon, Teflon or polypropylene. Add exactly 50 ml of absorbing reagent to the calibrated absorption tube. Disconnect funnel, insert calibrated flow-meter, and measure flow before sampling. If flow rate before sampling is not between 180 and 220 cm^3 min^{-1} replace the flow control device and/or check the system for leaks. Start sampling only after obtaining an initial flow rate in this range. Sample for 24 h and measure the flow after the sampling period.

Analysis:

Replace any water lost by evaporation during sampling by adding distilled water up to the calibration mark on the absorption tube. Pipette 10 ml of the collected sample into a test tube. Pipette in 1 ml hydrogen peroxide solution, 10 ml sulphanilamide solution and 1.4 ml NEDA solution with thorough mixing after the addition of each reagent. Prepare a blank in the same manner using 10 ml of unexposed absorbing reagent. After a 10 min colour development interval, measure the absorbance at 540 nm against the blank. Read μg NO_2^- ml^{-1} from the calibration curve. Samples with an absorbance greater than 1.0 must be reanalysed after diluting an aliquot (less than 10 ml) of the collected sample with unexposed absorbing reagent.

Calibration and efficiencies

Sampling:

(a) Calibration of flow-meter. Using a wet test meter and a stopwatch, determine the rates of air flow (cm^3 min^{-1}) through the flow-meter at a minimum of four different ball positions. Plot ball positions versus flow rates.

(b) Calibration of absorption tube. Calibrate the polypropylene absorption tube by first pipetting in 50 ml of water or absorbing reagent. Scribe the level of the meniscus with a sharp object, go over the area with a felt-tip marking pen and rub off the excess.

(c) Calibration curve. Dilute 5.0 ml of the 1000 μg NO_2^- ml^{-1} solution to 200 ml with absorbing reagent. This solution contains 25 μg NO_2^- ml^{-1}. Pipette 1, 1, 2, 15 and 20 ml of the 25 μg NO_2^- ml^{-1} solution into 100-, 50-, 50-, 250- and 250-ml volumetric flasks and dilute to the mark with absorbing reagent. The solutions contain 0.25, 0.50, 1.00, 1.50 and 2.00 μg NO_2^- ml^{-1} respectively. Run standards as instructed above, including the blank. Plot absorbance versus μg NO_2^- ml^{-1}. A straight line with a slope of 0.48 ± 0.02 absorbance units/μg NO_2^- ml^{-1}, passing through the origin, should be obtained.

(d) Efficiencies. An overall average efficiency of 82% is obtained over the range of 40 to 750 $\mu g/m^3$ NO_2.

Calculation

Sampling:
Calculate volume of air sampled.

$$V = \frac{F_1 + F_2}{2} \times T \times 10^{-6}$$

V = volume of air sampled, m^3; F_1 = measured flow rate before sampling, cm^3 min^{-1}; F_2 = measured flow rate after sampling, cm^3 min^{-1}; T = time of sampling, min; 10^{-6} = conversion of cm^3 to m^3.

Uncorrected volume. The volume of air sampled is not corrected to S.T.P. because of the uncertainty associated with 24-h average temperature and pressure values.

Calculate the concentration of nitrogen dioxide as μg NO_2 m^{-3} using

$$\mu g\ NO_2\ m^{-3} = \frac{(\mu g\ NO_2^-\ ml^{-1}) \times 50}{V \times 0.82}$$

50 = volume of absorbing reagent used in sampling, ml; V = volume of air sampled, m^3; 0.82 = collection efficiency.

Analysis of NO

Principle

After removal of NO_2 from the gas stream, NO is quantitatively oxidized to NO_2 by solid chromic oxide [67]. The NO_2 is then determined by absorption in arsenite reagent (see previous section).

This method is useful for the manual determination of atmospheric NO in the range 0.005 to 5 ppm. In addition to the factors affecting the measurement of NO_2, the precision of the method depends on the efficiency of the $NO \rightarrow NO_2$ converter, which, under controlled conditions, varies between 98 to 100%.

Apparatus and reagents

The NO_2 absorber consists of a 2 cm i.d. × 5 cm long polyethylene tube filled with 10 to 20 mesh inert porous material such as firebrick, alumina, zeolite, etc., coated with triethanolamine. (The granules are soaked in aqueous ethanolamine (20%) and dried for 0.5 to 1 h.) For the efficient operation of the chromic oxide oxidizer the r.h. of the sample should be between 40 and 70%. The r.h. is controlled by a humidity regulator consisting of a 2 cm i.d. × 5 cm long tube filled with crystals of a mixture of anhydrous and hydrated sodium acetate (50/50); 13 ml of H_2O are added dropwise to 40 g anhydrous sodium acetate with stirring.

The oxidizer consists of a 15 mm i.d. glass tube filled to a depth of 7 to 10 cm with pellets of chromic oxide adsorbed on an inert support. Glass, firebrick or alumina, mesh size 15 to 40, is soaked in a solution of 17 g chromium trioxide in 100 ml of H_2O. The pellets are drained, dried at 110° C and exposed to 70% r.h. in a desiccator containing saturated sodium acetate. The reddish colour should change to golden orange when equilibrated. Glass wool plugs should be used to hold the solid reagents in place. The NO_2 produced from NO is collected in a fritted bubbler containing arsenite reagent as described in Section 6.3.2.1.

Procedure

Assemble a sampling train comprising, in order, rotameter, NO_2 absorber, humidity regulator, oxidizer, absorber moisture trap, filter and pump. Pipette 10 ml of arsenite reagent into the absorber and draw sample air through the system at 0.2 litre min^{-1} for 24 h. Record the total air volume sampled and correct to 25° C and 760 Torr. Analyse the arsenite solution according to the procedure in Section 6.3.5.1 above.

Calibration and calculation

Calibration may be carried out using standard $NaNO_2$ solution as described for NO_2 in Section 6.3.5.1. Alternatively a gaseous standardization technique may be employed using standard mixtures of NO in air. These can be obtained by dilution of high concentration NO/N_2 mixtures (100 to 1000 ppm) either dynamically (into a flowing air stream) or in using the bag method (see Section 6.3.8.1). NO concentrations should be chosen to lie within the expected range of the samples. The calculation of the NO concentration is analogous to that for NO_2 described in Section 6.3.5.1.

Triethanolamine method for determination of average (24 h) ambient NO_2 concentration

Principle

This manual procedure involves the absorption of NO_2 in a triethanolamine solution or a triethanolamine impregnated molecular sieve followed by colorimetric analysis using a modified Griess–Saltzman reagent [55]. The absorber has a trapping efficiency of over 95% at NO_2 concentrations at the pphm level. The factor for the conversion of NO_2 (gas) to nitrite ion was 0.85 for the procedure outlined below, when evaluated against the Saltzman factor of 0.72. The method may be used for short duration sampling and samples may be stored for up to 4 weeks without significant loss of NO_2.

The presence of SO_2 has a slight interference at the 0.1 ppm level with the solid absorber but has no effect on the liquid absorber at the levels normally present in ambient air. NO does not interfere. The effect of O_3 has not been extensively investigated but field trials do not indicate serious interference.

Apparatus

For the liquid absorber use fritted bubblers (70–100 μm frits) with 163×32 mm polypropylene tubes. The solid absorbent is packed in glass tubes $2 \times 3/16$ inch) preferably with a standard luer lock fitting at one end. The solid can be held in place with glass wool plugs. A pump capable of maintaining at least 0.5 atm vacuum and a critical orifice flow-meter comprising a 27 gauge hypodermic needle is used to aspirate the samples.

Reagents

(a) Liquid absorber: Dissolve 15.0 g of the triethanolamine in approximately 500 ml distilled water, add 3.0 ml *n*-butanol and dilute to 1 litre (0.1 N).

(b) Solid absorber: In a 250 ml beaker, place 25 g of triethanolamine, 4.0 g of glycerol, 50 ml of acetone and sufficient distilled water to dissolve. Dilute to 100 ml with distilled water and add about 50 ml of 12 to 30 mesh molecular sieve 13X. Stir and allow to stand for 30 min, decant the liquid and transfer the molecular sieve to a porcelain pan. Remove the bulk of the water by drying the molecular sieves under a heating lamp and then oven dry at 110° C for 1 h. Store in an air-tight bottle.

(c) H_2O_2: Dilute 0.2 ml 100 vol. H_2O_2 to 250 ml with distilled water.

(d) Sulphanilamide: Dissolve 10 g of sulphanilamide in 400 ml distilled water. Add 25 ml of concentrated H_3PO_4, mix well and dilute to 500 ml.

(e) *N*-1-naphthylethylene diamine (NEDA): Dissolve 0.5 g NEDA dihydro-chloride salt in 500 ml distilled water.

(f) Standard nitrite stock: Dissolve 0.1500 g of analytical grade sodium nitrite in distilled water and dilute to 1 litre (i.e. 100 μg ml^{-1}).

Procedure

Attach a sampling probe (PTFE) to either the bubbler or the absorption tube and connect to the critical orifice flow-meter. 50 ml of absorbing solution is required in the liquid absorber. Switch on the pump and rapidly check the flow rate entering. Record the flow rate which should be between 150 to 200 ml min^{-1}. Sample for 24 h then recheck the flow rate.

To the liquid absorber, add distilled water to 50 ml to replace evaporation losses. Transfer a 10 ml aliquot to a 25 ml graduated tube. To another tube add 10 ml of unexposed reagent. To each add 1.0 ml of the H_2O_2 solution, 10 ml of the sulphanilamide solution, 1.4 ml of NEDA solution with thorough mixing after each addition. After 10 min measure the colour at 540 nm against the prepared blank.

Transfer the solid absorber and glass wool plugs to a 50 ml test tube. Wash the glass tube with approximately 10 ml of water and add washings to the sieve. Make up to 50 ml with absorbing solution. Cap and shake vigorously for about 1 min. Allow to stand and shake again after 10 min. Allow the solids to settle and transfer a 10 ml aliquot to a 25 ml graduated tube. Develop colour as for the liquid absorber.

Standardization and calculation

Dilute stock standard 50:1 with absorbing solution and add 0, 1, 3, 5 and 7 ml aliquots to a series of tubes. Make up to 10 ml with absorbing solution and develop colour as above. A plot of absorbance versus concentration is prepared so that the amounts of nitrite ion in the samples can be determined. The NO_2 concentration is then given by

$$NO_2 \text{ (ppm)} = \frac{\mu g \ NO_2^- \ \text{(in aliquot)} \times 5}{1.88 \times 0.85 \times V}$$

where V is the volume of air sampled (at 760 Torr and 25° C), 1.88 is the factor for converting $\mu g \ NO_2$ to $\mu l \ NO_2$ at 25° C and 0.85 is the nitrite ion (μg) equivalent to NO_2 gas (μg) i.e. stoichiometry factor. The system may also be calibrated using permeation tubes.

Instrumental chemical methods

A number of continuous analysers of oxides of nitrogen based on both electrochemical principles and colorimetric principles are commercially available.

A widely used instrument for ambient air monitoring is the Beckmann Model 910 NO_2 analyser (and model 909 NO analyser). The detector utilizes galvanic coulometry of the iodine liberated from buffered KI by NO_2. The recirculating cell is designed to give a claimed efficiency of 100% for NO_2 absorption, and oxidation and reagent replacement are not necessary. Selective scrubbers are used to reduce interferences from other oxidizing and reducing substances. In the NO analyser a small ozonizer produces sufficient O_3 to oxidize NO to NO_2 and the excess O_3 is removed by the scrubber. The minimum detectable concentration is 0.004 ppm with a useful range up to 1 ppm (3 switched ranges). Response time is 10 min to 90% of full scale. Concentrations of 1 ppm each of O_3, SO_2, H_2S, mercaptans will give a response equal to 0.1 ppm NO_2; these interferences will be significant in some environments.

Continuous colorimetric instruments for NO/NO_2 analyses are marketed in the US by Pollution Monitors Inc. (Model PM 100) WACO Ltd, and Precision Scientific Ltd (AERON). These instruments utilize a modified Griess–Saltzman method for the automatic colorimetric determination of NO_2; total oxides of nitrogen are measured with a separate oxidizer. Minimum detectable concentrations are of the order of 0.005 ppm with response times of the order of 4 min. A version of the Technicon Air-Monitor IV for analysis of NO and total oxides of nitrogen is also available. A minimum range of 0 to 50 ppb offers some advantage in sensitivity over other automated chemical systems.

6.3.5.2 *Physical methods*

A wide variety of physical methods have been employed for the analysis of NO and NO_2. Among the optical methods, chemiluminescence has emerged as the most sensitive, selective and practical instrumental system and is discussed in more detail below. Apart from some recent sophisticated developments (e.g. laser techniques, correlation spectrometry, second derivative spectrometry) all spectroscopic methods have too low a sensitivity for useful atmospheric monitoring. Furthermore, interference by water vapour and the consequent requirement of efficient drying of the sample gas, limits the usefulness of the more readily available optical techniques, such as non-dispersive i.r. spectroscopy, for the measurement of higher concentrations of NO.

There has been a substantial amount of effort toward the development of a gas chromatographic method for the analysis of oxides of nitrogen. The major problems encountered are, firstly, the difficulty in obtaining a column material which will quantitatively elute low concentrations of NO and NO_2. The high reactivity of these gases makes most chromatographic materials unsuitable, porous polymer packings offering the best prospects. Secondly, although many detectors have been tried, none has been shown to give a consistent performance at the required sensitivity and selectivity.

Chemiluminescence detectors for analysis of NO *and* NO_2

The chemiluminescent reaction between NO and O_3 (see Section 6.2.3.1) is accepted as the most reliable and precise method for detecting oxides of nitrogen. The response is linear from the limit of detection (< 0.001 ppm) up to at least 1000 ppm and the high sensitivity and relative freedom from interference by CO, CO_2, hydrocarbons, SO_2, O_3 and water vapour make this an excellent technique for the continuous analysis in both sources and the ambient atmosphere. Many commercial instruments based on this system and which are suitable for exhaust gas analysis [1 to 1000 volumes per million (vpm)] and ambient air monitoring (0.001 to 10 ppm) are now on the market.

Current models of these instruments employ catalytic decomposition of NO_2 to NO before detection, providing monitoring of NO and total oxides of nitrogen. Early models used a high carbon stainless steel converter operating at 650 to 750° C but there is a possibility of oxidation of NH_3 to NO in this system. This problem has been solved by the development of other catalytic materials which effect the conversion of NO_2 to NO at lower temperatures (e.g. molybdenum at 450° C). It should be noted, however, that any nitrogen compound which is thermally decomposed to NO at elevated temperatures (e.g. organic nitrites, oxyacids of nitrogen) will give a response in the 'total oxides of nitrogen' mode of a chemiluminescence NO analyser. In most environments NO_2 is the predominant nitrogen compound which is detected. A discussion of the operation of catalytic total oxides of nitrogen converters has been given recently by Breitenbach and Shelef [68]. Winer *et al.* [69] tested both carbon and molybdenum converters, finding near-quantitative positive interferences from PAN, ethyl and propyl nitrates and ethyl nitrite. Nitroethane gave a smaller positive interference. In most ambient air monitoring applications, these interferences will be of little significance.

Instruments for monitoring low concentrations in ambient air are normally equipped with a thermo-electrically cooled photomultiplier to minimize the thermal noise and dark current. For long-term monitoring, thermostating of the multiplier is necessary to eliminate thermal drift of the zero. Alternatively zero drift may be compensated by mechanical chopping of the chemiluminescence signal and phase sensitive detection. Both systems have been utilized in commercial instruments. When monitoring stack-gas or automobile exhaust gas, the chemiluminescence output is sufficiently high to dispense with elaborate control of the zero current; the need to prevent water condensation within the instrument, however, complicates the sampling of undiluted combustion products. All common desiccants have been found to remove NO_2 to varying degrees although the drying of gases containing NO is less of a problem.

Other chemiluminescence reactions of NO are suitable for use in methods for detecting oxides of nitrogen. O atoms undergo a fast reaction with NO

$$O + NO \rightarrow NO_2 \rightarrow NO_2 + hv$$

Furthermore NO_2 is rapidly converted to NO by O atoms

$$O + NO_2 \rightarrow NO + O_2$$

and these two reactions provide the basis of a very sensitive detector for $NO + NO_2$ [5]. The use of flame chemiluminescence has also been investigated in a possible detection method for NO and NO_2 [70]. The reactions, which take place in an oxygen rich H_2–O_2 flame, are:

$$H + NO_2 \rightarrow NO + OH$$

$$H + NO \rightarrow HNO \rightarrow HNO + hv$$

Commercial instruments based on these detection systems for determination of oxides of nitrogen are not, however, available at the present time.

An interesting indirect chemiluminescence method for the determination of NO and NO_2 in the atmosphere has been developed by Guicherit [71]. The equilibrium O_3 concentration produced by continuous u.v. irradiation of NO_2

$$NO_2 + O_2 \underset{\text{u.v. (360 nm)}}{\overset{}{\rightleftharpoons}} NO + O_3$$

is measured by chemiluminescence of O_3 with rhodamine B (see Chapter 7). The equilibrium O_3 concentration is a function only of wavelength, light intensity and temperature in the photolysis cell and by keeping these parameters constant, NO_2 was determined with a lower limit of detection of 0.002 ppm. NO was measured after oxidation to NO_2.

Operation of chemiluminescence analysers for estimation of oxides of nitrogen in the atmosphere

Any commercial chemiluminescence analyser with a maximum sensitivity of < 0.2 ppm FSD can be used. Suitable instruments are marketed by Thermo-electron Corporation (Models 12A and 14D), REM 'NO$_x$ Analyser', Monitor Labs Inc. Model 8440 and others. The operational specifications of these instruments are similar with response times of a few seconds or less, multiple ranges and various types of output circuitry for continuous analogue or stored response (for NO and total oxides of nitrogen). Some instruments operate at atmospheric pressure thereby dispensing with elaborate pumping and pressure controls but with some loss of sensitivity.

A continuous flow of sample gas is pumped into the instrument at a flow rate of the order of 1 litre min^{-1}. Sampling probes for NO and NO_2 should be of PTFE tubing or borosilicate glass. Tube lengths should be kept to a minimum for two reasons. Firstly, NO_2 has a tendency to absorb on even the most inert materials, particularly at high r.h. Secondly, the ratio of NO to NO_2 in ambient air when O_3 is present is a function of light intensity. In the dark NO is rapidly oxidized to NO_2 by ozone and therefore the NO/NO_2 ratio will change if the residence time in the sampling system is too long [72]. The half-life for oxidation of NO is given by

$$t_{1/2}(NO) \simeq 0.03[O_3]^{-1} \text{ min } ([O_3] \text{ in ppm})$$

and therefore for ambient air monitoring (O_3 concentration approximately 0.05 ppm) the residence time should not exceed a few seconds. For continuous monitoring it is desirable to insert a filter to prevent dust particles from entering the sample lines, where they may interfere with the flow control capillaries, etc. The filter holder should be of an inert material (e.g. PTFE) and a cellulose or Teflon membrane filter may be used with minimal loss of NO_2.

After start-up, a period of up to 2 h will be required for temperature equilibration of the multiplier cooler and the heater for conversion of oxides of nitrogen into NO. Before use it will be necessary to calibrate the response of the instrument using a source of NO of known concentration. The high linearity and stability of the detector make this a relatively simple procedure since calibration can be made at a single concentration. Initially, however, it may be desirable to check the linearity by controlled dilution of the standard gas. For calibration purposes, standard mixtures of NO in N_2, stored under pressure in corrosion resistant steel cylinders, are commercially available. If a regulator is used it should be of high quality stainless steel. The NO content of the calibration gas may be checked by chemical analysis after oxidation to NO_2 (not highly recommended in view of the uncertainties in the oxidation efficiency and the analysis) or by titration against a constant O_3 source which can in turn be calibrated by u.v. photometry (Chapter 7). The details of a method for gas phase titration of NO against O_3 are given by Hodgeson *et al.* [73], and suitable equipment is available commercially.

The efficiency of the conversion of NO_2 to NO can also be checked if an O_3 source of constant composition is available. A known flow of NO is injected into an air stream containing an excess of O_3. The concentrations and flow rates should be such that oxidation of NO_2 by excess O_3 does not occur (this will be important only at low flow rates and with a large excess of O_3). If the converter efficiency is 100% the response of the analyser in the total oxides of nitrogen mode should be independent of the relative amounts of NO and NO_2 in the sample.

After calibration and converter efficiency evaluation, the instrument is ready for monitoring operation. Calibration checks need not be more frequent than once per week providing operating conditions are maintained constant.

6.3.6   Analysis of NH_3

6.3.6.1   *Chemical methods*

Measurements of NH_3 in amounts of the order of the TLV value (25 ppm) are normally carried out acidimetrically, by absorption in dilute H_2SO_4 of known concentration [74]. For measurements in ambient air the presence of other acidic and basic components (principally SO_2) make this method impractical and more specific methods must be used. Another disadvantage of the use of a dilute acid collection medium is that the collection efficiency over lengthy sampling periods is rather low [75]. For this reason, impregnated filter papers are generally preferable. Shendrikar and Lodge [76] recommend a Whatman cellulose filter impregnated with 3% ethanolic oxalic acid and vacuum dried, which was found to give quantitative collection of ammonia from ambient air. An alternative impregnant is potassium bisulphate [77] which has a collection efficiency of >95% at face velocities up to 70 cm s^{-1}. The one possible problem with impregnated filters is that at very high ambient ammonia levels the filter may become saturated with a consequent loss in collection efficiency. Two colorimetric methods can be applied for measurement of NH_3 down to the ppb level. The first method [78] uses Nessler's reagent (formula $HgI_2 . 2KI$) which gives an intense colour when added to solutions containing the ammonium ion. However, a disadvantage is the interference of a number of compounds which may be present in ambient air (e.g. H_2S, other sulphides and formaldehyde). For accurate analysis it is necessary to carry out a time consuming distillation step in which the acid sample solution is made slightly alkaline and the NH_3 distilled into a second solution prior to addition of the colour reagent. These disadvantages are eliminated in the second method which is based on the indophenol reaction, i.e. the formation of a blue dye when NH_3 reacts in phenol-sodium hypochlorite solutions. A modified indophenol-blue method in which the reaction is catalysed by sodium nitroprusside is the recommended technique for the analysis of NH_3 in ambient air [79, 80].

When sampling the atmosphere for gaseous NH_3 the most likely interference is from ammonium compounds in particulate matter, which are nearly always present in ambient air. The particulate matter may be removed by filtration during sampling but certain complications may arise, particularly when sampling for long periods. Firstly, the collection of acid particulate matter on the filter may serve to remove gaseous NH_3 from the sample stream. Secondly, some particulate salts (e.g. ammonium nitrate, NH_4NO_3), which will be collected on the filter, have an appreciable vapour pressure at ambient temperatures and may volatilize into the collection solution. Thus, as a result of these sampling problems, the accuracy of the measurement of low concentrations of NH_3 in the atmosphere is subject to some uncertainty. In general, it is recommended that Teflon pre-filters are used, that they are maintained at atmospheric temperature (to avoid disturbance of dissociation equilibria) and cross-filter pressure changes are kept to a minimum. See also Sections 6.4.3 and 7.3.6.1.

Analysis of NH_3 by the catalysed indophenol-blue method

Principle

NH_3 is collected by aspiration of air through dilute H_2SO_4 in a standard impinger. The resultant ammonium ion is determined by spectrophotometric measurement of the blue indophenol dye formed in the sodium pentacyanonitrosyloferrate catalysed phenol-hypochlorite reaction with NH_3 in alkaline solution [80]. A filter placed upstream of the impinger prevents collection of particulate matter containing ammonium compounds which would interfere. Other common pollutant gases which may dissolve in dilute H_2SO_4, e.g. SO_2, O_3, NO_2 do not interfere at their normal atmospheric levels. The sensitivity of the method is approximately 0.02 μg NH_4^+ in the sample solution of 50 ml which corresponds to a detection limit of approximately 1 μg m^{-3} NH_3 (1.3 ppb at 760 Torr and 25° C) when sampling at the recommended rate (30 litres min^{-1}) for 30 min.

Apparatus

A standard impinger is used to contain the absorbing solution. A sampling probe with a filter assembly similar to that described for the sampling of SO_2 (Section 6.3.1) should be placed upstream of the impinger. The inlet probe should not be heated to avoid disturbance of the NH_4NO_3 dissociation equilibrium. A pump capable of drawing 30 litres min^{-1} air through the sample train is required. A calibrated rotameter and/or a gas meter (wet or dry) is used to measure the volume of air sampled to within 2%.

Reagents

Analytical grade reagents should be used. The colour developing solutions should be freshly prepared every 1 to 2 days and stored in the dark. Solutions should be prepared from distilled water which has been passed through a cartridge of ion exchange resin to remove the NH_4^+.

(a) Absorbing solution: 0.0025 M H_2SO_4. Prepare by dilution from 1 M H_2SO_4 made up by mixing 28 ml concentrated H_2SO_4 (sp. gr. 1.84) with 500 ml de-ionized H_2O and making up to 1 litre.

(b) Phenol-nitroprusside reagent: dissolve 5 g phenol plus 25 mg of sodium pentacyanonitrosyloferrate (sodium nitroprusside – $Na_4(Fe(CN)_5NO . 2H_2O)$ in 500 ml de-ionized H_2O.

(c) Alkaline hypochlorite: dissolve 5 g NaOH in 500 ml de-ionized water. Add 4.2 ml sodium hypochlorite solution (12% available chlorine) and make up to 1 litre with H_2O.

(d) Standard ammonium sulphate $[(NH_4)_2SO_4]$ solution (stock): weigh out 0.370 g $(NH_4)_2SO_4$ and dissolve in 1 litre de-ionized water in a volumetric flask. This solution contains 101 μg ammonium ion ml^{-1}.

(e) Dilute $(NH_4)_2SO_4$ solution in a volumetric flask dilute 10 ml of the stock solution to 1 litre with de-ionized water.

Procedure

Place 50 ml of the absorbing solution in the impinger and a fresh filter (Whatman 41, 'Microsorban' or Teflon) in the filter holder. Assemble the sampling train and aspirate the bubbler at 30 litres min^{-1} for 30 min. (Slower flow rates and/or longer times may be employed.) Note the volume of air sampled and correct to 760 Torr and 25° C. Appreciable evaporation of the absorbing solution may occur; this can be corrected by weighing before and after sampling.

For analysis transfer a 5.0 ml aliquot of the absorbing solution into a 25 ml stoppered flask and add from a pipette 10 ml of phenol-nitroprusside reagent. Shake in order to mix well. From a pipette add 10 ml of alkaline hypochlorite solution and mix well. Prepare a reagent blank in the same manner using unexposed absorbing solution. Allow 30 min at room temperature (which should be greater than 20° C) for colour development before measuring absorbance at 625 nm.

Calibration and calculation

Into a series of 25 ml volumetric flasks place 0, 0.5, 1, 2, 3 and 5 ml of the dilute ammonium sulphate solution. Add in turn exactly 10 ml each of the phenol nitroprusside reagent and the alkaline hypochlorite, shaking well after each addition. Make up to volume using absorbing solution, allow to stand for 30 min and measure the absorbance. Construct a calibration plot of μg NH_4^+ ml^{-1} versus absorbance.

The concentration of ammonium ion in the sample solutions can be determined from the calibration. The NH_3 concentrations in the air sample is given by

$$C_{NH_3} \text{ (ppm)} = \mu g \text{ ml}^{-1} \text{ } NH_4^+ \times 1.36 \times V_a/V_s$$

where V_a is the final volume of the absorbing solution after correction for evaporative losses, V_s is the volume of air (in litres) sampled at 760 Torr and 25° C and 1.36 is the factor for conversion of μg ammonium ion to μl NH_3.

Variations

NH_3 may be sampled on filter paper (discs or tape, Whatman 41) impregnated with 5% potassium bisulphate solution [77]. A flow rate of 10 litres min^{-1} through a 2.5 cm diameter filter gives a good collection efficiency. A prefilter must be used to remove particulate matter. The NH_3, which is trapped as NH_4^+, can then be extracted from the paper with H_2O at 70° C, and determined using the indophenol blue method. Potassium hydrogen sulphate ($KHSO_4$) is also removed from the tape during extraction and can cause attenuation in response. This can be taken into account by maintaining in all samples and standards a level of $KHSO_4$ approximately equivalent to that in the paper extracts (0.1 to 0.2 mg ml^{-1}).

6.3.6.2 *Physical methods*

At present, there are no widely used direct physical methods for the measurement of NH_3 in the atmosphere. NH_3 may be measured by i.r. or u.v. absorption spectrometry. The minimum detectable concentration of NH_3 by dispersive i.r. absorption in a 10 m path length is approximately 20 ppm using the 10.77 μm absorption maximum. Gaseous NH_3 exhibits several strong absorption bands in the u.v. between 190 and 230 nm. The determination of NH_3 in air by direct u.v. spectrophotometry at 204.3 nm in 10 cm quartz cells has been reported [81] with a lower detection limit of 7 ppm. The molar extinction coefficient of NH_3 at this wavelength is 2790 litres mol^{-1} cm^{-1} but many other compounds absorb strongly in this region (e.g. SO_2, H_2S, and organic compounds) and may interfere. These spectroscopic methods are only useful, therefore, for measuring NH_3 concentrations at the TLV level (TLV $NH_3 = 25$ ppm).

Above 800° C oxidation of ammonia to NO occurs in stainless steel tubing, and this may be used as an ammonia analysis method when the NO generated is analysed with a chemiluminescent analyser. The analyser is run with a permanently fitted converter and is cycled between analysis of $(NO + NO_2 + NH_3)$ and analysis of $(NO + NO_2)$, the latter mode requiring ammonia removal by an acidic dichromate [82] or phosphoric acid [83] pre-scrubber. In the latter case, measurements of NH_3 in the range 0.001–0.01 ppm are possible, although this sensitivity is not generally achievable. In an alternative system, ammonia is determined subtractively using two instrumental modes: an $(NO + NO_2)$ mode with a conventional NO_2 to NO converter, and an $(NO + NO_2 + NH_3)$ mode using a high temperature converter [84]. The method is effective for ammonia concentrations >5 ppb with a response time of 2 s. Commercially produced accessories for ammonia analysis with chemiluminescent analysers are available.

If ammonia is collected on Chromosorb T, it may be thermally desorbed at 100° C and detected by chemiluminescent or opto-acoustic analyser [85]. This allows detection at level of 0.5–22 ppb with a sampling time of 40 min, or less.

6.3.7   Miscellaneous N_2 compounds

The most important organo-nitrogen air pollutants, the peroxyacylnitrates, are discussed in Chapter 7. Other gaseous nitrogen compounds, which are encountered mainly in industrial environments, are amines, hydrazine and HCN. For industrial hygiene purposes specific colorimetric methods have been developed for the quantitative measurement of these compounds in air. Primary aliphatic amines can be determined colorimetrically using ninhydrin (triketohydrindine hydrate) [86]. A method utilizing collection in impingers containing HCl has been developed by Hantzsch and Prescher [87]. Photometric determination of small amounts of hydrazine in air using the colour reaction with dimethyl aminobenzaldehyde has been described [88, 89]. By absorption in a midget impinger a detection limit of approximately 3 ppb in a 100 litre air sample can be achieved. HCN has been determined by a number of techniques. A simple impregnated filter paper method for the determination of cyanide in air has been devised [90] based on the Prussian blue reaction. More sensitive methods have been developed by Hanker and co-workers [91, 92]. They found that the chelate compound formed between 8-hydroxy-7-iodoquinoline-5-sulphonic acid and palladous chloride, when converted to the potassium salt, reacts with ferric ion and cyanide to yield a blue green complex which can be estimated colorimetrically at 650 nm. A more sensitive variation of this palladium chelate procedure involves the measurement of the fluorescence of a co-ordination complex of 8-hydroxyquinoline-5-sulphonic acid with magnesium ion [93]. The action of cyanide on the non-fluorescent potassium salt of the above chelate compound liberates hydroxyquinoline-sulphonic acid which co-ordinates with magnesium ion to form a fluorescent chelate. Thus the intensity of fluorescence provides a measure of the amount of cyanide present.

Nitrous oxide (N_2O) is of importance as an air pollutant and is interesting because of its natural occurrence in low concentrations (~ 0.3 ppm), both in the atmosphere and in sea water. The analysis of this compound is characterized by its lack of any specific chemical reactions at room temperature and only physical methods are suitable for its determination at low concentrations. Several GC procedures for the measurement of atmospheric N_2O have used Porapak Q columns and Katharometer detection [94]; the N_2O having been previously concentrated on a solid absorbent. Lattue *et al.* [95] have developed a method giving an accuracy of 2% for the determination of N_2O in the atmosphere.

Nitrous and nitric acids are important secondary pollutants and are considered in Chapter 7.

6.3.8   Preparation of standard gas mixtures for calibration

Any method for the analysis of gaseous air pollutants is of doubtful validity unless the response can be reliably calibrated. Wet chemical methods involving colorimetric analysis can usually be accurately calibrated using standard solutions of the solvated species (usually an ion) which is being analysed. The response of coulometric systems can be calculated according to Faraday's law provided the redox reaction of the pollutant proceeds with close to 100% efficiency. These procedures for determining the response of the method do not, however, take into account possible errors resulting from variations in sampling efficiency, non-stoichiometric chemistry and losses by absorption on tubing, etc. These difficulties may be overcome by sampling standard mixtures containing known concentrations of the pollutant of interest at the concentration levels for which the method is to be employed. Furthermore, for many physical and instrumental methods particularly those involving chemiluminescence or GC, standard gas mixtures provide the only means of calibrating the method.

Mixtures containing accurately known concentrations of gases at the ppm level and below can be prepared by either static or dynamic methods.

6.3.8.1   *Preparation of standard mixtures by static methods*

Essentially this involves dilution of a measured volume of pollutant gas with a known volume of air or other diluent gas at a fixed pressure and temperature. This technique can be used satisfactorily for most of the gaseous N and S pollutants.

The dilution is most conveniently carried out in the laboratory at pressures near atmospheric using either a rigid (constant volume) container or a flexible bag. The container and fittings should be of a suitably inert material such that the pollutant is not absorbed rapidly on the walls, etc. Borosilicate glass, aluminium or plastic can be used for rigid containers and fluorocarbon plastic film (e.g. PTFE) for flexible bags. The type of material will be determined to some extent by the pollutant gas being handled. The volume of the container should be as large as possible, preferably of the order of 200 litres. A large size is particularly useful for rigid containers since the partial pressure (and hence the concentration) of the pollutant decreases as the sample gas is removed. The air sampled can be replaced by clean air to maintain constant pressure, and if the sample volume removed is small compared with the total volume only a small concentration change will result. The volume of a rigid container must be accurately determined. Rigid containers have an advantage in that a circulating fan can be installed to assist rapid mixing. Liquid pollutants can be introduced from a microsyringe and gases from a gas-tight syringe. Accurate amounts of gases and vapours can also be measured out from a vacuum system, if available. A known pressure of gas is measured into a previously evacuated glass cell (fitted with PTFE stopcocks if the gas is corrosive or soluble in vacuum grease).

When diluting in flexible bags the following procedure is adopted. The container is purged with diluent gas several times by filling and deflating through the entry port. The bag is then completely deflated by applying a slight vacuum and the diluent supply quickly connected. The bag is filled with a known volume of diluent gas by metering through an accurate rotameter (for a known time) or a gas-meter. During filling inject a measured volume of pollutant into the gas stream. When the container is approximately 0.75 full, stop the flow and mix the contents by kneading the bag several times. Leave for 15 min to allow complete mixing before sampling. The concentration in ppm is simply the mixing ratio as defined by

$$C \, (\text{ppm/v}) = V_P / V_D$$

where V_P is the volume of pollutant injected and V_D the total volume of gas in the reservoir, both corrected to 760 Torr and 25° C. If liquid pollutant is injected, the corresponding volume of vapour can be calculated from a knowledge of the density of the liquid at the laboratory temperature and the molecular weight of the vapour.

Calibration mixtures can also be prepared as pressurized gas mixtures in cylinders. Pressurized standard mixtures of a variety of pollutant gases are commercially available. Most of the N and S gases are, however, too reactive for storage under pressure in metal containers for long periods. An exception is NO which may be stored as a pressurized mixture in pure N_2. Standard mixtures of NO in N_2 which are prepared gravimetrically and which provide a useful source for the calibration of chemiluminescence NO–NO_2 analysers are commercially available. Pressurized standard mixtures of sulphur hexafluoride in air may be employed for the calibration of flame photometric detectors for sulphur compounds. If pressure regulators are used for dispensing these standard mixtures, they should be of the high quality, corrosion resistant type.

6.3.8.2   *Preparation of standard mixtures by dynamic methods*

Permeation tubes

The use of permeation tubes to standardize methods for the analysis of trace gases was first documented by O'Keefe and Ortman [96] in 1966. The principle of this device is based on diffusion of a gas or vapour through a plastic membrane at very slow rates. A sealed section of inert PTFE tubing containing a liquefied gas or volatile liquid can, when placed in a metered air stream, maintain a constant low concentration of the gas. By accurate measurement of the weight loss from the permeation tube over a period of time the system can be used as a gravimetric calibration standard. The diffusion rate is a non-linear function of temperature and therefore constant temperature conditions must be maintained during gravimetric standardization and use as a calibration source.

Certain compounds, e.g. SO_2 and H_2S react with O_2 or water vapour at the surface of the plastic tubing. Tubes containing these compounds should be stored and standardized under dry N_2. A relatively high concentration in dry N_2 carrier gas is then diluted stepwise in air to provide calibration mixtures.

Perhaps the most important parameter in the standardization of permeation tubes is the time factor required for the diffusion equilibrium to be reached. Two to five days at constant temperature ($\pm 0.1°$ C for 1% precision) under a flowing dry carrier gas stream is required for accurate standardization. The time is dependent on diffusion rate and accuracy of weighing. At intervals during the standardization, the tube can be quickly removed, weighed and replaced, and the process continued until a constant loss rate is achieved.

Permeation tubes may be purchased or prepared according to the method of O'Keefe and Ortman [96]. For the preparation of gas mixtures in the concentration range likely to be encountered in the atmosphere a loss rate of 0.1 to 0.2 μl min^{-1} is optimum. In addition to temperature, the loss rate depends on the pressure difference, wall thickness and surface area (and hence length) of the tube. Some tube dimensions and corresponding loss rates for S compounds [30] are given in Table 6.2.

An apparatus as shown in Fig. 6.9 is required to prepare standard

Table 6.2   PTFE permeation-tube characteristics for S gases [30]

Gas	Tube length (cm)	i.d. (in)	Wall (in)	Volume loss μl min^{-1} at 20.3° C and 1 atm.
CS$_2$	10.3	0.183	0.03	0.377
SO$_2$	2.3	0.183	0.03	0.364
H$_2$S	4.0	0.485	0.2	0.120
CH$_3$SH	7.2	0.183	0.03	0.087

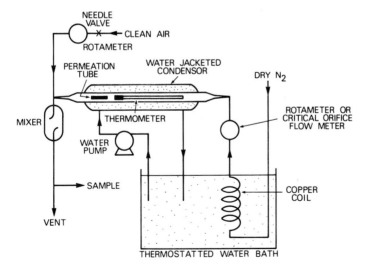

Fig. 6.9   Arrangement for producing standard gas mixtures using a permeation tube.

concentrations of a pollutant. Carrier gas (50 to 100 ml min^{-1}) is metered through a needle valve and flow-meter or a critical orifice and brought to temperature by passage through a 2 m copper coil immersed in a constant temperature water bath. The gas passes over the calibrated permeation tube which is housed in a water jacketed condenser through which water from the bath is circulated. The temperature should be the same as used for calibration of the tube. The gas stream is then diluted to the desired concentration by varying the flow rate of the diluent air which should be clean and dry. The minimum flow rate will be determined by the requirements of the sampling system and maximum flow rates of 15 litres min^{-1} or more may be used. This flow should be measured with an accuracy of 1 to 2%. The concentration at a given flow is simply given by

$$C \text{ (ppm)} = P/F$$

where F is the total flow rate (diluent+carrier) in litre min^{-1} and P is the permeation rate in μl min^{-1}, both corrected to 760 Torr and 25° C. The permeation rate P can be calculated from the rate of weight loss according to

$$P \text{ } (\mu l \text{ min}^{-1} \text{ at } 25° \text{ C}) = w \text{ } (\mu g \text{ min}^{-1}) \times \frac{22.4}{\text{mol. wt}} \times \frac{298}{273}$$

When working with low concentrations (i.e. ppb region) it is necessary to condition the system by allowing the trace gas to pass through the system for a period of day(s) before conducting calibrations. An accurate recording instrument is useful for monitoring the outlet gas concentration when attempting to calibrate manual techniques. All tubing into which the gas comes in contact should be of an inert material such as glass or PTFE.

Thermostatted ovens known as 'permeation tube systems' are available from a number of commercial suppliers. These generally incorporate flow control and measurement devices.

Exponential dilution method

This method produces a gas mixture in which the trace-gas concentration declines with time at a fixed exponential rate. It is useful, therefore, for calibration of instrumental methods which are either continuously recording or capable of giving a number of readings in reasonably rapid succession e.g. GC methods. The basic flow system is shown in Fig. 6.10. An aliquot of pure pollutant (or an accurately prediluted mixture) is injected into the dilution flask which is stirred for complete mixing. The flask is continuously flushed with a metered stream of pure, dry, carrier gas. For these conditions the concentration of the pollutant in the outlet decreases exponentially [97].

$$C = C_0 \exp(-Qt/V)$$

where C_0 = initial concentration, Q = volumetric flow rate at the flask pressure (1 atm), and V = the effective volume of the dilution flask, t = time elapsed after injection. The effluent gas is then mixed with a constant diluent flow to obtain a low concentration mixture, the composition of which is varying in a known

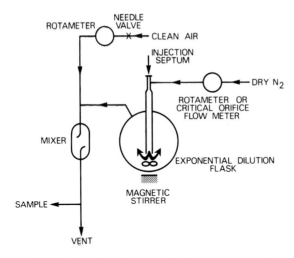

Fig. 6.10 Exponential dilution flask for preparing low concentration gas mixtures.

manner. By a judicious choice of flow rates and volumes, accurate mixtures in the sub-ppm range can be prepared. Thus with a 10 litre exponential dilution flask and carrier and dilution flows of 0.05 and 10 litres min^{-1} respectively, 1 ml of pure pollutant will give a concentration of 0.5 ppm with a half-time for decay of 140 min. The application of the exponential dilution technique for the calibration of a chemiluminescence NO detector [5] and the flame-photometric S detector [31] has been described.

6.4 Particulate compounds of S and N

6.4.1 Analysis of SO_4^{2-}

The principal natural source of SO_4^{2-} in the atmosphere is the oxidation of natural H_2S. Additionally, SO_4^{2-} enters the air in marine aerosol and in the erosion of rocks. Sulphates also arise directly from some industrial activities, and from atmospheric oxidation of anthropogenic SO_2. In general, significantly higher levels of SO_4^{2-} are measured in urban areas as compared to rural areas.

A review article by Tanner and Newman [98] discusses progress and problems in SO_4^{2-} analysis. The SO_4^{2-} must first be collected on a filter paper. Cellulose paper filters such as Whatman No. 41 have a low background of soluble SO_4^{2-} and are suitable, while Mitex or Fluoropore filters, made of Teflon, are recommended for speciation studies [98]. Glass fibre is unsuitable for sulphate sampling due to artifact formation from SO_2. This point is amplified in Chapter 1. In ambient atmospheres, particulate S is almost entirely SO_4^{2-} and X-ray fluorescence analysis of collected S may be used [98]. Alternatively, if a flame photometric S analyser is available, collected SO_4^{2-} may be volatilized into the analyser by rapid heating and this method has been shown to give results comparable with those obtained by XRF [99, 100]. More commonly, the sulphate is leached from the filter into an aqueous medium and determined by any sensitive specific technique for SO_4^{2-} such as those described in Section 6.3.1.1 for analysis of SO_2 when collected as sulphate in H_2O_2 solution.

Attempts to speciate airborne SO_4^{2-} depend upon selective volatilization or solvent extraction of acid SO_4^{2-}. Such techniques are of considerable interest due to the importance of sulphates in acid precipitation and methods are described in Chapter 11.

The development of ion exchange chromatography has greatly simplified anion analysis. Mulik *et al.* [101] describe the application of ion chromatography to the analysis of sulphate in ambient aerosols. A detection limit of $0.1 \mu g \, ml^{-1}$ and a precision (relative standard deviation) of 3% at the $5 \mu g \, ml^{-1}$ level were reported.

6.4.1.1   Experimental procedure for SO_4^{2-} (turbidimetric)

Principle

The method described is an alternative to that given in Section 6.3.1.1. SO_4^{2-} after leaching from filter paper, is mixed with a solution of barium chloride. The resultant precipitate of barium sulphate is measured turbidimetrically [102].

Sampling apparatus

A Teflon filter paper in almost any type of filter holder may be used. Membrane and Nuclepore filters have a higher soluble SO_4^{2-} background and are not recommended.

Reagents

(a) HCl 10 N.
(b) Glycerol–alcohol solution. Mix one volume glycerol with two volume 95% ethanol.
(c) Standard SO_4^{2-} solution. Dissolve 0.1479 g anhydrous sodium sulphate in de-ionized water and dilute to 1 litre to give a $100 \mu g \, ml^{-1}$ stock solution.
(d) Barium chloride. Analar grade, 20–30 mesh.

Procedure

Draw air through the filter at 10 litres min^{-1} for 24 h, or longer if required. Extract the exposed filter and a blank by warming to 70° C in de-ionized water (10 ml) for 10 min. Decant the liquid into a 25 ml volumetric flask and repeat with another aliquot of de-ionized water. Adjust to a final volume of 25 ml when cool, filtering if turbid before making the adjustment.

Prepare SO_4^{2-} standards by pipetting 0.5, 1.0, 1.5, 2.0 and 3.0 ml of stock SO_4^{2-} solution into 25 ml volumetric flasks and adjust the volume to 20 ml by addition of de-ionized water by pipette to give standards of 2.5, 5, 7.5, 10 and

15 μg ml^{-1}. Transfer 20 ml of sample and blank solutions to 25 ml volumetric flasks by pipette. To the sample, blank and standard solutions, add 10 N HCl (1 ml) followed by glycerol–alcohol solution (4 ml) and mix thoroughly. Add a constant measure (0.5 spoon spatula) of barium chloride crystals to each. Do not mix, but allow to stand for exactly 40 min. Carefully decant the turbid supernatant into 4 cm cells and determine the absorbance at 500 nm against a reagent blank.

Calculation

Draw a calibration curve of SO_4^{2-} concentration versus absorbance. This normally shows some curvature. Estimate the SO_4^{2-} concentrations of sample (x μg ml^{-1}) and filter blank (y μg ml^{-1}) solutions.

Then, atmospheric concentration of SO_4^{2-}

$$= \frac{(x-y) \times 25}{\text{Vol. of air sample (m}^3)} \mu g \ m^{-3}$$

6.4.2 Analysis of particulate NO_3^-

Oxidation of the nitrogen oxides in the atmosphere leads ultimately to formation of NO_3^-. This occurs naturally, and also at an accelerated rate in the highly oxidizing atmosphere of a 'photochemical smog'. Analysis for NO_3^- in ambient air is becoming increasingly frequent due to the importance of nitrates in acid rain.

Nitrates may be collected upon any type of filter material, but the use of Teflon papers is recommended since this paper has a low background of soluble NO_3^- and gives minimal artifact formation from NO_2 or HNO_3 [103]. A number of colorimetric procedures are available for NO_3^- but it is essential to establish the magnitude of possible interferences before adopting a method. Ion chromatography is also a popular means of nitrate analysis [101].

Recent work has highlighted the volatility of NH_4NO_3 (the main particulate nitrate compound in polluted air). It is in equilibrium with gaseous NH_3 and HNO_3, and may volatilize if heated, or form from the equilibrium gases if the air stream is cooled

$$NH_4NO_3 \text{ (s)} \rightleftharpoons NH_3 \text{ (g)} + HNO_3 \text{ (g)}$$

The equilibrium is highly temperature dependent and follows thermodynamic predictions over a wide range of temperatures [104–106]. Thus sampling of particulate nitrates should be carried out on a filter at ambient temperature and minimal pressure drop.

6.4.2.1 *Experimental procedure for nitrate (colorimetric)*

Principle

Nitrates are leached from the filter paper and then converted to nitrite by copper-catalysed reduction with hydrazine sulphate in alkaline solution [77]. The nitrite is then determined by a modification of the method of Saltzman [53]. After acidification with *o*-phosphoric acid, the nitrite is reacted with sulphanilamide (4-aminobenzene sulphonic acid) and the resultant diazonium salt is coupled with *N*-(1-naphthyl)ethylenediamine dihydrochloride to form an azo dye which is determined spectrophotometrically at 540 nm. Interference from nitrite is possible. The existence of this species in appreciable concentrations in ambient air is most improbable, but if its presence is suspected it may be determined separately by the above procedure, omitting the hydrazine sulphate reduction, and an appropriate correction made.

Sampling apparatus

A Teflon filter paper in any suitable holder.

Reagents

(a) Diazotization reagent. Dissolve sulphanilamide (10 g) and *N*-(1-naphthyl) ethylenediamine dihydrochloride (0.5 g) in 10% v/v *o*-phosphoric acid and make up to volume (1 litre) with the same acid.

(b) Copper sulphate solution. Make the stock solution by dissolving anhydrous copper sulphate (2.5 g) and making up to volume (1 litre) with water. Prepare fresh working solution daily by dilution of stock solution (12.5 ml) to 4 litres.

(c) Hydrazine sulphate. Prepare stock solution by dissolving hydrazine sulphate (27 g) and making up to volume (1 litre) with water. Prepare fresh working solution daily by dilution of stock (25 ml) to 1 litre.

(d) NaOH. Dissolve Analar NaOH (12 g) and make up to volume (1 litre) with water.

(e) NO_3^- standards. Dissolve Analar potassium nitrate (0.163 g) in water and make up to volume (1 litre to prepare a stock solution of 100 μg ml^{-1} NO_3^-. Prepare a working standard of 10 μg ml^{-1} by diluting stock solution (100 ml) to 1 litre.

Procedure

Draw air through the filter at 2.5 litres min^{-1} for 24 h, or longer if required. Extract the exposed filter and blank by warming to 70° C in de-ionized water (5 ml) for 10 min. Decant into a 10 ml volumetric flask, repeat the extraction with a further portion of water (4 ml), add the washings to the volumetric flask and when cool dilute to volume. The use of ultrasonic agitation may assist the wetting of the Teflon filter.

Dilute the stock standard solution with water to give standards 0, 1, 2, 3, 4 and 5 μg ml^{-1} NO_3^-. Pipette 2 ml aliquots of sample, blank and standard solutions into small flasks and add copper sulphate solution (6 ml), NaOH solution (6 ml) and hydrazine sulphate solution (4 ml). Shake gently to mix and then heat to 38° C using a water bath. Use conditions of subdued light at all times. Cool and add the diazotization reagent (6 ml). Using a spectrophotometer determine the absorbance of the solutions at 540 nm against a de-ionized water blank.

Calculation

Draw a calibration curve of absorbance versus NO_3^- concentration and estimate the concentration of the sample (x μg ml^{-1}) and filter blank (y μg ml^{-1}) solutions. Then, concentration of NO_3^- in air

$$= \frac{(x - y) \times 10}{\text{Vol. of air sample } (m^3)} \mu g \ m^{-3}$$

6.4.3 Analysis of NH_4^+ salts

NH_4^+ salts arise primarily from neutralization of NH_3 by acidic substances in the atmosphere. $(NH_4)_2SO_4$ is believed to be a major constituent of urban aerosols.

NH_4^+ salts may be collected from the atmosphere by filtration, and after leaching from the filter paper are determined by the same methods as are applicable to gaseous NH_3 after collection (see Section 6.3.6.1). One problem arises from the significant vapour pressures of salts such as NH_4NO_3 and chloride at ambient temperatures. A very small concentration of these salts may escape collection by filtration, and this may be of significance when determining the low levels present in unpolluted air. This point is discussed in Section 6.4.2. Another problem arises from reaction of gaseous ammonia with acid sulphates collected on a filter enhancing the ammonium loading of the filter. This problem can be overcome only by use of an ammonia denuder [107, 108], a device consisting of narrow glass tubes coated internally with oxalic acid [107] or phosphoric acid [108]. Gaseous ammonia is removed in the denuder, while particulate ammonium salts are unaffected and may be collected subsequently on a Teflon filter. If an appreciable level of acid sulphates is anticipated, the use of an ammonia denuder is essential to reliable measurement of NH_4^+.

6.4.3.1 *Experimental procedure for NH_4^+ (colorimetric)*

Principle

After collection upon a filter paper the NH_4^+ salts are leached into water, where they are determined by the catalysed indophenol-blue method.

Sampling apparatus

A Teflon filter is recommended, held in any convenient filter holder. This type of filter has a low background of soluble NH_4^+ ions. The filter holder is maintained at atmospheric temperature.

Procedure

Draw air through the filter at 2.5 litres min^{-1} for 24 h, or longer if required. Extract the exposed filter and blank by warming to 70° C in de-ionized water (10 ml) for 10 min. Ultrasonic agitation may assist wetting of the Teflon filter. Decant the liquid into a 25 ml volumetric flask and repeat with another aliquot

(10 ml) of de-ionized water. Adjust to volume when cool. Transfer an aliquot (5 ml) into a 25 ml stoppered flask and add phenol-nitroprusside reagent (10 ml) as described in Section 6.3.6.1 for analysis of gaseous NH_3, and complete the analyses as described. Calibration is performed in the manner described in Section 6.3.6.1, extending the range of standards to give concentrations of 0–5 μg ml^{-1} by increasing the concentration of the standard NH_4 solution five-fold.

Calculation

Draw a calibration curve of NH_4^+ ion concentration versus absorbance and estimate the concentration of the sample (x μg ml^{-1}) and filter blank (y μg ml^{-1}) solutions. Then, concentration of NH_4^+ in air

$$= \frac{(x-y) \times 25}{\text{Vol. of air sampled (m}^3)} \, \mu\text{g m}^{-3}$$

References

1. Environmental Protection Agency Publ. No. AP-50 (1969) *Air Quality Criteria for Sulphur Oxides.*
2. Environmental Protection Agency Publ. No. AP-84 (1971) *Air Quality Criteria of Nitrogen Oxides.*
3. Hersch, P. and Deuringer, R. (1963) *Anal. Chem.*, **35,** 897.
4. Brewer, A. W. and Milford, J. R. (1960) *Proc. Roy. Soc.*, A**256,** 470.
5. Fontjin, A., Sabadell, A. J. and Ronco, R. J. (1970) *Anal. Chem.*, **42,** 575.
6. Brody, S. S. and Chaney, J. E. (1966) *J. Gas Chromatogr.*, **4,** 42.
7. Okabe, H. (1971) *J. Am. Chem. Soc.*, **93,** 7095.
8. Moffat, A. J., Robbins, J. R. and Barringer, A. R. (1971) *Atmos. Environ.*, **5,** 511.
9. Williams, D. T. and Huger, R. N. (1970) *Appl. Opt.*, **9,** 1597.
10. Hodgeson, J. A., McClenny, W. A. and Hanst, P. L. (1973) *Science*, **182,** 248.
11. Stevens, R. K., Mulik, J. D., O'Keefe, A. E. and Krost, K. J. (1966) *Anal. Chem.*, **38,** 760.
12. Allen, J. D. (1973) *J. Inst. Fuel*, **46,** 123.
13. Driscoll, J. N. and Warneck, P. (1973) *J. Air Pollut. Control Assoc.*, **23,** 858.
14. Forrest, J. and Newman, L. (1973) *J. Air Pollut. Control Assoc.*, **23,** 761.
15. West, P. W. and Gaeke, G. C. (1956) *Anal. Chem.*, **28,** 1816.
16. Jacobs, M. B. and Greenburg, L. (1956) *Ind. Eng. Chem.*, **48,** 1517.
17. Huygen, C. (1962) *Anal. Chim. Acta*, **28,** 349.
18. Katz, M. (1950) *Anal. Chem.*, **22,** 1040.
19. Intersociety Committee for Manual Methods of Air Sampling and Analysis (1977). *Methods of Air Sampling and Analysis*, 2nd edn, American Public Health Association, Washington, DC.
20. Bostrun, C. E. (1965) *Int. J. Air Pollut.*, **9,** 333.
21. Pate, J. P., Ammons, B. E., Swanson, G. A. and Lodge, J. R. (1965) *Anal. Chem.*, **37,** 942.
22. Scaringelli, F. P., Saltzman, B. E. and Frey, S. A. (1967) *Anal. Chem.*, **39,** 1709.

23. Scaringelli, F. P., Frey, S. A. and Saltzman, B. E. (1967) *Am. Ind. Hyg. Assoc. J.*, **28**, 260.
24. Thomas, M. O. and Amtower, R. E. (1966) *J. Air Pollut. Control Assoc.*, **16**, 618.
25. Persson, G. A. (1966) *Int. J. Air Pollut.*, **10**, 845.
26. Volmer, W. and Frohlich, Z. H. (1944) *Anal. Chem.*, **126**(16), 414.
27. Kanno, S. (1959) *Int. J. Air Pollut.*, **1**, 231.
28. Lewin, E. and Zachau-Christiansen, B. (1977) *Atmos. Environ.*, **11**, 861.
29. Axelrod, H. D. and Hausen, S. G. (1975) *Anal. Chem.*, **47**, 2460.
30. Stevens, R. K., O'Keefe, A. E. and Ortman, G. C. (1969) *Environ. Sci. Technol.*, **3**, 652.
31. Hartmann, C. H. (1971) Proc. Joint Conference on Sensing of Environmental Pollutants, AIAA Paper No. 71-1046, Palo Alto, California.
32. Bruner, F., Liberti, A., Possanzini, M. and Allegrini, I. (1972) *Anal. Chem.*, **44**, 2070.
33. Thornsberry, W. L. (1971) *Anal. Chem.*, **43**, 452.
34. Jackson, P. J., Langdon, W. E. and Reynolds, P. J. (1970) *J. Inst. Fuel*, **43**, 10.
35. Fielder, R. S. and Morgan, C. H. (1960) *Anal. Chim. Acta*, **23**, 538.
36. Bond, R. L., Mullin, W. J. and Pinchin, F. J. (1963) *Chem. Ind. (London)*, **48**, 1903.
37. Jacobs, M. B., Braverman, M. M. and Hochheiser, S. (1957) *Anal. Chem.*, **29**, 1349.
38. Bamesburger, W. L. and Adams, D. F. (1969) *TAPPI*, **52**, 1302.
39. Bethea, R. M. (1973) *J. Air Pollut. Control Assoc.*, **23**, 710.
40. Pare, J. P. (1966) *J. Air Pollut. Control Assoc.*, **16**, 325.
41. Hochheiser, S. and Elfers, L. A. (1970) *Environ. Sci. Technol.*, **4**, 672.
42. Okita, T., Lodge, J. P. and Axelrod, H. D. (1971) *Environ. Sci. Technol.*, **5**, 532.
43. Natusch, D. F. S., Kloris, H. B., Axelrod, H. D. *et al.* (1972) *Anal. Chem.*, **44**, 2067.
44. Axelrod, H. D., Cary, J. H., Bonelli, J. E. and Lodge, J. P. (1969) *Anal. Chem.*, **41**, 1865.
45. Adams, D. F., Bamesburger, W. L. and Robertson, T. J. (1968) *J. Air Pollut. Control Assoc.*, **18**, 145.
46. Braman, R. S., Ammons, J. M. and Briker, J. L. (1978) *Anal. Chem.*, **50**, 992.
47. Sandalls, F. J. and Penkett, S. A. (1977) *Atmos. Environ.*, **11**, 197.
48. Pecsar, R. E. and Hartmann, C. H. (1971) *Anal. Instrum.*, **9**, H-2, 1.
49. Moore, H., Helwig, H. L. and Graul, R. J. (1960) *Ind. Hyg. J.*, **21**, 466.
50. Leithe, W. (1970) *The Analysis of Air Pollutants*, Ann Arbor-Humphrey, Ann Arbor, London, p. 228.
51. Okita, T. (1970) *Atmos. Environ.*, **4**, 93.
52. Ronkainen, P., Denslow, J. and Leppanen, O. (1973) *J. Chromatogr. Sci.*, **11**, 384.
53. Saltzman, B. E. (1954) *Anal. Chem.*, **26**, 1949.
54. Jacobs, M. B. and Hochheiser, S. (1958) *Anal. Chem.*, **30**, 426.
55. Levaggi, D. A., Siu, W. and Feldstein, M. (1973) *J. Air Pollut. Control Assoc.*, **23**, 30.
56. Nash, T. (1970) *Atmos. Environ.*, **4**, 661.
57. Saltzman, B. E. and Wartburg, A. F. (1965) *Anal. Chem.*, **37**, 1261.
58. Scaringelli, F. P., Rosenburg, E. and Rehme, K. A. (1970) *Environ. Sci. Technol.*, **4**, 924.
59. Margeson, J. H., Beard, M. E. and Suggs, J. C. (1977) *J. Air Pollut. Control Assoc.*, **27**, 553.
60. Christie, A. A., Lidzey, R. G. and Radford, D. W. F. (1970) *Analyst*, **95**, 519.
61. Margeson, J. H., Suggs, J. C., Constant, P. C. *et al.* (1978) *Environ. Sci. Technol.*, **12**, 294.
62. Mulik, J., Fuerst, R., Guyer, M. *et al.* (1974) *Int. J. Environ. Anal. Chem.*, **3**, 333.

63. Killick, C. M. (1976) Warren Spring Lab. Rept, LR 228 (AP).
64. Hartkamp, H. (1970) *Schr. Reihe Landesanst Imn-u Bodenutzungshutz Landes N. Rhein/Westfalen*, **18**, 55.
65. Ripley, D. L., Chingenpeel, J. M. and Hurn, R. W. (1964) *Int. J. Air Pollut.*, **8**, 455.
66. Thomas, M. D. (1956) *Anal. Chem.*, **28**, 1810.
67. Levaggi, D. A., Kothny, E. L., Belsky, T. *et al.* (1974) *Environ. Sci. Technol.*, **8**, 348.
68. Breitenbach, L. P. and Shelef, M. (1973) *J. Air Pollut. Control Assoc.*, **23**, 128.
69. Winer, A. M., Peters, J. W., Smith, J. P. and Pitts, J. N. Jr (1974) *Environ. Sci. Technol.*, **8**, 1118.
70. Krost, K. J., Hodgeson, J. A. and Stevens, R. K. (1973) *Anal. Chem.*, **45**, 1800.
71. Guicherit, R. (1972) *Atmos. Environ.*, **6**, 807.
72. Butcher, S. and Ruff, R. E. (1971) *Anal. Chem.*, **43**, 1890.
73. Hodgeson, J. A., Baumgardner, R. E., Martin, B. E. and Rehme, K. A. (1971) *Analyt. Chem.*, **43**, 1128.
74. Leithe, W. (1970) *The Analysis of Air Pollutants*, Ann Arbor-Humphrey, Ann Arbor, London, p. 170.
75. Okita, T. and Kanamori, S. (1971) *Atmos. Environ.*, **5**, 621.
76. Shendrikar, A. D. and Lodge, J. P. (1975) *Atmos. Environ.*, **9**, 431.
77. Eggleton, A. E. J. and Atkins, D. H. (1972) *Results of the Tees-side Investigation*, AERE R-6983, HMSO, London.
78. Buck, M. and Strathmann, H. (1965) *Z. Anal. Chem.*, **213**, 241.
79. Chaney, A. L. and Marbach, E. P. (1962) *Chim. Chem.*, **8**, 130.
80. Weatherburn, M. W. (1967) *Anal. Chem.*, **39**, 971.
81. Gunther, F. A., Barkley, J. H., Kolbezen, M. J. *et al.* (1956) *Anal. Chem.*, **28**, 1985.
82. Sigsby, J. E., Black, F. M., Bellar, T. A. and Klosterman, D. L. (1973) *Environ. Sci. Technol.*, **7**, 51.
83. Hodgeson, J. A. (1974) *Toxicol., Environ. Chem. Rev.*, **2**, 81.
84. Aneja, V. P., Stahel, E. P., Rogers, H. H. *et al.* (1978) *Anal. Chem.*, **50**, 1705.
85. McClenny, W. A. and Bennett, C. A. Jr (1980) *Atmos. Environ.*, **14**, 641.
86. Williams, D. D. and Miller, R. R. (1962) *Anal. Chem.*, **34**, 225.
87. Hantzsch, S. and Prescher, K. E. (1966) *Staub*, **26**, 332.
88. Price, J. G., Fenimore, D. C., Simmonds, G. P. and Patkiss, A. Z. (1968) *Anal. Chem.*, **40**, 541.
89. Porter, K. and Volman, D. H. (1962) *Anal. Chem.*, **34**, 748.
90. Gettler, A. O. and Goldbaum, L. (1947) *Anal. Chem.*, **19**, 270.
91. Hanker, J. S., Goldberg, A. and Witten, B. (1958) *Anal. Chem.*, **30**, 93.
92. Hanker, J. S., Gamson, R. M. and Klapper, H. (1957) *Anal. Chem.*, **29**, 879.
93. Leithe, V. W. and Hofer, A. (1968) *Allg. Prakt. Chem.*, **19**, 78.
94. Bock, R. and Schutz, K. (1968) *Z. Anal. Chem.*, **237**, 321.
95. Lattue, M. D., Axelrod, H. D. and Lodge, J. P. (1971) *Anal. Chem.*, **43**, 1113.
96. O'Keefe, A. E. and Ortman, G. C. (1969) *Anal. Chem.*, **38**, 761.
97. Lovelock, J. E. (1960) *Gas Chromatography, 1960* (ed. R. P. W. Scott), Butterworths, London, p. 26.
98. Tanner, R. L. and Newman, L. (1976) *J. Air Pollut. Control Assoc.*, **26**, 737.
99. Roberts, P. T. and Friedlander, S. K. (1976) *Atmos. Environ.*, **10**, 403.
100. Husar, J. D., Husar, R. B. and Stubits, P. K. (1975) *Anal. Chem.*, **47**, 2062.
101. Mulik, J., Puckett, P., Williams, D. and Sawicki, E. (1976) *Anal. Lett.*, **9**, 653.
102. Jacobs, M. B. (1960) *The Chemical Analysis of Air Pollutants*, Interscience Publishers, New York.

103. Spicer, C. W. and Schumacher, P. M. (1979) *Atmos. Environ.*, **13**, 543.
104. Doyle, G. J., Tuazon, E. C., Graham, R. A. *et al.* (1979) *Environ. Sci. Technol.*, **13**, 1416.
105. Stelson, A. W., Friedlander, S. K. and Seinfeld, J. H. (1979) *Atmos. Environ.*, **13**, 369.
106. Harrison, R. M. and Pio, C. A. (1983) *Tellus*, **35B**, 155.
107. Ferm, M. (1979) *Atmos. Environ.*, **13**, 1385.
108. Stevens, R. K., Dzubay, T. G., Russwurm, G. and Rickel, D. (1978) *Atmos. Environ.*, **12**, 55.

7   Secondary pollutants*

R. M. HARRISON

7.1   Introduction

The term secondary pollutants is applied to pollutants which are formed as a result of chemical reactions of primary gaseous pollutants within the atmosphere. Secondary pollutants may be either gaseous, or particulate aerosols. The gaseous pollutants are usually formed in homogeneous gas-phase reactions which in many cases are photochemically initiated. Aerosol pollutants may be formed within the atmosphere as a result of gas-phase reactions followed by condensation of the products, or by reactions taking place in the existing atmospheric aerosol phase.

Gaseous pollutants formed within the atmosphere include O_3, oxides and oxyacids of nitrogen (NO_2, N_2O_5, HNO_3, HNO_2), SO_2 (from oxidation of sulphides etc.), H_2O_2 and organic peroxides, aldehydes, organic acids and other partially oxidized organic compounds including nitrogen-containing esters such as the peroxyacylnitrates (PANs). The aerosol components which can be produced in atmospheric reactions are H_2SO_4 and sulphates, inorganic nitrates, chlorides, and their ammonium salts and also organic condensation aerosols. Some of the 'secondary' pollutants formed in the atmosphere are also emitted as primary pollutants, e.g. NO_2, SO_2, and sulphates; the classification is not, therefore, rigid. In this section only the gaseous secondary pollutants which have not been dealt with in other chapters will be considered. The aerosols produced by chemical transformations in the atmosphere contribute to the overall burden of particulate matter in the air. The analytical techniques for sampling, characterization and measurement of atmospheric particulate are described in Chapters 1 and 4.

Of the gaseous secondary pollutants, O_3 has undoubtedly received the most attention. O_3 has been known for a long time to be a natural constituent of the earth's atmosphere. Relatively high concentrations of O_3 exist in the stratosphere where it is produced by photolysis of molecular oxygen, O_2. O_3 from the high altitudes is transported to the lower atmosphere where it is removed by chemical reaction or by deposition at the earth's surface. Recognition of O_3 as a pollutant in urban air resulted from the discovery in the late 1940s by Haagen-Smit and his colleagues that abnormally high O_3 concentrations were present during air

---

* A revision of the chapter by R. A. Cox, *Handbook of Air Pollution Analysis*, first edition.

pollution episodes in the Los Angeles basin. It was subsequently found that the formation of O_3 and other 'oxidants' resulted from sunlight initiated photo-chemical reactions of nitrogen oxides and unburned hydrocarbons, emitted principally from motor vehicle exhausts. Other undesirable pollutants such as aldehydes, PANs and organic and inorganic aerosols are also produced in these reactions. This particular type of atmospheric pollution was termed 'photo-chemical smog', and although the phenomenon was originally thought to be confined to the west-coast cities of the United States, it has become apparent in recent years that the formation of photochemical oxidants in urban air is rather widespread. Severe photochemical smog episodes are confined, however, to certain large cities which are subject to special climatic conditions, i.e strong temperature inversions and prolonged sunshine. As a result of atmospheric transport processes, secondary pollutants are found at substantial distances from urban source areas and indeed regional scale pollution is common.

The chemistry of photochemical smog is very complex and the reader is referred to other sources for a detailed description [1, 2]. The phenomenon results from oxidation of organic pollutants by a chain reaction involving free radicals, which are generated photochemically following light absorption by certain primary and secondary pollutants (Chapter 3). The presence of nitrogen oxides is required to sustain the free radical chain reactions, and the primary pollutant, NO, is oxidized to NO_2 during the process. The hydrocarbons are oxidized firstly to aldehydes and acids which are converted to CO, and eventually to CO_2. This accounts for the large and complex array of partially oxidized organic compounds present in the smog. O_3 is formed as a result of photodissociation of NO_2 to give oxygen atoms, which combine with molecular oxygen:

$$NO_2 + \text{u.v. light} \rightarrow O + NO$$

$$O + O_2 \rightarrow O_3$$

Since NO is converted back to NO_2 in the free radical reactions, NO_2 is continuously replenished and the O_3 concentration builds up (see Chapter 3).

The sequence of chemical reactions occurring during smog formation gives rise to a characteristic pattern in the diurnal variation of the concentrations of primary and secondary pollutants in urban air [3]. In the early morning, the concentration of the primary pollutants, NO and hydrocarbons, rises due to the increased emissions from traffic. At this stage the O_3 concentration is usually very low due to overnight destruction at the ground and by reaction with other pollutants. During the morning, as sunlight intensity increases, photochemical oxidation of the primary pollutants causes the concentrations of NO_2 and other oxidation products to rise. The O_3 concentration also rises until a maximum is reached sometime after midday, and falls thereafter due to reactions with other pollutants. The concentrations of other secondary pollutants, both gaseous and particulate, also reach their maximum during the afternoon or early evening. This model diurnal pattern is determined by the chemistry and emission pattern of the pollutants. The diurnal variation in the concentration of the pollutants is also sensitive to changes in meteorological features and the model diurnal pattern may not always be observed in practice. Furthermore, even in unpolluted air, an afternoon maximum in the ground-level O_3 concentration can often be observed due to diurnal changes in the vertical mixing in the atmosphere, giving increased downward mixing of natural O_3 during daytime.

The rate of chemical transformation of primary to secondary pollutants normally depends on the air-concentration of primary pollutants. Thus the concentration of secondary pollutants will be influenced by the same gross atmospheric features, such as air-mass origin and degree of dispersion, as the primary pollutants. However the detailed variation in time and space of the concentration of the secondary pollutants will differ from their precursors. The transformation rates are also influenced by other parameters such as temperature, absence or presence of sunlight and its intensity, r.h., etc. The time scale of the overall chemical processes is normally of the order of a few hours and as a result the very high concentrations and rapid fluctuations associated with primary sources are not encountered with the secondary pollutants. Furthermore, the maximum concentration of the secondary pollutants may be observed some distance from the source of the precursors and a larger geographical region may be subjected to the effects of the pollutants.

Many of the secondary pollutants under discussion have known deleterious effects [3]. The 'cracking' of rubber by O_3 is well known and was in fact used as a diagnostic for O_3 formation in early work on photochemical smog. The most serious effect of elevated O_3 levels, however, is its phytotoxic effect on vegetation, in particular on agricultural crops such as tobacco, market-garden produce and citrus fruit trees. O_3 is also toxic to humans and the TLV of 100 ppb is often exceeded during photochemical episodes both in the United States and Europe. The peroxyacyl nitrates also show phytotoxic effects on some plants [4]. HNO_3 and H_2SO_4 produced by oxidation of nitrogen and sulphur oxides are both highly corrosive and can cause damage to many materials [5], particularly metals. H_2SO_4 aerosol may present a serious health hazard since it can be readily ingested into the lung by virtue of the small particle size [6]. Certain aldehydes and PANs are responsible for the severe eye irritation associated with photochemical smog [7].

The range of air concentrations likely to be encountered in an air sampling programme for secondary pollutants is indicated in Table 7.1, which shows typical levels of gaseous pollutants which have been observed in different locations. As noted above, the concentrations of the secondary pollutants do not

Table 7.1 Typical air concentrations in parts per billion of gaseous secondary pollutants

Pollutant	*Urban smog*	*Rural*	*Background*
O_3	150 to 500	50 to 100	20 to 50
Total aldehydes	~50	2 to 10	<1
Peroxyacylnitrate	3 to 30	<1	<0.1
Nitric acid and nitrates	10 to 100	–	–

fluctuate rapidly with time and, therefore, sampling time resolution of 1 to 2 h is adequate for obtaining a reasonable measure of the pollutant burden. In view of the diurnal variations in the pollutant levels, however, 24 h average measurements only give limited information and, in the case of O_3, may give an entirely false impression of the degree of pollution.

Analytical methods of sufficient sensitivity are now available for measurement of most of the above secondary pollutants in urban and rural air. In discussing the analytical techniques it is convenient to classify the pollutants into three groups i.e. oxidants, organic secondary pollutants and inorganic oxyacids.

Oxidants

These are broadly defined as those compounds present in the air which will oxidize a reference reagent that is not oxidized by atmospheric oxygen [3]. The widely adopted reference compound for the measurement of atmospheric-oxidants is KI, the oxidant concentration being equivalent to the amount of iodine released following exposure to a known volume of air. Using this technique a measurement of 'total oxidants' is obtained, without distinction as to the individual nature of the pollutants, which may include both inorganic and organic primary and secondary pollutants. Furthermore, reducing components such as H_2S, SO_2 and aldehydes, will decrease the degree of oxidation observed leading to a net oxidant concentration. O_3 is the main component giving a positive response, with minor contributions from peroxides, PAN and NO_2.

The 'total oxidants' measurement can be considered to represent the 'condition' of the atmosphere. Low (or negative) oxidants (<10 ppb) might indicate, for example, high levels of SO_2; oxidant readings of 20 to 50 ppb would be expected in clean air due to natural O_3; oxidant readings in excess of 100 ppb might indicate photochemical pollution. These measurements are, however, of limited use in air pollution studies and this has stimulated investigation into more specific methods for the measurement of individual oxidizing pollutants.

The organic secondary pollutants

These broadly comprise all partially oxidized hydrocarbon compounds, i.e. alcohols, aldehydes, organic acids and peroxides and organo-nitro compounds. The overall atmospheric burden of these compounds will contain, in addition to atmospheric reaction products, partially oxidized organic compounds from combustion and other sources. A distinction between the 'primary' and 'secondary' component is difficult. However some compounds, e.g. PANs have an exclusively atmospheric source and, therefore, serve as useful indicators of the overall photochemical reactivity in the atmosphere.

The inorganic oxyacids

These include HNO_2, HNO_3 and H_2SO_4. The precursor oxides NO_2 and other oxides of nitrogen (N_2O_3, N_2O_4, NO_3 and N_2O_5) and sulphur trioxide, which may also be considered as secondary pollutants, have been discussed in Chapter 6. H_2SO_4 only exists in the atmosphere as an aerosol and the analysis of particulate sulphate is described in Chapter 6, and speciation of sulphates in Chapter 11. HNO_3 exists in the gaseous phase under atmospheric conditions. Although there is evidence for the presence in air of both HNO_3 and HNO_2 vapours, analytical methods for these compounds are still at a relatively early stage of development.

7.2 Basic analytical techniques for the analysis of gaseous secondary pollutants

The basic techniques which have been used to measure the air concentrations of gaseous secondary pollutants are similar to those described in the previous chapter relating to gaseous nitrogen and sulphur compounds. Much of the earlier work on identification and measurement of oxidants, oxidized organic compounds, etc. was made using simple chemical methods. Infrared spectroscopic analysis has also played an important role in elucidating some of the unusual products and chemical reactions in photochemical smog, mainly in laboratory studies. Recently the application of physical methods, particularly chromatographic and chemiluminescence techniques, has become more widespread. The main advantages offered by the latter are improved specificity in the measurement of individual components in the complex array of substances present in the urban atmosphere. The general remarks regarding interferences, accuracy and precision, made in Chapter 6, also apply to the analysis of gaseous secondary pollutants.

7.2.1 Sampling methods

The basic techniques used for sampling trace gases from the atmosphere are surveyed in Chapter 1 and some additional comments relating to the design of samplers to minimize absorption losses and interferences, are given in Chapter 6. For the gaseous secondary pollutants, particular attention should be paid to the minimization of losses by absorption and decomposition during sampling, since these gases are generally themselves chemically reactive. Thus suitable inert materials should be used for sample probes, tubing, etc. Teflon (PTFE) equipment is recommended for the sampling of O_3 and other oxidants, but clean glass tubing may be used provided long sample probes are not used. Metals, rubber and PVC should be avoided. Organic vapours may be sampled with stainless steel, PTFE or glass equipment. However, since many organic secondary pollutants are water soluble it is essential that the sampling tubes be kept dry, to minimize absorption losses. In view of the reactive nature of most secondary pollutants, sample concentration by cryotrapping or adsorption on a porous support is not recommended for quantitative work.

7.2.2 Analytical techniques

7.2.2.1 Chemical methods

Classical chemical techniques involving colorimetric, coulometric or acidimetric analysis of air samples are still widely used for the measurement of gaseous secondary pollutants. In fact these are likely to remain the favoured methods for 'total oxidants' which, according to the definition given in Section 7.1 can only be measured by chemical methods.

Colorimetric analysis involves trapping of the gaseous air pollutant in a suitable medium until sufficient of the material of interest has been collected to give a measurable colour with a specific colour-forming reagent. In view of the unstable nature of many gaseous secondary pollutants, a liquid absorbing medium containing a reagent which reacts immediately with the gas(es) of interest to give an involatile stable product, is normally used. The colour-forming reagent may either be present in the absorbing solution or can be added to the sample after collection. Colorimetric analysis is most conveniently applied to batch samples but automated colorimetric instruments are available (e.g. Technicon Mk IV) which have been developed to give a semi-continuous measurement of a number of pollutants including oxidants and aldehydes. Autoanalysers also allow rapid and precise measurement of large numbers of batch samples. This aid is invaluable for any routine survey work.

The main problem with colorimetric analysis and with chemical methods generally is the lack of specificity of the colour-forming reactions. For example, the liberation of iodine from KI solution, which has been widely used for the colorimetric determination of O_3 and total oxidants, is subject to a strong negative interference from reducing gases such as SO_2 and H_2S, which are commonly present in urban air. Chemical methods for the analysis of organic secondary pollutants generally utilize a specific colour reaction of a characteristic functional group in the organic compound. Thus selective determination of individual compounds of a certain class is not normally possible. Notable exceptions are the aldehydes, formaldehyde and acrolein and also H_2O_2, for which specific colorimetric methods have been developed.

Coulometry using the I_2/I^- redox system is widely used for the measurement of total oxidants and O_3. The air sample containing oxidant is aspirated continuously through an electrochemical cell containing KI solution. The liberated I_2 causes current to flow between the anode and the cathode. Two methods have been used to monitor the electrochemical reaction. The galvanic coulometric O_3 sensor, first described by Hersch and Deuringer [8], utilizes a cyclic oxidation–reduction process and requires no reagent replacement or applied external voltage. The operating principle of this cell is described in detail in Chapter 6. The second type has been described as an amperometric coulometer and was originally developed as a balloon-borne instrument for measurement of O_3 in the upper atmosphere [9]. A commercially available form of amperometric coulometric detector for atmosphere oxidants (Mast Development Corporation, model 724-2) has been widely used for measurements in

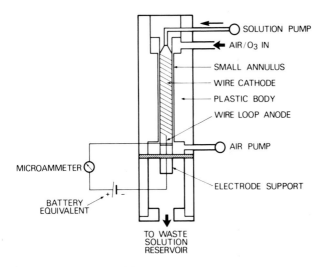

Fig. 7.1 Detector cell for Mast oxidant meter.

polluted air [10–12]. In this detector, which is shown diagrammatically in Fig. 7.1, a platinum wire helical cathode and a wire loop anode are wetted in turn by a solution containing 2% KI and 5% KBr, which is continuously pumped through the cell. Air is drawn through the cell at a constant flow rate of 140 cm^3 min^{-1} and, if O_3 is present in the sampled air, I_2 is liberated in the electrolyte. The liberated I_2 is continuously reduced by an applied cathodic potential of 0.25 to 0.3 V and the current flowing between the cathode and the wire loop anode is proportional to the amount of I_2 liberated. The current is assumed, therefore, to be a linear function of the O_3 concentration and the flow rate. In common with all methods based on oxidation of I_2 solutions, coulometric methods for measurement of O_3 suffer interferences from other oxidizing and reducing substances present in the sample.

7.2.2.2 *Physical methods*

Chemiluminescence methods
The basic principles of chemiluminescence analysis of atmospheric pollutants have been described in Chapter 6. The only secondary pollutant for which chemiluminescence methods have been developed to the instrumental stage is O_3. However, the development of these methods, which provide a relatively simple, sensitive and highly specific measurement of O_3, has been timely in view of the difficulties inherent in all chemical methods for this pollutant. A detailed survey of the techniques which have been used for chemiluminescence analysis of O_3 is given in Section 7.3.2.2.

Gas chromatography

Gas chromatography is by far the most widely used technique for the analysis of organic atmospheric pollutants. A useful review of the application of gas chromatography to atmospheric analysis has been given by Altshuller [13], and Fishbein [14] has surveyed selected chromatographic procedures for a variety of environmental pollutants including atmospheric gases. The use of gas–liquid or gas–solid chromatography with flame ionization detection (FID) for the analysis of hydrocarbon compounds is discussed in detail in Chapter 8. Measurement of secondary organic pollutants, i.e. oxygenated and nitrogen-containing organic compounds, by the application of normal GC–FID techniques presents two major difficulties. Firstly, the high reactivity and polar nature of these compounds leads to problems of material losses or modification during sample handling and chromatographic separation. Secondly, the response of the FID to low molecular weight O and N containing compounds is considerably less than for the corresponding hydrocarbons. The lower sensitivity for O and N containing organics makes sample concentration necessary for the detection of these compounds at the levels present in ambient air using FID. Furthermore, the chromatographic separation of aldehydes, etc., from the vast range of hydrocarbons of different volatility which are present in urban air presents a formidable problem.

For the GC analysis of aldehydes Levaggi and Feldstein [15] have developed a novel method of obtaining the necessary selectivity and sample stability. This was achieved by trapping the aldehydes in aqueous sodium bisulphite solution when a stable aldehyde–bisulphite complex is formed. Hydrocarbons, etc. are not trapped in this medium. The aldehydes are then quantitatively recovered by thermal decomposition of the complex in a specially designed heated injection port on the gas chromatograph.

In recent years, a variety of new types of column material have been developed which possess greatly improved elution characteristics and separating ability for trace amounts of reactive substances. These include modified forms of porous polymeric stationary phases (e.g. Porapak, Durapak, Porasil) and supports with chemically bonded stationary phases. The use of silanized support materials (e.g. HDMS treated Chromosorb) which provide a more inert chromatographic medium, has enabled chromatographic separation and quantitative elution of unstable compounds such as peroxyacetylnitrate (PAN). Unfortunately there have been rather few reports in the literature on the performance of these new columns, as applied to the analysis of organic compounds in ambient air samples.

For the analysis of organo-nitrogen compounds the inadequate sensitivity of the FID has been largely overcome by the application of the electron capture detector (ECD) [16]. The extreme sensitivity of this detector for strongly electron absorbing compounds such as PAN and alkyl nitrates allows direct measurement of their concentration at the ppb level in 1 to 10 ml air sample. Review articles on the ECD have been given by Wentworth [17] and Aue and Kapila [18] but a brief description will be given here.

Electron capture detection utilizes the drastic reduction in the electrical conductivity of a gaseous mixture in an ionization chamber when electrophilic compounds are present. The detector consists of an ionization chamber containing a radioactive β emitting source with a stream of inert gas flowing through it. The β activity causes ionization of the gas liberating free electrons. By the application of a low voltage potential, the electrons are caused to migrate to the anode and a constant 'standing' current through the detector results. On the introduction of a trace gas with a high electron-capture cross-section, e.g. compounds containing halogen atoms, nitro or amino groups and some sulphur and oxygen compounds, the electrons undergo reaction to form low mobility ions and the detector current drops. The fall in current is directly proportional to the amount of trace contaminant present, provided the fall does not exceed 20 to 30% of the standing current. When approaching saturation, i.e. the electron concentration is severely depleted by reaction with the electron absorbing component, the response becomes very non-linear. The d.c. potential is usually applied in a pulsed mode to prevent polarization in the detector leading to anomalous response particularly under conditions of detector contamination. In the pulsed mode the voltage is applied in short pulses (0.5 to 1 μs) at intervals of 50 to 500 μs, i.e. the voltage is off for most of the time. In this way not only are polarization effects minimized but the electron density in the detector will be greater. Thus the likelihood of electron capturing events and hence the sensitivity is increased. The average electron concentration increases with the interval between pulses up to 500 μs but above this, natural ion recombination and the presence of trace contaminants limits further increases in sensitivity. The linear response range of the ECD can be increased by operating in the pulse frequency feedback mode [19]. In this mode the electron concentration is maintained essentially constant as the electron absorbing component passes through the detector, by changing the pulse frequency in response to the change in standing current. The change in pulse frequency required to maintain the current is directly related to the amount of electron absorbing component present.

The molar response of the detector to a given electron absorbing component, in the linear region, depends on the rate constant for electron attachment to the component. This parameter varies greatly among the various electron absorbing species. Lists of relative response factors for halogenated compounds and methods of predicting them have been given [20, 21]. Calibration of the detector by normal methods of injecting measured quantities of the component is only practical for the more weakly absorbing compounds. The response is so high for strong absorbers that difficulty is encountered in injecting quantitatively the extremely small amounts of material required to calibrate the detector in the linear range. A novel approach to this problem is the application of gas phase coulometry [22]. If the detector is constructed in the form of a long tube then, for strong absorbers, the proportion ionized in the detector approaches 100% provided an appropriate flow rate is employed. The integrated response in As (i.e. peak area, A) is related to the mass of substance entering the detector (m) in grams as follows:

$$m = \frac{MA}{9.65 \times 10^4} \ (M = \text{mol. wt})$$

This gives an absolute calibration of the detector independent of ambient variables of temperature and pressure. This technique has not been applied so far to the organo-nitro compounds of interest in the present context.

The main practical problem in the operation of electron capture detectors arises from unwanted contaminants entering the detector. Impurities, present in the carrier gas or on the detector surfaces, serve to reduce the standing current, thereby decreasing sensitivity and the linear range and also increasing the noise in the detector. For optimum performance with commercial ECDs great care should be taken to follow the recommended operating procedure, with particular attention to the standing current in the detector.

Spectroscopic methods

Spectroscopic techniques whether in emission or absorption offer a means for the direct and continuous detection of trace gases in the gas phase. They have been used for many years to identify and measure pollutant gases both in the laboratory and in the atmosphere. A brief introduction to the use of spectroscopic techniques as applied to gaseous pollutants is given in Chapter 6. Historically absorption spectroscopy has been of considerable significance in the identification and measurement of secondary gaseous pollutants. The presence of elevated O_3 levels in Los Angeles smog was convincingly demonstrated in the mid-1950s using both u.v. [23] and long-path i.r. absorption spectroscopy [24]. The organic peroxyacylnitrates were first discovered at about the same time by use of i.r. absorption spectroscopy for studying the photochemical reactions of hydrocarbons and nitrogen oxides in laboratory smog chambers [25]. The presence of peroxacetylnitrate in the Los Angeles atmosphere was subsequently verified by means of long-path i.r. absorption spectroscopy [24].

Since these early experiments, considerable advances have been made in the development of spectroscopic techniques to provide the sensitivity, specificity and practicability necessary for measuring air pollutants. Two basic innovations have enabled those advances to be made. Firstly, the advent of lasers has removed some of the limitations in terms of energy, spectral purity and coherency of classical radiation sources. A laser can serve as a light source in a non-dispersive analyser if one of the laser emission lines falls on one of the absorption lines of the pollutant to be measured. Measurements can be made either in absorption or emission. Current research is focused on improving sensitivity and selectivity of laser systems for the measurement of pollutants both in discrete gas samples and by remote sensing. The second innovation is the application of computer techniques for processing complex spectral data to extract the required information. This has enabled the development of a particularly promising technique for the detection of pollutants in the i.r., i.e. Fourier transform spectroscopy with a Michelson interferometer. This technique has been applied both to remote sensing of trace gases in the upper atmosphere [26] and also to measurement of pollutant concentrations in the surface atmosphere [27, 28]. Progress in the field of advanced spectroscopic techniques has been summarized in several review articles [29–31].

At the present time, these advanced spectroscopic techniques can only be regarded as research instruments; the cost and expertise required for their operation precludes their use for routine monitoring. Furthermore, development has been largely focused toward measurement of the major primary pollutants i.e. NO, NO_2, SO_2, CO, hydrocarbons, etc. and it is for these gases that routine application of new optical techniques is envisaged in the near future. O_3 is the only secondary gaseous pollutant for which measurements on a routine basis using spectroscopic techniques are now feasible. However, detection of the less familiar secondary pollutants such as formaldehyde, H_2O_2, formic acid and HNO_3 using new spectroscopic techniques has already been reported [28] and their use in atmospheric research is likely to become more widespread.

7.3   Experimental section

7.3.1   Analysis of 'total oxidants'

Since oxidants are broadly defined as those compounds present in the air which will oxidize a chemical reagent which is not oxidized by molecular oxygen, 'total oxidants' can only be determined by chemical methods. The degree of oxidation of a chosen reference compound is measured, either by colorimetric or coulometric analysis. The results are usually expressed in terms of O_3 which is normally the most abundant atmospheric oxidant.

7.3.1.1   *Discussion of analytical methods*

Methods based on I_2

The most widely used reference compound for total oxidant determination is KI. The I_2 liberated, in either a neutral or alkaline solution of KI, can be measured colorimetrically from the optical extinction of the I_3^- complex at 352 nm [32, 33]. The alkaline KI method has the advantage that delay is permissible between sampling and analysis. However, the neutral KI procedure has greater simplicity, accuracy and precision and is, therefore, the preferred method.

For many years the stoichiometry (I_2/O_3) of the neutral buffered KI method was believed to be 1:1. When independent means of ozone analysis became available, it was found that the stoichiometry is variable, sometimes being as high as 1.5:1. Typically the neutral phosphate buffered KI method gives results which are 10–25% too high and dependent upon relative humidity [34]. Flamm has developed an alternative reagent consisting of 1% KI/0.1 M H_3BO_4 which gives consistent agreement with u.v. photometry in the range 0.1–3.5 ppm ozone [35].

Possible oxidizing pollutants which comprise 'total oxidant' are O_3, H_2O_2, organic hydroperoxides and peroxides, peracids, peroxyacylnitrates, NO_2 and Cl_2. Cohen *et al.* [36] have investigated the response of various reagents, including KI, to some of these compounds. Neutral KI shows immediate response to O_3 and peracids but a slow response to other peroxides. The response of NO_2 is rapid but it is only approximately 10% of O_3 on a molar basis (1 ppm

$NO_2 \equiv 0.10$ ppm O_3 in terms of I_2 liberated). An immediate response to H_2O_2 together with O_3 and peracids is obtained when ammonium molybdate is added as catalyst to the neutral KI reagent. (1.0 ml of a 10^{-3} M ammonium molybdate solution is added to KI absorbing reagent after sampling.)

Coulometric measurement of the I_2 released from KI or mixed KI–KBr neutral solutions has also been widely used for measurement of total oxidants, e.g. Mast oxidant meter. These instruments only respond to the oxidants which give rapid liberation of I_2. In the Mast instrument, the current required to reduce the liberated I_2 is assumed to be a linear function of the oxidant concentration and the sample flow rate. Several groups [11, 36, 37] have made careful evaluations of the response of Mast instruments to O_3 under laboratory conditions and in the field. In all cases, the response was less than that obtained using the colorimetric neutral KI method as a standard, and in some cases deviations of up to 50% were observed. Clearly results obtained with this type of instrument can only be semi-quantitative and caution should be exercised in comparing total oxidant or O_3 concentrations by this and other methods.

Reducing gases such as SO_2 and H_2S all give serious negative interference with oxidant measurements using KI (probably on a mol–mol basis). The procedures are also sensitive to reducing dusts which may be present in the air or on glassware. Elimination of the interference from SO_2 has been accomplished with a sample pre-filter containing glass fibre impregnated with chromium trioxide [38].

Other methods

An alternative method for manual total oxidant determination, developed by Cohen *et al.* [36] is based on the oxidation of ferrous ammonium sulphate followed by colorimetric measurement of the ferric ion by the addition of ammonium thiocyanate to form the highly coloured $Fe(CN)_6^{3-}$ ion. This method gives instant response for O_3 and most peroxy compounds and is more sensitive than the methods based on I_2 liberation. The high sensitivity and low selectivity combined with good reagent stability make this an ideal method for the determination of total oxidants. However, since most of the measurements of oxidants have been made using KI as the reference compound, the latter should be used if comparison with previous data is contemplated.

The most sensitive method for determining atmospheric oxidants is the NO_2 equivalent method, devised by Saltzman and Gilbert [33]. This method responds primarily to O_3 and involves the addition of a dilute mixture of NO in N_2 to the sample air flow. Any O_3 present reacts rapidly in the gas-phase with NO, converting it to NO_2, which is subsequently trapped and measured colorimetrically in a modified Griess reagent [39]. A second sample is taken to determine concentration of NO_2 initially present in the air (i.e. without NO addition) and the O_3 determined by difference. Difficulties may be encountered with the NO_2 equivalent method in environments where the NO_2 concentrations are greater than that of O_3; small changes in NO_2 levels could be misinterpreted as O_3 responses. Also peroxyacylnitrates give a positive interference to the colorimetry.

7.3.1.2   Neutral KI method for manual analyses of 'total oxidants'

Principle

This method [40] is intended for the manual determination of O_3 and other oxidants in the range 0.01 to 10 ppm. O_3, Cl_2, H_2O_2 and organic peroxides, when absorbed in a neutral buffered (pH $= 6.8 \pm 0.2$) solution of KI, liberate I_2, which is measured spectrophotometrically by determination of the absorption of the tri-iodide ion at 352 nm. It must be borne in mind that the stoichiometry of the method is open to question (see above) and its use can be justified only in terms of the production of data comparable with earlier measurements. The borate-buffered KI method [35] overcomes many of the problems and may be a more suitable choice.

SO$_2$ and sulphides produce a 100% negative interference on a molar basis. Up to 100-fold ratio of SO_2 to oxidant may be eliminated without loss of oxidant by incorporating a chromic acid paper absorber in the sampling stream. The absorber also oxidizes NO to NO_2, however, and NO_2 gives a positive interference equivalent to 10% of O_3 with neutral KI. Therefore, when SO_2 is less than 10% of the NO concentration the use of the chromic acid absorber is not recommended. NO_2 interference can be corrected for by concurrent analysis of the NO_2 concentration and subtracting one-tenth of this concentration from the total oxidant value. Peroxyacyl nitrates give a response equivalent to 50% of an equimolar concentration of O_3.

The precision of the method is approximately $\pm 5\%$ deviation from the mean. The major error is loss of I_2 by volatilization during longer sampling periods; this error can be reduced by the use of a second impinger. The calibration is based on the assumed stoichiometry for the reaction:

$$O_3 + 3KI + H_2O \rightarrow KI_3 + 2KOH + O_2 \tag{7.1}$$

Apparatus

Absorber

All glass midget impingers with a graduation mark at 10 ml are used. Other bubblers with nozzle or open-ended inlet tubes may be employed. Fritted bubblers are not recommended since they produce less I_2. Impingers must be kept clean and dust free. Cleaning should be done with laboratory detergent followed by liberal rinsing with tap and distilled water.

Air metering device

A glass rotameter capable of measuring a flow of 1 to 2 litres min^{-1} with an accuracy of $\pm 2\%$ is required.

Air pump

Any suction pump capable of drawing the sample flow for up to 30 min through a needle valve or critical orifice is suitable. A trap placed downstream from the absorber is recommended to protect the pump, etc. from accidental flooding by the reagent.

Reagents

All reagents are made from analytical grade chemicals and double distilled water. The latter is obtained by distillation in an all glass still with a crystal each of potassium permanganate and barium hydroxide added.

Absorbing reagent

Dissolve 13.61 g of potassium dihydrogen phosphate, 14.20 g of anhydrous disodium hydrogen phosphate (or 35.8 g of the dodecahydrate salt) and 10.00 g of KI successively and dilute to exactly 1 litre with double distilled water. Keep at room temperature for at least 1 day before use. This solution may be stored for several weeks in a glass stoppered brown bottle in a refrigerator. Do not expose to sunlight.

Standard I_2 solution, 0.05 M

Dissolve successively 16.0 g of KI and 3.173 g I_2 in doubly distilled water and make up to 500 ml. Keep for at least 1 day at room temperature before using. Standardization is unnecessary if the weighing is carefully done although, if desired, the solution may be standardized by titration with sodium thiosulphate solution, using starch indicator.

SO_2 absorber

Flash-fired glass fibre filter paper is impregnated with chromium trioxide as follows: drop 15 ml of aqueous solution containing 2.5 g chromium trioxide and 0.7 ml conc. H_2SO_4 uniformly over 400 cm^3 of paper and dry in an oven at 80 to 90° C for 1 h; store in a tightly capped jar. Half of this paper serves to pack one absorber. Cut the paper into 6 × 12 mm strips each folded into a V shape. Pack into an 8.5 ml U tube and condition by drawing dry air through overnight. The absorbent has a long life (at least 1 month). If it becomes visibly wet from sampling humid air, it must be dried with dry air before further use.

Procedure

Assemble the sampling train consisting, in order, of the SO_2 absorber (optional), impinger, rotameter and air pump. The sample probe should preferably be of PTFE but glass or stainless steel may be used for short probes. PVC should be avoided except for butt jointing of glass tubing.

Pipette exactly 10 ml of absorbing reagent into the impinger and sample at a flow rate of 1 to 2 litres min^{-1} for up to 30 min. Sufficient air should be sampled to collect the equivalent of 0.5 to 10 μl O_3 in the absorber. Measure the volume of air sampled and correct to 760 Torr and 25° C. Do not expose the reagent to direct sunlight.

For analysis, add distilled water to the impinger to make up to the 10 ml graduation mark (i.e. if evaporation losses have occurred). Within 30 to 60 min of sampling, transfer a portion of the exposed reagent directly to a curvette and measure the absorbance at 352 nm against a reference of double distilled water. Samples having a colour too dark to read may be quantitatively diluted with absorbing reagent. Also measure the absorbance of the unexposed reagent against the reference and subtract the blank from the sample absorbance.

Calibration and calculation

Firstly, prepare 0.001 M I_2 standard by pipetting exactly 4 ml of the 0.025 M standard I_2 solution into a 100 ml volumetric flask and diluting to the mark with absorbing reagent. Discard after use. For calibration purposes, exactly 4.09 ml of the 0.001 M I_2 solution is diluted with absorbing reagent just before use to 100 ml, i.e. making the final I_2 concentration equivalent to 1 μl O_3 ml^{-1} according to the stoichiometry of Equation (7.1).

$$1 \; \mu l \; O_3 \; ml^{-1} \equiv \frac{100}{24.47} \; \mu mol \; I_2 \; in \; 100 \; ml (= 4.09 \times 10^{-6} \; mol)$$

In order to obtain a range of concentration values, add graduated amounts of the above calibrating solution up to 10 ml to a series of 10 ml volumetric flasks and dilute to volume with absorbing reagent. Read the absorbances and plot them against the equivalent concentration of O_3 in μl O_3 (10 ml)$^{-1}$ absorbing reagent. The plot follows Beer's law. Draw the straight line through the origin giving the best fit and read off the total μl O_3 (10 ml)$^{-1}$ for the samples from the calibration graph. The concentration of O_3 in the gas phase in μl litre^{-1} or ppm is given by

$$O_3 \; (ppm) = \frac{Total \; \mu l \; O_3 \; per \; 10 \; ml}{Total \; volume \; of \; air \; sampled \; in \; litres}$$

Effects of storage

O_3 liberates 90% of the iodine from the buffered reagent immediately and the remaining 10% through a slow set of reactions. Some of the other oxidants cause a slow formation of iodine. Some indication of the presence of such oxidants and of gradual fading due to reducing agents can be made by making several measurements over a period of time.

7.3.1.3 *Instruments for measurements of total oxidants*

A number of commercial instruments for the measurement of total oxidants, using both colorimetric and coulometric principles, are available. All use either aqueous iodide or mixtures of iodide and bromide as absorbing medium for the oxidizing substances present in the air sampled.

Automatic colorimetric instruments are based on the continuous spectrophotometric measurement at 352 nm of the I_2 liberated in a flowing reagent stream. The air sample stream is exposed to the reagent in a wetted-wall absorber. Fig. 7.2 shows a schematic diagram of a typical system. The reagent reservoir contains approximately 4 litres reagent (neutral buffered 10% KI) which is forced

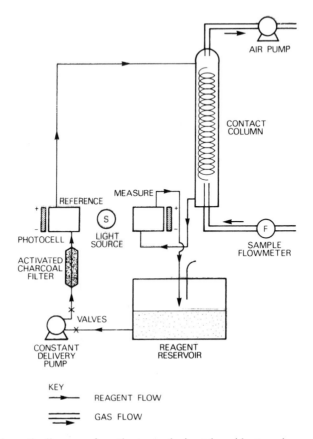

Fig. 7.2 Schematic diagram of continuous colorimetric oxidant analyser.

by a constant delivery pump at approximately 4 ml min^{-1} through a charcoal column (to remove I_2 from the solution) and a reference colorimetric cell, to the top of the contact column. Sample air is drawn countercurrently through the contact column at approximately 4 litres min^{-1}. The exposed reagent then flows by gravity through the optical cell where the absorbance is continuously measured. The flow rates of both the air sample and the liquid reagent must be carefully controlled for accurate recording of oxidant concentration. Calibration of the instrument response is carried out by means of a constant known O_3 source (see Section 7.3.2). The method is subject to the same interferences as discussed for the manual colorimetric determination of oxidants (Section 7.3.1), although on some instruments pretreatment of the sample is used to eliminate some of the interfering substances.

The most widely known coulometric instrument for the measurement of oxidants is the Mast 'O_3 meter'. Other instruments operating either on the amperometric or galvanic principle (see Section 7.2) are available. Many of these incorporate sample pretreatment to remove reducing substances and oxidants other than O_3 and are marketed as 'O_3' meters.

7.3.2 Analysis of O_3

O_3 is the most important and abundant 'oxidant' present in the atmosphere and a considerable amount of effort has been devoted to the development of specific analytical methods, both chemical and physical, for the measurement of this gas. Although O_3 is a highly reactive gas and will interact with a variety of chemical reagents, no truly specific chemical analytical method has yet been developed for the measurement of atmospheric O_3.

7.3.2.1 *Chemical methods*

Iodide and related methods

The most widely used chemical methods have utilized the liberation by O_3 of I_2 from solutions containing KI. The KI method, however, is not specific for O_3. All oxidizing and reducing agents can potentially interfere, as discussed in Section 7.3.1, and procedures for measurement of O_3 must involve elimination of these possible interferences. A number of sample pretreatment systems have been devised to minimize the effects of the major interferences likely to be encountered in atmospheric measurements, i.e. NO_2, SO_2 and H_2S. Thus the chromic acid scrubber developed by Saltzman and Wartburg [38] (described in Section 7.3.1.2) will eliminate interference of up to a 100-fold ratio of SO_2 to O_3 without loss of O_3.

The liberated I_2 may be measured by titration, by colorimetry or electrochemically using an amperometric cell [9] or a galvanic cell [8]. An interesting electrochemical sensor based on differential galvanic measurement, for atmospheric O_3 measurement which is virtually free of interference, has been described by Lindqvist [41]. Halogen is liberated from an aqueous electrolyte containing NaBr, NaI, Na_2HPO_4 and NaH_2PO_4 at concentrations of 3.0, 0.001, 0.1 and 0.1 M respectively, contained in a recirculating galvanic cell containing two cathodes and one counter electrode (Fig. 7.3). The air sample containing O_3 is divided into two equal streams. By selectively removing reducing agents in both sections and O_3 in only one section, the differential galvanic current is directly related to the concentration of O_3. This current is independent of the presence of other oxidants provided they are not removed by the O_3 scrubber and the O_3 concentration can be calculated directly by Faraday's law, precluding the necessity of calibrating the sensor. No interference of NO_2, Cl_2, SO_2, H_2S, NH_3, HCl, C_2H_4 and 1-C_4H_8 could be observed at concentrations up to 1 ppm. A 12% interference by peroxyacetylnitrate, in mol equivalents of O_3, was observed in the laboratory but this would have little significance in practice. Under laboratory conditions O_3 measurements agreed well with those using the neutral KI method.

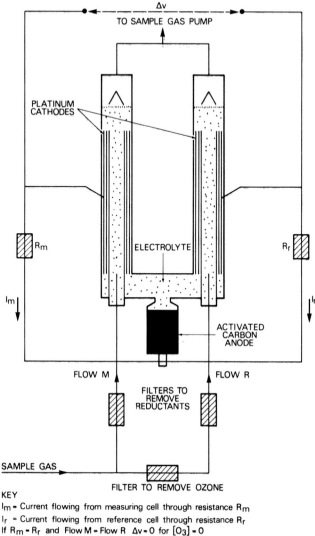

Fig. 7.3 Differential galvanic detection of O_3 in air.

In operation the instrument showed good stability and a reasonable time constant of about 40 s.

Both instrumental and manual methods based upon oxidation of KI are of variable stoichiometry and cannot be recommended for analysis of ozone in ambient air, or calibration of ozone sources. Only the borate-buffered method [35] appears to offer any real potential.

Other methods

Several colorimetric procedures, which are claimed to be specific for the determination of O_3, have been reported. The measurements have been based either on the bleaching action (i.e. decrease in absorbance) following the reaction of O_3 with a colour reagent or on the formation of ozonolysis products which undergo sensitive colour reactions.

Use of the bleaching action of O_3 on a buffered solution of indigo sulphuric acid in water was first described by Dorta-Shaeppi and Treadwell [42]. Guicherit *et al.* [37] recently evaluated a slightly modified indigo H_2SO_4 method and found a good correlation between O_3 concentrations determined by this method and those determined using the differential galvanic detector of Lindqvist [41]. At pH 6.85 the reaction proceeds stoichiometrically and the decrease in extinction follows Beer's law over a concentration 0.05 to 1.5 μg O_3 ml^{-1} solution. The method is simple, rapid, inexpensive and sensitive (5 ppb O_3 can be measured in a 60 litre air sample). Of the components which are most frequently present in ambient air, only NO_2 interferes. This interference, however, is only 4% in O_3 equivalents and will not normally give rise to serious error.

A basically similar method for O_3 determination based on bleaching of Diacetyl-dihydro-lutidine (DDL) has been described by Nash [43]. This method apparently has minimal interference from O_2 and oxidants other than O_3, SO_2 and oxides of nitrogen. However, the DDL reagent is less stable and the collection efficiency lower than the indigo H_2SO_4 reagent.

Specific methods for O_3 determination based on spectrophotometric determination of ozonolysis products have been described by Bravo and Lodge [44] and Hauser and Bradley [45]. Both methods are slow, complicated and use corrosive non-aqueous collection media. The latter method, which involves spectrophotometric determination (with 3-methyl-2-benzothiazolone hydrazone hydrochloride) of the pyridine-4 aldehyde formed when O_3 reacts with 1,2 di-(4-pyridyl) ethylene (DPE), has non-unit stoichiometry (reported values vary from 1.24 to 1.33). A field comparison study in which the DPE method was tested against an O_3-selective galvanic detector, showed deviations of more than a factor of 2 [37].

7.3.2.2 *Physical methods*

Chemiluminescence methods

In recent years, chemiluminescence detection has been widely adopted as a technique for the measurement of atmospheric O_3. The first chemiluminescence O_3 detectors were developed by Regener and co-workers for use on balloon-borne O_3 sondes [46]. These instruments utilized the chemiluminescent reaction of O_3 with an organic dye, rhodamine B. The solid dye was adsorbed on a silica-gel impregnated surface which was positioned close to the window of a photomultiplier tube. Sample gas is aspirated over the active surface and any O_3 present in the sample reacts with the dye with the emission of light, which is detected by the photomultiplier. This technique is extremely sensitive and highly selective for O_3 (no other components of the atmosphere have been observed to give chemiluminescence with the active surface). The main problems arise from instability and irreproducibility of the chemiluminescent surface, non-linearity of the response and effects of relative humidity changes in the sample gas.

A good deal of development work has been carried out with a view to utilizing the rhodamine B system for continuous automatic detection of O_3. Hodgeson *et al.* [47] eliminated the desensitizing effect of water vapour on the chemilumines-cent surface by treatment of the silica-gel with a hydrophobic agent, silicone resin, prior to adsorbing the dye. The response was then independent of relative humidities between 0 and 80% r.h. The lifetime of the reactive surface was limited by the slow oxidation of rhodamine B by O_3. Since sensitivity could be sacrificed in the system, the analyser was run over an extended period with a 1:6 dilution of the incoming sample with dry air. In this way the useful lifetime of a surface would be increased to 2 to 3 months. However, frequent calibration using a known O_3 source is necessary to allow for the small daily decrease in sensitivity.

Guicherit [48] increased the stability of the chemiluminescent surface by protecting the rhodamine B from oxidation, with another compound, which reacts with O_3 more easily, i.e. gallic acid. Following the reaction with O_3, the gallic acid or one of its oxidation products transfers energy to the dye, giving chemiluminescence. A stable surface was produced with no serious loss in sensitivity. The stability was further increased using an intermittent sampling procedure in which sample air was admitted for 15 s followed by the introduction of dry O_3 free air for about 45 s. The dry air admission eliminated moisture effects and also gave a 'zero-point' reading. Monitoring outdoor air over a period of 52 days showed no decay in sensitivity of the surface, but after this time the response fell off rapidly. The minimum detectable concentration of O_3 using the dye method is <1 ppb which is more than adequate for monitoring ambient levels.

The gas-phase chemiluminescent reaction between O_3 and ethylene (C_2H_4), which was originally utilized by Nederbragt *et al.* [49] for the detection of O_3 in air, is now widely used for ambient O_3 measurement. A large number of commercial instruments based on this method are now available. Air containing O_3 is mixed with a slow flow of C_2H_4 in an injector positioned close to the end window of a photomultiplier, and the low-level light emission resulting from the O_3–C_2H_4 reaction observed. Kummer *et al.* [50] have investigated the nature of the light-emitting products of this reaction. Emission from excited formaldehyde in the 350–550 nm region and also from vibrationally excited hydroxyl radicals has been detected. Other aliphatic olefins also give chemiluminescence with O_3 and with a much higher emission intensity, at least at low pressures. However, an advantage of the O_3–C_2H_4 system is that sufficient light emission for measurement is obtained at pressures near 1 atm, and therefore elaborate pumping and pressure control facilities are not required. The photocurrent is a linear function of the O_3 concentration in the sample gas and the detection limit is of the order of 1 ppb of O_3. As with all chemiluminescence methods, the electrical response must be calibrated with a known O_3 source.

Other chemiluminescence methods have been used for O_3 analysis. For example, the chemiluminescent reaction of NO with O_3, which is commonly used for the analysis of NO (see Section 6.2.3) can be used for analysis of O_3 if NO is used as the reactant gas instead of O_3. Finlayson *et al.* [51] have described a chemiluminescent reaction between O_3 and triethylamine. The latter compound also reacts with peroxyacetylnitrate to give luminescence but at a different wavelength to the emission following reaction with O_3.

Ultraviolet absorption

The O_3 absorption of the 253.7 nm mercury resonance line was the first spectrometric concentrations approach employed for monitoring pollutant concentration in the lower atmosphere. However, early instruments had insufficient sensitivity and stability for routine measurements of ambient concentrations of O_3. Sensitivity and stability has been greatly improved in a commercial O_3-photometer offered by DASIBI Corporation of Glendale, California, and more recently by other manufacturers. In this instrument the absorption by O_3 of the 253.7 nm line is measured in a 70 cm folded path cell which is alternatively filled with unfiltered and filtered (O_3 free) ambient air. A schematic diagram of the system is shown in Fig. 7.4. The light source is a stabilized miniature low pressure mercury arc. The light beam is split, a fraction being directed on to the incident light detector photocell and the remainder passing through the cell to a second detector photocell. The output of the photocells is digitized and stored and the detectors are linked electronically so that the second photocell measures the integrated transmitted light for a fixed amount of incident radiation. This operation is carried out alternatively on the direct and filtered air sample, the latter providing a unique internal reference system, allowing direct measurement of the fraction of the incident light absorbed by O_3. The O_3 concentration can then be determined using the Beer–Lambert relationship for weak absorption:

$$\log \frac{I_0}{I} = \varepsilon_{253.7} \times l \times [O_3]$$

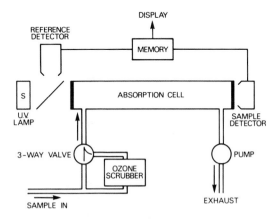

Fig. 7.4 Schematic diagram of O_3 photometer (Dasibi).

The absorption data is processed electronically to provide a direct reading of O_3 concentration in ppb. Minimum detectable concentration is of the order of 1 to 2 ppb.

Two features of the system may be disadvantageous under some circumstances. Firstly, the sequential mode of operation and signal integration system give the instrument a relatively long time-constant for measuring a change in O_3 concentrations. In practice a digital output, updated every 10 or 22 s, is obtained, giving a 90% rise or fall time of 30 s. While this is adequate for most ambient monitoring applications, it is not satisfactory for many laboratory applications. Furthermore, the sample flow requirement (1 to 4 litres min^{-1}) is quite large, which again may be a disadvantage in laboratory work. A more serious disadvantage is the potential interferences from some organic compounds e.g. carbonyl and aromatic compounds which also absorb in the u.v. region employed. Mercury vapour also interferes by virtue of its extremely strong absorption of the resonance 253.7 nm line. These interferences are, however, unlikely to be troublesome when making measurements in outdoor air. Since the ozone scrubber does not affect these potential interferents, their effect on the signal is only transient, lasting a few measurement cycles [52].

Infrared absorption

Infrared absorption spectrometry has been used for the measurement of O_3 under laboratory conditions, e.g. in smog-chambers. Long path lengths have to be employed since the extinction coefficient of O_3 in the i.r. ($\varepsilon = 3.80 \times 10^{-4}$ ppm^{-1} m^{-1} at 1054 cm^{-1}) is lower than that in the u.v. ($\varepsilon = 1.3 \times 10^{-2}$ ppm^{-1} m^{-1} at 254 nm). The advent of i.r. laser sources offers possibilities for remote sensing of O_3 in ambient air by i.r. absorption spectrometry. Hanst [29] discusses the possibility of using a CO_2 laser, wavelength shifted to 9500 nm with propane, for O_3 measurement. An estimated detection limit of 0.05 ppm in a 1 km path is given.

7.3.2.3 Measurement of O_3 by the C_2H_4-chemiluminescence method

Principle

O_3 reacts rapidly in the gas phase with C_2H_4 with an accompanying chemiluminescence emission in the 350 to 600 nm wavelength region. By monitoring this light emission with a sensitive photomultiplier, when air containing O_3 is mixed with C_2H_4 in a flow cell at atmospheric pressure, a signal is obtained which is proportional to the O_3 concentration.

The method is suitable for the measurement of O_3 concentrations in the range 0.001 to 100 ppm and the response is linear in this range provided the sample and C_2H_4 flow rates are maintained constant. At the lower end of the concentration range the response is limited by the thermal noise and drift of the photomultiplier tube. The latter may be minimized by thermoelectric cooling of the photomultiplier.

There are no known interferences with this O_3 method. An interfering substance would either have to produce chemiluminescence by reaction with C_2H_4 at atmospheric pressure or interfere with the reactions giving rise to the chemiluminescence from O_3 reaction. No known compounds present in atmospheric air have been demonstrated to produce these effects.

The precision of the method relies on maintaining constant gas flow rates and suppression and compensation for the 'dark' current of the photomultiplier tube. A precision of $\pm 2\%$ at a 50 ppb O_3 concentration can be readily achieved. The accuracy is dependent on the validity of the calibration which is carried out using a source of known O_3 concentration. Standard O_3 sources should be subject to independent calibration by gas phase titration or u.v. photometry. A relative simple laboratory-based u.v. photometer has been described by de More and Patapoff [53].

Apparatus

The apparatus consists of a reaction cell-photomultiplier detector assembly, a high voltage supply, amplifier and recorder and ancillary equipment for providing controlled flows of sample gas and C_2H_4.

Detector assembly: The arrangement for a suitable detector assembly is illustrated in Fig. 7.5. The body of the reaction cell may be either of glass (as in the design of Warren and Babcock [54] or of aluminium, with the quartz end window sealed on [either with cement or an 'O' ring] to give a gas tight fit. The Pyrex-glass injector is sealed into the cell so that the tip is between 5 and 10 mm from the end window. It is necessary to exclude all stray light from the detector and this is most conveniently achieved by adapting the reaction cell assembly to fit on to a standard photomultiplier housing. Alternatively, the whole assembly

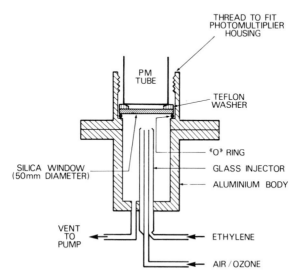

Fig. 7.5 Detector assembly for chemiluminescence O_3 analyser.

may be enclosed in a light-tight box. The inlet and outlet gas lines should be covered with black sleeving to prevent light entry.

A suitable photomultiplier is EMI 9635 QD (low dark current 13 dynode tube with quartz window and S2 cathode). This tube has a dark current of approximately 0.5 nA at 1000 volts (20° C) and gives a response of 50 nA (ppm O_3^{-1}). The photocurrent can be amplified with a conventional d.c. amplifier with an input range of 10^{-9} to 10^{-5} A. The high voltage supply unit should be capable of giving a constant voltage of up to -1000 V d.c. The output can be displayed on a recorder.

The ancillary equipment consists of:

(a) a PTFE sample probe with a filter to prevent entry of extraneous dust
(b) an air pump and metering system capable of drawing a constant flow of 1.0 ± 0.02 litre min^{-1} through the system and
(c) a C_2H_4 metering system consisting of a capillary orifice flow meter and rotameter capable of maintaining a flow of C_2H_4 of approximately 10 ml min^{-1} constant to $\pm 2\%$. A good quality pressure reducing valve fitted to the C_2H_4 supply cylinder can be used to maintain the required pressure at the capillary orifice. It may be advisable to incorporate a dust filter in the C_2H_4 line upstream from the capillary. Commercial purity grade C_2H_4 can be used. The flow system is shown diagrammatically in Fig. 7.6.

Operation

Before the instrument can be used for measurement of O_3 concentration it is necessary to establish and compensate for the photomultiplier 'dark' current and also to calibrate the electrical response as a function of O_3 concentration.

The 'dark' current is measured by either supplying the instrument with O_3 free air or by operating with the pump off so that no air enters the reaction cell. It is

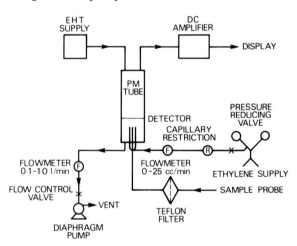

Fig. 7.6   Flow diagram for chemiluminescence O_3 analyser.

inadvisable to check the 'dark' current by stopping the C_2H_4 flow as a low level irreproducible response can be observed when O_3/air mixtures are present in the reaction cell even if C_2H_4 is nominally absent. A suitable EHT of, say, 800 V, is chosen and the 'dark' current measured after the photomultiplier has reached temperature equilibrium. The instrument is then supplied with air containing a constant known O_3 concentration of approximately 0.2 to 1.0 ppm, supplied from a suitable O_3 generator (see below) and the signal allowed to stabilize. The response may then be adjusted up or down to give the desired sensitivity by adjusting the EHT. The dark current must be remeasured after adjusting the EHT. The calibration factor F is then given by

$$F = \frac{I - I^0}{C_{O_3}} \text{ nA ppm}^{-1}$$

where I^0, I are respectively the dark current (or equivalent recorder deflection) and the total observed photocurrent (or recorder deflection) with the known concentration of O_3 (C_{O_3}) entering the instrument.

If the instrument is equipped with a thermostatted photomultiplier, the dark current may be backed off electrically to give a zero recorder reading when no O_3 is present. If no temperature control is used, it will be necessary to check the dark current periodically to compensate for zero drift due to temperature fluctuations. For continuous monitoring applications, this may be achieved conveniently by incorporating a programmed solenoid valve which vents the pumping line for a short period (e.g. 2 min) at intervals (e.g. every 30 min) thereby stopping the flow of O_3/air into the reaction cell. If it is proposed to operate in an environment with large ambient temperature variations $> 10°$ C, a thermostatted, cooled detector assembly is desirable.

When the instrument is calibrated initially, the linearity should be checked by measuring the response at different O_3 concentrations within the operating range. For atmospheric measurements, response over the range 0.02 to 1.0 ppm should be checked; at the lowest concentrations some O_3 losses in the sample line may occur giving rise to a fall-off in response and/or longer periods to reach a steady reading. This is often most noticeable when a dust filter is incorporated in the sample line. It is advisable to condition the system with higher O_3 concentrations for a few hours before calibration to minimize any effects of this kind. After this initial check for linearity, subsequent routine calibration need only be carried out at a single O_3 concentration. For accurate monitoring, a weekly check on the response should be carried out.

C_2H_4 chemiluminescence O_3 detectors have been operated continuously for long periods. When operating in urban locations, the filter should be replaced fairly frequently and it may be necessary to clean the reaction cell periodically to remove any accumulated deposits.

7.3.2.4 *Preparation of O_3/air mixtures for calibration purposes*

Preparation of mixtures containing ppm concentrations of O_3 in air is conveniently achieved by exposing air or O_2 to a mercury vapour lamp emitting short wavelength (185 nm) radiation. O_3 formation results from the photodissociation of oxygen: $O_2 + h\nu\, (\lambda = 185\ nm) \rightarrow 20;\ O + O_2 \rightarrow O_3$. Two suitable types of light sources are available commercially. The filament type operates on a low d.c. voltage (e.g. Philips OZ4) and the low pressure arc type (e.g. 'Pen ray' lamp, Ultraviolet Products, Inc.) is operated on a.c. supplied from a high voltage leak transformer.

Figure 7.7 shows a schematic diagram of a simple form of O_3 generator suitable for producing constant O_3 concentrations in the range 0.1 to 1.0 ppm in an air stream. Pure, dry air is passed continuously, at a constant flow rate, through a quartz tube which is positioned alongside a miniature mercury arc lamp. The ozonized air passes into a sampling manifold. Tubes downstream from the generator should preferably be of Teflon but clean borosilicate glass may be used. The O_3 concentration can be varied by adjusting either the lamp current, the air flow rate or the length of the arc tube exposed. A sliding shield may be used to alter the length of the arc tube.

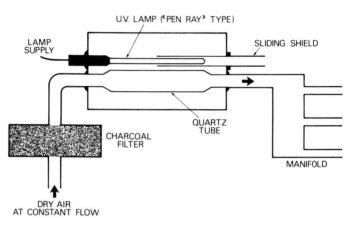

Fig. 7.7 O_3 generator.

The O_3 concentration from the generator can be determined using u.v. photometry, or possibly the borate-buffered KI method [55]. After allowing the generator to stabilize by operating under constant conditions for several hours samples are taken from the manifold. The initial calibration should be carried out at several different concentrations, so that the characteristics of the generator can be established. Provided operating conditions i.e. temperature, flow rate, supply voltage are carefully controlled, reproducible O_3 concentrations can be obtained with this type of generator. In fact, it has been shown that an O_3 source of this type, once initially calibrated, provides a more reliable routine calibration for O_3 detectors than manual iodometric analysis [56].

As an alternative to the above dynamic method, a static system may be used to obtain O_3/air mixtures for calibration. In this case the O_3/air mixture from the generator is collected in large bags fabricated of inert plastic film (e.g. Teflon or

Tedlar) where it can be diluted with clean air if required. It may be necessary to precondition the bag by exposure to higher O_3 concentrations to prevent surface destruction of the O_3. The O_3 concentration in the bag should be determined by an independent method prior to use for calibration purposes. The general procedures for the preparation of gas mixtures in static systems are described in Section 6.3.8.1.

7.3.3 Analysis of H_2O_2

Although it has been suggested that H_2O_2 may constitute a significant fraction of the oxidants present in polluted air very little progress has been made in the development of techniques for the measurements of air concentrations of H_2O_2. Gaseous H_2O_2 is well known for its instability, particularly in the presence of metal surfaces. This, combined with the high affinity for H_2O of H_2O_2, makes quantitative sampling by conventional techniques very difficult.

7.3.3.1 *Chemical methods*

H_2O_2 is an oxidant and can be measured by the neutral KI technique (Section 7.3.1) if ammonium molybdate is added as a catalyst to assist rapid colour development [36]. Methods specific for H_2O_2 and organic peroxides which yield H_2O_2 on acid hydrolysis, based on the formation of a titanium (4^+) peroxysulphate complex have been reported [36, 57]. Sensitivity is hardly adequate for the measurement of H_2O_2 in polluted air, however. (The molar absorptivity of the Ti (4^+)–peroxide complex at 407 nm is only 777 compared to 2.4×10^4 for the I_3^- complex formed when O_3 reacts with neutral KI [36]. Cohen and Purcell [58] have reported another method for the determination of microgram quantities of H_2O_2. The analytical procedure uses the coloured complex obtained upon extraction of titanium–H_2O_2 mixtures with 8-quinolinol in chloroform. Air containing H_2O_2 is sampled through a fritted bubbler containing 10 ml of a solution of titanous sulphate. The absorbing solution is then extracted with an 0.1% solution of 8-quinolinol in chloroform and the absorbance of the complex measured at 450 nm. The method is claimed to be very specific and to have a sensitivity at least 4 times greater than that of the titanous sulphate method. (Molar absorptivity $\simeq 3 \times 10^3$ at 450 nm.) Neither method appears to have been widely utilized, although Bufalini *et al.* [59] report measurement of H_2O_2 concentrations of up to 150 ppb in polluted air, using the Ti (4^+)-8-quinolinol method.

Kok and co-workers [60] developed a chemiluminescent procedure based upon reaction with luminol in the presence of a Cu(II) catalyst. A detection limit of 0.4 ppb was reported for a 15–20 min air sample. In a comparative study [61], three methods gave rather poor agreement but all showed concentrations of the order of 10–30 ppb in the Southern Californian atmosphere.

These methods must be regarded as unproven as yet, and in need of considerable further development.

7.3.3.2 *Physical methods*

The only physical method which has been applied to the measurement of H_2O_2 in air is i.r. absorption spectroscopy. Hanst *et al.* [28] have used a 417 m path length Fourier transform i.r. instrument to determine H_2O_2 by its absorption at 1250 cm^{-1} ($\varepsilon = 9 \pm 3$ atm^{-1} cm^{-1}). In atmospheric spectra, interference from water vapour and methane resulted in a rather high detection limit of the order of 100 ppb. In polluted air in Pasadena, California, no H_2O_2 could be detected at this concentration but by comparison of spectra recorded at different times during the day, evidence was found on one occasion for the accumulation of approximately 70 ppb H_2O_2 during the development of smog. This method seems to offer the best prospects at the present time for selective measurement of this difficult gas in polluted air.

7.3.4 Analysis of aliphatic aldehydes and oxygenated compounds

7.3.4.1 *Chemical methods*

Chemical methods for the analysis of aldehydes are based on the chemical reactivity of the carbonyl group. The results from analyses by these methods are often reported as 'total aldehydes' with the response to ketones ignored. Formation of the bisulphite addition complex, followed by titration of the 'trapped' bisulphite, has been widely used for industrial hygiene applications and source measurements, but the method is not sensitive enough for ambient air analysis. Aqueous sodium bisulphite is a useful trapping reagent for collecting low molecular weight aldehydes for measurement using more sensitive colorimetric techniques. However, trapping efficiency for higher molecular weight aldehydes and ketones is less satisfactory.

A useful method for the determination of total aliphatic aldehydes in air was first described by Sawicki *et al.* [62]. The aldehydes are determined colorimetrically by their reaction with 3-methyl-2-benzothiazolone hydrazone hydrochloride (MBTH) in the presence of ferric chloride to form a blue cationic dye in acid media. Air is sampled directly into the MBTH reagent and the sensitivity is sufficient for measurement of ppb concentrations of aldehydes in a 100 litre air sample. Because the contribution to the measured extinction, and also the sampling efficiency for the higher aldehydes is less than for formaldehyde (HCHO), the overall molar response depends on the relative proportions of the individual aldehydes present. For normal outdoor air samples, in which 60 to 80% of the aldehydes are HCHO, Altshuller and Leng [63] have suggested that the measured concentration calculated as HCHO should be multiplied by 1.125 ± 0.10 to obtain the real concentration of total aliphatic aldehydes in outdoor air. This is based on the molar absorptivities of 65 000, 50 000 and 23 000 for HCHO, straight chain aliphatic aldehydes and branched or unsaturated aldehydes respectively [62].

Specific colorimetric methods have been developed for HCHO and acrolein. Procedures involving the use of chromotropic acid for the analysis of HCHO have been widely investigated [64–67]. There is no significant interference from higher aliphatic aldehydes, ketones, alcohols, etc., and the interference from olefins and aromatic hydrocarbons can be largely eliminated by the use of appropriate sampling conditions. Thus, for higher concentrations of HCHO as found for example in combustion effluent, etc., collection in aqueous bisulphite is recommended. For measurements in the ambient atmosphere, where HCHO concentrations are unlikely to exceed a few tenths of a part per million, sampling directly into chromotropic acid/H_2SO_4 reagent is most advantageous. The sensitivity is sufficient for the detection of 10 ppb HCHO in a 100 litre air sample, sampled directly into the reagent. A slight interference from aromatic hydrocarbons and olefins may occur if the concentration of these compounds exceeds that of HCHO by a factor of 5 to 10.

Acrolein can be measured colorimetrically using 4-hexylresorcinol [68]. The reaction between these two compounds in an ethyl alcohol–trichloro-acetic acid solvent medium in the presence of mercuric chloride results in a blue coloured product with a strong absorption maximum at 605 nm. The sensitivity is adequate for the determination of ppb concentrations of acrolein in a 60 litre air sample, with no significant interference from common inorganic and organic pollutants. Air is sampled directly into the mixed sampling reagent contained in 2 fritted bubblers in series and the colour developed by heating to 60° C for 15 min.

7.3.4.2 *A colorimetric analysis of total aliphatic aldehydes in air (MBTH method) [62, 69]*

Principle

The aldehydes in ambient air are collected in an 0.05% aqueous solution of 3-methyl-2-benzothiazolone hydrazone hydrochloride (MBTH). The resulting azine is then oxidized by ferric chloride–sulphamic acid solution to form a blue cationic dye in acid solution which can be measured at 628 nm. The concentration of total aldehydes is calculated in terms of HCHO. Normally between 60 to 80% of the aldehydes occurring in outdoor air are HCHO and acrolein. Other aldehydes which may be present are higher aliphatic aldehydes and, to a lesser extent, aromatic aldehydes. Because the contribution to the measured extinction and also the sampling efficiency for the higher aldehydes is less than that for HCHO, the calculated concentration should be multiplied by 1.25 ± 0.10 to obtain a real concentration of total aliphatic aldehydes in ambient air [63].

The method is relatively free from interferences. Thus, none of the variety of organic and inorganic materials present in photochemical smog, generated by laboratory irradiation of diluted vehicle exhaust (e.g. hydrocarbons, ketones, nitrogen oxides, O_3, PAN) gave detectable interference [63].

From 0.03 to 0.7 μg ml^{-1} of HCHO can be measured in the colour developed solution (12 ml). This corresponds to a minimum detectable concentration of 0.03 ppm aldehyde (as HCHO) in a 25 litre air sample. The reproducibility of the method is to within ± 5%.

Apparatus

The samples are collected in all glass bubblers with a coarse fritted inlet. It is normally necessary to sample ambient air for quite long periods, to obtain sufficient aldehyde for analysis and a pump capable of drawing at least 0.5 litre min^{-1} for 24 h is required. A critical orifice, a rotameter or a gas meter can be used to meter the flow.

Reagents

Analytical reagent grade chemicals should be used.

(a) 3-methyl-2-benzothiazolone hydrazone hydrochloride absorbing solution (0.05%): Dissolve 0.5 g MBTH in distilled water and dilute to 1 litre. The reagent may become turbid either in storage or during sampling. If this occurs, filter by gravity. The solution is stable for a week or more if stored in the cold in a dark bottle.

(b) Oxidizing reagent: Dissolve 1.6 g sulphamic acid and 1.0 g ferric chloride in distilled water and dilute to 100 ml.

(c) Formaldehyde stock solution (1 mg ml^{-1}): Dilute 2.7 ml of 37% formalin solution to 1 litre with distilled water. This solution must be standardized by the addition of excess sodium bisulphite to an aliquot of the solution, followed by iodometric titration of the formaldehyde bisulphite addition product after the excess bisulphite has been oxidized by the addition of I_2. Alternatively, sodium formaldehyde bisulphite can be used as a primary standard. Dissolve 4.470 g in distilled water and dilute to 1 litre. Stabilize for at least 3 months.

(d) Dilute standard formaldehyde solution (10 μg ml^{-1}): dilute 1 ml of standard stock solution to 100 ml with distilled water. Prepare a fresh solution daily.

Procedure

Sample a measured volume of ambient air at a rate of 0.5 litre min^{-1} through 35 ml of MBTH solution in the absorber until sufficient aldehyde for analysis has been collected. In clean air a period of up to 24 h may be required. An average collection efficiency of 84% has been determined for sampling under these conditions although higher efficiencies have been reported for slightly modified conditions. Altshuller and Leng [63] obtained efficiencies of between 90 and 95% using a 10 ml of 0.2% aqueous MBTH in a bubbler aspirated at 1 litre min^{-1}.

After sampling, make up the volume of the absorbing solution to exactly 35 ml with distilled water (to compensate for evaporation losses) and allow to stand for 1 h. For analysis, pipette 10 ml of the sample solution to a glass stoppered tube and an equal volume of unexposed reagent to a second tube to serve as a blank. To each, add 2 ml of oxidizing solution and mix well. After allowing to stand for at least 12 min, determine the absorbance of the sample at 628 nm against the reagent blank in 1 cm cells. The aldehyde content (expressed as μg

HCHO ml^{-1}) can be determined from the calibration plot prepared as described below.

Calibration

The dilute standard HCHO solution, freshly prepared, is used to calibrate the method. Pipette 0, 0.5, 1.0, 3.0, 5.0 and 7.0 ml of this solution into a series of 100 ml volumetric flasks and dilute to volume with 0.05% MBTH solution. Allow these solutions to stand for 1 h and then transfer 10 ml aliquots of each solution to a stoppered test tube, add 2 ml of oxidizing solution and mix well. After 12 min determine the absorbance of each solution at 628 nm against the blank in 1 cm cells. Plot the absorbance against μg HCHO ml^{-1} of solution.

The air concentration of total aliphatic aldehydes (as HCHO) is given by

$$\text{ppm (volume)} = \mu\text{g ml}^{-1} \text{ HCHO} \times \frac{35 \times 24.45}{V \times M \times E}$$

where V = volume of air sampled (at 760 Torr and 25° C); M is the molecular weight of HCHO ($= 30.03$) and E is the collection efficiency of the bubbler which can be established by using 2 bubblers in series. If a coarse-fritted bubbler is used, E may be taken to be 0.84. In order to obtain the real concentration of aldehydes in ambient air the above result should be multiplied by 1.25 to allow for the lower molar extinction of the higher aldehydes when determined using MBTH.

7.3.4.3 *Colorimetric analysis of HCHO (chromotropic acid method)*

Principle

HCHO reacts with chromotropic acid (1,8 dihydroxynaphthalene-3,6-disulphonic acid) in concentrated H_2SO_4 solution to form a purple coloured dye, which is determined spectrophotometrically at 580 nm. The HCHO may be collected either in water, in aqueous bisulphite or directly in the chromotropic acid–H_2SO_4 solution.

Using the procedure given below, involving collection in water, from 0.1 to 2.0 μg ml^{-1} HCHO can be measured in the final solution. This corresponds to a minimum detectable quantity of 0.1 ppm in a 40 litre air sample collected in 20 ml water. The sensitivity can be increased ten-fold by sampling directly into chromotropic acid–sulphuric acid solution (see below).

The chromotropic acid method has very little interference from other aldehydes (<0.01% from saturated aldehydes and a few % from acrolein). Alcohols, phenols, aromatic hydrocarbons and olefins all show a negative interference but the concentrations of these classes of compound in ambient air are too low to lead to serious error. Possible interference from this source should be considered when analysing for HCHO in combustion effluent, etc., however. The reproducibility of the method is to within ±5%.

Apparatus

Two coarse-fritted bubblers in series are used to collect the samples. A pump capable of maintaining a metered flow of 1 litre min^{-1} through the sampling train for up to 24 h is required.

Reagents

(a) Chromotropic acid reagent: Dissolve 0.10 g of chromotropic acid disodium salt in water and dilute to 10 ml. Filter if necessary and store in the dark. Make up a fresh solution weekly.
(b) Concentrated H_2SO_4 (sp. gr. = 1.86).
(c) Standard HCHO solutions – prepare as for MBTH method, Section 7.3.4.2.

Procedure

Place 20 ml distilled water in each bubbler, connect up sampling train, and aspirate at 1 litre min^{-1} for a suitable time to collect enough HCHO for analysis. The collection efficiency in a single bubbler is approximately 80%, i.e. 95% of the HCHO is trapped in the 2 bubblers. For long period sampling (e.g. 24 h) evaporative losses may be considerable. In this case, it may be preferable to use larger volumes of water (e.g. 35 ml and 25 ml respectively in the two bubblers).

After sampling adjust volumes to 20 ml and pipette a 4 ml aliquot of each, to stoppered test tubes. A blank is also prepared using 4 ml distilled water. Add 0.1 ml of 1% chromotropic acid solution and mix. Then add, cautiously, 6 ml concentrated H_2SO_4 from a pipette. Allow the solution to cool and read the absorbance at 580 nm in a 1 cm cell. The colour is remarkably stable but there is a small increase in absorbance of the solutions over a period of a few days. The amount of HCHO present in each aliquot taken can be determined from the calibration curve obtained as described below.

Calibration

A freshly prepared standard HCHO solution containing 10 μg HCHO ml^{-1} is used to prepare calibration solutions. Pipette 0, 0.1, 0.3, 0.5, 0.7, 1.0 and 2.0 ml of this solution into a series of glass stoppered graduated tubes. Dilute each to 4 ml with distilled water and develop colour as described above. Plot the absorbance against HCHO concentration (μg ml^{-1}) in the colour developed solution.

The air concentration of HCHO is given by

$$\text{ppm (volume)} = \text{total } \mu\text{g HCHO} \times \frac{24.47}{V \times M}$$

where the total HCHO is the sum of the amounts collected in each bubbler; V is the volume of air sampled (at 760 Torr and 25° C) and M = mol. wt of HCHO (= 30.03).

Variation

Altshuller *et al.* [63] have described a more sensitive procedure in which air is sampled directly at 1 litre min^{-1} through an impinger containing 10 ml 0.1% chromotropic acid in concentrated H_2SO_4. Collection efficiency was essentially

100% and, therefore, only a single collection vessel is required. The disadvantages of this method for some applications are the increased absorption efficiency of interfering organic compounds and the handling of the strong acid solutions during sampling and analysis.

7.3.4.4 *Physical methods*

Although GC has been widely used for the measurement of hydrocarbons, relatively little progress has been made in applying this technique to the analysis of oxygenates in emissions and particularly in atmospheric samples. One of the difficulties arises from the tendency of these compounds to adsorb on containing surfaces, particularly in moist environments. Thus, losses in sampling, storage and transfer can be serious. However, a useful procedure for the GC measurement of aliphatic aldehydes in air has been developed by Levaggi and Feldstein. Air samples are collected in 1% sodium bisulphite solution which serves to separate the aldehydes from hydrocarbons (which are not trapped at all in the aqueous solution) and other oxygenated compounds (alcohols, esters or ketones) which are only trapped with poor efficiency. An aliquot of the absorbing solution is injected on to a GC fitted with a heated injection port containing a short column of solid sodium carbonate. The aldehyde–bisulphite addition compounds are decomposed on this column and the volatile aldehydes carried on to the analytical column where they are separated and detected on a flame ionization detector. HCHO is not detected by this method. Sensitivity is sufficient for the determination of 0.02 to 0.03 ppm acetaldehyde in a 100 litre air sample. Large volume air samples are required for the measurement of aldehydes in relatively clean air by this method.

Acrolein may be collected using activated molecular sieve type 13X. After desorption, chromatographic analysis allows measurement at sub-ppm levels [70]. Acetaldehyde can be collected by freeze-out on 25% tris (2-cyanoethoxy) propane on Shimalite support maintained in liquid oxygen. After elution at 100° C, analysis is by gas chromatography [71].

Although the carbonyl C–H stretching vibration at 3.5 to 3.7 μm should be useful for i.r. analysis of aldehydes, no studies appear to have been reported on the quantitative analysis of aldehydes, either in combustion effluents or in the atmosphere by this technique. Aldehydes have been determined in laboratory 'smog chamber' studies using long-path i.r. spectroscopy, and Fourier transform i.r. spectroscopy has been successfully used to detect HCHO in polluted atmospheres [28].

7.3.5 Analysis of PAN and related compounds

7.3.5.1 *Chemical methods*

The PANs absorb readily in aqueous sodium hydroxide where they are hydrolysed to yield nitrite ion in quantitative yield [72]. The nitrite ion can be determined colorimetrically using a Griess type reagent (see Section 6.2.2). This method has been used as a check on mixtures containing PAN at high concentrations which are analysed using i.r. absorption spectrometry. It is of little practical use for atmosphere measurements because of interference from NO_2, which also yields NO_2^- in alkaline solution, and which is normally present at much higher concentration than the peroxyacyl-nitrates.

7.3.5.2   Physical methods

Infrared spectroscopy has been used to identify and analyse PAN and its higher homologues in smog chamber studies and also in the atmosphere in the Los Angeles basin using long-path length absorption methods. These compounds show characteristic bands at 5.4, 7.7 and 12.5 μm and PAN also shows a band at 8.6 μm. The 8.6 and 12.6 μm bands are normally used for analytical work, and molar absorptivities of the various bands for PAN, peroxypropionyl nitrate and peroxybutyryl nitrate have been reported [73]. The 8.6 μm band (absorptivity = 14.3×10^{-4} ppm^{-1} m^{-1}) of PAN is used as a primary standard for measurement of calibration mixtures of this gas in the 200 to 1000 ppm range. The discovery and identification of peroxybenzoyl nitrate (PBzN) in laboratory studies of photochemical reactions involving aromatic hydrocarbons has been made by means of its i.r. absorption spectrum [7], but the concentrations of PBzN in polluted air are likely to be far too low for measurement by i.r. absorption.

The minimum detectable concentration of PAN using 120 m path length i.r. absorption spectroscopy is approximately 0.03 ppm. This sensitivity is insufficient for normal atmospheric concentrations which are of the order of a few ppb or less. The very high sensitivity of the electron capture detector enables direct GC measurement of PAN and related compounds in ambient air samples of a few millilitres.

Techniques for the measurement of PAN by GC were first described by Darley *et al.* [74] and have subsequently been developed by other workers [75, 76]. Since the peroxyacylnitrates are rather reactive compounds and tend to absorb or decompose on column materials, only a limited number of columns have been found that are of practical use. Suitable stationary phases are Carbowax E400 [74] or PEG 400 (polyethylene glycol) [76] with silanized Chromosorb W or Gas Chrom Z as a support. Column and sampling tubes should be of Teflon although glass and stainless steel have been successfully used. Samples cannot be stored before analysis for PAN. A typical ECD chromatogram of polluted air sample using a 1.8 m × 3 mm Pyrex column containing 10% PEG on Gas Chrom Z is shown in Fig. 7.8.

Columns used for the analysis of atmospheric PAN have normally been shorter in length than used in the analysis depicted above. Thus Darley *et al.* [74] used a 9 inch Carbowax E400 and Penkett *et al.* [76] a 1.3 ft PEG 400 column. These give retention times of approximately 1.5 to 2.5 min for PAN with satisfactory separation from other electron absorbing species.

The main difficulty in the quantitative analysis of PAN by ECD gas chromatography lies in calibration of the detector response. This can only be achieved using air mixtures containing known concentrations of PAN of the

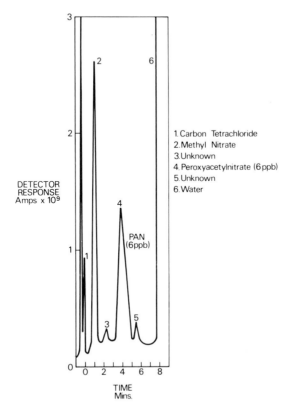

Fig. 7.8   Chromatogram of electron capturing substances in Cincinnati air [13, 75]. Column: 6 ft × 3 mm Pyrex tube packed with 10% PEG on 60/80 mesh Gas Chrom Z at 22° C. Carrier: argon + 5% methane at 60 ml min^{-1}.

order of 10 ppb (at concentrations above about 50 ppb in a 2 ml sample the detector response becomes non-linear). Preparation of these mixtures requires accurate dilution of mixtures containing 200 to 1000 ppm PAN which can be standardized by i.r. absorption spectroscopy. Furthermore, high concentration PAN mixtures are not available commercially at the present time, and PAN must be prepared by photolysis of ethyl nitrite in oxygen and purified by preparative GC. The techniques for preparing these mixtures, which have been described by Stephens *et al.* [77] are rather laborious and required extensive laboratory facilities. The pure PANs are violently explosive compounds and suitable precautions should be taken in handling these substances.

Holdren and Rasmussen [78] have described an effect of water vapour on the gas chromatographic analysis of PAN in which sensitivity was decreased substantially at low relative humidities. This effect could not be repeated by Lonneman [79], and Watanabe and Stephens [80] concluded that PAN losses occur only in dry acid-washed flasks, due probably to adsorption on active sites. The use of moist air as a calibration medium [80] was recommended.

In addition to PAN and its higher homologues, alkyl nitrites, nitrates and nitroalkanes have good electron capture responses and can be analysed by electron capture GC. Several alkyl nitrates have been identified in irradiated motor vehicle exhaust and detection of these compounds at the ppb level in the atmosphere is possible.

Efforts to detect PBzN in the atmosphere by electron capture GC have so far been unsuccessful, possibly due to absorption losses and interference from unknown components. A new GC procedure for PBzN has been described [81] which involves conversion of PBzN to methyl benzoate by trapping the former in basic methanol solution at $0°$ C. The methyl benzoate is then determined by GC with flame ionization detection. With this technique, atmospheric concentrations of PBzN of less than 0.1 ppb should be detectable.

7.3.5.3   *Analysis of PAN by electron capture GC*

Principle

PAN is measured in a discrete air sample by an electron capture detector after GC separation from other electron absorbing components in the sample. The procedure given here is based on the techniques used by Darley *et al.* using modifications recommended by I.S.C. [82].

The method is extremely sensitive, the detection limit being less than 1 ppb on a 2 ml air sample. With conventional type ECD detectors, the concentration range used is limited by the non-linear response when the detector current is reduced by about 25 to 30%. The maximum concentration measurable depends on sample size, column length, temperature and carrier flow rate but under typical operating conditions this occurs at considerably less than 1 ppm PAN. A much extended linear range can be obtained using a detector operating in the pulse frequency feedback mode. There are no known interferences with the analysis. An interfering substance would have to meet three conditions:

   (a) It must have a high electron capture cross-section
   (b) It must have a chromatographic retention time very close to that of PAN on the analytical column and
   (c) It must be present in the sample at a concentration detectable by this procedure.

These conditions eliminate virtually everything.

The detector response is calibrated by analysis of mixtures containing known concentrations of PAN and the accuracy of the method depends on the accuracy with which these mixtures are prepared. The reproducibility of measurement of a given PAN mixture is reported to be within 2% at the 50 ppb level. The reported overall accuracy of within 5% is probably over optimistic considering the difficulties in calibration for such a reactive substance at very low concentrations.

Apparatus

A flow diagram of the GC system for analysis of PAN is shown in Fig. 7.9. Three major components are required for the analysis.

(1) Sample injection system: A stainless steel (or Teflon) 6-way gas sampling valve (GSV) with an external sample loop (1 to 5 ml) is used to inject air samples into the GC. A manual GSV can be used for laboratory work but for atmospheric monitoring automatic operation of the GSV is necessary. A number of automatic GSVs are available commercially from chromatographic suppliers. These valves are usually operated by a compressed air supply controlled by solenoid valves. The solenoid valves are activated with a suitable cam-timer so that the GSV is switched between the 'sample' and 'inject' position at the required times. A 15 min interval between samples will give a satisfactory record of the ambient concentration of PAN. A small pump is required to draw the air sample through the sample loop.

(2) Chromatographic column: A column is required to separate PAN from O_2 and from other electron absorbing components which may be present in the sample, e.g. higher homologues of PAN, alkyl nitrites and nitrates, organic halogen compounds. This separation may be effected using a 9 inch long column of 1/8 inch i.d. Teflon tubing packed with 5% Carbowax E400 on 100–120 mesh

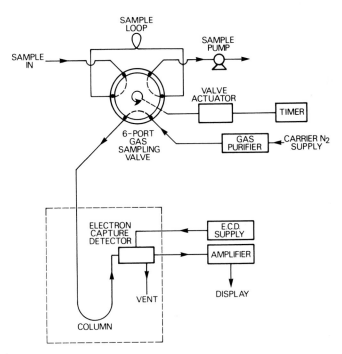

Fig. 7.9 Schematic diagram of GC system for automatic analysis of PAN.

HMDS treated Chromosorb W. When operated at 25° C with a nitrogen carrier gas flow of 40 ml min^{-1} the retention time for PAN on this column is 60 s.

(3) Electron capture detector: The column can be mounted in any commercial chromatograph equipped with an electron capture detector, with associated amplifier and recorder. In order to avoid excessive use of recorder chart in continuous monitoring, it is preferable to arrange for the recorder chart drive to be on only during the period necessary to record the chromatogram (i.e. 3 min). A second cam-timer can be used for this purpose.

Procedure

Dry nitrogen carrier gas is passed continuously through the sample loop of the GSV, the column and the detector at 40 ml min^{-1}. The standing current in the ECD should be checked according to the manufacturers' procedures to ensure that the detector is clean and functioning satisfactorily. The automatic sample injection system is then operated and chromatogram recorded. The large air peak appears during the first 30 s following injection followed by the PAN peak at approximately 60 s. A small shoulder on the tail of the PAN peak, assumed to be the higher homologue, peroxypropionylnitrate, may also appear.

Calibration

Calibration is based on comparison of the sample peak heights with those obtained by injection of mixtures containing known concentrations of PAN. The latter are prepared by dilution of mixtures containing high concentrations of PAN measured by i.r. spectrometry. Since concentrations of a few hundred ppm are required for accurate i.r. measurement, quantitative dilution by a factor of 10^4 must be carried out.

PAN is synthesized, purified as described by Stephens *et al.* [77] and stored in stainless steel cylinders. The cylinders, pressured to 100 psig with N_2, contain 500 to 1000 ppm PAN and are stored at 16° C. An i.r. spectrophotometer with a 10 cm cell is employed to determine the concentration of PAN in the cylinder gas using the absorptivity of 13.9×10^{-4} ppm^{-1} m^{-1} at 8.6 μm (1161 cm^{-1}), i.e.

$$\text{Concn (ppm)} = \text{Absorbance}/(0.1 \text{ m} \times 13.9 \times 10^{-4})$$

A flow dilution method can be used to reduce the concentration of PAN accurately to a level suitable for calibrating the GC system (10 to 50 ppb). Figure 7.10 shows a suitable system involving two 100:1 dilutions in activated-charcoal-filtered air. The concentration of PAN in the effluent can be varied by adjusting the dilution at either stage. Samples from the diluted effluent are injected on to the chromatograph and the peak height recorded for a series of calculated concentrations.

Alternatively an exponential dilution system may be employed (see Section 6.3.8). In this case a single aliquot of the i.r. calibrated gas mixture is required. This enables an alternative simplified method to be used for the preparation of PAN. Small quantities of PAN are prepared directly in a 10 cm i.r. cell with a Pyrex glass body by photolysing traces of ethyl nitrite (C_2H_5ONO), in pure O_2. The PAN formed is then measured in the cell without separation from

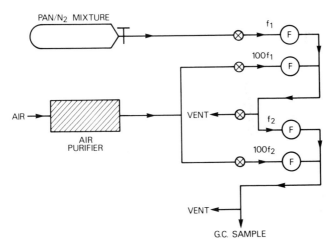

Fig. 7.10 Flow dilution system for preparing low concentrations of PAN.

unreacted nitrite or other products. The cell is first flushed with pure O_2 and then small amounts (50 μl) of vapour from the above liquid C_2H_5ONO are added at 15 min intervals whilst the mixture is irradiated with 'Blacklight' fluorescent u.v. lamps (radiation of 300 to 400 nm is required). The C_2H_5ONO concentration must be kept low to maximize the yield. When sufficient PAN has accumulated (as measured from the i.r. absorption at 8.6 μm), an aliquot of the mixture which now contains a known concentration of PAN is injected into an exponential dilution flask connected to the appropriate flow system (Chapter 6). Since the concentration of PAN in the effluent from the flask decreases in a known manner, a calibration may be carried out by repeated injection of samples from the effluent.

The relationship between peak height and PAN concentration should be linear up to approximately 50 ppb and, therefore, the concentration in unknown samples within this range can be calculated by multiplying the peak height by a constant. The constant varies from one instrument to another and varies inversely with changes in the detector standing current. The automatic system can be operated continuously over periods of months. Occasional cleaning of the electron capture detector is desirable since contamination of the detector causes a slow decline in standing current which leads to a reduction in sensitivity. It is desirable to calibrate the instrument about once a week to compensate for the gradual reduction in sensitivity. When the detector is thoroughly cleaned the original sensitivity is regained.

7.3.6 Analysis of oxyacids of N

Although the oxyacids of N, in the gaseous state, are widely recognized as being important products of photochemical reactions, there has been relatively little progress in developing specific analytical techniques for these compounds until recently. Both HNO_2 and HNO_3 have been detected in ambient air by FTIR methods (see Chapter 10). These are outside the capability of most laboratories and alternative methods are required.

Both HNO_2 and HNO_3 are extremely reactive water soluble vapours and absorption losses can present serious problems during sampling and analysis by conventional techniques.

7.3.6.1   *Chemical methods*

Laboratory measurements of low concentrations of HNO_2 were first described by Nash [83]. HNO_2 was determined as nitrite ion (NO_2^-) using a modified Griess–Saltzman reagent, following its absorption in dilute aqueous NaOH contained in a simple impinger. By deliberately using a fast sampling flow rate ($\geqslant 1$ litre min^{-1}) absorption of NO_2, which also yields NO_2^- in aqueous alkali, was made inefficient. Less than 10% of the NO_2 was absorbed at the 0.1 ppm level whereas HNO_2 was trapped essentially quantitatively in a single bubbler. Thus HNO_2 could be measured in the presence of NO_2. However, any other nitrogen compounds which yield nitrite or alkaline hydrolysis (e.g. peroxyacylnitrates) can potentially interfere. Nash [84] has also reported measurements of HNO_2 in the atmosphere using this technique.

Nitric acid vapour may be pre-separated from particulate nitrate by passage through an inert filter of acid-washed quartz or of Teflon (the latter is generally preferred). It is important to maintain the filter at ambient temperature and to avoid high pressure drops so to maintain the ammonium nitrate dissociation equilibrium (see also Chapter 6). The nitric acid vapour may then be collected on a nylon filter, or a sodium chloride impregnated filter, where the following reaction takes place:

$$HNO_3 + NaCl \rightarrow NaNO_3 + HCl$$

After leaching with water, the collected nitric acid is analysed as nitrate (Chapter 6). Alternatively, nylon and Teflon filters are run in parallel; after analysis for nitrate, the HNO_3 concentration is given by difference as the nylon filter collects both particulate nitrate and nitric acid vapour.

Spicer and co-workers [85] carried out a comparative study of techniques involving both chemical and physical measurement methods. Both nylon and NaCl impregnated filters were found to be suitable for analysis of HNO_3 in ambient air.

It should be noted that since HNO_3 concentrations are affected by ambient temperatures and NH_3 levels, in cooler climates very low concentrations of HNO_3, generally <1 ppb, are encountered.

Method for analysis of HNO_3 vapour in air [86]

Impregnated filters are prepared by immersing Whatman No. 41 cellulose filters in 5% aqueous Analar NaCl solution, and drying under an infrared lamp after draining off excess solution.

Air is sampled 4 litres min^{-1} through a 0.45 μm Teflon membrane filter maintained at ambient temperature followed by the impregnated filter. After 24 h, sampling is terminated and exposed and blank impregnated filters are extracted into de-ionized water (10 ml) and analysed for nitrate by an appropriate technique (Chapter 6). The detection limit is approximately 0.04 ppb for a 24 h air sample.

7.3.6.2 *Physical methods*

Commercial chemiluminescence oxides of nitrogen analysers employing carbon, molybdenum or stainless steel converters are non-specific for the determination of NO_2 (see Chapter 6). The instruments not only respond to NO_2 but also to PAN and a variety of other organic nitrates and nitrites [87]. The response of chemiluminescence oxides of nitrogen analysers to HNO_2 [88] and HNO_3 [87, 89] has also been investigated.

HNO_2 gives a quantitative response in an 'NO_x' analyser employing a stainless steel converter although some evidence of absorption in the sample lines was found giving rise to a longer than normal time constant for measurement of 'total NO_x' when HNO_2 was present. By passing the sample through a bubbler containing aqueous alkali HNO_2 could be selectively removed, and its concentration could be determined in the presence of NO and NO_2 by difference. The response of commercial NO_x analysers to HNO_3 appears to be non-quantitative. When HNO_3 vapour is sampled into a chemiluminescence analyser, a low and irreproducible response is observed. Adsorption and chemical attack on the instrument components is responsible for these effects.

The most promising technique at the present time for measurement of low concentrations of HNO_3 in the gas phase is i.r. absorption spectroscopy. The HNO_3 absorption band most sensitive for atmospheric analysis is centred at about 880 cm^{-1} with two distinctive peaks located at 879 and 896 cm^{-1}. At these frequencies water vapour interference is not serious. Using a Fourier transform i.r. spectrometer with a scanning Michelson interferometer and a folded path of 400 m, Hanst *et al.* [27] claim a detection limit at least as low as 10 ppb for HNO_3 in ambient air and a 1 m path length FTIR gave a detection limit of 6 ppb [85] (see also Chapter 10). This level of sensitivity is adequate for measurements only in highly polluted and hot environments.

References

1. Leighton, P. A. (1961) *Photochemistry of Air Pollution*, Academic Press, New York.
2. Demerjian, K., Kerr, J. A. and Calvert, J. G. (1974) *Adv. Environ. Sci. Technol.*, **4**, 1.
3. US Dept of Health, Education and Welfare (1970) *Air Quality Criteria for Photochemical Oxidants*, Natl Air Pollut. Control Admin. Publication No. AP-63.
4. Taylor, O. C. (1969) *J. Air Pollut. Control Assoc.*, **19**, 34.
5. US Dept of Health, Education and Welfare (1969) *Air Quality Criteria for Particulate Matter*, Natl Air Pollut. Control Admin. Publication No. AP-50.

6. US Dept. of Health, Education and Welfare (1969) *Air Quality Criteria for Sulphur Oxides*, Natl Air Pollut. Control Admin. Publication No. AP-49.
7. Huess, J. M. and Glasson, W. A. (1968) *Environ. Sci. Technol.*, **2**, 1109.
8. Hersch, P. and Deuringer, R. (1963) *Anal. Chem.*, **35**, 897.
9. Brewer, A. W. and Milford, J. R. (1960) *Proc. R. Soc. London, Ser. A*, **256**, 470.
10. Mast, G. M. and Saunders, H. E. (1962) *ISA Trans.*, **1**, 325.
11. Wartburg, A. F., Brewer, A. W. and Lodge, Jr., J. P. (1964) *Int. J. Air Water Pollut.*, **8**, 21.
12. Potter, L. and Duckworth, S. (1965) *J. Air Pollut. Control Assoc.*, **15**, 207–9.
13. Altshuller, A. P. (1966) *Adv. Gas Chromatogr.*, **5**, 229.
14. Fishbein, L. (1973) *Chromatography of Environmental Hazards*, Vol. 2, Elsevier Scientific, Amsterdam, The Netherlands.
15. Levaggi, D. A. and Feldstein, M. (1970) *J. Air Pollut. Control Assoc.*, **20**, 312.
16. Lovelock, J. E. (1963) *Anal. Chem.*, **35**, 474.
17. Wentworth, W. E. (1971) *Recent Advances in Gas Chromatography* (eds I. I. Domsky and J. A. Perry), Marcel Decker, New York, p. 185.
18. Aue, W. A. and Kapila, S. (1973) *J. Chromatogr. Sci.*, **2**, 255.
19. Maggs, R. J., Jaynes, P. L., Davies, A. J. and Lovelock, J. E. (1971) *Anal. Chem.*, **43**, 1966.
20. Boettner, E. A. and Dallos, F. C. (1965) *Am. Ind. Hyg. Assoc. J.*, **26**, 289.
21. Sullivan, J. J. (1973) *J. Chromatogr.*, **87**, 9.
22. Lovelock, J. E., Adlard, E. R. and Maggs, R. J. (1971) *Anal. Chem.*, **43**, 1962.
23. Renzetti, N. A. (1956) *J. Chem. Phys.*, **24**, 209.
24. Scott, W. E., Stephens, E. R., Hanst, P. L. and Doerr, R. C. (1957) *Proc. Am. Petroleum Inst., Series III*, **37**, 171.
25. Stephens, E. R., Hanst, P. L., Doerr, R. C. and Scott, W. E. (1956) *Ind. Eng. Chem.*, **48**, 1498.
26. Harries, J. E. (1973) *Nature*, **241**, 515.
27. Hanst, P. L., Lefohn, A. S. and Gay, Jr., B. W. (1973) *Appl. Spectrosc.*, **27**, 188.
28. Hanst, P. L., Wilson, W. E., Patterson, R. K. *et al.* (1975) *A Spectroscopy Study of California Smog*, EPA Environmental Monitoring Series, EPA-650/4-75-006, US Environmental Protection Agency, Research Triangle Park, NC.
29. Hanst, P. L. (1970) *Appl. Spectrosc.*, **24**, 161; also in (1971) *Adv. Environ. Sci. Technol.*, **2**, 92.
30. Menzies, R. T. (1971) *Appl. Opt.*, **10**, 1532.
31. Hodgeson, J. A., McClenny, W. A. and Hanst, P. L. (1973) *Science*, **182**, 248.
32. Byers, D. H. and Saltzman, B. E. (1958) *Am. Ind. Hyg. Ass. J.*, **19**, 251.
33. Saltzman, B. E. and Gilbert, N. (1959) *Anal. Chem.*, **31**, 1914.
34. Pitts, Jr., J. N., Sprung, J. L., Poe, M. *et al.* (1976) *Environ. Sci. Technol.*, **10**, 794.
35. Flamm, D. L. (1977) *Environ. Sci. Technol.*, **11**, 978.
36. Cohen, I. R., Purcell, T. C. and Altshuller, A. P. (1967) *Environ. Sci. Technol.*, **1**, 247.
37. Guicherit, R., Jeltes, R. and Lindqvist, F. (1973) *Environ. Pollut.*, **3**, 91.
38. Saltzman, B. E. and Wartburg, A. F. (1965) *Anal. Chem.*, **37**, 779.
39. Saltzman, B. E. (1954) *Anal. Chem.*, **26**, 1949.
40. Intersociety Committee (1970) Tentative method for the manual determination of oxidising substances in the atmosphere, 44101-02-70T. *Health Lab. Sci.*, **7**, 152.
41. Lindqvist, F. (1972) *Analyst*, **97**, 549.
42. Dorta-Shaeppi, Y. and Treadwell, W. D. (1949) *Helv. Chim. Acta*, **32**, 356.
43. Nash, T. (1967) *Atmos. Environ.*, **1**, 679.

44. Bravo, H. A. and Lodge, Jr., J. P. (1964) *Anal. Chem.*, **36**, 671.
45. Hauser, T. R. and Bradley, D. W. (1966) *Anal. Chem.*, **38**, 1529.
46. Regener, V. H. (1964) *J. Geophys. Res.*, **65**, 3975.
47. Hodgeson, J. A., Krost, K. J., O'Keefe, A. E. and Stevens, R. K. (1970) *Anal. Chem.*, **42**, 1975.
48. Guicherit, R. (1971) *Z. Anal. Chem.*, **256**, 177.
49. Nederbragt, G. W., Van der Horst, A. and Van Duijn, J. *Nature*, **206**, 87.
50. Kummer, W. A., Pitts, Jr., J. N. and Steer, R. P. (1971) *Environ. Sci. Technol.*, **5**, 1045.
51. Finlayson, B. J., Pitts, Jr., J. N. and Akimoto, H. (1972) *Chem. Phys. Lett.* **12**, 495.
52. Huntzicker, J. J. and Johnson, R. L. (1979) *Environ. Sci. Technol.*, **13**, 1414.
53. De More, W. B. and Patapoff, M. (1976) *Environ. Sci. Technol.*, **10**, 897.
54. Warren, G. J. and Babcock, G. (1970) *Rev. Sci. Instrum.*, **41**, 280.
55. Paur, R. J. (1977) *Nat. Bur. Stand. (US)*, *Spec. Publ.* 464, 15.
56. Hodgeson, J. A., Stevens, R. K. and Martin, B. E. (1971) *ISA Trans.*, **11**, 161.
57. Pobiner, H. (1961) *Anal. Chem.*, **33**, 1423.
58. Cohen, I. R. and Purcell, T. C. (1967) *Anal. Chem.*, **39**, 131.
59. Bufalini, J. J., Gay, Jr., B. W. and Brubacker, K. L. (1972) *Environ. Sci. Technol.*, **6**, 816.
60. Kok, G. L., Holler, T. P., Lopez, M. B. *et al.* (1978) *Environ. Sci. Technol.*, **12**, 1072.
61. Kok, G. L., Darnall, K. R., Winer, A. M. *et al.* (1978) *Environ. Sci. Technol.*, **12**, 1077.
62. Sawicki, E., Hauser, T. R., Stanley, T. W. and Elbert, W. (1961) *Anal. Chem.*, **33**, 93.
63. Altshuller, A. P. and Leng, L. J. (1963) *Anal. Chem.*, **35**, 1541.
64. Bricker, C. E. and Johnson, H. R. (1945) *Ind. Eng. Chem. Anal. Edn*, **17**, 400.
65. West, P. W. and Sen, B. (1956) *Z. Anal. Chem.*, **153**, 177.
66. Altshuller, A. P., Miller, D. L. and Sleva, S. F. (1961) *Anal. Chem.*, **33**, 621.
67. Altshuller, A. P., Leng, L. J. and Wartburg, A. F. (1962) *Int. J. Air Water Pollut.*, **6**, 381.
68. Cohen, I. R. and Altshuller, A. P. (1961) *Anal. Chem.*, **33**, 726.
69. Intersociety Committee on Manual Methods of Air Sampling and Analysis (1972) *Methods of Air Sampling and Analysis*, American Public Health Association, Washington, DC, p. 199.
70. Gold, A., Dube, C. E. and Perni, R. B. (1978) *Anal. Chem.*, **50**, 1839.
71. Hoshiba, Y. (1977) *J. Chromatogr.*, **137**, 455.
72. Nicksik, S. W., Harkins, J. and Mueller, P. K. (1967) *Atmos. Environ.*, **1**, 11.
73. Stephens, E. R. (1964) *Anal. Chem.*, **36**, 929.
74. Darley, E. F., Kettner, K. A. and Stephens, E. R. (1963) *Anal. Chem.*, **35**, 589.
75. Bellar, T. A. and Slater, R. W. (1965) Proc. 150th Am. Chem. Soc. Meeting, Atlantic City, NJ.
76. Penkett, S. A., Sandalls, F. J. and Lovelock, J. E. (1975) *Atmos. Environ.*, **9**, 131.
77. Stephens, E. R., Burleson, F. R. and Cardiff, E. A. (1965) *J. Air Pollut. Control Assoc.*, **15**, 87.
78. Holdren, M. W. and Rasmussen, R. A. (1976) *Environ. Sci. Technol.*, **10**, 185.
79. Lonneman, W. A. (1977) *Environ. Sci. Technol.*, **11**, 194.
80. Watanabe, I. and Stephens, E. R. (1978) *Environ. Sci. Technol.*, **12**, 22.
81. Appel, B. R. (1973) *J. Air Pollut. Control Assoc.*, **23**, 1042.
82. Intersociety Committee (1971) *Health Lab. Sci.*, **8**, 1.
83. Nash, T. (1968) *Ann. Occup. Hyg. (Cambridge)*, **11**, 235.
84. Nash, T. (1974) *Tellus*, **26**, 175.

85. Spicer, C. W., Howes, Jr., J. E., Bishop, T. A., Arnold, L. H. and Stevens, R. K. (1982) *Atmos. Environ.*, **16,** 1487.
86. Harrison, R. M. and Pio, C. A. (1983) *Tellus*, **35B,** 155.
87. Winer, A. M., Peters, J. W., Smith, J. P. and Pitts, Jr., J. N. (1974) *Environ. Sci. Technol.*, **8,** 13.
88. Atkins, D. H. F. and Cox, R. A. (1974) AERE/R7615, HMSO, London.
89. Stedman, D. H. and Niki, H. (1973) *J. Phys. Chem.*, **77,** 2604.

8 Hydrocarbons and carbon monoxide

A. E. McINTYRE and J. N. LESTER

8.1 Introduction

All industrial processes contribute to the pollution of our environment. The greatest man-made contributions to atmospheric pollution (in terms of bulk of pollutants) are, however, processes of combustion in power generation and industrial plants, in automobile and aircraft engines, and in domestic heating. Such processes are responsible for the generation of a wide range of pollutants; amongst those pollutants which have received considerable attention in recent years are unburned and partially oxidized gaseous and particulate species. More recently, interest in this general field has been concentrated on investigations into the identification and measurement of levels for groups of compounds or even individual pollutants.

In this chapter such combustion products are examined in three sections: volatile hydrocarbons; hydrocarbon content of particulate matter; and carbon monoxide. No sampling nor analytical techniques of general applicability to all environments and the extensive range of potential pollutants have as yet been reported. In this chapter, sampling and analytical procedures are considered separately. It should be remembered, however, that the quantification of results will depend on the efficiencies of both procedures, and on their compatibility.

When considering volatile hydrocarbons in the atmosphere, the analyst is concerned primarily with alkanes or with lower alkenes and aromatics as a group of unsaturated hydrocarbons. The analysis of such compounds has been the subject of a review [1].

Methane, which is the only hydrocarbon found naturally occurring world-wide, has a background concentration in the atmosphere of 1.3 to 1.4 ppm. Other hydrocarbons found in air are derived from a wide variety of sources, the most significant of which include: oil and petroleum refineries and storage depots; chemical production and oil burning industries; commercial and geogenic gas leaks; biological processes [2]; and agricultural and forest burning programmes [3]. In certain forest areas terpenes are emitted by the living trees [4, 5].

At the ppb* concentrations in which they are normally encountered in the atmosphere the hydrocarbons are relatively harmless to mammals. It has been shown [6], however, that ethylene at concentrations as low as 0.01 ppm, and to a

---

* 1 ppb = 1 part in 10^9.

lesser extent other hydrocarbons produce deleterious effects in various species of plant. A more serious problem is the so-called 'photochemical smog' [7–9], which is produced under certain meteorological conditions by the reactions of nitrogen oxides with other air pollutants including hydrocarbons, leading to the formation of compounds such as ozone, aldehydes, peroxyacylnitrates and alkyl nitrates. These reaction products can cause severe irritation of the eyes and mucous membrane. Visibility may also be reduced by the formation of aerosols of polymer molecules originating from photochemical reactions of hydrocarbons.

Specific hydrocarbons derived from different sources will have varying degrees of impact on the environment. Thus ideally it would be desirable to determine levels of individual hydrocarbons in the atmosphere. Many surveys, however, have been carried out measuring only total hydrocarbon levels, in for example, St Louis [10], and comparing Paris with other cities [11]. A more significant survey was carried out measuring total hydrocarbon and calculating non-methane hydrocarbon levels in the USA [12], using an empirical relationship derived from previous measurements. Thus measurements of non-methane hydrocarbon levels are to be preferred to those of total hydrocarbon levels. More data may be obtained by monitoring specific compounds and a study of diurnal patterns for several aliphatic and aromatic hydrocarbons has resulted in their occurrence being attributed to automobile exhaust [13]. A similar range of compounds has been investigated using an alternative method of evaluating vehicular emissions [14], which involved measurements in New York and New Jersey and included calculations to demonstrate the effect of wind direction on hydrocarbon levels. A study in Los Angeles [15] has attempted to correlate hydrocarbon levels measured with possible sources of such species, and the relative significance of sources was quoted as: automobile exhausts 47%, evaporative losses of gasoline 31%, geogenic gas leaks 14% and leaks in commercial natural gas supply systems 8%. An attempt has been made [16] to produce a distribution profile for individual hydrocarbons (benzene and toluene) over a large area (Toronto) by sampling at twelve sites simultaneously. Overall average concentrations were 13 and 30 ppb respectively but the variation in levels between individual sites reached a factor of 40. The toluene to benzene ratio has been suggested as an indication of the proportion of hydrocarbon pollution in the atmosphere produced by automobiles. Other works [17, 18] have found similar ratios to those recorded in Toronto. In Zurich, however, a far lower toluene to benzene ratio has been measured [19]. Benzene and toluene have been measured [20] at levels of 23 ppm and 11 ppm respectively in the vicinity of a chemical reclamation plant. A three-month survey of a rural community in Eastern Pennsylvania showed that EPA limits for hydrocarbons [21] were not exceeded [22]. Other measurements have been reported covering several molecular weight ranges and include: C_1–C_5 by an automatic method [23], C_1–C_5 amongst other gases associated with peat and coal [24]; C_8–C_{18} in Paris air [25]; C_6–C_{20} in Zurich [19], C_9–C_{20} in Italy [26]; and C_6–C_{10} in the Netherlands [27].

The research outlined above has demonstrated the ubiquity of hydrocarbons in the environment. They occur in great structural variety as contributions from many sources, each source differing in the relative proportions of its component hydrocarbons. Their stabilities vary greatly but most hydrocarbons are sufficiently inert to enter into complex pathways of dispersal, involving transport by air, water, particulates, and by the food chain. Much of the research into the origins and fates of hydrocarbons in the environment has centred on the lower molecular weight alkanes and alkenes, principally because of their relatively facile analysis by gas chromatography, alone or in conjunction with mass spectrometry. Combustion chemistry is notoriously complex and the combustion of fossil fuels for heat, power and transportation produces most of the organic fraction of the atmospheric particulate matter commonly encountered in urban environments.

Some attention has been paid to the higher molecular weight aliphatic compounds in particulate matter, and compounds in the C_{11} to C_{33} range have been tentatively identified in an atmospheric extract [28]. Organic materials in airborne particles have also been studied [29] by high resolution mass spectrometry to determine whether hydrocarbon distributions typical of major sources of pollution are readily identified. Thin layer chromatography (TLC) has been used to separate alkane fractions prior to gas chromatographic analysis [30]. The majority of such studies have involved other fractions and have concentrated on the identification of specific compounds, particularly suspected carcinogens of the polycyclic aromatic hydrocarbon (PAH) type. The increasing use of fossil fuels, whose pyrolysis produces complex mixtures of PAH, poses many environmental and public health problems. Environmental samples may contain complex PAH mixtures derived from many sources and it has been stated [31] that:

'No existing method separates and resolves adequately the entire PAH fraction on a mg to μg scale and on samples that contain a very large excess of nonhydrocarbons. Even in petroleum analysis, where large samples are available, a complete analytical resolution of the PAH fraction exceeds the capability of any existing combination of analytical techniques.'

Analyses of PAH mixtures using methods developed by Sawicki *et al.* [32] typically entail various partitioning sequences followed by column chromatography to separate the organic fraction of the particulate matter into subfractions prior to spectrophotometric examination of the subfractions. Other methods involve fluorescence techniques [33], TLC [34] and high-pressure liquid chromatography (HPLC) [35]. Specialized techniques of this type enable the separation and identification of PAH subfractions and have been extensively reported [36]. Typical examples are the analyses of groups of eleven PAH in particulate samples from Baltimore [36] and a characteristic urban area of Budapest [37].

Carbon monoxide (CO) is a poisonous gas, but it owes its place in the hierarchy of pollutants to different considerations. Although it is not a stable end-product of balanced combustion, unless the supply of O_2 is insufficient, it represents the penultimate stage in combustion through which all the carbon species must pass. Combustion cannot take place without the intermediate formation of CO. As a combustion-generated pollutant, CO need not be a problem under normal operating conditions except in the spark ignition engine. There, however, it is a serious problem, especially under idling and decceleration conditions. Thus, it is in the context of monitoring motor vehicle emissions that the analysis of CO has received considerable recent attention.

Classically the analysis of CO has been performed by non-dispersive infrared spectroscopy (Ndir) [38], which is sensitive enough to measure the levels normally encountered in street air of 1 to 50 ppm. A gas chromatographic technique, although not allowing continuous measurement of CO, has a substantially higher sensitivity. The basis of the method is catalytic reduction of the CO to methane and detection by flame ionization. This principle is used in commercial instruments which measure methane, CO and total hydrocarbons [39]. An analyser for continuous determination of carbon monoxide at levels down to 1 ppm using an electrochemical cell has been described [40].

Air quality standards for particulate matter, hydrocarbons and carbon monoxide established by the United States Environmental Protection Agency (USEPA), Commission of the European Communities (CEC) and the World Health Organization are included in Table 8.1.

Table 8.1   Air quality standards and guidelines established by national and international bodies

Pollutant		Averaging time	Primary standard levels	Secondary standard levels
Particulate matter	NAAQS	Annual (geometric) mean	75 μgm^{-3}	60 μgm^{-3}
	NAAQS	24 hours[a]	260 μgm^{-3}	150 μgm^{-3}
	WHO	Annual (arithmetic)	40–60 μgm^{-3}	—
	WHO	24 hours	100–150 μgm^{-3}	—
	EC	Annual (geometric) mean	68 μgm^{-3}	—
Hydrocarbons (non-methane)	NAAQS	3 hours[b] (6 to 9 am)	160 μgm^{-3} (0.24 ppm)	160 μgm^{-3} (0.24 ppm)
Carbon monoxide	NAAQS	8 hours[a]	10 mgm^{-3} ppm (9 ppm)	10 mgm^{-3}
	NAAQS	1 hour[a]	40 mgm^{-3} (35 ppm)	40 mgm^{-3}

[a] not to be exceeded more than once a year
[b] not to be exceeded more than once a year
NAAQS = US National Ambient Air Quality Standards
   WHO = World Health Organization Health Criteria
      EC = European Community Directive

8.2 Volatile hydrocarbons

8.2.1 Sampling procedures

The accuracy and precision of a sampling technique is of the utmost importance and such a technique must allow one to relate readily the values measured to those which were present in the environment. These requirements are of course true of all analyses and in this application inert materials such as glass, polytetrafluoroethene (PTFE) and stainless steel should be used to reduce the adsorption effects on the walls of containers and transfer lines. Particular attention should be given to the cleaning of all apparatus and to the selection and application of any sealing or lubricating materials in order to avoid contamination.

Both the duration of sampling and the time at which samples are taken should be related to the purpose of a particular investigation. Studies on short duration pollution incidents require rapid sampling techniques, whereas in investigations into ambient levels of pollutants time-averaged results over as long a period as is practicable are desirable. German States developing legislation in respect of smog incidence are specifying half-hour sampling periods and in the USA a 3 h mean sample is prescribed [41].

The requirement of gas chromatography for a small (<10 ml), discrete injection necessitates the use of a point-sampling technique. Analytical systems employing gas chromatography are therefore not area monitors. Multiple-point sampling may be used to simulate area monitoring.

At ambient levels normally encountered, preconcentration (of the volatile components from a highly diluted sample) prior to analysis is usually required, however, some methods in which a gas sample is analysed directly have been reported [13, 23]. The high volatility of the low molecular weight hydrocarbons requires the use of cryogenic trapping techniques to ensure 100% retention, whereas for the higher molecular weight compounds ($>C_6$) adsorption on to solid adsorbents is the preferred technique [42–45]. When using solid adsorbents it is essential to ensure that a valid sample is collected. Particular attention should be directed towards the determination of the 'breakthrough volume', the sampled air volume at which the compound being collected begins to elute from the tube. The breakthrough volume is dependent upon the gas chromatographic retention time of the compounds at ambient temperature, the adsorbent acting as the stationary phase and the sampled air as the carrier gas. An equilibration position will be reached after the breakthrough volume is exceeded where if the ambient concentration of the compound remains constant, the amount of the compound being adsorbed will be equal to the amount eluting from the tube.

The choice of adsorbent for use in a particular situation depends upon the chemical properties of the compounds to be sampled and subsequently analysed. The following criteria must be considered: quantitative collection efficiencies and recovery of trapped vapours, high breakthrough volumes, minimal decomposition or polymerization of sample constituents during collection and recovery, low background contribution from the adsorbent and little or no affinity of the adsorbent for water. The parameters involved in determining the performance of an adsorbent can be divided into two categories: those related to the sampling environment such as sample flow rate, sampling time, air temperature and humidity and those related to the physico-chemical properties of the adsorbent such as surface area, particle size and porosity, solute capacity, sorption mechanism and degree of solute affinity. Furthermore, some of the factors which influence adsorbent performance are not independent of each other.

8.2.1.1   *Cryogenic systems*

A battery operated system for sampling in remote areas has been reported [46], and requires approximately 3 litres of liquid N_2. This system was originally developed for use in the tropical forests of Panama [47] where very high relative humidities are encountered [48], but has exhibited variable freeze-out efficiency and requires constant attention [49]. A portable cryocondenser has been shown [52] to have a high enrichment capacity for large air volumes and, when coupled via sub-ambient GC to a flame ionization detector, to give a detection limit at the sub-ppb level. A similar system has been used for the sampling of industrial air pollutants at various altitudes [53]. A more sophisticated application of a freeze-out technique to ultratrace analysis has been described [54], in which a temperature gradient is applied to a tube packed with a GLC stationary phase on an appropriate support. The main problem with freeze-out techniques is the presence of water vapour in the air. The application of drying agents prior to condensation has reduced this problem and systems based on this principle have been described [55]. Separations of the contents of cryogenic traps prior to mass spectral analysis have been accomplished by thermal analysis [56] or by preparative gas chromatographic techniques [55].

A recently developed cryostat (Fig. 8.1) is at present being used for the analysis of individual hydrocarbons in street air [52]. This instrument has a temperature range of -196 to $+300°$ C. Temperatures may be maintained within this range by the balancing of the cooling and heating effects of a liquid nitrogen reservoir and a 240 W cartridge heater.

The instrument basically consists of two independent vacuum chambers. The inner chamber contains the liquid N_2 can and an integral carbon adsorption pump. The outer vacuum space contains the 'U' tube, heat exchanger, heater and temperature sensors. The use of two independent vacuum chambers interconnected via an isolation valve facilitates the changing of the 'U' tube without removal of the refrigerant. The 'U' tube is only packed with adsorbent along the length of the cartridge heater. Once the sampling temperature has been achieved and the can refilled with liquid N_2, the supply will last for approximately 4 h, during which time the desired temperature may be maintained.

In remote areas with no electricity supply, sampling is carried out at $-196°$ C. Prior to analysis, liquid O_2 that has been condensed out in the 'U' tube, can, with controlled heating, be slowly evaporated so as to avoid interference with the subsequent chromatographic separation of the adsorbed material.

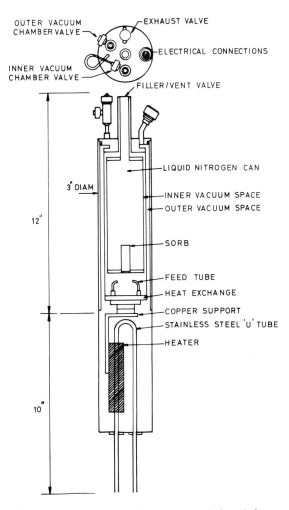

OUTER VACUUM CHAMBER VALVE

EXHAUST VALVE

ELECTRICAL CONNECTIONS

INNER VACUUM CHAMBER VALVE

FILLER/VENT VALVE

LIQUID NITROGEN CAN

3" DIAM

INNER VACUUM SPACE

OUTER VACUUM SPACE

12"

SORB

FEED TUBE

HEAT EXCHANGE

COPPER SUPPORT

STAINLESS STEEL 'U' TUBE

HEATER

10"

Fig. 8.1 Schematic diagram of a portable cryostat designed for sampling C_1–C_5 hydrocarbons.

This cryostat has several advantages over other sampling techniques that are currently available. It is easily portable, which is advantageous when sampling in remote areas, and the results obtained with it are both reproducible and consistent. By selection and control of the sampling temperature it is possible to ignore certain of the more volatile compounds. If required, however, the complete range of C_1–C_{12} compounds may be analysed in a single sample. A certain amount of separation of the adsorbed material prior to introduction into the gas chromatograph is achieved by temperature programming the 'U' tube. With the cryostat it is possible to sample large volumes of air with the confidence that a truly representative sample is being collected and without the limitations of relatively small breakthrough volumes. This is an important consideration when a GC/MS analysis is to be employed and ng quantities of each compound to be identified are required. After sampling at sub-ambient temperature (to take advantage of the greater breakthrough volumes) the sample tube can be stored at ambient temperature without loss of the higher boiling hydrocarbons (C_5 and above).

Although cryogenic sampling is the only technique presently available which produces, directly, a true sample composition in the low molecular weight range of air pollutants, procedures involving adsorption on solid adsorbents or in liquids, coated on inert supports and acting as stationary phases, are simpler and more convenient for field sampling. Adsorption techniques suffer from high percentage losses of volatile compounds. The chemical nature of a solid adsorbent has a considerable influence on the adsorption/desorption characteristics for specific compounds, and losses due to irreversible adsorption are common, particularly when a thermal desorption step is used.

8.2.1.2   *Solid adsorption systems*

Activated carbon is particularly prone to irreversible adsorption of certain organic species. This material has, however, been used widely as an adsorbent followed by either thermal desorption [25] or by various methods for solvent extraction [29, 26, 53] of the adsorbed species. Carbon disulphide has been shown to be an effective solvent for the extraction of activated carbon [19]. Recovery of solutes by solvent extraction is more complete than thermal desorption but is also a more time-consuming and complex process.

The usefulness of silica gel as an adsorbent for the collection of environmental samples is greatly reduced by its affinity for water and the consequent influence of relative humidity on the adsorption curves for hydrocarbons. Sampling on to silica gel followed by thermal desorption of the adsorbed species, their oxidation to CO and determination by acidimetric and coulometric measurements has recently been reported [54]. The procedure was limited by the number of adsorption/desorption cycles that could be carried out, by inadequate desorption temperatures and by variations in retention volumes for specific organic species present. This method is limited to atmospheric pollutants of relatively low volatility but could be improved by sub-ambient operation.

Porous polymers with high capacities for the sampling of organics and low affinities for water provide extremely useful adsorbent species and have been extensively investigated [2, 49, 56–62] utilizing both thermal desorption and solvent extraction. An evaluation of various organic adsorbents and their relative dynamic capacities for certain classes of compound has been reported [42]. Chromatographic packing materials studied were Porapak P, a porous polymer of styrene and divinyl-benzene; Carbosieve, prepared by thermally cracking polyvinylidine chloride, and Tenax GC, a porous polymer, 2-6-diphenyl-*p*-phenylene oxide. In this, as in other recent publications [57, 63, 64], Tenax GC was shown to be superior as a general adsorbent, having excellent temperature stability to 380° C and proving easy to handle. In these studies relative humidity, ubiquitous background levels, repetitive use, transportation and storage were all considered. Mieure and Dietrich [61] classified the adsorption/desorption characteristics of various column packings:

Chromosorb 101 adsorbs and desorbs acidic and neutral components
Chromosorb 105 adsorbs and desorbs low-boiling components
Tenax GC adsorbs and desorbs basic, neutral and high-boiling components.

Recently, a great deal of research effort has been directed towards the detailed evaluation of solid adsorbents for sampling of volatile hydrocarbons, including determination of sampling efficiency, safe sampling volumes, retention and breakthrough volumes [65–68]. In addition, numerical expressions have been derived on both practical and theoretical bases, for determining such variables [69].

Prior to the use of porous polymers such as Tenax GC or Tenax TA as adsorbents for volatile hydrocarbons, purification by solvent extraction and/or purging with an inert gas (helium or nitrogen) at elevated temperatures is necessary in order to remove trace contaminants which may interfere with subsequent analyses, following solvent or thermal desorption. A suitable procedure would involve extraction with superior grade methanol in a Soxhlet apparatus with two changes of fresh solvent and then thermal conditioning at 350° C in a metal column under a stream of helium carrier gas. After cooling, the purified adsorbent may then be packed into sampling tubes (typically 74 mm × 3 mm i.d. stainless steel), which should then be thermally conditioned prior to use.

8.2.1.3   *Gas sampling systems*

Gas tight syringes [16, 45] and glass aspirators and PTFE bags [16, 23, 70, 71] have all been used as sample containers. Ambient gas has been sampled using a modified gas pipette with Teflon coated rubber seals [72]. The use of Saran bags for the sampling of industrial atmospheres in the 200–1000 ppm range has been reported [70] and losses after 3 h of 15% and 40% for benzene and toluene respectively were recorded. A remote sequential sampler has been described [73], in which spring loaded syringes are activated by a timer and the samples obtained have been stored for 18 h without leakage. A small lightweight gas sampler for time integrated samples has been described [74]. An evacuated container equipped with a critical orifice is used to collect the sample, and the sampling period is dictated by the area of the orifice and the volume of the container. The system has been evaluated for the sampling of methane, ethane, benzene and hexane. Pressurized stainless steel containers have also been used for sampling of hydrocarbons in ambient air in Sydney [75]. Ideally, the sampling vessel should be of an inert nature, to minimize possible degradation reactions and should be shielded from light in order to prevent photochemical reactions. The suitability of various materials for sampling and storage of certain hydrocarbon compounds has been evaluated by a number of workers, some of whom reported that Tedlar bags were found to contaminate air samples with acetaldehyde and acetone [76]. Clearly, with any system rigorous contamination testing will be essential.

8.2.2   Analytical methods

8.2.2.1   *Continuous instrumental analysers*

The flame ionization detector, widely used in gas chromatography, has been adapted for use in instruments designed specifically for the measurement of ambient air concentrations of hydrocarbons. An important feature of such analysers is the ability to discriminate between total and non-methane hydrocarbons in the sample, for two reasons. Firstly, naturally-generated methane (typically present at concentrations of between 1.3 and 1.4 ppm in the atmosphere) is of little importance with respect to photochemical smog formation, owing to its low reactivity, and secondly, it is often important to obtain an independent measure of methane concentrations, particularly in the vicinity of such installations as sewage treatment plants or agricultural premises.

A number of instruments are currently available for determining total hydrocarbons, which operate upon the principle of feeding a continuous small sample stream of sample air, via a capillary restriction in order to damp variations in pressure and flow, to an FID, the electrical response of which is linearly proportionate to the concentration of hydrocarbon components in the sample stream. A schematic diagram of a typical continuous hydrocarbon analyser with non-methane facility is included in Fig. 8.2. Recent developments in the design of such monitors have included the incorporation of 'scrubber' systems for the removal of all non-methane hydrocarbons from the sample stream such that methane concentrations may be measured periodically at preselected intervals in order to derive non-methane hydrocarbon concentrations. However, such an instrument is incapable of providing continuous measures of both components. Several other types of non-methane hydrocarbon monitors are currently available, including those operating upon the gas-filter correlation principle, where continuous measures of both non-methane and total hydrocarbon concentrations are obtained [77].

Several commercial instruments equipped with FID are available for the continuous measurement of hydrocarbons at ambient levels. The FID is capable of providing a continuous response which is approximately linear for hydrocarbons over a wide range of concentrations. Carbon atoms bonded to hetero atoms (e.g. O, N or Cl) give a decreased FID response, and measurement of total hydrocarbons can be distorted by the presence of organic compounds which are not purely hydrocarbons. Comparative evaluations of the available instruments have been reported [78, 79] and their sensitivity, accuracy, portability and ease of operation discussed. Calibration techniques for such instruments have also been reported [80] using both standard calibration gases from cylinders and diffusion from permeation tubes.

The response of the FID to hydrocarbons is a function of the rate of introduction of carbon atoms into the flame. Thus for a given instrument setting,

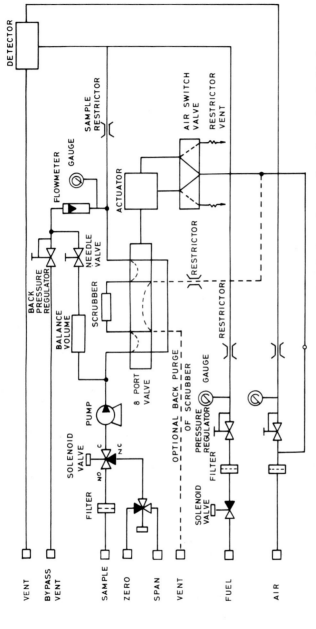

Fig. 8.2 Flow diagram of a commercially available non-methane hydrocarbon analyser (reproduced with the permission of Analysis Automation Ltd).

1 ppm ethane gives a response equal to 2 ppm methane or 0.5 ppm butene. For simplicity, therefore, the analyser is calibrated with a standard atmosphere of methane and ambient concentrations of total hydrocarbons are reported as ppm (as CH_4) or ppm C. In order to facilitate comparisons of total hydrocarbon and specific hydrocarbon concentrations, the latter are commonly reported as ppm C, derived by multiplying the concentration (in ppm v/v) by the number of carbon atoms in the molecule.

FID instruments have been operated under reduced pressure [81, 82] with the sample passing through the column and detector prior to entering the pump. This method has been applied to corrosive or unstable compounds. As previously noted, non-methane hydrocarbon measurements are of greater significance than methane measurements as methane contributes little to photochemical reactions.

There are now many commercial non-methane hydrocarbon monitors available, the majority of which operate by producing a value for the difference between total hydrocarbon and methane levels [83]. Methane and non-methane hydrocarbon levels may be monitored continuously and simultaneously using a dual flame instrument [84]. A system has been described [85] which measures methane and/or total hydrocarbons by selective combustion. Selective combustion has also been used, with the incorporation of a water sorption detector, to measure hydrogen, methane, reactive and non-reactive hydrocarbons. The air is first dried and then burnt, the reactive hydrocarbons being distinguished from the non-reactive species by the relative ease of their combustion. In this context the term reactive refers to the ease with which a particular hydrocarbon will undergo photochemical reaction with NO or O_3. All hydrocarbons appear to be capable of taking part in such processes, but the reaction rates are extremely dependent on molecular structure [86]. Methane is the least reactive and its estimated atmospheric residence time is up to 16 years [87], thus in terms of urban air quality it may be considered as photochemically inert. The most reactive of such species probably only have residence times of hours in ambient air and strong sunlight [88].

Instruments have been described [89–91] for the measurement of CO and CH_4 in which the CO is first reduced to CH_4 and subsequently detected by FID and detection limits of 'a few ppb' have been quoted [92]. Interference free measurements of CO, CH_4 and total hydrocarbons in the range 0–1 ppm have been obtained automatically with a cycle time of 5 min [93], similarly an automatic timing device has been incorporated into a total hydrocarbon analyser to enable repetitive measurements of CH_4, total hydrocarbons and reactive hydrocarbons [94]. An automatic gas chromatographic system, which measured CO, CH_4 and non-methane hydrocarbons in a single sample, has recently been described [95].

Instruments are also available for measurement of CH_4, C_2H_4, CO and total hydrocarbons [39, 96–98] with detection limits in the region of 200 ppb for total hydrocarbons and 1 ppm or better for individual components. An air quality chromatograph (AQC) system has been reported [99] and evaluated for the measurements of CH_4, C_2H_4, C_2H_2, propene, propane, iso- and normal-butane. CH_4 has been separated from the other hydrocarbons in air by the use of a cryogenic trap [100]. The device traps more than 95% of all non-methane hydrocarbons and when used in conjunction with a total hydrocarbon analyser makes the continuous determination of methane possible. A schematic diagram of a commercially-available air quality chromatograph is shown in Fig. 8.3.

Reactive hydrocarbons can also be estimated utilizing the chemiluminescent reaction of O atoms with unsaturated hydrocarbons at pressures of 1 mb which produces intense emission in the 700–900 nm region of the spectrum [101]. Owing to the high background level of methane, techniques for the determination of total hydrocarbons in air do not require exceptional sensitivity (merely a capability for precise operation at the ppm level). Although only of barely adequate sensitivity for ambient measurements, non-dispersive i.r. (Ndir) [102] has been a popular technique. The sensitivity and selectivity of the method has been improved by refinements in instrumental design [103].

High resolution i.r. spectroscopy has been used to monitor methane in the Pyrenees [104]. A problem in using i.r. for the measurement of trace levels of organics has been its limited sensitivity. The sensitivity has been improved [105] by using a scanning Michelson interferometer, cooled solid state detector, fast minicomputer and multiple pass long path length cells. The spectra are analysed by the computer which minimizes the interference from H_2O and CO and has enabled the measurement of organics at ppb levels.

Laser Raman spectroscopic techniques have been employed for the remote sensing of organics [106] and in particular for levels of CH_4 [107, 108], C_2H_4 [109] and C_2H_6 [110]. The detection limits for hydrocarbons using Laser Raman methods have been quoted as 1–7 ppm with a range resolution of 10 m [111]. Direct current discharge emission spectra have been reported for selected organics [112]. An advanced theoretical treatment of remote air pollution measurement has been published [113], in which several techniques made possible by the advent of high energy tunable laser sources are discussed and compared.

8.2.2.2 *Gas–liquid chromatography (GLC)*

There is a need for high specificity in measuring the components of an environmental sample but it is also desirable that the analysis may be performed routinely, rapidly and with precision. The apparatus should be simple if it is required for use in a mobile laboratory; thus permitting rapid analysis in a field situation. The application of GLC to the analysis of complex environmental air samples is the only really viable system, with FID being the most commonly used detector. Conventional packed columns remain adequate in many applications, but in order to achieve the required separation of the broad range of components detectable, and to enable accurate measurement of retention times, capillary columns have gained increasing use. There are no accepted standard methods available for the analysis of hydrocarbons; and it is difficult to envisage the establishment of any in the immediate future because of the large variation in the analytical requirements of specific applications.

Several GLC systems have been described and compared [114]; the authors

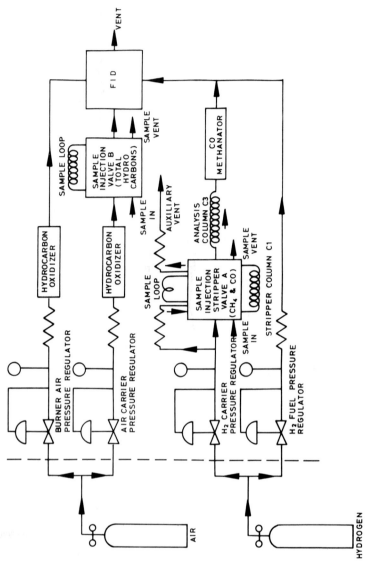

Fig. 8.3   Flow diagram of a commercially available air quality chromatograph for the measurement of total hydrocarbons, methane and carbon monoxide (courtesy of Beckman Instruments Ltd).

concluded that there is a need for better chromatographic methods designed to resolve the 50 to 100 or more readily measurable hydrocarbons in urban atmospheres. They added that methods should be developed, standardized and made generally available in order to enable more meaningful comparative and collaborative studies. There is also a requirement to reduce experimental error in identification and estimation of individual compounds, especially in the correlation of the photochemical reactivity of specific hydrocarbons with the resulting oxidant formation.

Rasmussen *et al.* [114] described 'some of the present-best GC column methods used in this laboratory'. They preferred a packed capillary column for the analysis of light hydrocarbons having compared its performance with both 3.1 mm o.d. packed columns and support coated open tubular (SCOT) columns. The actual column used was 6.1 m, 1.6 mm o.d., 0.8 mm i.d. packed with Durapak (n-octane, Porasil C). Sub-ambient temperature programming was found to give excellent separation of the C_2–C_6 hydrocarbon fraction. The construction of a suitable cryogenic GC oven has been described [115]. When considering methods for the analysis of hydrocarbons it is convenient to treat the compounds in groups specified by a range of molecular weights. Thus Rasmussen *et al.* [114] considered intermediate hydrocarbons, aromatic and aliphatic (C_6–C_{12}) compounds, and discussed their comparative chromatographic characteristics on several types of column: on open tubular columns of relatively large diameter capable of accepting high sample loadings, *m*-bis(*m*-phenoxyphenoxy)benzene (91.5 m, 1.5 mm o.d.) on a system using three packed 3.1 mm o.d. columns in series; and on small diameter 0.25 mm i.d. open tubular columns in conjunction with temperature programming designed to produce optimum resolution. They found a 0.5 mm i.d., 61 m OV-101 SCOT column to be most satisfactory. A recent publication [116] describes the application of photoionization detection (PID) coupled with FID for the analysis of C_4–C_{10} saturated, unsaturated and aromatic hydrocarbons in ambient air. Prior gas chromatographic separation was achieved using an SE-30 fused silica capillary column, temperature programmed from $-50°$ C to $80°$ C at $4°$ C min^{-1}. Absolute PID detection limits for the range of compounds considered were found to vary from 0.6 pg for aromatics to 15 pg for alkanes. The authors concluded that the dual-detector system provided reliable information on the various alkane, alkene and aromatic fractions in atmospheric samples. PID have also been used in conjunction with electron capture detectors (ECD) [117] for the analysis of organics in ambient air samples.

Jeltes and Burghardt [23, 118] have reviewed methods available for the analysis of the C_1–C_5 hydrocarbon fraction present in air and described an automatic GC method claiming it to be 'a relatively simple, direct and inexpensive method for sensitive and quick measurement'. Having considered both the spectroscopic and chromatographic methods available they chose to develop an automatic system for measurement combining GSC and FID. They listed the requirements for a monitor for measurement of hydrocarbons:

1. Maximum information; separated measurement of individual hydro-

carbons, measurement of ethylene (phototoxic), 'fingerprinting' of hydro-carbon fraction (source identification)
2. Relatively comfortable automized mode
3. As inexpensive as possible
4. Simplicity of the whole set-up
5. Quick sampling and analysis
6. Quick results for guarding and 'alarms'
7. Specificity
8. Sensitivity.

The merits of methods involving preconcentration as opposed to direct analysis for the two distinct hydrocarbon fractions (C_1-C_5) and (C_5-C_9) were also discussed. A 2 m $\times$ 4 mm i.d. column packed with 80–100 mesh alumina type F1 was employed for the analysis with air as carrier gas. The detection limits at a signal to noise ratio of 2 were 2–5 ppb for C_1-C_3 hydrocarbons and 7–20 ppb for the C_4-C_5 fraction. An Aquasorb filter was used to remove water from the air samples. The cycle time for this system was 45 min including n-C_5 compounds. Points 2 and 6 appear to lose a little clarity in the translation but the concepts of facile automation and rapid results are obviously desirable. An automatic system for C_1-C_5, total C_6 and benzene working on a 30 min cycle has also been described [119].

The problems associated with the condensation of water vapour are even more significant in cryogenic sampling systems. In order to minimize such difficulties samples have been passed through a desiccant prior to freeze-out: potassium carbonate has been reported [50, 120] to be extremely effective in this application owing to its very low affinity for organics. K_2CO_3 has been used prior to the trapping of organics in a cryogenic U tube containing glass beads [50], and the sensitivity was quoted as 0.2 ppb where the compounds were concentrated from 20 litres of air with 100% efficiency. A range of desiccants has been examined [120], including both anhydrous salts such as potassium carbonate, calcium sulphate and sodium sulphate and adsorbents of the type Sephadex (G-100) or Linde molecular sieve 3A. The same authors considered the removal of water by its conversion to acetylene and hydrogen by calcium carbide and calcium hydride.

A cryogenic sampling system incorporating a freeze-trap containing stainless steel gauze and employing potassium carbonate as desiccant has been reported [44] to give complete recovery of all hydrocarbons boiling above $-90°$ C with a detection limit of 0.001 parts in 10^8. Air samples (50–100 litres) were taken and liquid O_2 was used as coolant. The transfer of the concentrate from the trap was a two-stage process: initially by pressure equilization into evacuated containers and secondly by Toepler pumps to give a residual pressure in the trap of less than 1 Torr. Subsequent analysis was by GC using retention data from several columns or, with suitable samples, by GC/IR.

Light hydrocarbons have been measured [13] directly using a 5 ml sample, while C_2-C_5 aliphatic hydrocarbons (excluding C_4 alkenes) have been investi-gated [14] by a more sophisticated technique involving a silica gel packed column at liquid nitrogen temperature. The latter analysis employed a 2.4 m × 3.1 mm o.d. stainless steel analytical column packed with the same silica gel substrate and operated at 30° C. The same workers have measured C_4–C_8 aliphatic hydrocarbons and C_6–C_{10} aromatic hydrocarbons using different trapping and analytical columns for each group. The C_4–C_8 aliphatic hydrocarbons were trapped on a 0.3 m × 3.1 mm o.d. stainless steel column packed with 10% Carbowax 1540 on 60–80 mesh Gas Chrom Z, and separated on a 91.5 m × 1.5 mm i.d. open tubular column coated with dibutylmaleate at 0° C. The C_6–C_{10} aromatic hydrocarbons were trapped on a 0.3 m × 3.1 mm o.d. stainless steel column packed with 60–80 mesh glass beads, and analysed on a 91.5 m × 1.5 mm i.d. open tubular column coated with *m*-bis(*m*-phenoxyphenoxy)benzene at 70° C. The CO and methane species were separated on a 2.4 m × 3.1 mm o.d. stainless steel column packed with 60–80 mesh 13X molecular sieve (Wilken Instrument Co.) at 70° C. The three gas chromatographic procedures produced 52 measurable hydrocarbon peaks representing the C_1–C_{10} hydrocarbons. The higher molecular weight hydrocarbons (C_{11}–C_{12}) were observed on occasion but at very low concentrations. Some peaks contained unresolved shoulders. The overlap of alkane and alkene species was not a significant problem. There were interferences between benzene and some of the C_7 and C_8 alkanes, particularly methylcyclohexane. A direct comparison has been made of the resolution obtained for similar ambient air samples on a 2 m × 3.1 mm o.d. packed analytical column and on a 0.5 mm × 30.5 m SCOT column – the latter proving far superior [61].

Benzene and toluene have been analysed with other compounds [16] by trapping on a 3.1 mm i.d. × 203 mm stainless steel column packed with 20% Dow-Corning SF-20 on 60–80 mesh Columpak at dry ice/acetone temperature followed by analysis on 3.1 mm i.d. × 1.8 m stainless steel column packed with 9% SF96 on 60/80 Chrom W at 50° C.

Acetylene is one of the most dangerous hydrocarbon traces present in air as it may be accumulated and lead to explosions. This type of hazard is at a maximum where vast amounts of air are involved in industrial processes, typically in the steel industry (particularly in the LD basic oxygen process) and obviously in the production of liquid air. Acetylene was detected and quantified colorimetrically before the advent of gas chromatography, which is a more rapid and sensitive technique and also more readily automated. Several commercial instruments similar to those described elsewhere in this chapter for the analysis of total and non-methane hydrocarbons have been adapted for the monitoring of acetylene levels merely by the incorporation of a chromatographic system producing separation of acetylene from other volatile hydrocarbons. The work reported by Klein [121] who used a 3.4 m × 4 mm i.d. column packed with silica gel (grain size 0.3–0.5 mm) is typical. This readily separated acetylene from other hydrocarbons at 50° C using nitrogen carrier gas and reliably detected acetylene down to 5 ppb in 10 ml air samples.

Grob and Grob [19], in agreeing with a previous worker [122] that high molecular weight hydrocarbons in the atmosphere represent a complex mixture of hundreds of compounds, considered the analysis of such mixtures as three separate components. These three components were: high resolution trace analysis on capillary columns as a necessary prerequisite; appropriate sampling of volatile organics for qualitative and/or quantitative analysis; and gas chromatographic–mass spectrometric (GC/MS) design maintaining the full separation power of a high-resolution column and working in the ng range. They emphasized that the key to the possibility of injecting very dilute solutions on to capillary columns is the avoidance of stream splitting, referencing their own work (see also [123]). They also illustrated that direct sampling was not viable for all but the most abundant species, as the majority of components were present in the 0.1–10 ppb range and 0.1 ng of an individual compound was required for positive identification. 108 volatile organic substances were identified in the C_6–C_{20} range, the majority of which were aliphatic or aromatic hydrocarbons. The samples were trapped at ambient temperature in a glass tube containing 25 mg of wood charcoal as used for cigarette filters with an average particle size of 0.08 mm. Extraction of the filters was with carbon disulphide. The analyses were carried out on two GC columns, a 120 m × 0.33 mm Ucon HB 5100 for the full spectrum (C_6–C_{20}) and an 80 m × 0.33 mm Ucon LB550 for the volatile fraction (benzene to C_3-substituted benzenes) with temperature programming from ambient to 190 and 120° C respectively. A more recent application of this type of technique (carbon adsorption) allied to semi-automatic sampling and chromatography on highly polar columns has been reported [124] for the analysis of C_6–C_{20} fractions.

Using a similar approach but substituting a thermal desorption step for liquid extraction of the sampling trap Bertsch *et al.* [125] have recognized several hundred substances in the C_5–C_{16} range and identified almost 100. The sampling traps used were 100 × 8 mm i.d. stainless steel containing Tenax GC and the GLC was carried out on 100 m × 0.5 mm i.d. nickel open tubular columns coated as described elsewhere [25] with Emulphor ON-870.

Tenax GC has been used as an adsorbent [60] with the subsequent desorption step achieved by the passage of current through a heating tape attached to the sampling tube. The tube was heated to 260° C for 5 min, the desorbed components being retained at the front of the analytical capillary column at ambient temperature prior to temperature programmed elution.

Raymond and Guiochon have discussed the merits of graphitized carbon black as an adsorbent for a wide range of organic compounds [126] and subsequently described [25] its use as the sampling component of a GC/MS system, with which they identified more than 70 components in the air of Paris. These compounds were in the C_8–C_{18} range and predominantly aliphatic or aromatic hydrocarbons. Sampling was carried out at ambient temperature using tubes 5 cm × 4 mm i.d. packed with 0.25–0.40 g of 200–250 μg sieved graphitized carbon black. Desorption was achieved by heating the trap to 400° C within 15 s using an electric wire (Thermocoax) soldered around it. The analytical column was maintained at ambient temperature so that the compounds eluted from the sampling tube were retained at the column inlet; desorption usually took approximately 5 min. The analytical column used was 100 m × 0.4 mm i.d.

coated with OV-101 and was operated under temperature programming between ambient and 230° C.

A rather complex two-stage extraction/injection system has been described [57] which was designed to overcome problems of water condensation from the air sampled. The air sample was first concentrated on to Chromosorb 102 at ambient temperature; the subsequent 'thermal' desorption step was somewhat unusual in that the sampling tube was warmed only slightly above ambient but the collecting vessel (part of the injection port) was cooled to liquid N_2 temperatures. After condensation of the organic compounds the injection port was rapidly heated and analysis was carried out on a 30.5 m support-coated Carbowax 20 M column temperature programmed between 60 and 180° C.

Several reports [61, 127] have appeared describing the connection of the sample collection tube directly to the analytical column. An interesting extension of this approach was presented by Mieure and Dietrich [61], who designed their sample tube such that it could be incorporated into the injection port of their chromatograph. This design has considerable potential for less volatile hydrocarbons as it enables independent and rapid heating of sample tubes to higher temperatures than those required for analytical columns.

Thermal desorption has recently become the preferred method for transfer of adsorbed organic compounds from sampling tubes to gas chromatographic columns for analysis. A number of specifically-designed thermal desorbers are now commercially available, including both single- and two-stage models, for packed column and capillary column gas chromatography. In the simplest form, a thermal desorber consists of an oven or heating block, into which an adsorbent sampling tube is sealed, capable of attaining temperatures in excess of 300° C. During a period of approximately 5 min, while the sampling tube achieves the desired preselected temperature, the carrier gas flow bypasses the sampling tube. When the desired temperature is reached, the carrier gas is diverted through the sampling tube, by manual operation of a switching valve and the volatilized sample components are swept on to the gas chromatographic column for analysis. A simplified schematic diagram of a typical single-stage thermal desorber is shown in Fig. 8.4. This type of instrument can only be used successfully in conjunction with packed column gas chromatography. With the development and widespread use of glass and fused silica capillary columns in gas chromatography, two-stage thermal desorption instruments have been produced. These thermal desorbers contain, in addition to the primary sample tube heating oven, a secondary cold trap of a small volume, in which volatile components desorbed from sample tubes are cryogenically trapped and concentrated in a narrow band, suitable for injection on to narrow bore capillary columns by flash-heating to a temperature of up to 300° C with carrier gas flow. In this way, the superior resolution characteristics of capillary columns are used to advantage, particularly in the analysis of complex samples. A 'thermo-desorption cold trap injector' of this type has recently been described [128] and a schematic diagram is included in Fig. 8.5. A more comprehensive automated system for sequential unattended analysis of up to 50 sampling tubes, with

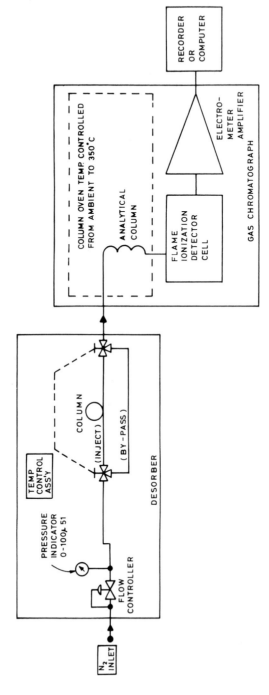

Fig. 8.4 Flow diagram of a typical single-stage thermal desorber.

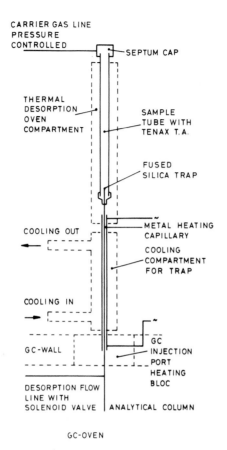

CARRIER GAS LINE
PRESSURE
CONTROLLED

SEPTUM CAP

THERMAL
DESORPTION
OVEN
COMPARTMENT

SAMPLE
TUBE WITH
TENAX T.A.

FUSED
SILICA TRAP

COOLING OUT

METAL HEATING
CAPILLARY

COOLING
COMPARTMENT
FOR TRAP

COOLING IN

GC-WALL

GC
INJECTION
PORT
HEATING
BLOC

DESORPTION FLOW
LINE WITH
SOLENOID VALVE | ANALYTICAL COLUMN

GC-OVEN

Fig. 8.5 Schematic representation of a commercially available two-stage thermal desorber (courtesy Chrompack (UK) Ltd).

integral pressure testing and fault finding facilities has been available for a few years and is compatible with both packed and capillary columns, utilizing either single or two-stage desorption. A diagram of this system is detailed in Fig. 8.6.

8.2.2.3   *Mass spectrometry and gas chromatography/mass spectrometry*

Mass spectrometry is almost mandatory if definitive identification of organic compounds in environmental samples is required. Even with this powerful technique it is often difficult to differentiate between the many similar compounds which may be present, and thus considerable effort is currently being directed towards improvement of MS techniques in this application.

GC/MS systems, many of which are interfaced to computerized data handling systems, are now in widespread use [51, 53, 57, 126, 129–133]. A high resolution

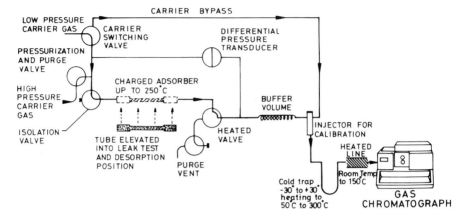

Fig. 8.6   Flow diagram of two-stage thermal desorber, capable of analysing up to 50 samples in automatic, unattended operation (reproduced with the permission of Perkin-Elmer Ltd).

mass spectrometer coupled to a computer has been used for the analysis of multi-component air samples [134].

A technique [135] which has found application in the estimation of refinery pollutant dispersion and in air quality control utilizes a mass spectrometer modified by the incorporation of a static condensation system and gives quantitative results for hydrocarbons of $\pm 10\%$ at the 5 ppm level and $\pm 20\%$ at 1 ppm.

A technique has been described by which it is claimed that micropollutants present in the air within the concentration range of 10^{-4} to $10^{-8}\%$ by volume may be separated and identified routinely on commercial equipment [136]. Enrichment was carried out on a microgradient tube followed by separation on a glass thin-film open tubular column by linear-programmed low temperature gas chromatography and identified by mass spectrometry. The authors reported that significant MS data was obtained for concentrations down to 0.02 ppb.

A portable MS based vapour detection system with complete digital control has recently been described [137] which was reported to give detection limits in the ppb range with a reproducibility of approximately 5% when used with a compatible GC system. The 'advantage' of cryogenic preconcentration of volatile trace components in air samples has recently been exploited [138] in chemical ionization (CI) mass spectrometry by using the condensed water as a secondary CI reagent gas. Further refinements in the cryogenic trap and heater system should produce lower detection limits.

A transportable MS system designed for either ambient air or workplace monitoring, with ppb detection limits for various components, has recently been described. Preconcentration of the air sample and injection into the quadrupole analyser is achieved by use of a selective membrane [139], giving concentration factors of up to 500 for selected compounds.

The development of fast-scanning analytical MS instruments which are compatible with capillary column gas chromatography has enabled the detection of a wide range of hydrocarbon-based materials in ambient air. Over 300 hydrocarbons and related compounds have been identified to varying degrees of certainty in air samples taken from the vicinity of a road tunnel [140], using capillary column GC–MS. Several other reports describe GC/MS analysis of various hydrocarbon compounds in ambient air [141–144].

8.2.2.4 *Calibration methods*

The types of analytical method described in this chapter do not determine the absolute amount of compound present but do measure a property of that compound which may be related to its concentration by an appropriate calibration procedure. Gas chromatographic techniques require particular care in calibration as the response of many components of a complex mixture may vary non-linearly with relatively slight changes in operating parameters.

Gas mixtures containing very low but accurately known levels of organic compounds are extremely difficult to prepare. The techniques which are available for the generation of such standard gas mixtures fall into two categories, static and dynamic procedures. Static procedures are usually necessary in the preparation of multi-component systems while dynamic methods are applied to the introduction of trace levels of single species into carrier or diluting gas streams.

A major problem in preparing very dilute mixtures is adsorption of trace components on to the walls of cylinders and tubing; this effect may be reduced by rigorous cleaning and a careful choice of materials. Mixtures of saturated hydrocarbons in clean glass or metal containers are relatively stable. Unsaturated compounds, however, especially C_2H_2 and butadiene, are slowly decomposed, particularly in the presence of water vapour and O_2.

Mixtures may be prepared in glass containers or purchased as 'standard gases' in cylinders; it is often unwise to place too much reliance on commercially available standards without first checking them against their specification. Mixtures for less rigorous work can be prepared in plastic bags but these must be used rapidly to avoid excessive diffusion losses.

Gas mixing pumps are available and the use of a two-stage apparatus has enabled the production of standards down to 1 ppm concentration levels. Another system, suitable for work not demanding the highest accuracy, employs motor driven syringes.

A more sophisticated and extremely accurate dynamic procedure employs permeation tubes, in which the component of interest, as the liquid or gas phase, is enclosed in a cylinder ampoule made of plastic (often PTFE). After a certain induction period the material permeates slowly through the plastic at a uniform rate by the diffusion controlled process. An apparatus based on this principle requires precise and stable control of two physical parameters, temperatures and dilution gas flow rate. These parameters are readily controlled to within 1% and thus the overall precision of the system is high. The absolute amount of a compound released has been determined in several ways; for liquid hydrocarbons gravimetric methods are accurate and simple to perform. Such techniques allow the confident production of standards at the ppb level. Diffusion tubes may also be used to produce mixtures at similar concentrations. Although not specifically related to their application to hydrocarbons, the principles of operation of diffusion tube devices have been extensively discussed [145–149].

In the capillary restricted flow method the calibration component to be diluted is dosed into a carrier gas stream through a restricting capillary. The concentration is calculated in accordance with the flow laws of Hagen-Poiseuille. As the capillary radius enters these calculations raised to the 4th power it must be known precisely. This procedure has application as an absolute primary standard with an accuracy of 1–2% [150].

8.2.3   Methods for specific compounds

8.2.3.1 C_2-C_5 hydrocarbons

Since these compounds (with the exception of n-pentane) are in the gaseous form at ambient temperatures, special sampling procedures are required for quantitative analysis, at the concentrations normally encountered in ambient air, usually involving a cryogenic preconcentration step originally developed by Westberg *et al.* [45]. The method described here has been utilized previously for the analysis of low molecular weight compounds in ambient air in the vicinity of industrial complexes [151].

Reagents

   (a) Anhydrous potassium carbonate, 20/60 mesh
   (b) 60 mesh glass beads
   (c) Liquid oxygen or liquid nitrogen
   (d) Standard gas mixtures C_2-C_5 hydrocarbons.

Equipment

   (a) Copper drying tubes, 45 cm × 6.2 mm o.d., 'U'-shaped
   (b) Stainless steel sampling trap, 15 cm × 1.66 mm o.d., 'U'-shaped
   (c) Solvent-washed glass wool
   (d) Gas-tight syringes 1–5 ml capacity
   (e) Liquid oxygen/nitrogen bath
   (f) Potassium permanganate crystal pre-filter
   (g) Six-port sampling valve
   (h) FEP-Teflon sampling bags (10 l) enclosed in rigid containers
   (i) Gas chromatograph equipped with flame ionization detector (FID) and 6 m × 1.65 mm stainless steel column packed with Durapak n-octane/Porasil C. A sub-ambient temperature programming unit is necessary to achieve the required initial column temperature

(j) Water bath
(k) Constant flow pump.

Preparation of sample traps and drying tube

The glass beads should be suitably cleaned prior to use, preferably by soaking overnight in chromic acid, rinsing copiously with distilled water and drying in an oven at 120° C. These should then be packed into the 'U' tube sampling trap under gentle vacuum and the ends plugged with glass wool. The drying tube should be packed with anhydrous potassium carbonate in a similar manner and purged overnight with an inert dry gas (helium) at 120° C prior to use.

Sampling

The FEP-Teflon bags, sited at the appropriate locations, are filled with the sample air in the following manner. The bags are first evacuated and sealed using a toggle valve. Air is then drawn out of the surrounding container until a vacuum persists which is sufficient to allow the bag to expand fully. The toggle valve is then opened and air flows into the bag as it expands to fill the container. The sampling period may be controlled by the vacuum inside the container and the valve orifice. Samples taken in FEP-Teflon bags contained within glass vessels should be protected from light.

The use of the $KMnO_4$ crystal prefilter while sampling ambient air is recommended in order to remove traces of ozone, which may react with hydrocarbons.

Analysis

The expanded FEP-Teflon bag is connected through the drying tube into the six-port valve, through the sample trap, which is immersed in liquid oxygen, and thus to atmosphere where a measured volume of the sample may be drawn through the trap. C_2–C_5 hydrocarbons are then immobilized on the cooled trap. The six-port valve is then operated such that helium carrier gas (7 ml min^{-1}) flows through the sampling trap (which is warmed with a hot (60° C) water bath) and on to the gas chromatographic column maintained at $-70°$ C. After 4 min, the column temperature is programmed to 65° C and held for 25 min. Component gas chromatographic peaks are identified on the basis of retention time.

Calibration

This is achieved by the use of proprietary gas mixtures of known concentration, which may be further diluted if necessary to obtain calibrations in the desired range. These standard mixtures may be directly injected on to the sampling trap by gas syringe or, if desired, may be injected into FEP-Teflon bags and analysed in a similar manner to ambient air samples.

8.2.3.2 C_6–C_9 hydrocarbons

The method detailed below involves sampling by adsorption using Tenax GC sampling tubes and analysis by thermal desorption/gas chromatography with a packed column and FID. The compounds determined range from benzene to isopropyl benzene in ambient air. Further details and application of the method may be found elsewhere [152].

Reagents

   (a) Tenax GC 35–60 mesh size, purified and conditioned as described in Section 8.2.1.2
   (b) Benzene, toluene, xylene, trimethylbenzene, isopropyl benzene, all of analytical reagent standard (>99% purity).

Equipment

   (a) Stainless steel sampling tubes 74 mm long × 4 mm i.d.
   (b) Solvent washed glass wool
   (c) Gas-tight syringes 1–5 ml
   (d) Constant flow sampling pump
   (e) Thermal desorption system
   (f) Gas chromatograph with FID containing a 1.8 m × 3.1 mm o.d. stainless steel column, packed with 10% TCEP on 80/100 mesh chromosorb WHP.

Preparation of sampling tubes

Tenax GC (0.13 g previously purified) is slowly added to a stainless steel tube, with gentle tapping or vibration, to ensure close packing of the particles, and the ends are packed tightly with glass wool plugs. The tubes are then conditioned at 300° C in the gas chromatograph oven with a carrier gas flow of 15–30 ml min^{-1} nitrogen for 16 h. After cooling, the ends of the tubes are tightly sealed with Swagelok blanking plugs prior to use.

Sampling

A sampling tube is connected to the sampling pump via PTFE or PVC tubing and ambient air is drawn through the tube at a flow rate for a specified time period such that a total sample volume of 6.3 l is not exceeded, this being the retention volume for benzene [152]. A typical sampling period and flow rate would be 20 min at 200 ml min^{-1}. The tube is then resealed and taken for analysis.

Analysis

The tubes containing adsorbed materials are transferred to the thermal desorber and allowed to attain a temperature of 250° C. The GC carrier gas is then diverted through the heated tube and the desorbed sample components are swept on to the analytical column which is maintained isothermally at 70° C with a carrier gas flow rate of 40 ml min^{-1}. Component gas chromatographic peaks are identified on the basis of retention times. This procedure has been used previously for monitoring aromatic hydrocarbon concentrations in ambient air at urban, rural and motorway locations [153, 154].

Calibration

This is achieved by loading known quantities of each of the compounds to be analysed on to sampling tubes, using one of the methods described in Section 8.2.2.4. The most accurate procedure would involve the use of a permeation tube oven with accurate temperature and gas flow control, although an alternative method using a permeation oven coupled to an exponential dilution flask has been successfully used for calibrating a thermal desorption/GC analytical system for a range of aromatic hydrocarbons [152].

8.3 Hydrocarbon fraction of airborne particulate matter

Methods for the analysis of particulate organic air pollutants have been reviewed [155]. Such analyses are not simple – over 100 PAH species alone have been separated and identified by GC/MS techniques in a single study [54]. In this study airborne particulate samples were collected on glass fibre filters using a high volume sampler; the filters were Soxhlet-extracted using cyclohexane and extractable material was subjected to a Rosen separation. The PAH fraction was concentrated and injected into a GC/MS system. Samples of less than 100 μg produced good MS data for the individual components emerging from the GC column in ng quantities. A number of other procedures have been used for the analysis of PAH in air. Most accepted methods including the ASTM Standard (1971) consist of an extraction process, a fractionation by column chromatography and the spectrophotometric analysis of the eluted fractions. The complexity of the samples involved, however, continues to promote research interest, often directed towards achieving a more convenient analytical procedure for an extended range of compounds. GC and LC have provided avenues of approach to this problem.

8.3.1 Sampling procedures

Most methods for the analysis of organic particulates in ambient air depend on filter collection and some form of chromatographic separation. In a recent survey at locations throughout Los Angeles County [156] conventional high volume samplers as described by the Intersociety Committee [157] were used on an intermittent schedule. The full 'tentative method' [56] is a lengthy process involving particulate sample collection on glass fibre filters, Soxhlet extraction, column chromatography and u.v.-visible spectrophotometry, and owes much to the early work of Sawicki and co-workers [158–161]. Many people have attempted to apply more direct chromatographic techniques to the problem and an authoritative series of papers relating to the analysis of PAH in the environment has been published by Lao and co-workers [162–164]. These workers also found the use of commercial high volume samplers, using glass fibre filters, adequate for the collection of airborne particulate samples. Filters are weighed before and after sampling to determine the weight of entrained particulate material. It has long been suspected that the volatility of PAH might prevent their complete collection by particulate filter techniques and also cause loss after collection. If equilibrium between solid and vapour (particulate on filter and vapour in passing air respectively) is established, then losses during collection depend on the equilibrium vapour concentration (EVC). These effects have been considered [163–165] and the question posed [163]: do the high volume particulate filter techniques which are presently used yield meaningful measurements of the levels of these pollutants? The authors concluded that qualitative evidence indicated that considerable losses occur during ambient air sampling of compounds having EVCs of 500 ng m^{-3} or higher at ambient temperatures and noted that such species included pyrene and benzo(a)anthracene.

The problems associated with quantitative sampling of both particulate and vapour phase PAH and 'blow-off' from high volume filters during sampling may be resolved by the use of porous polymer or cryogenic back-up traps in order to achieve gas-phase enrichment of PAH [166]. These measures have been recommended following determination of the gas phase–particulate matter distribution of PAH [167, 168]. Other solutions to this problem have included the use of polyurethane foam plugs [169] and high volume filters coated with glyceroltricaprylate [170, 171]. An additional complication with extended high volume sampling is the possible reaction of entrained PAH with reactive gases in the atmosphere such as ozone, oxides of nitrogen and sulphur dioxide [172] although this is only normally a problem with samples of vehicle exhaust or stack gases, where relatively high concentrations of these reactant gases may be present.

8.3.2 Extraction and clean-up procedures

PAH are soluble in many organic solvents including cyclo-hexane, benzene, chloroform, acetone and alcohols, but the efficiency of such solvents in the extraction of PAH is largely dependent on the nature of the particulate matter under investigation. So strongly are PAH adsorbed by carbon black [173], for example, that many hours are required to solvent extract samples of airborne particulates and other source materials containing carbon black. There is by no means agreement on the length of time for complete extraction, and the various workers in the field have suggested 2 h [174], 12 h [175] and 20–30 h [176] using cyclohexane. The Intersociety Committee has adopted 24 h with cyclohexane for one method [33] and 6–8 h for another [34], and also 6–8 h with benzene [32]. A quantitative investigation into the extraction efficiencies of these solvents has been carried out [177], in which 76% and 95% recoveries of benzo(a)pyrene from enriched air samples were obtained after 6 h refluxing with cyclohexane and benzene respectively. Possible losses of PAH as the result of decomposition during long periods of Soxhlet extraction have been investigated [178]. The possibility of decomposition and the length of time involved in solvent extraction methods makes the use of alternative extraction procedures very attractive. Such a procedure has been reported [179] using ultrasonic vibration at ambient temperature and requiring only 30 min. Vacuum sublimation has also been used with a certain degree of success for extraction of PAH from carbonaceous materials, but this procedure was found to be particularly time consuming for the higher molecular weight compounds [180]. Having extracted the organic matter from the particulate sample, the aromatic fraction is usually separated by liquid–liquid partition followed by benzene elution from a silica gel chromatographic column [181]. Tabor *et al.* [182] chromatographed the neutral fraction of the organic material in air pollution samples on silica gel, eluting the aliphatics with iso-octane and the aromatics with benzene. A quantitative method for PAH based on this procedure required additional chromatography, on alumina [183]. These types of procedure enable the production of simplified mixtures of PAH species from the more complex arrays of organics present in the particulate extracts.

More recent clean-up procedures have involved the use of alumina column chromatography followed by gel permeation chromatography with Sephadex LH-20 [184] which has the advantage of increased sample capacity, improved separation from interfaces and reproducibility.

8.3.3 Analytical methods

8.3.3.1 *Thin-layer chromatography*

In general thin layer chromatography as applied to air pollutants has been concerned with qualitative analysis, although some semi-quantitative work has been done. Most analyses have been carried out on benzene extracts of particulate fractions, often following clean-up on a chromatographic column. Identification of separate species has normally involved fluorescence spectrophotometry or occasionally absorption spectrophotometry.

The use and relative merits of three TLC systems for the analysis of 20 PAH extracted from airborne particulate matter has been described [185]. The systems described were: (i) aluminium oxide with pentane-ether (19:1, v/v); (ii) cellulose with *N,N*-dimethylformamide–water (1:1, v/v); (iii) cellulose acetate with ethanol–toluene–water (17:4:4, v/v). The third system has enabled the separation of benzo(a)pyrene and benzo(k)fluoranthene [186]. Quantitative determinations of benzo(a)pyrene have been made using aluminium oxide–cellulose (2:1, w/w) with ethanol–toluene–water (17:4:4, v/v) [178]. A two-dimensional system has proved satisfactory in the separation of benzo(a)pyrene and benzo(k)fluoranthene employing aluminium oxide–cellulose acetate (2:1, w/w) with pentane (first dimension) and ethanol–toluene–water (17:4:4, v/v) (second dimension) [187]. The significance of PAH in the environment is their carcinogenic health hazard. The hazards posed by individual members of this group of compounds vary over several orders of magnitude and most researchers are seeking a convenient quantitative technique, particularly for the analysis of the most potent carcinogens. It is unlikely that TLC will provide such a technique in the near future.

8.3.3.2   *High-performance liquid chromatography*

The distinctive features of gas chromatography are the use of long pressurized columns, a long-lived adsorbent, a single mobile phase throughout the procedure and the highly reproducible retention volumes that the system provides. Many of the criteria applied to GC have been adapted to various forms of liquid chromatography, liquid–solid, liquid–liquid, gel and ion exchange. GC has often been considered inappropriate to the analysis of PAH because of the thermal instability of that class of compounds and the low volatility of the higher molecular weight species. During the relatively short and rapidly advancing history of modern liquid chromatography many workers have been of the opinion that some variation of this technique will provide the most convenient procedure for the rapid analysis of an extensive range of PAH, and particularly for the separation of high molecular weight compounds. Since Karr *et al.* [188] used a 7.6 m column of 6.2 mm i.d. copper (providing a length:diameter ratio of 1200) packed with 80/100 mesh alumina (4% water content), and cyclo-hexane at 50 lb inch^{-2} to separate a synthetic mixture of PAH, applications of LC techniques to such analyses have been numerous.

A significant advance in the application of HPLC to PAH analysis was reported by Schmit *et al.* [189], who permanently bonded octadecyl silane to spherical siliceous particles with a porous surface of specific thickness and pore size [190]. Such columns have been used to give, for example, near baseline separation of the carcinogenic benzo(a)pyrene and its non-carcinogenic isomer benzo(e)pyrene. 'Pressure-assisted reverse-phase LC' has been described for the analysis of PAH in engine oil [191] using Corasil/C_{18} and methanol–water solvents. It seems probable that such a system would be effective for the analysis of particulate air samples. The separation of benzo(a)pyrene from its isomers has been reported [192] using cellulose acetate columns and ethanol/dichloromethane solvent (2:1).

A recent publication has compared the performance of eight commercially available reverse phase C_{18} HPLC columns [193] and has indicated that only two (Zorbax-ODS and Vydac-201TP) are particularly satisfactory for separation and resolution of PAH. Although this technique does not offer as efficient resolving power as capillary GLC, the wider selection of column materials and solvent systems allows the separation of many isomeric compounds not normally separable by GLC. Detection systems which have been used in the analysis of PAH include u.v. absorption and u.v. fluorescence, the latter technique being typically 10–100 times more sensitive than the former. Detection of picogram and sub-picogram amounts of certain PAH has been achieved by use of a fluorescence detector incorporating a small-volume (3 μl) flow-through cell [194]. A comprehensive analytical scheme for the determination of 13 PAH compounds in airborne particulate samples has been described recently [195]. Samples (1 g) were Soxhlet extracted for 48 h with methylene chloride (450 ml). The extracts were then evaporated, redissolved in cyclohexane and partitioned into nitro-methane. This solution was then analysed by HPLC using a 5 μm polymeric C_{18} column (Vydac 201TP) with a linear solvent gradient from 40% acetonitrile in water to 100% acetonitrile in 45 min at a flow rate of 1.5 ml min^{-1}. Detection was achieved using a programmable wavelength fluorescence unit, such that the excitation and emission wavelengths could be changed automatically during the chromatographic analysis in order to optimize sensitivity and selectivity for particular PAH compounds. Initial separation of the PAH compounds into five fractions, based on carbon number, using an aminosilane HPLC column, further simplified the final analysis.

8.3.3.3   *Gas–liquid chromatography*

The analysis of the C_{15}–C_{36} n-alkane range in benzene-soluble fractions from airborne particulates has been achieved on a 6.1 m × 3.1 mm o.d. stainless steel column packed with 3% SE30 on Chromosorb W (100/200 mesh) [195]. This study demonstrated that petrol and diesel fuel combustion processes can generate this range of aliphatic compounds in particulate matter, although they are not present in the fuels. A similar range of n-alkanes (C_{11}–C_{33}) was analysed on a glass column, 5.4 m × 6 mm o.d. packed with 1% OV-7 on Chromosorb W (80/100 mesh) by Lane *et al.* [28] who also separated 20 PAH.

Lao *et al.* [162] have concluded a feasibility study into the quantitative measurement of sub-μg quantities of PAH in polluted air using GC/MS by stating that such techniques 'can provide the definitive approach to complete PAH analysis which to date has been impossible'. The same authors subsequently [163] add that 'the GC-FID-Quadrupole-MS-Computer system is, at present, the method of choice for PAH analysis in all types of environmental samples. In their first study both packed and SCOT Dexsil-300 columns were used, with the packed columns proving more satisfactory. The claims made by these workers may appear extravagant but they have recorded some very impressive chromatograms. Despite such claims it seems unlikely that any one system will provide adequate separation of all the PAH species present in airborne particulates – but specialized techniques have enabled resolution of some extremely stubborn fractions. For example [196], GLC in the nematic region of *N*,*N'*-bis(p-methoxybenzylidene)-α,α'-bis-toluidine, utilizing the unique selectivity of this liquid phase, based upon differences in the molecular length to breadth ratio of solute geometric isomers, has enabled the separation of 16 3–5 ring PAH compounds. Gas chromatography has been applied extensively to the analysis of PAH although the low volatility of these compounds presents a problem. The use of packed columns has achieved only partial separation and resolution of individual PAH, most success being obtained with non-polar liquid phases such as OV-17 and high-temperature phases, e.g. Dexsil 300 [197]. By far the greatest success has been obtained with the use of glass, and, more recently, fused silica capillary columns. High-efficiency separation and resolution of the common PAH from naphthalene to coronene are possible with only the benzfluoranthenes presenting particular separation problems [198]. The typical detection limits for individual PAH by capillary column gas liquid chromatography with flame ionization detection are of the order 0.1–0.5 ng [199].

A review concerned with the performance and applications of capillary columns in PAH analysis has been published recently [200]. A typical scheme for the analysis of PAH in airborne particulate matter, utilizing capillary column gas–liquid chromatography with flame ionization detection has been described [201]. Particulate matter obtained by high volume sampling was extracted by Soxhlet for 24 h using methylene chloride.

8.3.3.4   *Gas–liquid chromatography/mass spectrometry*

The use of this technique for the identification and quantification of PAH in samples of airborne particulate matter constitutes an extremely sensitive and reliable analytical method, particularly where capillary columns are employed, although packed column GC/MS has been used to considerable effect [202]. Under normal electron impact ionization (EI) conditions, mass spectra of isomeric PAH pairs are virtually identical, such that confirmatory retention time data are required for positive identification. Since intense molecular ions are produced under EI conditions, multiple ion monitoring techniques may be employed to good advantage and detection limits in the picogram (10^{-12} g) range may be achieved.

8.4   Carbon monoxide

Several instrumental methods are available for the measurement of CO at the concentrations normally encountered in ambient atmospheres, including electrochemical, non-dispersive infrared (Ndir) and gas chromatographic. Of these, the Ndir and gas filter correlation continuous analysers have proved to be most sensitive, stable and accurate.

CO exhibits i.r. adsorption of sufficient strength to be detected over reasonable path lengths at normal urban atmospheric concentrations. The use of a 10 m path length i.r. cell enables the measurements of concentrations in excess of 10 ppm to within 10%. The actual measurement made is of the i.r. absorbance at 4.67 μm. The vast majority of published data on CO levels have been obtained using Ndir methods, and such measurements usually have an uncertainty of about 1 ppm.

The essence of the Ndir method is the design of the detector, which is a two-sided cell containing equal concentrations of mixture of CO and an i.r. transparent gas. Infrared radiation is incident on both halves of the detector, one beam having passed through a sample cell and the other through a reference cell containing no CO. There is an integrated absorption of i.r. energy over all wavelengths passed by the optical system by the CO present in both halves of the detector. This absorbed energy is converted into heat producing a change in pressure or volume of the gases in the detector. If CO is present in the sample cell it will absorb radiation, reducing that incident on the detector and producing a decreased heating effect. It is normally the unequal volume change between the sample and reference sides of the detector which is converted into an electrical signal directly related to the CO concentration. Principal interferences are from H_2O and CO_2 but these may be greatly reduced by the use of filter cells. The

Intersociety Committee has produced a full tentative method together with a calibration procedure [148]. Infrared gas filter correlation instruments with a detection limit of 0.1 ppm are now available.

References

1. Leinster, P., Perry, R. and Young, R. J. (1977) *Talanta*, **24**, 205.
2. Leggett, D. C., Murrmann, R. P., Jenkins, T. J. and Barriers, R. (1972) *US N.T.I.S. AD Rep.* 745125.
3. Darley, E. F., Biswell, H. H., Miller, G. and Goss, J. (1973) *J. Fire Flammability*, **4**, 74.
4. Rasmussen, R. A. (1970) *Environ. Sci. Technol.*, **4**, 667.
5. Rasmussen, R. A. (1972) *J. Air Pollut. Control Assoc.*, **22**, 537.
6. Abeles, F. B. and Heggestad, H. F. (1973) *J. Air Pollut. Control Assoc.*, **23**, 517.
7. Perry, R. and Slater, D. H. (1975) *Chemistry and Pollution* (eds F. R. Benn and C. A. McAuliffe), Macmillan, London.
8. Huess, J. M. and Glasson, W. A. (1968) *Environ. Sci. Technol.*, **2**, 1109.
9. Glasson, W. A. and Tuesday, C. S. (1971) *Environ. Sci. Technol.*, **5**, 151.
10. Breeding, R. J., Klonis, H. B., Lodge, J. P. *et al.* (1976) *Atmos. Environ.*, **10**, 181.
11. Nerat, G. (1973) *Ingenieursblad*, **42**, 109.
12. Altshuller, A. P., Lonneman, W. A. and Kopczynski, S. L. (1973) *J. Air Pollut. Control Assoc.*, **23**, 597.
13. Altshuller, A. P., Lonneman, W. A., Sutterfield, F. D. and Kopczynski, S. L. (1971) *Environ. Sci. Technol.*, **5**, 1009.
14. Lonneman, W. A., Kopczynski, S. L., Darley, P. E. and Sutterfield, F. D. (1974) *Environ. Sci. Technol.*, **8**, 229.
15. Mayrsohn, H. and Crabtree, J. M. (1976) *Atmos. Environ.*, **10**, 137.
16. Pilar, S. and Craydon, W. F. (1973) *Environ. Sci. Technol.*, **7**, 628.
17. Lonneman, W. A., Bellar, T. A. and Altshuller, A. P. (1968) *Environ. Sci. Technol.*, **2**, 1017.
18. Perry, R. and Twibell, J. D. (1974) *Biomed. Mass Spec.*, **1**, 73.
19. Grob, K. and Grob, G. (1971) *J. Chromatogr.*, **62**, 1.
20. Smoyer, J. C., Shaffer, D. E. and De Witt, I. L. (1971) *Inst. Environ. Sci. Tech. Meet. Proc.*, **17**, 339.
21. Miller, S. S. (1971) *Environ. Sci. Technol.*, **5**, 503.
22. Grob, R. L., Schuster, J. L. and Kaiser, M. A. (1974) *Environ. Lett.*, **6**, 303.
23. Jeltes, R. and Burghardt, E. (1972) *Atmos. Environ.*, **6**, 793.
24. Kim, A. G. and Douglas, L. J. (1973) *J. Chromatogr. Sci.*, **11**, 615.
25. Raymond, A. and Guiochon, G. (1974) *Environ. Sci. Technol.*, **8**, 143.
26. Ciccoli, P., Garetti, G., Liberti, A. and Passanzini, L. (1974) *Ann. Chim.*, **64**, 753.
27. Burghardt, E. and Jeltes, R. (1975) *Atmos. Environ.*, **9**, 935.
28. Lane, D. A., Moe, H. K. and Katz, M. (1973) *Anal. Chem.*, **45**, 1776.
29. Shultz, J. L., Sharkey, A. G., Friedel, R. A. and Nathanson, B. (1974) *Biomed. Mass Spec.*, **1**, 137.
30. Brocco, D., Dipalo, V. and Possanzini, M. (1973) *J. Chromatogr.* **86**, 234.
31. Giger, W. and Blumer, M. (1974) *Anal. Chem.*, **46**, 1663.
32. Sawicki, E., Carey, R. C., Dooley, A. E. *et al.* (1970) *Health Lab. Sci.*, **7**, 31.
33. Sawicki, E., Carey, R. C., Dooley, A. E. *et al.* (1970) *Health Lab. Sci.*, **7**, 45.
34. Sawicki, E., Carey, R. C., Dooley, A. E. *et al.* (1970) *Health Lab. Sci.*, **7**, 60.

35. Fallick, G. J. and Walters, J. L. (1972) *Am. Lab.*, **4**(8), 21.
36. Fox, M. A. and Staley, S. W. (1976) *Anal. Chem.*, **48**, 993 and references therein.
37. Kertesz-Saringer, M. and Morlin, Z. (1975) *Atmos. Environ.*, **9**, 831.
38. US Federal Registry (1971) 36 (228) 22384.
39. Villalobos, R. and Chapman, R. L. (1971) *Anal. Instrum.*, **9**, D-6, 1.
40. Perry, R. and Harrison, R. M. (1976) *Chem. Brit.*, **12**, 185.
41. Bongers, W. (1976) *Stichting Cowcawe, Hague*, 3/76.
42. Zlatkis, A., Lichtenstein, H. A. and Tishbee, A. (1973) *Chromatographia*, **6**, 67.
43. Rasmussen, R. A. and Holden, M. W. (1972) *Chromatogr. Newslett.*, **1**, 31.
44. Narain, C., Marron, P. J. and Glover, J. H. (1972) *Proc. Int. Symp. Gas Chromatogr.*, **9**, 1.
45. Westberg, H. H., Rasmussen, R. A. and Holdren, M. W. (1974) *Anal. Chem.*, **46**, 1852.
46. Rasmussen, R. A. and Hutton, R. S. (1972) *Bio. Science*, **22**, 294.
47. Rasmussen, R. A. (1972) *Am. Lab.*, **4**(7), 19.
48. Rohrscheider, L., Jaeschke, A. and Kubik, W. (1971) *Chem. Ing. Tech.*, **43**, 1010.
49. Kaiser, R. E. (1973) *Anal. Chem.*, **45**, 965.
50. Tyson, B. J. and Carle, G. C. (1974) *Anal. Chem.*, **46**, 610.
51. Schubert, R. (1972) *Anal. Chem.*, **44**, 2084.
52. Leinster, P. (1977) PhD Thesis, University of London.
53. Rollet, M. and Moisson, M. (1972) *Rev. Inst. Pasteur Lyon*, **5**, 439.
54. Crecelius, H. J. and Forweg, W. (1975) *Staub-Reinhalt. Luft*, **35**, 330.
55. Aue, W. A. and Teli, P. M. (1971) *J. Chromatogr.*, **62**, 15.
56. Perry, R. and Twibell, J. D. (1973) *Atmos. Environ.*, **7**, 929.
57. Bertsch, W., Chang, R. C. and Zlatkis, A. (1974) *J. Chromatogr. Sci.*, **12**, 175.
58. Dravnieks, A., Kiotoszynski, B. K., Whitfield, J. *et al.* (1971) *Environ. Sci. Technol.*, **5**, 1220.
59. Bourdin, M., Badre, R. and Dumas, C. (1975) *Analusis*, **3**, 34.
60. Russell, J. W. (1975) *Environ. Sci. Technol.*, **9**, 1175.
61. Mieure, J. P. and Dietrich, M. W. (1973) *J. Chromatogr. Sci.*, **11**, 559.
62. Pellizzari, E. D., Bunch, J. E., Berkley, R. E. and McRae, J. (1976) *Anal. Lett.*, **9**, 45.
63. *Idem* (1976) *Anal. Chem.*, **48**, 803.
64. Novotny, M., Lee, M. L. and Bartle, K. D. (1974) *Chromatographia*, **7**, 333.
65. Vejrosta, J., Roth, M. and Novák, J. (1981) *J. Chromatogr.*, **219**, 37.
66. Vejrosta, J., Roth, M. and Novák, J. (1981) *J. Chromatogr.*, **217**, 167.
67. Janák, J., Ruzicková, J. and Novák, J. (1974) *J. Chromatogr.*, **99**, 689.
68. Piecewicz, J. F., Harris, J. C. and Levins, P. L. (1979) *Further Characterization of Sorbent Resins for Use in Environmental Sampling*, EPA Report-600/7-79-216, Research Triangle Park, NC.
69. Senum, G. I. (1981) *Environ. Sci. Technol.*, **15**, 1073.
70. Desbaumes, E. and Imhoff, C. (1971) *Staub-Reinhalt. Luft*, **31**, 257.
71. Dimitriades, B. and Seizinger, D. E. (1971) *Environ. Sci. Technol.*, **5**, 223.
72. Schneider, W. and Fronhne, J. C. (1975) *Staub-Reinhalt. Luft*, **35**, 275.
73. Griffith, G. A., Drivas, P. J. and Shair, R. F. (1974) *J. Air Pollut. Control Assoc.*, **24**, 776.
74. Williams, F. W., Stone, J. P. and Eaton, H. G. (1976) *Anal. Chem.*, **48**, 442.
75. Nelson, P. F. and Quigley, S. M. (1982) *Environ Sci. Technol.*, **16**, 650.
76. Cox, R. D., McDevitt, M. A., Lee, K. W. and Tannahill, G. K. (1982) *Environ. Sci. Technol.*, **16**, 57.

77. Burch, D. E. (1980) EPA report EPA-600/2-80-201, Environmental Protection Agency, Cincinnati, Ohio.
78. Dennison, J. E., Viscon, R. E. and Broyde, B. (1973) *West Electr. Eng.*, **17**, 3.
79. Derwent, R. G. and Stewart, H. N. M. (1974) *Meas. Control*, **7**, 101.
80. Decker, C. E., Royal, T. M. and Taummerdahl, J. B. (1974) *Gov. Rep. Announce (US)*, **74**, 68.
81. Frostling, H. and Brantte, A. (1972) *J. Phys. A: Gen. Phys.*, **5**, 251.
82. Fischer, H., Neafelder, Koh, G. and Pruggmayer, D. (1974) *Fachz. Lab.*, **18**, 214.
83. Saena, O., Boldt, C. A. and Tarazi, D. S. (1972) *Gov. Rep. Announce (US)*, **72**, 64.
84. Poli, A. A. and Zinn, T. L. (1973) *Anal. Instrum.*, **11**, 135.
85. King, W. H. (1974) *Environ. Sci. Technol.*, **4**, 1136.
86. Altshuller, A. P. and Bufalini, J. J. (1971) *Environ. Sci. Technol.*, **5**, 39.
87. Kamens, R. M. and Stern, A. C. (1973) *J. Air Pollut. Control Assoc.*, **23**, 592.
88. Kopczynski, S. L., Lonneman, W. A., Sutterfield, F. D. and Darley, P. E. (1972) *Environ. Sci. Technol.*, **6**, 342.
89. Stevens, R. K. (1971) *Gov. Rep. Announce (US)*, **71**, 184.
90. Todd, T. M. (1971) *Am. Lab.*, **3**(10), 51.
91. Smith, R. C., Bryan, R. J., Feldstein, M. *et al.* (1972) *Health Lab. Sci.*, **9**, 58.
92. Stevens, R. K., O'Keefe, A. E. and Ortman, G. C. (1972) *Air Qual. Instrum.*, **1**, 26.
93. Fee, G. G. (1971) *Anal. Instr.*, **9**, D-4, 1.
94. McCann, R. B. (1971) *J. Air Pollut. Control Assoc.*, **21**, 502.
95. Burgett, C. A. and Green, L. E. (1976) *Am. Lab.*, **8**(1), 79.
96. Villalobos, R. and Chapman, R. L. (1974) *Chimia*, **28**, 411.
97. Villalobos, R. and Chapman, R. L. (1971) *ISA Trans.*, **10**, 356.
98. Villalobos, R. and Chapman, R. L. (1972) *Air Qual. Instrum.*, **1**, 114.
99. Karten, M. P. H. (1972) *Chem. Weekbl.*, **68**, L19.
100. Cooper, J. C., Birdseye, H. E. and Donnelly, R. J. (1974) *Environ. Sci. Technol.*, **8**, 671.
101. Krieger, B., Malki, M. and Kummler, R. (1972) *Environ. Sci. Technol.*, **6**, 742.
102. Houben, W. P. (1971) *Anal. Instrum.*, **9**, D-3, 1.
103. Hodgeson, J. A., McClenny, W. A. and Hanst, P. L. (1973) *Science*, **182**, 248.
104. Bargues, P. (1973) *Off. Nat. Etud. Rech. Aerosp. (Fr.) Note Tech.*, 213.
105. Hanst, P. L., Liefoh, A. S. and Gray, B. W. (1973) *Appl. Spectrosc.*, **27**, 188.
106. Robinson, J. W. and Guagliardo, J. L. (1974) *Spectrosc. Lett.*, **7**, 121.
107. Kobayashi, T. and Inaba, H. (1971) *Rec. Symp. Electron Ion Laser Beam Technol.*, 11th, 385.
108. Lidholt, L. R. (1972) *Opto-electronics*, **4**, 133.
109. Katayana, N. and Robinson, J. W. (1975) *Spectrosc. Lett.*, **8**, 61.
110. Loane, J. and Krishman, K. (1974) *Gov. Rep. Announce (US)*, **74**, 72.
111. Hirschfeld, T., Schildkraut, E. R., Tannenbaum, H. and Tannenbaum, P. (1973) *Appl. Phys. Lett.*, **22**, 38.
112. Braman, R. S. (1971) *Atmos. Environ.*, **5**, 669.
113. Byer, R. L. (1975) *Opto-electronics*, **7**, 147.
114. Rasmussen, R. A., Westberg, H. H. and Holdren, M. (1974) *J. Chromatogr. Sci.*, **12**, 80.
115. Giannovaria, J. A., Gondek, R. J. and Grob, R. L. (1974) *J. Chromatogr.*, **89**, 1.
116. Kapila, S. and Vogt, C. R. (1981) *J. High Resolut. Chromatogr. Commun.*, **4**, 233.
117. Cox, R. D. and Earp, R. R. (1982) *Anal. Chem.*, **54**, 2265.
118. Jeltes, R. and Burghardt, E. (1972) *Chem. Weekbl.*, **68**, 16.

119. Siegel, D., Mueller, F. and Neuschwander, K. (1974) *Chromatographia*, **7**, 399.
120. Heatherbell, D. A., Wrolstad, R. E. and Libbey, L. M. (1971) *J. Agric. Food Chem.*, **19**, 1069.
121. Klein, G. (1964) *Linde Rep. Sci. Technol.*, **17**, 24.
122. Altshuller, A. P. (1968) *Adv. Chromatogr.*, **5**, 229.
123. Grob, K. and Grob, G. (1970) *J. Chromatogr. Sci.*, **8**, 635.
124. Burghardt, E. and Jeltes, R. (1975) *Atmos. Environ.*, **9**, 935.
125. Bertsch, W., Shunbo, F., Chang, R. C. and Zlatkis, A. (1974) *Chromatographia*, **7**, 128.
126. Raymond, A. and Guiochon, G. (1973) *Analusis*, **2**, 357.
127. Shadoff, L., Kallos, G. and Woods, J. (1973) *Anal. Chem.*, **45**, 2341.
128. Dooper, R. P. M. (1983) *Chrompack News*, **10**, 1.
129. Snyder, R. E. (1971) *J. Chromatogr.*, **9**, 638.
130. Mignano, M. J., Rony, P. R., Grenoble, D. and Purcel, J. E. (1972) *J. Chromatogr.*, **10**, 637.
131. Nicrotra, C., Cornu, A., Massot, R. and Perilhon, P. (1972) *Proc. Int. Symp. Gas Chromatogr. (Eur.)*, **9**, 9.
132. Lao, R. C., Oja, H., Thomas, R. S. and Monkman, J. L. (1973) *Sci. Total Environ.*, **2**, 223.
133. Pebler, A. and Hicham, W. M. (1973) *Anal. Chem.*, **45**, 315.
134. Schuetzle, D., Crittenden, A. L. and Charlson, R. J. (1973) *J. Air Pollut. Control Assoc.*, **23**, 704.
135. Rasmussen, D. V., Fisher, T. P. and Rowan, J. R. (1972) *Can. J. Spectrosc.*, **17**, 79.
136. Bergert, K. H., Betz, V. and Pruggmayer, D. (1974) *Chromatographia*, **7**, 115.
137. Evans, J. E. and Arnold, J. T. (1975) *Environ. Sci. Technol.*, **9**, 1134.
138. Wang, I. C., Swafford, H. S., Price, P. C., Martinsen, D. P. and Butrill, S. E. (1976) *Anal. Chem.*, **48**, 491.
139. Vacuum Generators Company (1977) Technical Publication 02/521, VG Gas Analysis Ltd, Cheshire, UK.
140. Hampton, C. V., Pierson, W. R. and Harvey, T. M. (1982) *Environ. Sci. Technol.*, **16**, 287.
141. Pellizzari, E. D. (1982) *Environ. Sci. Technol.*, **16**, 781.
142. Termonia, M., Monseur, X., Alaerts, G. and Dourte, P. (1980) *Pergamon Ser. Environ. Sci.*, **3**, 135.
143. Possanzini, M., Cicioli, P., Brancaleoni, E. *et al.* (1982) CEC Report EUR 7624, Commission of the European Communities, Brussels, Belgium, p. 76.
144. Knoeppel, H., Versino, B., Schlitt, H. *et al.* (1980) CEC Report EUR 6621, Commission of the European Communities, Brussels, Belgium, p. 25.
145. Scarengelli, F. P., O'Keeffe, A. E., Rosenberg, E. and Bell, J. P. (1970) *Anal. Chem.*, **42**, 871.
146. Blacker, J. H. and Brief, R. S. (1971) *Am. Ind. Hyg. Assoc. J.*, **32**, 668.
147. Saltzmann, B. E., Burg, W. R. and Ramasaway, G. (1971) *Environ. Sci. Technol.*, **5**, 1121.
148. Homolya, J. B. and Bachman, J. P. (1971) *Int. Lab.*, **5**, 37.
149. Dietz, R. M., Cote, E. A. and Smith, J. D. (1974) *Anal. Chem.*, **46**, 315.
150. Savitsky, A. C. and Siggia, S. (1972) *Anal. Chem.*, **44**, 1712.
151. Harrison, R. M. and Holman, C. D. (1980) *Environ. Technol. Lett.*, **1**, 345.
152. Clark, A. I., McIntyre, A. E., Lester, J. N. and Perry, R. (1982) *J. Chromatogr.*, **252**, 147.

153. Clark, A. I., McIntyre, A. E., Lester, J. N. and Perry, R. (1984) *Environ. Pollut. (Ser. B)*, **7**, 141.
154. Clark, A. I., McIntyre, A. E., Lester, J. N. and Perry, R. (1984) *Sci. Total Environ.*, **39**, 265–79.
155. Sawicki, E. (1970) *Crit. Rev. Anal. Chem.*, **1**, 275.
156. Gordon, R. J. (1976) *Environ. Sci. Technol.*, **10**, 370.
157. Intersociety Committee (1972) *Methods of Air Sampling and Analysis*, American Public Health Association, Washington, DC, 11104-01-69T.
158. Sawicki, E., Elbert, W. C., Stanley, T. W. *et al.* (1960) *Int. J. Air Pollut.*, **2**, 273.
159. Sawicki, E., Elbert, W. C., Stanley, T. W. *et al.* (1960) *Anal. Chem.*, **32**, 810.
160. Sawicki, E., Fox, F. T., Elbert, W. C. *et al.* (1962) *Am. Ind. Hyg. Assoc. J.*, **23**, 482.
161. Sawicki, E. (1964) *Chem. Anal.*, **53**, 24, 28, 56 and 88.
162. Lao, R. C., Thomas, R. S., Oja, H. and Dubois, L. (1973) *Anal. Chem.*, **45**, 908.
163. Pupp, C., Lao, R. C., Murray, J. J. and Pottie, R. F. (1974) *Atmos. Environ.*, **8**, 915.
164. Lao, R. C., Thomas, R. S. and Monkman, J. L. (1975) *J. Chromatogr.*, **112**, 681.
165. Murray, J. J., Pottie, R. F. and Pupp, C. (1974) *Can. J. Chem.*, **52**, 557.
166. Van Cauwenberghe, K. and Van Vaeck, L. (1983) in *Mobile Source Emissions Including Polycyclic Organic Species* (eds D. Rondia, M. Cooke and R. K. Haroz), D. Reidel, Dordrecht, Netherlands, pp. 327–47.
167. Van Vaeck, L., Broddin, G. and Van Cauwenberghe, K. (1980) *Biomed. Mass. Spectrom.*, **7**, 473.
168. Thrane, K. E. and Mikalsen, A. (1981) *Atmos. Environ.*, **15**, 909.
169. Lao, R. C. and Thomas, R. S. (1980) in *Polynuclear Aromatic Hydrocarbons: Chemical and Biological Effects* (eds A. Bjorseth and A. J. Dennis), Battelle Press, Columbus, Ohio, pp. 120–40.
170. Brockhaus, A. (1974) *Atmos. Environ.*, **8**, 521.
171. König, J., Funcke, W., Balfanz, E. *et al.* (1980) *Atmos. Environ.*, **14**, 609.
172. Hughes, M. M., Natusch, D. F. S., Taylor, D. R. and Zeller, M. B. (1980) Chemical Transformations of particulate polycyclic organic matter. In *Polynuclear Aromatic Hydrocarbons*, Battelle Press, Columbus, Ohio, pp. 232–74.
173. Falk, H. L. and Steiner, P. E. (1952) *Cancer Res.*, **12**, 30 and 40.
174. Lindsey, A. J. and Stanbury, J. R. (1962) *Int. J. Air Water Pollut.*, **6**, 387.
175. Clearly, G. J. and Sullivan, J. L. (1965) *Med. J. Aust.*, **52**, 758.
176. Del Vecchio, V., Valori, P., Melchiorri, C. and Grella, A. (1970) *Pure Appl. Chem.*, **24**, 739.
177. Stanley, T. W., Meeker, J. E. and Morgan, M. J. (1967) *Environ. Sci. Technol.*, **1**, 927.
178. Sawicki, E., Stanley, T. W., Elbert, W. C. *et al.* (1967) *Atmos. Environ.*, **1**, 131.
179. Chatot, G., Castegnaro, M., Roche, J. L. *et al.* (1971) *Anal. Chim. Acta*, **53**(2), 259.
180. Stenberg, U. R. and Alsberg, T. E. (1981) *Anal. Chem.*, **53**, 2067.
181. Rosen, A. A. and Middleton, F. M. (1955) *Anal. Chem.*, **27**, 790.
182. Tabor, E. C., Hauser, T. R., Lodge, J. P. and Burtschell, R. H. (1958) *Arch. Ind. Health*, **17**, 58.
183. Moore, G. E. and Katz, M. (1960) *Int. J. Air Pollut.*, **2**, 221.
184. Oelert, H. H. (1969) *Fresenius Z. Anal. Chem.*, **224**, 91.
185. Sawicki, E., Stanley, T. W., Elbert, W. C. and Pfaff, J. D. (1964) *Fresenius Z. Anal. Chem.*, **36**, 497.
186. Sawicki, E., Stanley, T. W., Pfaff, J. D. and Elbert, W. C. (1974) *Chem. Anal.*, **53**, 6.
187. Sawicki, E., Stanley, T. W., McPherson, S. and Morgan, M. (1966) *Talanta*, **13**, 619.
188. Karr, C., Childers, E. E. and Warner, W. C. (1963) *Talanta*, **35**, 1290.

189. Schmit, J. A., Henry, R. A., Williams, R. C. and Dieckman, J. F. (1971) *J. Chromatogr. Sci.*, **9**, 645.
190. Kirkland, J. J. (1969) *Anal. Chem.*, **41**, 218.
191. Vaughan, C. G., Wheals, B. B. and Whitehouse, M. J. (1973) *J. Chromatogr.*, **78**, 203.
192. Klimisch, H. J. (1973) *Anal. Chem.*, **45**, 1960.
193. Wise, S. A., Bonnett, W. J. and May, W. E. (1980) in *Polynuclear Aromatic Hydrocarbons: Chemical and Biological Effects* (eds A. Bjorseth and A. J. Dennis), Battelle Press, Columbus, Ohio, p. 179.
194. Hatano, H., Yamamoto, Y., Saito, M. *et al.* (1973) *J. Chromatogr.*, **83**, 373.
195. May, W. E. and Wise, S. A. (1984) *Anal. Chem.*, **56**, 225.
196. Janini, G. M., Johnston, K. and Zielinski, W. L. (1975) *Anal. Chem.*, **47**, 670.
197. Lee, M. L., Novotny, M. V. and Bartle, M. D. (1981) in *Analytical Chemistry of Polycyclic Aromatic Compounds*, Academic Press, London, pp. 188–239.
198. Stenberg, U., Alsberg, T., Blomberg, L. and Wännman, T. (1979) in *Polynuclear Aromatic Hydrocarbons* (eds P. W. Jones and P. Leber), Ann Arbor, Michigan, pp. 313–25.
199. Blomberg, L. and Wännman, T. (1979) *J. Chromatogr.*, **168**, 81.
200. Lee, M. L., Novotny, M. and Bartle, K. D. (1978) *Anal. Chem.*, **48**, 1566.
201. Bjorseth, A. (1977) *Anal. Chim. Acta*, **94**, 21.
202. Van Vaek, L. and Van Cauwenberghe, K. (1978) *Anal. Lett.*, **6**, 214.

9 Halogen compounds

P. W. W. KIRK and J. N. LESTER

9.1 Fluorides

Sources of fluorides in ambient and urban air are varied. They include emissions from volcanoes, dust generated by the weathering of fluoride-containing soils and outcroppings of fluoride-containing minerals, ocean spray, smoke from the combustion of coal and effluents from a variety of industrial processes. Major industrial sources of fluorides include primary aluminium manufacture, open hearth steel production, coal burning for power, the fertilizer industry and cement, brick, tile and ceramic industries. Less significant sources of emission include glass manufacture, hydrogen fluoride production and high octane fuel production.

Fluoride present in waste gases occurs in the gaseous form as HF or less commonly as SiF_4 and in the particulate form as metal fluorides. Elemental F is only found in the atmosphere in trace amounts. A summary of fluoride species emitted from stationary sources is included in Table 9.1 [1].

Fluoride compounds in the atmosphere are generally considered to be toxic to plants and animals [2, 3]; the present threshold limit value (TLV) for total

Table 9.1 Summary of emission species of fluoride from stationary industrial sources

Industry	Fluorides	
	Gaseous	Particulate
Primary aluminium	HF	Na_3AlF_6 AlF_3 $Na_5Al_3F_{14}$
Iron and steel	HF SiF_4	CaF_2
Glass	SiF_4, HF, F_2, BF_3, H_2SiF_6	CaF_2, NaF, Na_2SiF_6 PbF_2
Phosphate rock acidulation	HF, SiF_4 F_2	$Ca_{10}(PO_4)_6F_2$ CaF_2
Electric furnace	—	$Ca_{10}(PO_4)_6F_2$

fluorides as F in the air is 2.5 mg m^{-3}. Fluorides can affect man directly through air pollution and indirectly through contaminated plants and through animals feeding on contaminated vegetation. The highest concentrations tend to occur in green vegetables and fruit grown in the vicinity of emissions. Among the common air pollutants, fluoride is ranked fifth in importance with respect to the amount of plant damage produced in the United States. The four pollutants with higher rankings are ozone, sulphur dioxide (SO_2), oxidants other than ozone (peroxyacyl nitrates, nitrogen oxides) and pesticides [2]. However, fluoride is the most phytotoxic of these pollutants. Unlike SO_2, fluorides do not take part in plant metabolic reactions, but tend to accumulate in concentrations over 50 ppm both on the inside and outside of leaves. Thus, it may cause injury to susceptible plant species at atmospheric concentrations 10 to 1000 times lower than the others (less than 1 ppb or *c*. 0.8 μg F m^{-3}). Plants most sensitive to fluorides are tulips, gladioli, conifers and stone fruits such as apricot and prune. In the presence of SO_2, HF produces synergistic effects on these plants.

Fluoride is ubiquitous; detectable traces occur in almost all substances. According to Martin and Jones [4] an individual living and working in central London could be expected to inhale 0.001 to 0.004 mg of fluoride per day. In heavily industrialized cities this intake may rise to 0.01 to 0.04 mg. Chronic fluoride poisoning in man can result in abnormal hardening of bones (osteosclerosis) and spots on teeth, caused by disturbances in Ca metabolism and a lowered resistance to illness, especially amongst young children. The threshold for chronic effects has been the subject of considerable debate. Documented evidence exists of the result of industrial exposure but far fewer studies have been reported for exposure by the general population in the vicinity of fluoride-emitting industries [1]. In one study radiological evidence of skeletal fluorosis was associated with an atmospheric fluoride concentration as low as 0.05 mg m^{-3} (as F) [5]. At higher concentrations, acute effects include bronchitis, pneumonia, laryngitis and tracheitis, again, predominantly amongst the young and old. Fluorides also affect animals, especially grazing stock, through plants rather than by direct inhalation, causing mainly chronic, and, in some instances, acute fluorosis.

9.1.1   Sampling procedures

Sampling for airborne fluorides is complicated by the very low concentrations of these compounds generally present in the ambient atmosphere (from <0.1 to generally <10 μg F m^{-3}) and by the occurrence of both gaseous and particulate forms. In fact fluoride may occur in four possible states in the atmosphere: free gaseous fluoride, particulate fluoride, aerosol and gaseous fluoride adsorbed on dust. Depending on the species of interest (see Table 9.1) the gaseous or particulate fraction may be sampled, or alternatively continuous separation and sampling of both gaseous and particulate species may be undertaken. Moreover, since gaseous fluoride compounds, such as HF or SiF_4, are more toxic to vegetation than most particulate fluoride compounds, it is important that the method used for collection be capable of separating these two forms where potential injury to vegetation is of concern [6].

Sampling for total fluorides, although of limited value, is relatively easy and may be of use for environmental surveillance, particularly when the distribution of fluoride species is known. Several methods for total fluoride exist which can be usefully employed. The most common is the wet impingement technique, where either water or an alkaline solution is used to collect or trap fluorides. In general, water is preferable to alkaline solution due to its ready availability in a purified form, stability and easy dissolution of HF gas. Recovery of HF through a standard Greenburg–Smith impinger containing water was found on average to be 99.8% in 40 tests [7] while recovery of particulates above 1 μm in diameter may be 90–98% or lower depending on impingement velocity [8]. Using a Greenburg–Smith impinger or one of similar design flow rates of 20–30 litres min^{-1} are recommended for predominantly gaseous sampling while for particulates, flow rates of 50 litres min^{-1} should be adopted. Depending upon the subsequent analytical procedure alternative impinger solutions include NaOH–KCl solution [9] and total ionic strength adjustment buffer (TISAB) [10]. West *et al.* [11] described an impinger method where a prescrubber containing hot concentrated H_2SO_4 dissolved particulates and aerosols, releasing F^- which was then collected in a fritted glass bubbler containing sulphamic acid to remove nitrite interference prior to the subsequent spectrophotometric determination. Flow rates up to 8 litres min^{-1} should be adopted. Where appropriate the collection efficiency of any impinger system may be improved by connecting several impingers in series. Advantages of this method include simplicity and disadvantages include delicacy of the impingers, evaporation or freezing of the collection liquid and the relatively large collection volume required making the analysis of low-fluoride-containing samples difficult [6].

Buck and Stratmann [12] have described a method where fluorides are adsorbed in a quartz tube containing sodium carbonate-coated silver beads of 3 mm diameter, at a sampling flow rate of 60 litres min^{-1}. The inlet of the column is fitted with a separator unit (Herpetz cap) which excludes large particulates. Small particulates are admitted but not quantitatively collected, making this system of greatest value for predominantly gaseous fluoride sampling. Reported collection efficiencies are up to 99% from air containing between 5 and 500 μg m^{-3} fluorides. An apparatus for the automatic elution and electrometric analysis of air samples using this technique has been described [13]. The coated bead method of collection and electrometric analysis has been recommended as a Standard Method in the German Federal Republic [14].

Collection of fluoride as an integrated sample is relatively simple but may not provide the information required to account for specific chemical or biological effects. Since there is no satisfactory method of separating gaseous and particulate fluoride after collection it is necessary to undertake a separation as part of the collection process. Two general approaches have been applied to this problem: firstly removal of the particulates with a filter or electrostatic precipitator followed by collection of the gaseous component or alternatively allowing the reactive HF to adsorb by diffusion to non-impinged surfaces, followed by a particulate collector [8].

A relatively simple but effective technique for separating particulate from gaseous fluoride is the prefilter-impinger method. This consists of a standard or modified Greenburg–Smith impinger and a citric acid-treated prefilter [15] to remove particulate fluoride. Despite its simplicity of operation, however, this method suffers from the inherent problems associated with impingers as outlined above. The prefilter and impinger method has been adopted as a Standard Method by the ASTM [7] and may be used as described or with the modification of a membrane filter instead of the citric acid-treated filter. Certain membrane filters have been reported to be efficient collectors of particulate fluorides with only low retention of gaseous forms [16, 17].

An alternative procedure to separate, collect and concentrate particulate and gaseous fluorides involves the use of a prefilter and alkali-treated filter. As in the prefilter–impinger method an acid treated or membrane filter may be used as a prefilter to remove particulate fluorides. In one modification, the prefilter and holder is heated to desorb gaseous fluorides from the filter, from the surfaces of the filter holder or from collected particulate matter [18]. The prefilter is followed by an alkali treated filter to retain gaseous fluorides. Many alkaline compounds have been used to treat filters including calcium, potassium and sodium salts of acetate, carbonate, formate and hydroxide. Indeed where separation of gaseous and particulate fluoride is not required, alkali-treated membrane or cellulose filters have been used to collect total fluorides [19–21].

The prefilter and alkali-treated filter arrangement forms the basis of an automatic method for the collection and storage of individual air samples obtained over periods from several minutes to several hours by means of a double paper tape sampler [22]. Ambient air is drawn through an inlet tube and passes through a citric acid-treated prefilter tape to remove particulate fluoride and then through an alkali-treated tape to remove acidic gaseous fluoride. The effluent gas is then filtered through soda lime and glass wool and is used to prevent contamination of the paper tapes by fluoride in the ambient air. Advantages of the device include portability and automatic unattended operation while disadvantages include inability of the prefilter to retain aerosol particulates of < 1 μm and potential sorption of HF on collected particulate matter. The double paper tape sampler method forms the basis of a Standard Method adopted by the ASTM [23] and the Intersociety Committee [24].

The second approach to the separation and collection of gaseous and particulate fluorides is exemplified by the bicarbonate-coated tube and membrane filter method. This device consists of a borosilicate glass tube, 7 mm i.d. and 122 cm in length, coated on the inside with sodium bicarbonate [22]. The tube is mounted vertically with a polypropylene filter holder and membrane filter at the lower end. Sampled air is drawn downwards through the tube where gaseous fluorides are removed by chemical reaction with the sodium bicarbonate while particulate fluorides are collected on the membrane filter. In field use the sampler is normally fitted with a weather cap and is heated to prevent condensation within the tube. Advantages of the system include low capital cost while disadvantages include fragility. The bicarbonate-coated tube method has been adopted by the ASTM [25] and the Intersociety Committee [24] as a Standard Method. Results of air analyses by the double paper tape sampler and bicarbonate-coated tubes have been compared under field conditions [26].

With the exception of the double paper tape sampler, which is only a collection unit, there are a number of automatic methods available for the determination of gaseous fluoride. None of the available instruments separate and analyse gaseous and particulate fluorides. Commercially available instruments generally rely on liquid absorption or a semi-continuous version of the coated tube principle in which the coating is periodically removed by a liquid wash and subsequently recharged with coating solution. Invariably these monitors incorporate an ion specific electrode to determine the fluoride concentration in the absorbent or wash liquid.

The methods discussed above are only applicable to air sampling where fluoride levels are relatively low. For higher concentrations, such as are to be found in stack gases, alternative methods must be employed. Dorsey and Kemnitz [27] and Smith and Martin [28] have developed a method for this purpose, in which HF is converted to the less reactive SiF_4, followed by the separation of particulates by filtration and collection of the gaseous phase in two impingers containing solutions of NaOH. Filters used to trap the particulates may be of sodium carbonate-impregnated paper or membrane type, but SiF_4 may be retained unless the filter is maintained at a temperature high enough to prevent moisture condensation.

9.1.2   Analytical procedures

The analytical procedure adopted is determined both by the nature of the sample and the analytical method to be used for fluoride determination. Pretreatment techniques vary according to whether the sample is soluble or particulate and according to the fluoride species present. All analytical methods commonly employed for fluoride determination, with the exception of activation analysis, require that fluoride be in the soluble inorganic form and that the sample be free from interfering cations and anions, or that any interfering fluoride complexes present are destabilized.

9.1.2.1   *Pretreatment for particulates*

Particulate fluorides must be converted to soluble form prior to fluoride determination, and preferably this step should be concomitant with the removal of interfering cations and anions. Problems associated with interference are particularly prevalent in colorimetric analysis. The most commonly utilized pretreatment technique for particulate fluorides is fusion with sodium hydroxide followed by steam distillation, as originally developed by Willard and Winter [29] and subsequently modified [30]. If the distillation is carried out with adequate care, recoveries of up to 99% of the amount of fluorides present may be achieved. This technique, although time consuming in its manual form, should be regarded as the standard against which other pretreatment procedures must be checked prior to their adoption. Integrated samples containing both gaseous and particulate fluorides, which have been collected on glass fibre filters, are not, however, amenable to fusion and should be transferred directly to the distillation flask. A semi-automated microdistillation method has been developed and adopted as a Standard Method by the ASTM [31].

Fluorides may also be recovered from interfering substances by preferential adsorption on to an ion exchange resin [32], particularly when the interfering agents are cationic. The fluoride is then recovered as hydrofluoric acid on an anionic exchange resin which is then removed as sodium fluoride by elution with NaOH solution. Newman [33] used De-Acidite FF to remove both cationic and anionic interferences successfully. The efficiency of the method depends on the resin used, and overall use of these resins is limited to samples containing low concentrations of interfering ions. Several authors have reported that samples containing high concentrations of Al, phosphate or Fe^{3+} ions may not be suitable for treatment by ion exchange. Ion exchange as a pretreatment procedure has been adopted by the ASTM [30] as a Standard Method.

Singer and Armstrong [34] introduced the diffusion technique for separating fluoride from interfering substances; subsequently there have been many modifications proposed. However, the basic method has not changed: an aliquot of prepared sample is heated gently with a strong acid in a sealed or unsealed container, and the liberated HF is sorbed in an alkaline coating on the lid of the vessel. ASTM [30] have adopted a microdiffusion method for this separation using plastic Petri dishes.

9.1.2.2   Ion-selective electrode determination

The most commonly employed method of determining fluoride in prepared air samples is the ion-selective electrode [9, 10, 19, 35–37]. The electrode consists of a LaF_3 crystal doped with europium [38], one face of which is in contact with a reference solution of chloride and fluoride ions, the other face coming into contact with the sample solution. A potential difference is generated when a sample containing fluoride ions (F^-) is introduced, and its magnitude is related to the ratio of the fluoride ion activities of the sample to that in the reference solution. The potential developed is measured either against a standard single junction reference electrode or now more commonly against a double junction reference electrode. Since the electrode responds to the free fluoride ion activity, the observed potential is affected by the ionic strength of the solution and by the presence of any fluoride-complexing components or conditions which reduce the dissociation of fluoride. In order to minimize such effects a buffer is normally added to the sample prior to F^- determination. This buffer usually contains a pH buffer, a fluoride decomplexing agent and some additional electrolyte to adjust the ionic strength. Potential decomplexing or 'masking' agents have been investigated by Tanikawa *et al.* [35], who compared the masking ability of nine complexing agents in the presence of aluminium, the most serious common cationic interferent. Of the nine agents examined, tiron (disodium 4,5-dihydroxybenzene-1,3-disulphonate) was the most effective at relatively high aluminium concentrations, and especially at high pH, although the sensitivity is then decreased by the presence of OH^-. Citrate was found to be almost as satisfactory under these conditions.

However, such concentrated buffer solutions can cause problems at low fluoride concentrations. The complexing agent not only complexes interfering cations, such as Al^{3+} and Fe^{3+}, but also attacks the LaF_3 sensor with the La^{3+} slowly dissolved from the sensor surface and complexed. At the same time, small amounts of fluoride may go into solution. This results in a decreased sensitivity and rather slow response of the electrode [39]. In addition fluoride may be introduced by the reagents as an impurity, particularly when using highly concentrated buffer solutions [40].

Nicholson and Duff [41] proposed the use of an organic complexing agent immiscible with water to remove aluminium prior to determination by ion-selective electrode. The aqueous sample is shaken with oxine (quinolin-8-ol) as a 10% solution in chloroform, centrifuged and the organic layer discarded. An equal volume of buffer containing CDTA (1,2-cyclohexanediaminetetracetic acid) and triammonium citrate is then added. Recoveries of fluoride were 96% or greater up to 300 mg litre^{-1} of Al, while no pretreatment at this Al concentration produced a recovery of 76.6%.

Alary *et al.* [19] studied the recovery of fluoride in crude industrial products, dust samples originating from emissions and dust samples collected at various distances from an alumina reduction plant using alkali-treated cellulose filters. In order to assess recovery of fluoride from these samples, prior to determination by ion-selective electrode, alkaline fusion with NaOH followed by distillation in phosphoric acid (reference technique) was compared with the Willard–Winter distillation from perchloric acid, extraction for 24 h with sodium citrate buffer and extraction for 1 h with dilute sulphuric acid. It was apparent that anhydrous AlF_3 was not recovered by any of the non-reference procedures, although the low proportion of AlF_3 in dust emissions suggest that this compound does not constitute an important contribution to the total fluoride content of air. The citrate extraction and distillation procedures are time consuming and the former gives incomplete recovery. It was concluded that a rapid and simple sulphuric acid extraction procedure is the preferred pretreatment prior to fluoride determination by ion-selective electrode. For the recovery of total fluoride in dust samples collected in the vicinity of fertilizer plants extraction with dilute hydrochloric acid is recommended.

Where the concentration of Al is considerably lower than that of fluoride the use of the buffer CDTA for aqueous samples has been recommended [39]. CDTA does not attack the LaF_3 crystal at the concentration used in the buffer solution, which is sufficient to complex about 10 mg litre^{-1} of Al. Samples containing higher concentrations of Al require dilution below this range bearing in mind that the electrode may only be used down to concentrations of 20 μg litre^{-1}. Incomplete recovery may be accommodated by standard additions with an accompanying increase in analytical time. The manual determination of fluoride in air samples using a fluoride ion-selective electrode with an appropriate pretreatment procedure (Willard–Winter distillation, ion exchange or diffusion) has been adopted as a Standard Method by the ASTM [30].

9.1.2.3 Colorimetric determination

Colorimetric procedures either involving titration or spectrophotometry represent the main alternative to the use of the ion-selective electrode. The basis of the method is either the removal of free dyestuff from a metal dye lake as in zirconium–alizarin or on the formation of a coloured complex between fluoride and alizarin complexan. The earlier methods were based on the former principle and modifications of these are now used in the analysis of fluorides. When fluoride ions react with alizarin–thorium or alizarin–zirconium colour lakes, the metals are removed from the lake to form stable fluorides, initiating a change in colour, which can be measured with a spectrophotometer at low concentrations and measured visually at high concentrations. The determination of fluorides using the latter principle involves the titration of fluorides against a zirconium or thorium salt using alizarin as the indicator, until the yellow alizarin changes to the complex red colour. A revision of this method has been published by the ASTM [30] to cover a wide range of fluoride in the sample from 10 mg down to about 5 μg. The pH and composition of the solution must be carefully controlled and interfering substances removed by prior separation of fluoride.

There are other modifications and revisions of these basic methods, notably those of Bellack and Schouboe [42], who introduced the use of SPADNS (sodium-2-(p-sulphophenylazo)-1-8-dihydrocynapthalen-3,6-disulphate) instead of alizarin, applicable for concentrations between 0.05 and 1.5 mg litre^{-1}. The sensitivity of SPADNS to interference is minimal. Instead of SPADNS, zirconium–erichrome cyanine R may be used [43, 44]. Here the colour of the reagent is bleached out by the fluoride ions in solution and the intensity of the colour is measured using a spectrophotometer.

Barney and Hensley [45] developed a spectrophotometric method which is based on the reaction between fluoride and thorium chloroanilate in aqueous methyl cellulose at pH 4.5. For solutions of low fluoride concentrations (0–2 mg litre^{-1}) a methyl cellulose to water ratio of 1:9 is recommended and optical density measured at 330 nm. For higher concentrations, a 1:3 ratio is used at 540 nm. However, interferences by both cations and anions are common. Belcher et al. [46] proposed a method where an alizarin blue complex is produced by reaction between fluoride ions and a red complex consisting of a lanthanum salt and alizarin complexan (3-aminomethyl-(dicarboxymethyl)-1-2-di-hydroxy-anthranquinone). The reaction occurs at pH 4.3 to form a red metal lake which gives a measurable blue colour under the specific action of fluoride ions. The most sensitive pH range for the spectrophotometer is 5.0–5.2 and at this pH, the lanthanum complex is most sensitive to fluoride ions. The method can be used for a concentration range of 5 to 50 μg per unit volume of final solution and can be extended down to 1.0 μg by further modification [47].

This method has been further developed by West et al. [48] and Liiv and Luiga

[49]. They adopted a lanthanum to alizarin complexan molar ratio of 2:1 for the visible range, which reportedly gives a considerable increase in sensitivity towards fluoride ions. Liiv and Luiga [49] have further shown that for determinations in the u.v. range, a lanthanum to alizarin molar ratio of 1:1 gave an increased sensitivity towards fluoride. The optimum concentration of the reagent was 6×10^{-5} M and the absorbance was measured at 280 nm.

The colorimetric determination of fluoride has been incorporated in automated procedures for the analysis of fluorides in solution either for continuous monitoring or for discrete samples in the laboratory. A typical arrangement consists of a sampler, proportioning pump, gas/liquid separator, a colorimeter and a recorder. In the automated procedure, fluorides react with the red lanthanum alizarin complexan to form a lilac-blue species and absorbance is measured by a colorimeter at a suitable wavelength [50]. The continuous monitoring mode uses NaOH to absorb HF only, particulates being removed by filtration at the sampling stage, and cannot take account of effects produced by interfering agents. For discrete samples, the autoanalyser can be linked to a distillation apparatus, which removes interfering ions. This method forms the basis of a Standard Method adopted by ASTM [31].

Colorimetric methods are in general time consuming and require sample pretreatment for the removal of interfering ions. Alternative procedures which avoid pretreatment problems but which are not yet widely used include ion-chromatography [39] and activation analysis [51, 52].

9.1.3 Recommended experimental procedures

9.1.3.1 *Sampling*

The volume of air sampled depends generally upon the location selected for sampling and the type of sample required. The sampling volume needed to give a sufficient quantity of fluoride for analysis may be estimated as in Table 9.2. The sampling rate adopted depends on the system used. For a simple prefilter and impinger system air is drawn through an inlet tube, which is heated to prevent condensation, and is first passed through an acid-treated prefilter to remove particulates and then through an impinger to remove water-soluble fluorides. Two impingers are used in series, the first wet and the second dry to act as a demister prior to a gas meter, control value and vacuum pump. Prior to use the prefilter is immersed in alcoholic citric acid solution (0.1 M citric acid in 95%

Table 9.2 Sampling volume required for fluoride quantification

Location	Sampling volume (m^3)
Inside manufacturing plant or emitting sources, stack gases after pretreatment	0.1
Outside emission sources	1.0
Ambient away from emission sources	10

ethanol), dried under an infrared lamp and stored in a sealed container to prevent contamination. Filter changes are recommended after 12 m^3 have been sampled. De-ionized water (75–150 ml) should be used in the first impinger which prior to sampling contains less than 0.005 mg litre^{-1} of fluoride. For short-term sampling (3 h or less) an unmodified Greenburg–Smith impinger can be used to collect gaseous fluoride at sampling rates of around 28 litres min^{-1}, while for longer sampling times (3–72 h) a modified Greenburg–Smith impinger is required which maintains the level of water in the impinger automatically, and hence makes up any losses due to evaporation and carryover to the demister. Further details of this sampling procedure can be obtained by reference to the ASTM Standard [7]. Alternative sampling procedures which should be considered are the bicarbonate-coated glass tube and filter technique [25], and the double paper tape sampler method.

9.1.3.2   *Pretreatment and clean-up*

Particulate matter collected during air sampling generally requires fusion with sodium hydroxide for conversion into soluble form prior to the separation or masking of fluoride. This is achieved by soaking the filter paper in a slurry of limewater prepared from fluoride free lime in a nickel or resistant crucible until the mixture is alkaline to phenolphthalein, then evaporating to dryness and ashing in a muffle furnace at 550–600° C for 1 h until all the organic matter is completely oxidized. When membrane (cellulose ester) filters have been employed they should be drenched with ethanolic sodium hydroxide solution and ignited with a small gas flame to control their combustion. Integrated samples collected on glass fibre papers are also not amenable to fusion and should be transferred directly to the distillation flask (see below). Integrated samples collected in impingers should be transferred to nickel beakers, evaporated to dryness in an alkaline condition and ashed if organic matter is present.

Following one of the previous treatments, the contents of the crucible are fused with 2 g of NaOH and the cold melt dissolved in 10–15 ml of de-ionized water, followed by the addition of a few drops of H_2O_2 solution (30%) to convert sulphites to sulphates. Excess peroxide is then destroyed by boiling. The sample solution is then ready for fluoride separation or masking.

Ion exchange

For the removal of interfering ions using an ion exchange procedure a chromatographic column(s) is required made of borosilicate glass tubing 10 mm i.d. and 160 mm in length, having a fritted glass disc fused into the constricted base, and a reservoir of about 100 ml capacity at the top. A short piece of PVC tubing attached to the bottom and closeable with a screw hose clamp permits adjustment of flow rates and prevents the complete drainage of the column; while the quartz sand (−60 to +120 mesh), purified by hot extraction with 20% NaOH solution followed by hot 10% HCl, is used as a protective layer at the top of the resin bed.

Reagents

Intermediate-base of the granular aliphatic polyamine type Anolite A41, A43; Permutit A; Rexyn 204(OH) (-60 to $+100$ BS sieve); 2.0 and 1.0 M HCl; and 2.0, 0.1, and 0.01 M NaOH.

Prepare a resin bed 100–120 mm in height and add a 20 mm layer of quartz sand on top. Treat the resin with 200 ml of 2 M HCl, rinse with distilled water, then 200 ml 2 M NaOH followed by a final rinse of 200 ml of de-ionized water.

Precondition the resin by passing 400 ml of hydrofluoric acid (1 mg litre^{-1}) and then 400 ml of sodium fluoride (1 mg litre^{-1}) through the column at a rate of 5 ml min^{-1}. Follow this with 50 ml of 0.1 M NaOH, then 25 ml of 0.01 M NaOH. Discard the eluate. The resin is now ready for use.

The treated sample is acidified with 0.5 ml 1 M HCl per 100 ml (maximum 3 ml HCl) and then filtered to remove suspended matter. It is then added to the column and allowed to pass through at 10 ml min^{-1}, followed by a de-ionized water rinse of 10 ml. Fluoride ions are then eluted with 25 ml of 0.1 M NaOH, followed by 25 ml of 0.01 M NaOH.

When interferences are present in both cationic and anionic forms a modification of the previous procedure is used [33]. The procedure and apparatus is basically the same, but with the following changes. Strongly basic anionic exchange resin: polystyrene quaternary ammonium type: De-Acidite FF (Chloride form) ($-60+100$ BS sieve); 1 M HCl; 1 M NaOH (Analar); 1 litre 0.025 M ammonium chloride, pH 9.2; and EDTA solution.

The column is prepared as follows: wash the resin with 100 ml 1 M HCl, 100 ml water rinse, 100 ml NaOH and a final rinse with 100 ml de-ionized water, followed by 50 ml of the eluting solution, ammonium chloride.

After treating the sample with 2 ml EDTA solution and filtering to remove suspended matter, add it to the column, allowing it to pass through at 5 litres min^{-1} and wash the column with 10 ml de-ionized water. Elute the fluoride ions with 50 ml 0.025 M ammonium chloride solution.

Steam distillation

Apparatus is set up as shown in Fig. 9.1 and the sample is distilled from either perchloric or sulphuric acid solution in the presence of silica as fluorosilicic acid.

Reagents

Concentrated H_2SO_4 (98% by weight); concentrated perchloric acid (72% by weight); 50% silver perchlorate solution; and 0.2 M NaOH solution.

Transfer the prepared sample to the distilling flask, limiting the total volume to 50 ml. Rinse the sample container with 50 ml perchloric acid, add 1 ml silver perchlorate solution and transfer to the distilling flask plus washings. Commence distillation, heat the contents of the distilling flask to 135° C and maintain for about 1 h, until 250 ml of distillate is collected for analysis. If all interferences are not removed, a double distillation may be adopted. In this case concentrated H_2SO_4 is used for the first distillation at 165° C instead of perchloric acid, and

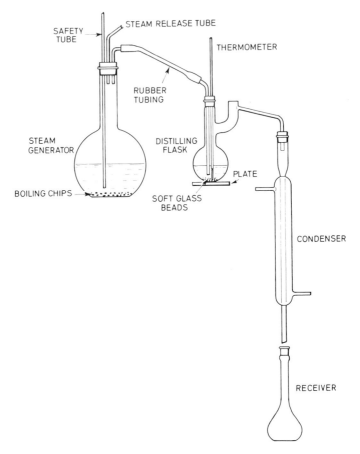

Fig. 9.1   Steam distillation apparatus for fluoride determination.

375 ml of distillate is collected over 1.5 to 2 h. NaOH solution (10 g litre^{-1}) is then added to the distillate until it is alkaline to phenolphthalein. The alkaline distillate is evaporated to 10 to 15 ml by heating below the boiling point and the concentrated distillate redistilled using perchloric acid as previously described.

Microdiffusion technique

   Apparatus

Polystyrene Petri dish, 50 mm in diameter, 8 mm deep; polythene micropipette, 0.1 ml capacity (0.01 ml subdivisions); and thermostatically controlled oven, 55–60° C range.

   Reagents

Perchloric acid solution (72%); methanolic NaOH (4 g of NaOH in 5 ml of water, add 95 ml methanol). Coat the underside of the Petri dish lid with 0.05 ml of methanolic NaOH to an area about 3 cm in diameter and dry in a clean, fluoride free atmosphere. Transfer 1.0 ml of prepared sample solution to the dish bottom, add 2.0 ml of perchloric acid solution, immediately replace the prepared lid and place in an oven at 60° C for 16–20 min. Remove the Petri dish and allow to cool. Transfer the alkaline coating from the lid quantitatively to a 25 ml flask and make up with distilled water.

9.1.3.3　*Analytical methods*

Ion-selective electrode

Apparatus

A combination fluoride ion electrode or separate fluoride ion and reference electrodes; an electrometer or expanded scale pH meter with a millivolt scale for measurement of potential; a compatible chart recorder attached to the electrometer; PTFE coated magnetic stirring bars and air driven magnetic stirrers or alternatively a thermostatted jacket at 25° C to avoid heating effects from motor-driven stirrers.

Reagents

A combined buffer, ionic strength adjuster and complexing agent solution (TISAB): to approximately 500 ml of distilled water in a 1 litre beaker add 57.0 ml glacial acetic acid, 58.0 g NaCl and 4.0 g CDTA. Bring to room temperature in a water bath and adjust the pH to 5.9 to 6.1 with 5 M NaOH. Pour into a 1 litre volumetric flask and make to volume with distilled water.

Sodium fluoride standard solution: dissolve 0.222 g of dry Analar grade sodium fluoride in 500 ml of distilled water and dilute to 1 litre with TISAB. The stock solution should then be stored in a plastic bottle in the cold and allowed to come to room temperature prior to use.

For calibration of the electrode, place 50 ml TISAB and 50 ml distilled water into a plastic beaker containing a PTFE coated stirring bar. Pipette 0.1 ml of stock fluoride solution into the stirred solution and record the potential. Repeat this procedure with 1.0 ml and 10 ml of stock solution added successively to the same beaker. Analyse one set of standards at 0.1, 1.09 and 9.99 mg litre^{-1} of fluoride before and after analysing each set of samples. Prepare a calibration curve on semilog paper. Reproducibility of each point should be ± 1 mV. A linear calibration is obtained in the range 0.1 to 10 mg litre^{-1} with a slope of between 57 and 59 mV per ten-fold change in fluoride concentration. If solutions containing less than 0.1 mg litre^{-1} of fluoride are measured, use additional standards since the calibration curve is not linear below 0.1 mg litre^{-1}.

For measurement of dissolved fluoride dilute samples $1+1$ with TISAB solution in plastic beakers. Insert the fluoride electrode into each beaker and record the potential with constant stirring after 2 min.

Measurement of fluoride in particulate matter may be accomplished by fusion with NaOH or simple acid treatment. It is recommended that any pretreatment procedure adopted be checked against fusion with NaOH and distillation to assess recovery prior to use.

Titrimetric method

Here, the sample solution is titrated against thorium nitrate using alizarin sulphuric acid as the indicator.

Reagents

Alizarin solution; 0.1 N thorium nitrate solution; 0.25 M HCl; 0.25 M NaOH; and chloracetate buffer solution.

Pipette an aliquot (10–15 ml) of the sample solution into a 250 ml flask and dilute to 100 ml with de-ionized water. Add 2 drops of indicator followed by NaOH dropwise until a pink colour appears. Carefully add HCl until the pink colour just disappears, add 5 drops buffer solution and titrate against thorium nitrate solution to a permanent faint pink end-point. A blank should be carried out using de-ionized water in place of the sample, and a calibration curve should be constructed for the determination of F^- ion concentration. By determining the concentration of fluoride ions in the sample aliquot and working backwards the original atmospheric concentration can be determined, and expressed as either mg F m^{-3} or ppm by volume of F or HF.

Spectrophotometric methods

Alizarin complexan forms a red lake with lanthanum which in the acetate buffer solution at pH 4.3 or 5.2 yields a blue water-soluble complex with fluoride ions. The principle is suitable for samples of low fluoride concentration.

Interfering cations may be removed by the use of masking agents such as potassium cyanide and sodium sulphide or by extraction with 8-hydroxy quinoline.

Reagents

5×10^{-4} M alizarin complexan; 0.02 M lanthanum nitrate; buffer solution (pH 4.3); buffer solution (pH 5.2); acetone (Analar); glacial acetic acid; and stock fluoride solution.

Visible region determination

Transfer an aliquot of the sample solution to a 100 ml flask, add 10 ml 5×10^{-4} M alizarin complexan and 2 ml buffer solution (pH 5.2), followed by 10 ml 5×10^{-4} M lanthanum nitrate solution and shake. Finally add 25 ml acetone and dilute to 100 ml with water and mix. Allow to stand for 90 min and measure the absorbance against a reagent blank at 618 nm in a 1 cm cell. Determine the fluoride concentration in a sample by reference to a suitably constructed calibration curve. The modification of this method (described below) gives an increase in sensitivity and is basically the same, but uses a lanthanum nitrate to alizarin complexan ratio of 2:1.

Ultraviolet region determination

Repeat the procedure as described for the visible region determination above, but with the following variations:

(a) Use 2 ml pH 4.3 buffer solution
(b) Do not add acetone
(c) Measure the absorbance at 281 nm in a 1 cm cell.

There are a number of variations on the basic spectrophotometric determinations, notably those of Belcher and West [47], Bellack and Schouboe [42], and Megregian [44]. It is not intended to go into detail concerning these methods as the basic principles are similar, variations occurring in the use of reagents and wavelength measurements only.

Semi-automated and automated analysis

As previously mentioned in Section 9.1.2 these methods involve the use of colorimetric determination of an alizarin–fluorine blue complex. In the semi-automated mode, the analysers can be used with a microdistillation unit to remove any interferences for discrete, operator-supplied samples (see Figs 9.2 to 9.5). In the automated mode, the microdistillation unit is not used and therefore

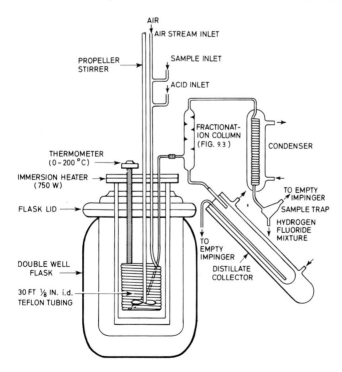

Fig. 9.2 Microdistillation apparatus.

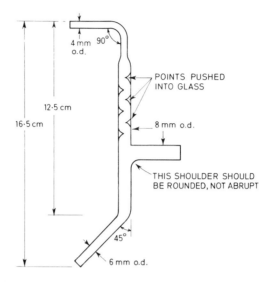

Fig. 9.3   Microdistillation column.

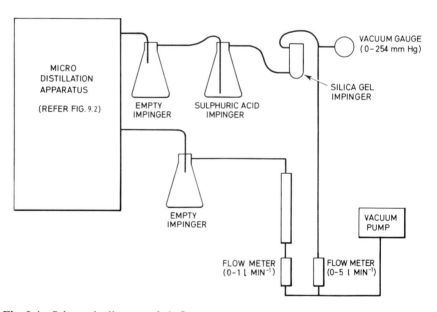

Fig. 9.4   Schematic diagram of air flow system.

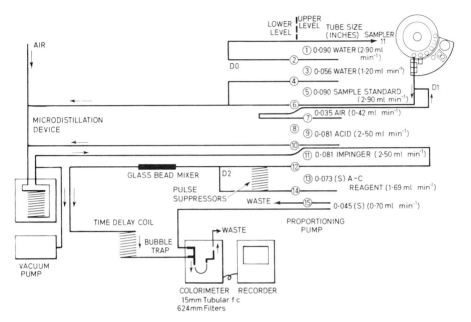

Fig. 9.5 Flow diagram of semi-automated procedure.

autoanalysers should only be used when interfering substances are low or absent, if accurate results are required.

9.2 Chlorine

Chlorine (Cl_2) is emitted to the atmosphere mainly as a result of its own manufacturing process, manufacture of associated compounds or as a by-product or waste from other industrial activities. Since the late 1920s, when chlorinated solvents and plastics started to become important, chlorine production has increased at a rate of about 7% per annum to the present world capacity of about 40 Mt annum^{-1}. The principal use of chlorine is in the manufacture of vinyl chloride (27% of production) followed by chlorinated solvents (14%), inorganic chemicals (13%), chloromethane production (11%), propylene oxide (7%), pulp and paper production (6%) while other uses constitute 22%. Less significant sources include emissions from sewage and potable water treatment plants, where it is used as a disinfectant.

During the manufacture of Cl_2 gas by electrolysis of brine in a mercury or diaphragm cell, the product is subjected to a number of repurification steps in various plants, and the relevant emissions are shown in Table 9.3 [53, 54]. Where Cl_2 is used in the pulp and paper industry for bleaching purposes, emissions of up to 0.08–0.19 kg t^{-1} chlorine used have been reported [55].

Table 9.3 Typical emissions from a chlorine-producing plant (USA)

Source	Chlorine (kg $(100\ t)^{-1}$ of liquefied chlorine)
Mercury cell plant (uncontrolled)	17.87–71.50
Diaphragm cell (uncontrolled)	8.94–44.68
Water absorber	1.79
Carbon tetrachloride absorber	0.40
Sulphur monochloride	0.13
Caustic/lime scrubber	0.0004
Tank car vents	2.01
Storage car vents	5.36
Air blowing of mercury cell	2.23

Chlorine gas is toxic to plants and animals, the suggested maximum concentration in working areas in the UK and USA being 1 ppm or 3 mg m^{-3}. At low levels, typical symptoms in man include irritation of the eyes, nose and throat, while acute symptoms are permanent lung damage and bronchitis. Prolonged exposure to severe levels may result in fatal pulmonary oedema. The effects on man are summarized in Table 9.4 [56].

Table 9.4 Effects of chlorine exposure on man

Concentration of chlorine in air (ppm v/v)	(mg m^{-3}) (20° C)	Effects
1	3	Mild smarting of eyes and
4	12	irritation of nose and throat
10	29	Severe coughing and eye irritation within 1 min
> 10	> 29	Immediate and delayed serious effects

There is little conclusive data available regarding the damage to vegetation due to gaseous Cl$_2$, reflecting its infrequent occurrence in significant concentrations in ambient air. This is in part due to a relatively low number of possible emission sources and the fact that most waste gases are recovered for economic reasons. The most commonly reported damage to plants includes marginal and bifacial neurosis and interveinal lesions, with the most sensitive species being Eastern cotton wood, silver maple and Bartlett pears. Defoliation was evident in walnut, willow, sycamore and peach with premature fruit drop in apple, peach and pear [57]. Injury may occur at Cl$_2$ concentrations of greater than 0.3 ppm in the atmosphere, the lower levels in the ppb range not causing damage because chlorine is not accumulated in the leaves, as reported by Brennan *et al.* [58].

9.2.1 Sampling procedures

Unlike HF, Cl_2 gas poses few problems in sampling due to its low interaction with particulate matter in the air. Therefore, the two can easily be separated by use of a suitable filter or glass wool plugs. The gas is then drawn through two midget impingers containing an absorbing medium. Midget impingers are used as they give better recovery efficiencies than the normal types. Several types are available, their use having been discussed in detail in Chapter 1.

The choice of absorbing medium is governed by the type of analytical method to be employed, but the most commonly used medium is NaOH solution, in which Cl_2 is converted to sodium hypochlorite and NaCl. Alternative solutions used react with Cl_2 gas to produce a colour change which is measured spectrophotometrically [24, 59].

The sampling rates adopted depend on the reactivity of the absorbing media and the type of media used. These will be discussed in more detail in the specific analysis sections.

9.2.2 Analytical procedures

Although Cl_2 is rarely found in toxic concentrations in the atmosphere, determination in the 0.1–1.0 ppm range is important as toxic effects occur both in plants and man at these levels. There are several methods of analysis available, but none sufficiently sensitive nor specific for accurately determining atmospheric concentrations. Almost all the available methods rely upon the strong oxidizing properties of Cl_2 and therefore are not selective for Cl_2, and are subject to interference from other strong oxidizing agents such as O_3, NO_2, bromine and chlorine dioxide.

In a standard procedure recommended by the Intersociety Committee [24] methyl orange is used as the detecting reagent which at pH 3.0 is decolorized, and monitored spectrophotometrically. This method is claimed to measure chlorine in the range 0.05–1.0 ppm (145–2900 μg m^{-3}) but is subject to a number of limitations. A modification of the standard method [60] involves the collection of Cl_2 by passage through NaOH solution, and subsequent addition of aliquots of the collecting solution to standard methyl orange reagent until bleaching of the colour is sufficient for accurate spectrophotometric measurement. Using this modified procedure sampling times of up to 4 h may be adopted.

A method for the determination of Cl_2 in ambient air has been described by Gabbay *et al.* [59], which consists of a scrubber containing chromic acid impregnated paper to eliminate interference from sulphur compounds, ammonia and hydrogen peroxide followed by two midget impingers containing an absorbing solution of alkaline-4-nitroaniline. A sampling rate of 0.5 to 2 litres min^{-1} was recommended for ambient air. The absorption of Cl_2 develops an orange-brown colour in the solution, with a molar absorptivity of 19 000 at 485 nm. The collection efficiency of the absorbing reagent was found to exceed 99% in the first impinger. This method was found to be suitable for the analysis of Cl_2 in the range from a few pphm to about 20 ppm, with high sensitivity and stability of the coloured reaction product. A study of common interferents demonstrated that the chromic acid impregnated paper scrubber removed problems associated with SO_2, NO_2, H_2O_2, NH_3 and methanethiol while high concentrations of HCl had negligible effect. Interference due to high O_3 concentrations was reduced by the addition of sodium nitrite to the absorbing reagent, while Br_2 interference could be accounted for by the relative instability of the reaction products (stable for 90 min) compared with those formed with Cl_2 (stable up to 24 h).

An amperometric procedure for the determination of Cl_2 in air has been proposed by Kuempel and Shults [61]. When a stream of Cl_2 containing air is passed through an electrolytic cell having one polarizable Pt electrode, chlorine is reduced to chloride if the appropriate potential is applied to the electrode. The cathodic current produced by the reaction is proportional to the quantity of chlorine present in the gas stream. Other electroactive substances present in the air will generally be reduced or oxidized at a different potential and hence should not interfere. This procedure has been compared with the methyl orange photometric method and was found to be superior [62]. The limit of detection of the amperometric procedure was found to be 50 μg m^{-3}, compared with 300 μg m^{-3} for the photometric method.

Collection of Cl_2 on filters, extraction with water and analysis by neutron activation has been reported as a suitable technique for the determination of Cl_2 in the stratosphere during balloon flight sampling [63], while for monitoring contaminated air, flow injection analysis [64] or collection in NaOH and subsequent reaction with *o*-tolidene [65] have been recommended.

9.2.3   Recommended experimental procedure

Methyl orange method

The reaction is carried out at pH 3, which is the most sensitive level for Cl_2. Substances such as manganese (3^+), bromine, iron (3^+), nitrates and SO_2 interfere in the determination, of which SO_2 is critical, being generally found in similar or greater concentrations than Cl_2 in the atmosphere. Presence of SO_2 can cause up to 30% reduction in the Cl_2 reading. Nitrates give possible interferences of the order of 20% of the true Cl_2 concentration.

Reagents

Stock methyl orange solution; and stock Cl_2 solution [24]. The air sample is drawn through a 250 ml fritted tip bubbler containing 100 ml of sample solution at a rate of 1–2 litres min^{-1} over a suitable time period to give at least 5 μg free Cl_2 in the solution. The solution is then transferred to a 5 cm cell and the absorbance measured at 505 nm using distilled water as a reference solution. The amount of Cl_2 in the aliquot can be determined by reference to a suitably constructed calibration curve. By working back to the original sample volume, the Cl_2 concentration in the air can be calculated.

9.3   HCl and particulate chloride

HCl is found in the atmosphere in three different forms: as a gas, acid mist or in chloride-containing particulates. HCl gas emitted to the atmosphere will react with water droplets to produce a HCl_{aq} mist which will in turn adsorb on to particulate matter. Sometimes, adsorption is followed by a reaction which produces metal chlorides, depending on the nature of the particulates. HCl is rarely found in the atmosphere as a gas. It is emitted to the atmosphere from metallurgical and chemical production processes and combustion processes. The sources most generally contributing to HCl emissions, particularly on a wide geographical basis, are combustion processes, and especially those involving the burning of coal [66] and municipal refuse [67]. HCl emissions from incinerators burning plastics and polyvinyl chlorides [68] have also been the subject of concern.

In the manufacture of HCl, Cl_2 is burned in an excess of H_2, and the atmospheric emissions from this process are usually less than 0.5% of total flue gas volume. It is also a by-product of the chlorination of organic compounds generally and specifically from the production of chlorinated hydrocarbons, but emissions from these sources have been virtually eliminated out of the necessity for process efficiency. Emissions from the combustion of coal containing 0.2% Cl by weight can result in 230 ppm HCl in the air surrounding the source [69]. Stack gases from refuse incinerators have been reported to give levels as high as 2000 ppm [70]. Other sources of emissions include steel pickling plants and glass manufacture.

Recently, concern has been expressed over the role of chlorine in the catalytic destruction of O_3 in the stratosphere [71]. It has been proposed that a significant source of stratospheric chlorine is HCl through its reaction with the hydroxyl radical, and therefore it has become important to determine nanogram quantities of HCl [71].

Like Cl_2, HCl is a phytotoxic pollutant, causing injury by acid reaction in plant tissues resulting in necrosis, strong decolorization and whitening of leaves. The most sensitive plants include sugar beet, cherry, larch, maple and tomato. Injury has been caused to plants by concentrations of greater than 5 ppm, particularly at high humidities. (The present TLV for HCl is 5 ppm or 7 mg m^{-3}.)

The strong acidity of this compound causes irritation of the eye membrane and upper respiratory tract in man, chronic effects are manifested as lowered resistance to illness and bad general health while severe, acute exposures can result in pulmonary oedema and laryngeal spasm.

9.3.1   Sampling procedures

HCl presents problems in sampling, as it may exist in either gaseous, particulate or aerosol form, as shown by Russian workers. Mists pose a special problem in sampling because it is often difficult to separate them from particulate and gaseous HCl. For general air pollution studies it is often considered sufficient to differentiate between so-called particulates (solid particles and mists) and gaseous

HCl. The separation step involves filtration, usually with a membrane or fibre glass filter, followed by wet impingement of the HCl gas. Sampling rates of 10 litres min^{-1} or greater should be adopted, depending on the type of filter used.

The filter must be heated to prevent condensation of moisture, as this would retain gases, but heating may cause loss of HCl as gas from the mist. Cheney and Duke [72] recently evaluated the interaction of HCl gas with a number of collecting filter types in order to explain variation in particulate HCl concentration observed in field sampling. Standardized HCl in N$_2$ cylinder gas was diluted to 40 ppm (v/v) HCl with laboratory air and drawn through heated filters (105° C) at 22.6 litres min^{-1} for one and three minutes. After sampling each filter was extracted separately and analysed for chloride. The results demonstrated that glass fiber filters strongly absorb gaseous HCl and hence that the subsequent chemical analysis of the filter resulted in a chloride level not attributable to particulate chloride. A cellulose acetate membrane filter also produced erratic results while PTFE and carbonate filters represented a reduced but still significant problem. Midget impingers are normally used in ambient air sampling, with 0.01 or 0.001 M NaOH as the absorbing medium, which rapidly achieves better than 99% recovery of gaseous HCl. The sampling rate is not particularly critical, the ambient air sampling rate normally used being between 15 and 35 litres min^{-1}. Recently the use of an alkali-treated filter, either with [73, 74] or without a prefilter [75] has been adopted for the collection of atmospheric chloride. Baechmann *et al.* [76] have recommended the use of silica gel as an adsorbent for HCl in air.

9.3.2   Analytical procedures

Several classical analytical methods are available for the determination of HCl as chloride ions Cl$^-$, and in air pollution studies, the method employed depends upon the quantity of Cl$^-$ collected by the absorbing media. For the determination of high Cl$^-$ concentrations, such as in stack gases, and for daily averages, the more common precipitation titration methods are suitable. The method involves precipitation of Cl$^-$ ions by AgNO$_3$ as AgCl$_2$, the end-point of the titration being determined by potassium chromate (Mohr), thiocyanate (Volhard) or fluorescence (Fajan). Of these, Volhard's back titration using thiocyanate is more accurate and is generally preferred to the other two methods. Volhard's method has been used successfully to determine Cl$^-$ concentrations down to about 25 μg Cl$^-$ ml^{-1} of solution. Leithe [77] modified the Mohr method and is reported to give a detection limit of 1 μg Cl$^-$ ions in the total sample analysed.

Chlorides can also be determined by titration of mercuric nitrate in the presence of a mixture of diphenyl carbazone and bromophenol blue [78]. In this method, chlorides are precipitated as mercuric chloride and bromophenol blue, yellow at pH 3.5, gives a blue-violet coloration due to reaction of excess mercuric ions with diphenyl carbazone. The procedure is applicable for absorber solutions containing at least 2 μg Cl^{-1} ion ml^{-1} [24]. A modification of this procedure with improved sensitivity has been reported in the Russian literature [79]. HCl is collected by aspiration through water, then titrated with 0.0005 M $Hg(NO_3)_2$ at pH 1.5 with diphenyl carbazone in toluene as indicator; the end-point is reached when the colour of the toluene layer changes from yellow to violet. Average error is 4% using this modification and a detection limit of 0.3 μg Cl^{-1} ml^{-1} was achieved.

Cheney and Fortune [80] compared and contrasted the Fajan, Mohr and Volhard techniques and the mercuric nitrate procedure for samples of HCl collected during stack sampling. The Mohr titration had a variation of 20% from the mean, when dealing with standard chloride solutions, which was attributed to problems with the visual interpretation of the colorimetric end-point. This procedure was consequently discounted as being unsuitable.

Titrations with the Fajan procedure gave highly reproducible results and had a clear end-point. However, the detection limit achieved was not suitable for the concentrations envisaged. Evaluation of the mercuric nitrate and Volhard procedures on four primary KCl solutions produced relative standard deviations of 0.19% and 1.03% respectively. Since the mercuric nitrate procedure had the lowest standard deviation, exhibited better precision, required only a single standardized solution and offered a sharper end-point when performed according to the procedure of Clark [78] this procedure was adopted for stack samples.

Chloride in the ambient air at low concentrations (ppbv) has been determined by constant current coulometry, following collection by filtration [73, 81]. The sampling train used consisted of a prefilter and alkali-impregnated filter, housed in a 20.3 × 25.4 cm filter holder, operating at a sampling rate of 1 m^3 min^{-1}. Each filter was then cut into small pieces and extracted with 1 M KNO_3. Following decantation the solution was acidified with 1 M HNO_3 and made 40% ethanol and then titrated using constant current coulometry with an Ag wire as anode. Titrations of standard chloride solutions at the detection limit had a relative error of +0.8% with a relative standard deviation of 2.5%. Gaseous SO_2, HI and HBr, although collected, should not interfere in the analysis. The detection limit for the titration is 400 μg HCl. Since only half of each filter was analysed in one titration, a total of 800 μg HCl should be collected. Thus, if air which is 1 ppbv (1.46 μg m^{-3}) HCl is to be analysed roughly 550 m^3 must be sampled. Care should be exercised for sampling periods of less than 9 h, however, since contamination of the filters during processing may be significant. A microcoulometric procedure for monitoring ppm concentrations of HCl has also been described [82].

Chlorides in the lower concentration range can be effectively determined by spectrophotometric methods and a method has been developed by Hagino *et al.* and Leithe [83, 84] which depends upon the displacement of thiocyanate ions from mercuric thiocyanate by Cl^- ions. In the presence of ferric ions, a highly coloured ferric thiocyanate complex is formed and the intensity is measured spectrophotometrically. The method is applicable for quantities of Cl^- in the range 0–100 μg in final solution. Torrance [85] has reported a lower detection limit of 0.015 μg ml^{-1} at the 95% confidence level for this method.

A colorimetric procedure has been reported in the Japanese literature [86] for the determination of Cl^- in urban air. Air (360 litres) is passed through 20 ml of H_2O, 5 to 20 ml of which is used for the determination. The solution is made slightly alkaline and $NaNO_3$ solution is added. The mixture is then evaporated and the residue dissolved in H_2SO_4 containing sulphanilic acid. To this solution are added *N*-1-naphthylethylenediamine dihydrochloride and NaOH solutions and the absorbance is measured at 545 nm. A detection limit of 1 μg Cl^- is claimed with a coefficient of variation of 12% ($n=6$). Sulphite does not interfere at up to 50 μg and nitrite at up to 500 μg can be removed by the addition of ethanol.

The original method of West and Coll [87] was later adopted to measure chlorides in the atmosphere by drawing air through a solution of iron (3$^+$) perchlorate in perchloric acid [88]. The absorbance of the chloro-complex of iron (3$^+$) formed is then measured at 340 nm. Bromides, iodides and sulphides do not interfere with this method. Cl in small concentrations can also be measured by a nephelometric method [89–91] where the absorbing liquid is treated with silver nitrate and the turbidity of the precipitate is observed. Bromides, iodides and sulphides interfere with this method.

Russian literature [89, 92] refers to the frequent use of the nephelometric method to determine chloride ions in ambient air in the presence of HCl_{aq} aerosol. The HCl_{aq} is determined by micro-titration or spectrophotometry, where methyl red in ethyl alcohol is used as the indicator. Total chlorides are measured nephelometrically.

Increasing attention has been paid to the determination of HCl by derivatization and gas–liquid chromatography. Proposed derivatization re-agents include 7-oxabicylco(4.1.0)heptane [93, 94], epibromohydrin [71] and cyclohexene oxide [76]. When used in combination with a suitable sampling system the detection limits claimed would be suitable for the determination of HCl in tropospheric and stratospheric air, without the need to sample large volumes of air.

Few physical methods exist for the determination of chlorides, but the method developed by Belcher *et al.* [95] involves the use of atomic absorption spectrophotometry. Chloride ions are converted to phenyl mercury (2$^+$) chloride which is quantitatively extracted into chloroform and the determination is carried out in ethyl acetate. Belcher *et al.* [96] have used this method for the analysis of chloride by gas chromatography with flame ionization detection (FID) using 2.5% diethylene glycol adipate as the stationary or liquid phase. Biehl and Baechmann [97] have described a flow through system where an air sample containing HCl is injected and passes through a reaction zone (heated at 130 to 150° C) containing CrO_3. The CrO_2Cl_2 thereby produced passes through tubing at 200° C to a special graphite tube, where it is decomposed at $>700°$ C and the Cr determined by atomic absorption spectrophotometry. If atmospheric H_2O is removed initially, a detection limit of 2 to 3 ng of HCl is claimed.

In recent years, silver chloride-type solid state membrane electrodes have been widely used for the measurement of Cl^- ion activities in water. This can also be adapted for use in determining Cl^- levels in absorber solution or in filter extract [74, 98]. The common interfering anions (nitrates, sulphates, hydroxyl, fluorides)

Table 9.5 Maximum allowable ratio of interfering ions to chloride ions

Interfering ions	Maximum allowable ratio
Hydroxyl	80
Ammonium	0.12
Thiosulphate	0.01
Bromide	3×10^{-3}
Iodide	5×10^{-2}
Cyanide	2×10^{-2}
Sulphide	Must be absent

have little or no effect on the measurement. Ions that form stable silver complexes or insoluble silver salts, however, interfere with the response to chloride ions. Table 9.5 shows the maximum allowable ratio of interfering ions to chloride ions.

Bourbon et al. [74] extracted PTFE prefilters and alkali impregnated papers with K_2SO_4–acetic acid buffer solution at pH 2.6 and determined Cl^- with an ion selective electrode. With a 24 h sampling period at 500 litres h^{-1}, the detection limit was 1 μg HCl m^{-3}. The only likely interferents (sulphide and nitrite) were eliminated by adding peroxide and sulphamic acid solution to the extracted solution containing Cl^{-1}.

Interest in the remote sensing of pollutants has increased rapidly in recent years [99, 100]. The advantages of such systems include the determination of gaseous pollutants from emission sources at stack height, in stack plumes, and at great distances from major sources. These methods are also useful in determining pollutant gases at roadways, from emission sources such as motor vehicles and from aircraft, balloons and space platforms. Several reviews of the applications of remote sensing including the determination of HCl have been prepared [100, 101] and the reader should refer to Chapter 10 for further details.

9.3.3 Recommended experimental procedures

Mercuric nitrate method

Cl^{-1} ions displace nitrate from mercuric nitrate to form essentially un-ionized mercuric chloride. Excess Hg^{2+} reacts with the diphenylcarbazone indicator to form a blue-violet complex.

Reagents

Acidic 0.007 M $Hg(NO_3)_2 . H_2O$; 0.025 M NaCl; mixed indicator (0.5 g diphenylcarbazone and 0.5 g bromophenol blue powder in 100 ml methanol); 0.3% v/v HNO_3; 0.01 M, 0.25 M NaOH; 30% H_2O_2.

A known volume of air is drawn through an impinger or bubbler containing

50 ml 0.01 M NaOH solution at a rate of about 30 litres min^{-1}. Standardize the Hg(NO$_3$)$_2$ solution by titrating 10 ml of standard NaCl as described below. Pipette an aliquot of up to 50 ml, containing less than 12 mg Cl$^-$, into a 250 ml Erlenmeyer flask and dilute (if necessary) to 50 ml with de-ionized water. Add 5 drops of mixed indicator. If the solution is red or blue-violet, add HNO$_3$ until the colour becomes yellow, then add 1.0 ml in excess. The pH should now be 3.0 to 3.5. If the solution is yellow or orange, add NaOH dropwise until the colour changes to blue-violet, then repeat the previous procedure to obtain a pH of 3.0 to 3.5. Titrate the solution with standard Hg(NO$_3$)$_2$ to a blue-violet end-point. A blank should be carried out using de-ionized water in place of the sample, and the titrant volume subtracted. By determining the concentration of Cl$^-$ ions in the sample aliquot and working backwards, the original atmospheric concentration can be determined.

Potentiometric method using the chloride ion electrode

   Reagents

Acetic acid–ammonium acetate buffer solution; stock chloride solution; standard solution A, 1 μg Cl$^-$ ml; standard solution B, 10 μg Cl$^-$ ml; and absorber solution, 0.001 M NaOH.

   Air is drawn through a suitable volume of absorber solution at a predetermined rate and an aliquot (100 ml) is transferred to a 200 ml beaker followed by the addition of 4 ml buffer solution. The beaker is then stabilized at a temperature of 20–25° C for 10 min in a thermostatically controlled water bath, using a magnetic stirrer to mix the contents well. Immerse the indicator and reference electrodes in the solution and clamp them 1.5 cm above the magnetic stirrer, which should be kept running during measurement. Calculate the concentration of the chloride ions in the aliquot by reference to a suitably constructed calibration curve of e.m.f. plotted against log $(C+1)$, where C is the concentration of chloride ions ml^{-1}.

9.4 Bromides

The determination of bromides (Br) in the atmosphere has received considerable interest in recent years due to the association between vehicle-emitted lead and bromide concentrations of atmospheric particulates [102]. Lead is added to petrol as organic tetra-alkyl lead, an anti-knock agent, which during combustion reacts with ethylene dihalide 'scavengers' also added to petrol, with consequent emission of the lead in an inorganic particulate form predominantly PbBrCl [103]. Thus from an appreciation of atmospheric bromide, and the calculation of the Br/Pb ratio, the contribution of vehicle-emitted lead to the total atmospheric burden may be estimated.

   Additional sources of Br in the atmosphere include the burning of coal and emissions from works producing organohalogen compounds. Koppenhaal and Manahan [104] stated that the mean concentration of Br in coal is 15 mg kg^{-1} while agricultural sources include emissions due to the use of methyl bromide as a fumigant. There is also a low natural background concentration of Br emanating from marine aerosols.

Harrison and Sturges [102] have reviewed the literature on the measurement and interpretation of Br/Pb ratios in airborne particles, and this should be taken as indicative of sampling technique, sampling duration and method of analysis.

9.4.1 Sampling procedures

The instability of PbBrCl in the atmosphere and the progressive loss of the halogens in volatile form [105] has led some workers to separate gaseous and particulate forms of Br, although determination of particulate Br predominates. Duce *et al.* [106] separated gaseous and particulate Br with a crude potassium carbonate bubbler and prefilter arrangement. Their results showed a possible gaseous Br concentration of 0.008–0.040 μg m^{-3} for the unpolluted coastal atmosphere, although subsequent determinations in the same area suggests that these concentrations were below the actual figure of around 0.050 μg m^{-3} [107].

Gaseous Br is normally sampled using a high-purity granular activated charcoal bed following a suitable prefilter or electrostatic precipitator [107, 108]. The choice of prefilter is dependent not only on the retention required but also on the analytical technique employed. Whatman 41 cellulose filters have low analytical blanks for Br but may clog during prolonged sampling while Millipore membrane filters although having low analytical blanks tend to disintegrate during neutron irradiation [109]. A comparison of Br contamination of certain filters has been undertaken [109, 110] and is presented in Table 9.6.

Table 9.6 Bromide impurity levels of certain filter materials

Filter	Br (ng cm^{-2})
Polystyrene	1000
Whatman 41	5
Cellulose	20
Cellulose ester (Millipore)	4
Cellulose acetate (Millipore)	6
Cellulose triacetate (Gelman)	4
Glass fibre	< 10

9.4.2 Analytical procedures

A variety of analytical procedures have been applied to the determination of Br; the most common are listed in Table 9.7, together with their respective detection limits [102]. PIXE, XRF, IPAA and INAA normally require a minimum of sample preparation as the filters can be analysed directly whereas for CSV the filter must first be extracted and the Br brought into solution. Dennis *et al.* [111] recommend the dissolution of Br in 1.0 M HNO$_3$ prior to analysis by CSV. It has

Table 9.7 Detection limits for bromide using some common instrumental techniques

Method	Volume of air sampled (m^3)	Detection limit ng cm^{-2a}
INAA[b]	10	1.6
IPAA[b]	1000	250
XRF (Cr tube)[b]	50	500
XRF (Mo tube)[b]	50	80
PIXE[b]	50	0.25
CSV[c]	—	1000

INAA instrumental neutron activation analysis; IPAA instrumental photon activation analysis; XRF X-ray fluorescence; PIXE particle induced X-ray emission; CSV cathodic stripping voltammetry.
[a] Loading of Br on the filter.
[b] Assumes that the volume of air sampled passes through an area of the filter equal to 12 cm^2 (roughly equivalent to the exposed area of a 47 mm diameter filter).
[c] Assumes that the filter (see [b]) is extracted into 25 ml of solvent.

been reported that water soluble Br, as determined by ion chromatography, may only account for 80% of total Br [112].

The advantages of PIXE, XRF and INAA lie in the minimum pretreatment requirement coupled with high sensitivity, but PIXE and INAA require expensive particle accelerator or nuclear reactor facilities making XRF a popular alternative. Unfortunately analysis by XRF has its associated problems: in particular loss of Br due to volatilization from samples or standards during irradiation by X-rays, loss during sample storage and during preparation of standards. XRF is also unsuitable for Cl determinations and hence simultaneous determination of halogens is not possible. Regardless of the type of analysis to be applied, care must be taken to prevent loss of Br. Precautions which help minimize further reactions in the collected material include sealing the filter surfaces with plastic film (e.g. Mylar) and storing samples in the cool and dark.

The review by Harrison and Sturges [102] documents 33 literature references to the determination of Br in relation to the estimation of Br/Pb ratios, and provides short details of the techniques employed. It is evident that the order of application of the various analytical techniques for Br in this context is:

XRF > INAA > PIXE ≫ ISE + ion chromatography

with only one publication respectively adopting the latter two procedures.

9.5 Halogenated hydrocarbons

The halogenated hydrocarbon gases include a wide range of compounds containing fluorine, chlorine and, to a far lesser extent, bromine. The great variations in the nature, properties and uses of these compounds, and in the scale and mode of their production, ensure that no standard methods are applicable to their identification and measurement in ambient air. In recent years certain groups within this general class of compounds have been the subjects of intense interest and particular study.

The 'fluorocarbon* problem' came to prominence in mid-1974 with the publication by Molina and Rowland [113] of a paper, in which they suggested a possible adverse impact of fluorocarbon release on the steady-state ozone concentration in the atmosphere. Many far-reaching and conflicting publications on this topic have followed and the controversy has not, as yet, been resolved. Chlorofluorocarbons are the working fluids responsible for the safety and efficiency of almost all air-conditioning and refrigeration equipment and are used as propellants in many types of aerosol products.

For quite some time there has been concern about the potential toxic hazards of chlorinated organic compounds used as insecticides. More recently some degree of attention has been transferred to the lower molecular weight organo-halogen species which are produced and used industrially on a vast scale. The problems associated with the wide distribution of the chloroderivatives of methane, ethane and ethylene in the atmosphere at a background level in the ppb range (1 part in 10^9) have been discussed [114, 115].

Investigations into the presence of low molecular weight brominated compounds in the atmosphere have largely been confined to the study of the petrol additive 1,2-dibromoethane (EDB) which in combination with 1,2-dichloroethane (EDC) act as scavengers in the combustion process [116].

9.5.1 Fluorocarbons

Sampling procedures

Several systems have been described for the sampling and collection of atmospheric fluorocarbons. Indeed, depending on the fluorocarbon of interest direct injection of air samples, previously collected as grab-samples in stainless steel vessels, on to a suitable gas–liquid chromatograph with electron capture detection (ECD), may be adequate [117]. However, for the less common fluorocarbons and those halocarbons exhibiting reduced sensitivity to ECD, preconcentration techniques may be required.

The sampling techniques originally adopted for fluorocarbons included concentration from air cryogenically [118] or on activated charcoal [119]. The use of stainless steel vessels for the collection of samples using these procedures may present difficulties due to adsorption on to the vessel walls or decomposition catalysed by the metal (or metal oxide) surface. Small stainless steel sampling tubes containing a porous polymer (a non-catalytic surface) to entrain the fluorocarbons, allowing the 'fixed' gases to pass through, greatly reduces these

---

* The compounds most widely studied have been CCl_3F(F-11) and CCl_2F_2(F-12) which are usually referred to as fluorocarbons or chlorofluorocarbons.

problems [120]. Porapak N at $0°$ C has been applied with some success, although the collection of highly volatile fluorocarbons such as dichlorodifluoromethane is unlikely to be quantitative at this temperature [120]. The application of Porapak Q at $-50°$ C has also been adopted to improve sampling efficiency [121] while the use of Carbopack B (80–100 mesh) coated with 0.5% SP1000 and operated at $-93°$ C [122], and Carbochrome C at room temperature [123] have also been reported. In some cases a pre-column containing a desiccant has been added to prevent interferences from water in subsequent analysis [120].

Analytical procedures

The first direct measurements of halocarbon species in the atmosphere were made using gas chromatography and ECD. Grimsrud and Rasmussen [123] have reviewed this work and noted that, although the ECD was of adequate sensitivity to measure $CFCl_3$, CF_2Cl_2 could not be monitored adequately in non-urban air masses. These workers described a gas chromatographic–mass spectrometric (GC–MS) method which gave a detection limit for CF_2Cl_2 and $CFCl_3$ of 5 ppt (1 ppt is 1 part in 10^{12} by volume) in 20 cm^3 air samples. It should be noted that sensitivity and precision using GC–MS is not generally as good as that obtained by GC with ECD [124]. Recently Crescentini *et al.* [125] have described a GC-high-resolution MS method for the determination of all the C_1 and C_2 halocarbons in the atmosphere with the exception of CHF_2Cl.

A number of GC column packings have been recommended by various workers for the GC–ECD determination of halocarbons. Russell and Shadoff [120] described a technique where halocarbons were thermally desorbed from Porapak N collecting tubes on to a stainless steel column (6 ft × 0.081 in.) packed with 20% DC-200 silicone on 100–200 mesh Chromosorb W AW DMCS HP. Initially the oven temperature was maintained at $-40°$ C for 8 min using carbon dioxide cooling and then temperature programmed to $140°$ C. CCl_3F, $CHCl_3$, 1,1,1-trichloroethane, CCl_4, trichloroethylene and tetrachloroethylene were then determined in the range 0.03 to 0.13 ppb. The identities of the compounds were confirmed by GC–MS. Bruner *et al.* [122] used a Carbopack B (80–100 mesh) column coated with 0.5 or 1.0% SP1000, depending on the separation required, to determine 15 halocarbons in the atmosphere of a suburban area previously collected on Carbopack B. Vidal-Madjar *et al.* [121] recommended the use of a Carbopack C-HT column for the separation of the 8 halocarbons studied, while Lasa *et al.* [117] described a system for the direct determination of CCl_4, CCl_3F and $CHCl_3$ using a glass column (2 m × 4 mm) packed with 10% DC-200 silicone on Chromosorb W. Detection limits for these components of 19, 16 and 23 pg respectively, in 5 ml of air were obtained. Analysis by GC–ECD on a column (3 m × 5 mm) packed with 10% OV101 on Chromosorb W-HP (80–100 mesh) was used by Makide *et al.* [126] while Filatova *et al.* [127] used Carbochrome B modified with 0.05 mg m^{-2} of a non-polar or mildly polar liquid to study CCl_3F and 1,1,1,trichloroethane in the atmosphere.

Calibration

Discrepancies have been reported in interlaboratory comparisons of the determination of halocarbons, which range from 13% to 30% [128, 129]. This may be attributed to the lack of reliable sources of known concentrations of halocarbons. Calibration procedures which have become generally acceptable include exponential dilution [127], permeation tube [128] and a combination of these methods [130].

9.5.2 Chlorinated hydrocarbons

The methods described above for the analysis of fluorocarbons are also generally applicable to chlorinated species. The sensitivity of ECD increases with chlorine content and thus mass spectrometric detection is not required for the more common chlorinated solvents which have relatively high background levels as the result of extensive industrial use.

Sampling procedures

Nineteen chlorine-containing species including vinyl chloride and substituted benzenes have been detected in ambient air samples collected in either Houston or Los Angeles by adsorption concentration with Tenax GC [131]. The applicability of Tenax GC to the sampling of EDC has also been demonstrated [132]. Collection on Tenax GC is normally carried out at ambient temperature for the less volatile chlorinated organics using a sampling rate determined by the 'safe sampling volume' for the particular organics to be collected [132]. Collection efficiencies for a number of chlorinated solvents including CCl_4 on Tenax GC have been assessed [133]. Carbon tetrachloride and CH_3Cl may also be concentrated cryogenically from ambient air [134, 135]. Desorption from Tenax GC is normally carried out thermally directly into a GC or GC–MS.

Alternative adsorbents include activated carbon [136–138] and XAD-2 resin [139]. Chlorinated hydrocarbons are generally desorbed from activated carbon by solvent extraction often using CS_2 or from XAD-2 using CCl_4.

Chlorinated pesticides and polychlorinated biphenyls (PCB) have been efficiently adsorbed from large volumes of air on to plugs of polyurethane foam, at levels down to $pg\,m^{-3}$ [140, 141]. Recoveries of hexachlorobenzene were however poor using this procedure [140]. A comparison of the collection efficiencies of Tenax-GC, XAD-2 and polyurethane foam has been undertaken for high-volume sampling of PCB in air [142]. This confirmed that the foam was unable to retain the more volatile compounds (e.g. hexachlorobenzene and some PCB) quantitatively, and that the resins were superior.

Analytical procedures

Again analytical procedures for trace levels of chlorinated hydrocarbons are centred on gas chromatography with either EC or MS detection. Sixteen halocarbons (including methyl bromide and methyl iodide) were investigated in samples from the rural north-west of the USA [123]. It was found necessary to use three chromatographic columns to ensure resolution of all these species: 20 ft × 1/16 inch i.d. stainless steel Durapak n-octane/Porasil C (100–120 mesh);

50 ft SCOT OV-101; and 20 ft × 1/16 inch i.d. stainless steel Durapak Carbowax 400/Porasil F.

Various other column packings have been applied to the analysis of particular chlorinated hydrocarbons or groups of related compounds. For C_1–C_2 chlorinated hydrocarbons 0.1% SP1000 on Carbopack C has been recommended using a temperature programme from 60° C to 220° C with an FID [138] while a Carbowax 1500 column [137] or a Durapak n-octane on Porasil C (100–120 mesh) column temperature programmed from 0° C to 100° C [135] coupled with MS has been applied to the quantification of sub-ppb quantities of CH_3Cl. For the determination of chlorobenzenes, 0.2% Carbowax E-40 M plus 0.5% Synerg C on GLC-110 (130–140 mesh) has been recommended prior to quantification by a photoionization detector [139]. In order to separate monochlorobenzene from the solvent (CCl_4) this column was preceded by one comprising 0.5% Carbowax E-20 M over bonded E-20 M on Chromosorb W-AW (80–100 mesh) plus 5% Synerg C. The results indicated that the collection and desorption efficiencies for chlorobenzenes in concentrations between 5 ppb and 15 ppm in air averaged 95% with a coefficient of variation of 6%.

Watson *et al.* [143] have described a technique for the enhancement of an ECD by passage of the sample over NaI just before detection. The analyte reacts selectivity to form a product with a high electron affinity. In the analysis of CH_3Cl in air, the sensitivity was improved by a factor of 10^4 over that achieved by direct ECD analysis.

Chlorinated pesticides and PCB have been determined by solvent extraction of exposed polyurethane plugs with hexane-ethyl ether (9:1); clean-up of the extract with HPLC on silica; and GC–MS on a capillary column coated with OV-1, with negative ion chemical ionization (CH_4 as reagent gas) and multiple-ion detection, to achieve a detection limit of 0.1 to 1 pg [144]. Recoveries applying this technique to ambient air ranged from 80 to 115% for ng quantities [141].

The determination of vinyl chloride in air has been the subject of a review [145]. In the Russian literature a stainless steel column packed with 0.2% PEG 1500 on graphitized carbon (0.25–0.5 mm o.d.) coupled with FID has been recommended for this analysis while Hobbs [146] used a 0.5% Carbowax 1500 on Carbopack A coupled with a microwave plasma detector. Goldan *et al.* [147] describe a technique for increasing the sensitivity of a ^{63}Ni ECD by a factor of 10^3 to vinyl chloride, thereby giving a detection limit of 0.5 ng ml^{-1} in air. To achieve this sensitivity the N_2 used as carrier gas was doped with 10–50 ppm of N_2O.

Calibration

Similar calibration procedures to those described for fluorocarbons in Section 9.5.1 may also be applied to chlorinated hydrocarbons.

9.5.3   Brominated hydrocarbons

The methods described above for the analysis of fluorocarbons and chlorinated hydrocarbons are also generally applicable to brominated species. In recent years, particular emphasis has been placed on the quantification of the fuel additive EDB in the atmosphere.

Sampling procedures

Sampling and preconcentration of EDB is normally undertaken on Tenax-GC prior to extraction with hexane [148] or thermal desorption directly on to GC–ECD [132, 149, 150]. Suitable equipment and sampling procedures have been described for urban, rural and motorway locations [150]. Quantitative recovery of EDB has also been observed from charcoal extracted with 1% methanol in benzene [151].

Analytical procedures

Following preconcentration, quantitative recovery may be obtained by thermally desorbing the Tenax-GC sampling tubes at 230° C directly on to a stainless steel column (2.0 m × 3.1 mm i.d.) packed with 5% Carbowax 1500 on Chromosorb W HP (80–100 mesh), operated isothermally at 65° C with argon/methane (95:5) as carrier gas [150].

Direct injection has recently been made possible for the determination of EDB down to 1 ppb in air, by the development of a portable GC utilizing a photoionization detector [152, 153].

Calibration

Procedures for the preparation of calibration gas mixtures either for collection on sampling tubes or direct injection include static dilution [152] or the coupling of a permeation tube oven to an exponential dilution flask [130].

References

1. Smith, F. A. and Hodge, H. C. (1979) *CRC Crit. Rev. Environ. Control*, **8**, 293.
2. Weinstein, L. H. (1977) *J. Occup. Med.*, **19**, 49.
3. Suttie, J. W. (1977) *J. Occup. Med.*, **19**, 40.
4. Martin, A. E. and Jones, C. M. (1971) *HSMHA Health Rep.*, **86**, 752.
5. Agate, J. N., Bell, G. H., Boddie, G. F. *et al.* (1949) *Industrial Fluorosis. A Study of the Hazard to Man and Animals near Fort William, Scotland*, MRC Council Memo. 22, HMSO, London.
6. Jacobson, J. S. and Weinstein, L. H. (1977) *J. Occup. Med.*, **19**, 79.
7. ASTM (1980) ASTM Standards on Methods of Sampling and Analysis of Atmospheres, Part 26, D3267-80. In *Annual Book of ASTM Standards*, American Society for Testing and Materials, Philadelphia, Pa.
8. Farrah, G. H. (1967) *J. Air Pollut. Control Assoc.*, **17**, 738.
9. Dolegeal, A., Devilliers, D., Villard, G. and Chemla, M. (1982) *Analusis*, **10**, 377.
10. Hrabeczy, A., Toth, K., Pungor, E. and Vallo, F. (1975) *Anal. Chim. Acta*, **77**, 278.
11. West, P. W., Lyles, G. R. and Miller, J. L. (1970) *Environ. Sci. Technol.*, **4**, 487.
12. Buck, M. and Stratmann, H. (1965) *Brennst.-Chem.*, **46**, 231.
13. Svoboda, K. and Ixfeld, H. (1971) *Staub*, **31**, 1.

14. VDI 2452 Blatt 2 (1975) Messen gasformiger Immissionen Messen der Fluorkon-zentration. Silberkugel-Sorption ver fahren mit Vorabscheidung und elektrometris-chen Nachweiss. VDI-Handbuch Reinhalten der Luft.
15. Weinstein, L. H. and Mandl, R. H. (1971) *VDI Ber.*, **164**, 53.
16. Davison, A. W., Rand, A. W. and Betts, W. E. (1973) *Environ. Pollut.*, **5**, 23.
17. Okita, T., Kaneda, K., Yanaka, T. and Sugai, R. (1974) *Atmos. Environ.*, **8**, 927.
18. Jahr, J. (1972) *Staub*, **32**, 17.
19. Alary, J., Bourbon, P. and Segui, M. (1980) *Sci. Total Environ.*, **16**, 209.
20. Huygen, C. (1963) *Anal. Chim. Acta*, **29**, 448.
21. Elfers, L. A. and Decker, C. E. (1968) *Anal. Chem.*, **40**, 1658.
22. Mandl, R. H., Weinstein, L. H., Weiskopf, G. J. and Major, J. L. (1970) *Separation and Collection of Gaseous and Particulate Fluorides*, Proc. 2nd Int. Clean Air Congress (eds H. M. Englund and W. T. Beery), Academic Press, New York, pp. 450–8.
23. ASTM (1980) ASTM Standards on Methods of Sampling and Analysis of Atmospheres, Part 26, D3266-79. In *Annual Book of ASTM Standards*, American Society for Testing and Materials, Philadelphia, Pa.
24. Intersociety Committee (1977) *Methods of Air Sampling and Analysis*, 2nd edn (ed M. Katz), American Public Health Association, Washington, DC.
25. ASTM (1980) ASTM Standards on Methods of Sampling and Analysis of Atmospheres, Part 26, D3268-78. In *Annual Book of ASTM Standards*, American Society for Testing and Materials, Philadelphia, Pa.
26. Israel, G. W. (1974) *Atmos. Environ.*, **8**, 159.
27. Dorsey, J. A. and Kemnitz, D. A. (1968) *J. Air Pollut. Control Assoc.*, **18**, 12.
28. Smith, W. S. and Martin, R. M. (1969) Paper 86, 60th Annual Conference of Air Pollution Control Association, Ohio.
29. Willard, H. H. and Winter, O. B. (1933) *Ind. Eng. Chem., Anal Ed.*, **5**, 7.
30. ASTM (1980) ASTM Standards on Methods of Sampling and Analysis of Atmospheres, Part 26, D3269-79. In *Annual Book of ASTM Standards*, American Society for Testing and Materials, Philadelphia, Pa.
31. ASTM (1980) ASTM Standards on Methods of Sampling and Analysis of Atmospheres, Part 26, D3270-80. In *Annual Book of ASTM Standards*, American Society for Testing and Materials, Philadelphia, Pa.
32. Nielsen, J. P. and Dangerfield, A. D. (1955) *Arch. Ind. Health*, **11**, 61.
33. Newman, A. C. D. (1958) *Anal. Chim. Acta*, **19**, 471.
34. Singer, L. and Armstrong, W. D. (1954) *Anal. Chem.*, **26**, 904.
35. Tanikawa, S., Kirihara, H., Shiraishi, N. *et al.* (1975) *Anal. Lett.*, **8**, 879.
36. Mascini, M. (1976) *Anal. Chim. Acta*, **85**, 287.
37. Glabisz, U. and Trojanowski, Z. (1980) *Chem. Anal. (Warsaw)*, **25**, 291.
38. Belcher, R., West, T. S. and Leonard, M. A. (1959) *J. Chem. Soc. A*, 3577.
39. Oehme, M. and Stray, H. (1981) *Fresenius Z. Anal. Chem.*, **306**, 356.
40. Kauranen, P. (1977) *Anal. Lett.*, **10**, 451.
41. Nicholson, K. and Duff, E. J. (1981) *Analyst (London)*, **106**, 904.
42. Bellack, E. and Schouboe, P. J. (1958) *Anal. Chem.*, **30**, 2032.
43. Deeker, C. E. and Elfers, L. A. (1968) *Anal. Chem.*, **40**, 1658.
44. Megregian, S. (1954) *Anal. Chem.*, **26**, 1161.
45. Barney, J. E. and Hensley, A. L. (1960) *Anal. Chem.*, **32**, 828.
46. Belcher, R., Leonard, M. A. and West, T. S. (1961) *J. Chem. Soc. B*, 2390.
47. Belcher, R. and West, T. S. (1961) *Talanta*, **8**, 853.
48. West, P. W., Lyles, G. R. and Miller, J. L. (1970) *Environ. Sci. Technol.*, **4**, 487.

49. Liiv, R. and Luiga, P. (1972) Paper 72-1, 65th Annual Meeting of the Air Pollution Control Association, Florida.
50. Hitchcock, A. E., Jacobson, J. S., Mandl, R. H., McCune, D. C. and Weinstein, L. H. (1965) *Simplified Semi-automated Analysis of Plant Tissues*, Proc. Technicon Symp. Automation in Analytical Chemistry, New York.
51. Bakes, J. M. and Jeffery, P. G. (1966) *Analyst (London)*, **91**, 216.
52. Jones, K. W., Gorodetzky, P. H. and Jacobson, J. S. (1976) *Int. J. Environ. Anal. Chem.*, **4**, 225.
53. National Air Pollution Control Administration (USA) (1971) *Atmospheric Emissions from Chlor Alkali Manufacture*, Publication AP-80, Environmental Protection Agency, North Carolina.
54. US Public Health Service (1968) *Compilation of Air Pollution Factors*, US-PHS 999-AP-42.
55. Devitt, T. W. and Gerstle, R. W. (1971) Chlorine and hydrogen chloride emissions and control. Paper 71-5, 64th Annual Meeting of the Air Pollution Control Association, New Jersey.
56. Kubota, K. and Surak, J. G. (1959) *Anal. Chem.*, **31**, 283.
57. McCormac, B. M. Ed. (1971) *Introduction to the Scientific Study of Atmospheric Pollution*, D. Reidel Publishing Co., Dordrecht, Holland.
58. Brennan, E., Leone, I. A. and Daines, R. H. (1965) *Int. J. Air Water Pollut.*, **9**, 791.
59. Gabbay, J., Davidson, M. and Donagi, A. E. (1976) *Analyst (London)*, **101**, 128.
60. Dharmarajan, V. and Rando, R. J. (1979) *Am. Ind. Hyg. Assoc. J.*, **40**, 161.
61. Kuempel, J. R. and Shults, W. D. (1971) *Anal. Lett.*, **4**, 107.
62. D'Amboise, M. and Meyer-Grall, F. (1979) *Anal. Chim. Acta*, **104**, 355.
63. Janghorbani, M. (1983) *Development of Analytical Methodology for the Measurement of Chlorine and Bromine in the Stratosphere*, Report of Nucl. React. Lab., MIT, Cambridge, MA.
64. Ramasamy, S. M., Jabbar, M. S. A. and Mottola, H. A. (1980) *Anal. Chem.*, **52**, 2062.
65. Cooper, H. B. H. and Rossano, A. T. (1970) *Source Testing for Air Pollution Control*, Environmental Science Services Division, Wittan, Connecticut.
66. Hart, A. B. and Lawn, C. J. (1977) *Combustion of Coal and Oil in Power Station Boilers*, CEGB Research No. 5, Central Electricity Generating Board, London.
67. Robertson, A. M. (1974) *Solid Waste Manag.*, **3**, 64.
68. Kaiser, E. R. and Carotti, A. A. (1971) *Municipal Incineration of Refuse with 2% and 4% Additions of Four Plastics: Polyethylene Polystyrene, Polyurethane, and Polyvinylchloride*, a report to the Society of the Plastics Industry, New York.
69. Iapolucci, T. L., Demski, R. J. and Beinstock, D. (1969) *Chlorine in Coal Combustion*, US Bureau of Mines, Report of Investigation No. 7260, US Dept of Interior, Bureau of Mines, Washington, DC.
70. Hiyayama, N., Kazuo, H., Sadao, K. and Toshio, O. (1968) *Bull. Japanese Soc. Mech. Eng.*, **47**, 902.
71. Baechmann, K., Goldbach, K. and Vierkorn-Rudolph, B. (1979) *Mikrochim Acta (Wien)*, **I**, 17.
72. Cheney, J. L. and Duke, D. L. (1983) *Anal. Lett.*, **16**, 309.
73. Williams, K. R. and Jacobi, S. A. (1978) *Atmos. Environ.*, **12**, 2509.
74. Bourbon, P., Alary, J., Lepert, J. C. and Escalssan, J. F. (1979) *Analusis*, **7**, 186.
75. Takamine, K., Tanaka, S. and Hashimoto, Y. (1982) *Bunseki Kagaku*, **31**, 692.
76. Baechmann, K., Matusca, P., Vierkorn-Rudolph, B. and Kontos, M. (1982) *Fresenius Z. Anal Chem.*, **310**, 89.
77. Leithe, W. (1947) *Mikrochem-Michrochim Acta*, **33**, 167.

78. Clarke, F. E. (1950) *Anal. Chem.*, **22**, 553.
79. Getmanenko, E. N. and Perepletchikova, E. M. (1979) *Gig. Sanit.*, **10**, 48.
80. Cheney, J. L. and Fortune, C. R. (1979) *Sci. Total Environ.*, **13**, 9.
81. Vogel, A. I. (1975) *A Text-book of Quantitative Inorganic Analysis*, 3rd edn, Longman, London.
82. Bailey, R. R., Field, P. E. and Wightman, J. P. (1976) *Anal. Chem.*, **48**, 1818.
83. Hagino, K., Iwasaki, J., Ozawa, T. and Utsuiur, S. (1956) *Bull. Chem. Soc. (Japan)*, **29**, 860.
84. Leithe, W. (1971) *Analysis of Air Pollutants* (Trs. R. Kondor), Ann Arbor Science Publishers, Ann Arbor.
85. Torrance, K. (1971) *Anal. Chim. Acta*, **54**, 373.
86. Suzuki, Y. and Imai, S. (1983) *Bunseki Kagaku*, **32**, 313.
87. West, P. W. and Coll, H. (1957) *Spectrophotometric Determination of Chloride in Air*, Proc. Symp. Atmospheric Chemistry of Chloride and Sulphur Compounds, Cincinatti, Ohio.
88. Coll, H. and West, P. W. (1956) *Anal. Chem.*, **28**, 1834.
89. Alexseyeva, M. V. and Elfimova, E. V. (1960) Survey of USSR Literature on Air Pollution and Related Occupational Diseases, **3**, 31–3 (Trs. B. S. Levine), Technical Translation No. TT-60-21475.
90. Katz, M. (1968) *Air Pollution* (ed. A. C. Stern), 2nd edn, ch. 17, Academic Press, New York.
91. Denice, E. C., Luce, E. N. and Akerland, F. E. (1943) *Ind. Eng. Chem.*, **28**, 365.
92. Manita, M. D. and Melekhina, V. P. (1964) *Hyg. Sanit.*, **29**, 62.
93. Vierkorn-Rudolph, B., Savelsberg, M. and Baechmann, K. (1979) *J. Chromatogr.*, **186**, 219.
94. Vierkorn-Rudolph, B. and Baechmann, K. (1981) *J. Chromatogr.*, **217**, 311.
95. Belcher, R., Najafi, A., Rodriguez-Vazgne, J. A. and Stephen, W. I. (1972) *Analyst*, **97**, 993.
96. Belcher, R., Majer, J. R., Rodriguez-Vazgne, J. A. *et al.* (1971) *Anal. Chim. Acta*, **57**, 73.
97. Biehl, H. M. and Baechmann, K. (1980) *Mikrochim. Acta*, **II**, 357.
98. Converse, J. G., Fowler, L. and Emery, E. M. (1976) *ISA Trans.*, **15**, 220.
99. Griggs, M. and Ludwig, C. B. (1978) *J. Air Pollut. Control Assoc.*, **28**, 119.
100. Katz, M. (1980) *J. Air Pollut. Control Assoc.*, **30**, 528.
101. Patel, C. K. N. (1978) *Science*, **4364**, 157.
102. Harrison, R. M. and Sturges, W. T. (1983) *Atmos. Environ.*, **17**, 311.
103. Habibi, K. (1973) *Environ. Sci. Technol.*, **7**, 223.
104. Koppenhaal, D. W. and Manahan, S. E. (1976) *Environ. Sci. Technol.*, **10**, 1104.
105. Robbins, J. A. and Snitz, F. L. (1972) *Environ. Sci. Technol.*, **6**, 164.
106. Duce, R. A., Winchester, J. W. and Van Nahl, T. W. (1965) *J. Geophys. Res.*, **70**, 1775.
107. Moyers, J. L., Zoller, W. H., Duce, R. A. and Hoffman, G. L. (1972) *Environ. Sci. Technol.*, **6**, 68.
108. Gladney, E. S., Sedlacek, W. A. and Berg, W. W. (1983) *J. Radioanal. Chem.*, **78**, 213.
109. Dams, R., Rahn, K. A. and Winchester, J. W. (1972) *Environ. Sci. Technol.*, **6**, 441.
110. O'Connor, B. H., Kerrigan, G. C., Thomas, W. W. and Gasseng, R. (1975) *X-ray Spectrom.*, **4**, 190.
111. Dennis, B. L., Wilson, G. S. and Moyers, J. L. (1976) *Anal. Chim. Acta*, **86**, 27.
112. Countess, R. J., Cadle, S. H., Groblicki, P. J. and Wolff, G. T. (1981) *J. Air Pollut. Control Assoc.*, **31**, 247.

113. Molina, M. J. and Rowland, F. S. (1974) *Nature (London)*, **245**, 27.
114. McConnell, G., Ferguson, D. M. and Pearson, C. R. (1975) *Endeavour*, **34**, 13.
115. Singh, H. B., Salas, L. J. and Cavanagh, L. A. (1977) *J. Air Pollut. Control Assoc.*, **27**, 332.
116. Leinster, P., Perry, R. and Young, R. J. (1978) *Atmos. Environ.*, **12**, 2383.
117. Lasa, J., Rosiek, J. and Wcislak, L. (1979) *Chem. Anal. (Warsaw)*, **24**, 819.
118. Heidt, L. E., Lueb, F., Pollock, W. and Enhalt, D. H. (1975) *Geophys. Res. Lett.*, **2**, 445.
119. Murray, A. J. and Riley, J. P. (1973) *Anal. Chim. Acta*, **65**, 261.
120. Russell, J. W. and Shadoff, L. A. (1977) *J. Chromatogr.*, **134**, 375.
121. Vidal-Madjar, C., Gonnard, M. F., Benchah, F. and Guiochon, G. (1978) *J. Chromatogr. Sci.*, **16**, 190.
122. Bruner, F., Bertoni, G. and Crescentini, G. (1978) *J. Chromatogr.*, **167**, 399.
123. Grimsrud, E. P. and Rasmussen, R. A. (1975) *Atmos. Environ.*, **9**, 1014.
124. Cronn, D. R. and Harsch, D. E. (1979) *Anal. Lett.* Part B, **12**, 1489.
125. Crescentini, G., Mangani, F., Mastrogiacomo, A. R., Cappiello, A. and Bruner, F. (1983) *J. Chromatogr.*, **280**, 146.
126. Makide, Y., Yokohata, A. and Tominaga, T. (1982–3) *J. Trace Microprobe Tech.*, **1**, 265.
127. Filatova, G. N., Koropalov, V. M. and Kovaleva, N. V. (1983) *J. Chromatogr.*, **264**, 129.
128. Crescentini, G., Mangani, F., Mastrogiacomo, A. R. and Bruner, F. (1981) *J. Chromatogr.*, **204**, 445.
129. Rasmussen, R. A., Khalil, M. A. K., Crescentini, G. *et al.* (1983) *Anal. Chem.*, **55**, 1834.
130. Bruner, F., Canulli, C. and Possanzini, M. (1973) *Anal. Chem.*, **45**, 1790.
131. Pellizzari, E. D., Bunch, T. E., Berkley, R. E. and McRae, J. (1976) *Anal. Chem.*, **48**, 803.
132. Clark, A. I., McIntyre, A. E., Lester, J. N. and Perry, R. (1982) *J. Chromatogr.*, **252**, 147.
133. Seshadri, S. and Bozzelli, J. W. (1983) *Chemosphere*, **12**, 809.
134. Hanst, P. L., Spiller, L. L., Watts, D. M. *et al.* (1975) *J. Air Pollut. Control Assoc.*, **25**, 1220.
135. Cronn, D. R. and Harsch, D. E. (1976) *Anal. Lett.*, **9**, 1015.
136. Drugov, Yu, S. and Muraveva, G. V. (1980) *Zh. Anal. Khim.*, **35**, 1319.
137. Wilkes, B. E., Priestley, L. J. and Scholl, L. K. (1982) *Microchem. J.*, **27**, 420.
138. Otson, R., Williams, D. T. and Bothwell, P. D. (1983) *Am. Ind. Hyg. Assoc. J.*, **44**, 489.
139. Langhorst, M. L. and Nestrick, T. J. (1979) *Anal. Chem.*, **51**, 2018.
140. Billings, W. N. and Bridleman, T. F. (1980) *Environ. Sci. Technol.*, **14**, 679.
141. Oehme, M. and Stray, H. (1982) *Fresenius Z. Anal. Chem.*, **311**, 665.
142. Billings, W. N. and Bridleman, T. F. (1983) *Atmos. Environ.*, **17**, 383.
143. Watson, A. J., Ball, G. L. and Stedman, D. H. (1981) *Anal. Chem.*, **53**, 132.
144. Oehme, M. (1982) *TrAC, Trends Anal. Chem.*, **1**, 321.
145. Lande, S. S. (1979) *Am. Ind. Hyg. Assoc. J.*, **40**, 96.
146. Hobbs, J. S. (1978) *Int. Environ. Saf.*, 22.
147. Goldan, P. D., Fehsenfeld, F. C., Kuster, W. C. *et al.* (1980) *Anal. Chem.*, **52**, 1751.
148. Going, J. and Long, S. (1975) *Sampling and Analysis of Selected Toxic Substances Task II. Ethylene Dibromide*, EPA 560/6-75-001, Environmental Protection Agency, Washington, DC.

149. Tsani-Bazaca, E., McIntyre, A. E., Lester, J. N. and Perry, R. (1981) *Environ. Technol. Lett.*, **2,** 303.
150. Clark, A. I., McIntyre, A. E., Perry, R. and Lester, J. N. (1984) *Environ. Pollut.* (Series B), **7,** 141.
151. Mann, J. B., Freal, J. J., Enos, H. F. and Danauskas, J. X. (1980) *J. Environ. Sci. Health*, Part B, **15,** 507.
152. Dumas, T. and Bond, E. J. (1982) *J. Assoc. Off. Anal. Chem.*, **65,** 1379.
153. Collins, M. and Barker, N. J. (1983) *Int. Lab.*, 106.

10 Remote monitoring techniques*

R. H. VAREY

10.1 Introduction

Remote sensing is usually associated with observations of the earth from satellites using photography and other forms of imaging to obtain information on features such as land use, crop condition and meteorological phenomena. However, in studies of air pollution, satellite borne instruments have not yet played a significant role and remote sensing techniques are generally ground based. What is meant by the term is simply that instruments are used to make measurements not just at the particular location of the instrument itself but also at points remote from it. Such techniques are common in areas other than satellite surveillance, radar being one example. In pollution studies the importance of remote sensing is principally in providing measurements not just at ground level but at heights through the first few kilometres of the atmosphere, the region within which the major proportion of man-made and natural emissions are found. Of course the impact of pollutants is generally most significant at ground level where it can be measured by point samplers. The importance of measuring higher in the atmosphere lies in understanding and predicting the processes which control the dispersion of pollutants and the physical and chemical changes they undergo as they are carried downwind from their sources. In order to develop realistic theoretical models which can be used to describe the dispersion and subsequent ground level concentration of emissions from elevated sources, plumes must be measured in locations where they are still high in the atmosphere and have no impact at all at ground level. Moreover, in a monitoring role, it is sometimes necessary to measure the flux of emitted material, either from a single industrial source or an extended urban area. This can only be done by integrating measurements made through the whole cross-section of a plume, not just from ground level concentrations. Even if restricted to ground level measurements, remote sensing can still offer advantages. For example, it is sometimes more effective to scan an area rapidly with a single remote sensor than to deploy a network of point samplers. Also it can sometimes be useful to operate a remote sensor in a manner which determines an average concentration over a path of up to several hundred metres rather than just use a single point sensor which may be subject to rapid local variations.

* Section 10.6 (Long pathlength absorption spectroscopy) was contributed by Dr A. M. Winer.

Another application of remote sensing is in making meteorological measurements, a central feature of any pollution study. Methods are available for measuring the depth of the mixing layer and vertical profiles of temperature, humidity, wind speed and direction and atmospheric turbulence. Perhaps the most widely used remote sensing technique is that for wind measurements.

In pollution studies remote sensing can be achieved in two ways. One is to monitor solar radiation reaching the earth and measure modifications produced by pollutants or meteorological effects in the atmosphere. The other is to introduce radiation into the atmosphere from a laser or other source and again measure modifications produced in passing through the atmosphere. This can be done by directing the radiation along a horizontal path and having instruments at both ends. More conveniently a reflector can be used at one end with a transmitter and receiver at the other. The most efficient reflector is a mirror or retroreflector, which permits the use of the lowest power transmitter but requires the prior deployment of one or more reflectors along predetermined paths. Alternatively natural targets such as trees or rising ground can be used. This is much less efficient, typically five orders of magnitude lower in reflectivity, and thus requires a more powerful source of radiation but it does have a substantial advantage in versatility. Finally reflection can be detected from backscatter caused by molecules, aerosol particles, or other atmospheric parameters in the atmosphere. Such backscatter is generally very weak, typically about four orders of magnitude lower than topographic targets, so the technique requires the most powerful transmitters and most sensitive detectors. The use of a retroreflector or topographic target provides only an average value along the return path of the parameter being measured. On the other hand using the atmosphere to provide the return signal gives spatial resolution. This is achieved by using pulses of radiation and, as in radar, measuring the backscattered signal as a function of time, the delay between transmission and reception gives the range at which the backscatter occurs.

Compared with more traditional methods of point sampling the application of remote sensing to the measurement of air pollution is relatively new. It is an area of rapid development, see for example [1, 2], in which there are many interesting prospects. Perhaps the most far reaching are satellite borne instruments which have the potential to make rapid surveys over large distances. These may prove valuable in the study of the transport of pollutants and in understanding the long range transformation and deposition of species during the drift of air masses. Already satellite systems have been used to measure fluxes of particulates [3] and plans are well advanced to use laser systems in the Space Shuttle [4, 5]. There are other methods which are in their infancy and while showing considerable promise are still at the stage of laboratory prototypes. However, some methods have been used for several years and are finding increasingly wide application. The object of this chapter is to describe techniques which have reached the stage in development where they have found practical use in field work. Consequently the main emphasis is on correlation spectroscopy, sodar, lidar and long path absorption spectroscopy, although some less widely used techniques such as those for temperature measurement are also included. A further objective is to illustrate the type of results which remote sensing can offer and how they complement those obtained by traditional methods.

10.2   Correlation spectroscopy

10.2.1   Mode of operation

The principle of correlation spectroscopy is straightforward. As in absorption spectroscopy, light is selectively absorbed at wavelengths characteristic of the gas of interest. In absorption spectroscopy light from a standard source is passed through a volume of the gas being studied and from the attenuation of the light the concentration of the gas can be deduced. It is common practice to monitor attenuation at two adjacent wavelengths corresponding to weak and strong absorption in order to normalize the effects of extraneous losses in the optics and detection system. If the absorption coefficients of the gas are known at the two wavelengths then the ratio of the transmitted signals gives the number of molecules in the light path. Alternatively, the system can be calibrated by introducing cells containing known concentrations of the gas of interest into the light path.

The procedure with correlation spectroscopy is similar except that solar radiation is used instead of a standard source. It is not direct solar radiation which is monitored but scattered radiation from the bright sky background, so called skylight, the spectrometer being generally arranged to collect radiation from vertically above. This radiation is modified by pollutants in the atmosphere and its spectrum reveals the total absorption produced by molecules in the field of view of the instrument, i.e. the integrated effect, or total column content of molecules in the atmosphere above the instrument. Thus there is no discrimination between a narrow layer of high concentration or a thicker layer of lower concentration. Nor does the instrument give any indication of vertical distribution of the gas in question. However, it does have the attribute of a remote sensor in that it is sensitive to the presence of the gas being measured whether it is at ground level adjacent to the instrument or high in the atmosphere remote from it. The units used in correlation spectroscopy are usually micrometres, μm, i.e. a concentration of 1 ppm through a depth of 1 m. This is equivalent in terms of absorption to 0.5 ppm occupying a depth of 2 m or 1 ppb through 1 km, i.e. the integral of the product of concentration, in ppm, and depth in metres.

In practice the useful region of the spectrum is limited to the ultraviolet and visible and the method is restricted to the measurement of SO_2 and NO_2. However, this limitation is not as serious as it may seem since SO_2 and NO_2 are two of the major atmospheric pollutants. The absorption spectra of these two gases are shown in Fig. 10.1. The most convenient wavelengths for the measurement of SO_2 are in the region of 300 nm and for NO_2 440 nm.

The essential components of a correlation spectrometer are shown in Fig. 10.2. Light enters the system from a telescope, and is directed on to a diffraction grating which produces a skylight spectrum. This is not scanned, as it might be in

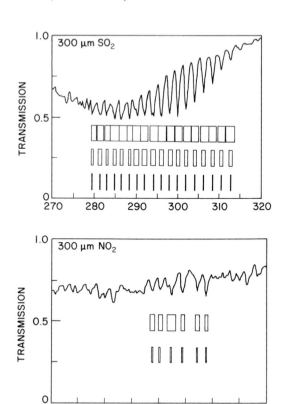

Fig. 10.1   Spectra of SO_2 and NO_2 with examples of arrangements of mask slits used to sample absorption. (Millán and Hoff [6].)

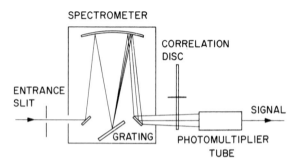

Fig. 10.2   Correlation spectrometer: Barringer Cospec IV – schematic.

absorption spectroscopy, but sampled sequentially at wavelengths of strong and weak absorption. This is achieved by placing masks in the plane of the spectrum, one mask with slits at positions of strong absorption and another at positions of weak absorption. The masks are mounted on a disc in such a way that as it rotates the two regions of the spectrum are measured alternately by the photomultiplier. The effectiveness of the instrument depends upon the number, width and spacing of the slits and the methods of signal analysis. This is dealt with in detail by Millán and Hoff [6].

10.2.2   Baseline drift, sensitivity and multiple scattering

Although it is the use of skylight which makes correlation spectroscopy such a convenient tool for measuring SO_2 and NO_2 it does present some serious problems. The ideal light source for absorption spectroscopy is one which does not change with time and does not vary with wavelength, at least not on a wavelength scale similar to that of the absorption spectra being used for the measurements. Skylight is far from ideal in that it has a marked variation during daylight hours both in intensity and in the structure of its spectrum. Figure 10.3 shows typical variations in the radiance of the zenith sky measured during the course of a summer day with clear skies [7]. The fine structure is due principally to Fraunhoffer lines in the solar spectrum with larger scale differences originating from ozone, scattering and the diurnal variation in the elevation of the sun. The result is that both the baseline and sensitivity of the instrument change during the course of the day. The effect of these changes can be reduced to some degree by careful instrument design and signal processing [8] but the best way to minimize

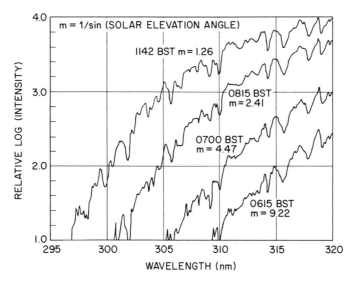

Fig. 10.3   Skylight spectra of the zenith sky measured at Leatherhead, UK, in the course of a July day with clear skies. (Hamilton *et al.* [7].)

errors is to calibrate the instrument frequently by the introduction of standard cells. Less easy to deal with are effects caused by rapid changes in skylight radiation at the appearance of a bright cloud edge in the field of view. These can be identified by monitoring the output from an automatic gain control unit incorporated into the instrument and in such circumstances care must be taken in interpreting results [9].

In general, changes in baseline and sensitivity are not a serious problem when the instrument is mounted in a vehicle and moved in a traverse underneath a well-defined plume. A typical result of such a traverse is shown in Fig. 10.4. Here the contribution of the plume to the reading of the correlation spectrometer is clear. In most situations of practical interest such a traverse will take no longer than ten to fifteen minutes during which changes in sensitivity are small. The plume may not be the only source of SO_2 or NO_2 in the atmosphere and other sources may contribute a background which may vary across the plume traverse. This will produce an apparent change in the instrument baseline between one edge of the plume and the other. The usual way to deal with this is to note the level on each side of the plume and to assume a linear variation in between. Monitoring extended source areas where overhead burden is relatively low, where plumes are less well defined and where traverses can be of a duration of an hour or more present more serious difficulties. These will be illustrated in Section 10.8 on monitoring with remote sensors.

The detection limits and accuracy of the instrument vary with skylight

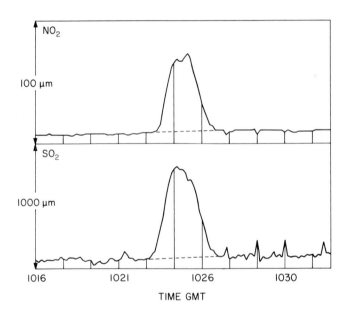

Fig. 10.4 Traverses under a plume with correlation spectrometers. Simultaneous measurements of SO_2 and NO_2 showing linear interpolation used to remove background.

radiance. Clearly, it does not work at all at night and is less effective in the early morning and late evening than in the middle of the day. Typically for several hours around noon on a spring or autumn day the noise on a carefully adjusted instrument is likely to be about 10 μm for SO_2 and 5 μm for NO_2. This means that the instrument can for example detect changes of around 5 ppb NO_2 distributed through a mixing layer 1000 m deep or 50 ppb in a layer 100 m deep. Such sensitivity can be maintained for a longer period in mid-summer and for considerably less in winter. Indeed, in winter in England in overcast cloudy conditions efficient operation can be limited to only an hour or so either side of noon. A further important feature of correlation spectroscopy is that measurements can be made with a short time constant, generally in the region of one second or so compared with typical ground level sensors which have a time constant of about one minute. This is an important factor when a plume is being traversed by an instrument mounted in a vehicle and particularly if an aircraft is used.

Practical applications of correlation spectroscopy will be reviewed and illustrated in Section 10.8 but it is clear that it is a powerful and convenient technique for measuring sources of SO_2 and NO_2. It has a short time constant, good sensitivity and can be used to measure plumes in the atmosphere whether or not they are at ground level. Furthermore, the instrument is compact, typically $0.6 \times 0.3 \times 0.2$ m and easy to operate. Limitations are that it operates only during daylight hours and does not provide vertical spatial resolution. Due to uncertainties and fluctuations in the skylight spectrum it is not an absolute instrument and operates most successfully in measuring plumes which are well defined against a background concentration and when the instrument is calibrated frequently by the use of internal standard cells.

Correlation spectrometers are available commercially. The early development of the technique and its application to air pollution studies were carried out in Canada by Barringer Research Ltd [10–12] and produced under the trade name Cospec. Recently a second commercial instrument has appeared, also produced by a Canadian company, Moniteq.

10.3 Single wavelength lidar

10.3.1 Principles of lidar

Unlike correlation spectroscopy lidar does not depend upon skylight. Thus it is not limited to daylight operation, indeed it works more effectively at night. It is not limited to measuring total overhead burden since it can provide measurements with spatial resolution. Furthermore, it does not suffer from problems of baseline drift and changing sensitivity, though it does have problems of its own as discussed below in this section and the next.

The principle of lidar is that a laser pulse is fired into the atmosphere and as it proceeds along its path, radiation is continuously scattered by molecules and aerosol particles back towards the laser where it is collected with a telescope and measured with a detector. The signal is analysed to provide information on the magnitude of backscatter and attenuation experienced by the pulse in its passage through the atmosphere. The technique is applied in two ways. The first is to use a laser operating at a single fixed wavelength to monitor the distribution of aerosol particles. This has applications in meteorology, Section 10.7, and in studies of plume rise and dispersion, Section 10.8. The second mode of operation is to use a tuneable laser in which the wavelength can be changed. This enables the distribution of selected molecules in the atmosphere to be determined. This section is concerned with some general principles of lidar and single wavelength systems and the next section with multi-wavelength operation.

Lidar has developed because lasers are particularly suitable for providing intense pulses of short duration with a narrow bandwidth and small beam divergence. High energy is important since the proportion of the pulse which is backscattered by the atmosphere is small. The narrow bandwidth is useful because it makes it possible to measure the concentration of particular molecules along the path of the pulse by probing their absorption spectra. It also makes it possible to exclude a high proportion of the skylight radiation, which would otherwise swamp the receiving system, by including in the optics a narrow band filter which allows only radiation close to the wavelength of the laser to reach the detector. The small beam divergence is useful because it improves the lateral spatial resolution of the system and also reduces skylight radiation reaching the detector by permitting the use of a narrow aperture telescope. As with radar (*R*adio *D*etection *a*nd *R*anging) the range from which the lidar (*L*ight *D*etection *a*nd *R*anging) signal is being received is provided by the time interval between the firing of the pulse and the reception of the signal. By continuously monitoring the backscattered signal, information is received out to a range at which the signal becomes too weak to be detected, the range resolution obtainable being related to the length of the laser pulse.

It is not appropriate here to treat in detail the theoretical aspects of lidar operation – that can easily be found elsewhere, as for example in references [13–16]. Nevertheless, in order to illustrate some of the practical aspects of the technique it is worth examining some features of the so-called lidar equation. As a laser pulse travels through the atmosphere it is attenuated as a result of being scattered and absorbed. Scattering has two components, one provided by particles much smaller than the laser wavelength, Rayleigh scattering, and the other by larger particles, Mie scattering. Rayleigh scattering is proportional to λ^{-4}, where λ is wavelength. Mie scattering is much more complex [17] and depends on the details of the size, shape and material of the particles. Absorption depends upon the laser wavelength and the absorption spectrum of the molecules involved. Scattering and absorption are grouped together in a coefficient α, the volume extinction coefficient [18]. Some of the scattered radiation returns along its path back towards the laser and telescope. The proportion of the total scattered radiation which does this varies with the size and shape of the scattering centres. It is described by a volume backscatter coefficient, β [18]. The energy, dE, received at the telescope after backscatter from an element dr at range r along the path of the laser pulse is given by

$$dE = E_0 TA \frac{\beta(r)dr}{r^2} \exp\left[-2 \int_0^r \alpha(r')\,dr' \right] \tag{10.1}$$

where E_0 is the initial laser pulse energy, T is related to the optical efficiency of the receiver system, A is the telescope area, $\beta(r)$ is the volume backscatter coefficient at range r and $\alpha(r')$ is the volume extinction coefficient at range r'. $\beta(r)$ describes the radiation which is backscattered towards the laser, the factor $1/r^2$ arising from the usual inverse square law.

The power received, P, is dE/dt where t is the interval between the start of the pulse from the laser and the return to the telescope of the radiation from range r. Since $t = 2r/c$ where c is the speed of light then Equation (10.1) becomes

$$P(r) = \frac{E_0 TAc\beta(r)}{2r^2} \exp\left[-2 \int_0^r \alpha(r')\,dr' \right] \tag{10.2}$$

which is the so-called lidar equation (see for example references [12–14]). The scattering coefficients α and β appear as functions of range since absorbing and scattering species are not in general distributed uniformly through the atmosphere. For example a power station plume contains a higher concentration of aerosol particles than the surrounding atmosphere and so provides an enhanced backscatter signal. It is this characteristic which enables a plume and its dimensions and position to be determined, as will be described in Section 10.8.

Spatial resolution depends upon the length of the laser pulse. At any instant the detector receives light simultaneously from a length of the path of the laser pulse equal to $c\tau/2$, where τ is the duration of the pulse. This defines the spatial resolution of the system since details of the structure within the length from which radiation is received simultaneously cannot be resolved, at least not without the use of deconvolution techniques. Clearly, for a given power the longer the pulse length the greater the received power. However, in practice with pulsed lasers it is not the instantaneous power which is fixed but the total energy in the pulse. This can, within limits, be used to give longer, lower power or shorter higher power pulses. There is thus nothing to be gained in using long pulses and it is best to use as short a pulse as possible because although this does not change the instantaneous power received it does give the best spatial resolution, provided of course that the electronics of the receiving system are fast enough to provide the time resolution τ.

10.3.2 Essentials of a practical system

The basic components of a lidar system comprise a laser with its associated optics, a receiving telescope and a detection system. For single wavelength operation two types of laser are commonly used, the ruby laser operating at a wavelength of 694.3 nm and the neodymium-YAG laser operating at its fundamental frequency with a wavelength of 1064 nm or frequency doubled at 532 nm. Pulse duration is in the region of 10 ns or so and energy up to about 1 J. The laser linewidth is a few tenths of a nanometre and the beam divergence about a milliradian. Such lasers can provide pulses at a rate of between about 1–20 per second, the higher pulse repetition rates being more important for differential lidar operation where many pulses have to be averaged in order to achieve an adequate signal-to-noise ratio in the backscattered signal, as described in Section 10.4.1. In the infrared in the region of 10 μm CO_2 lasers are commonly used [19–21]. These provide pulse energies, repetition rates and beam divergence similar to the ruby and neodymium-YAG lasers but generally with a longer pulse duration, typically 100 ns. Telescopes generally employ a parabolic reflector often with a Newtonian configuration although Cassegrain or Gregorian designs are also used. Ground based single wavelength systems used for the study of plume rise and dispersion are usually mounted in such a way that the direction in which the pulse is fired can be varied, both horizontally and vertically. This enables cross-sections of a plume to be measured from a fixed location. This is achieved most conveniently by rotating the complete laser/telescope assembly, an example of which is shown in Fig. 10.5. When mounted in a vehicle or aeroplane for mobile operation over greater distances the laser is fixed to fire vertically, upwards from a vehicle and generally downwards from an aircraft. Systems with the characteristics described above have a spatial resolution of a few metres and a range of operation, in conditions of moderate visibility, approaching 10 km. An example of a typical system, Fig. 10.5, is the Central Electricity Generating Board ruby lidar which can be used in a scanning mode or on the move pointing vertically upwards.

Fig. 10.5   Single wavelength lidar. (Central Electricity Generating Board.)

10.3.3   Signal processing

A significant component of most remote sensing techniques is the system used for analysing the collected data and presenting the results in a meaningful way. This is a substantial task and the effort needed to assemble the necessary hardware and, where it is necessary, as in the majority of applications, to produce the associated computer software should not be underestimated. A vehicle carrying a correlation spectrometer and a normal complement of point sensors collects a substantial quantity of data in even a modest pollution survey. The same is also true, but to a greater degree, of data collected from a lidar system. In a few hours even a simple system firing at a low repetition rate produces many thousand pulses each of which has to be examined in some detail. Furthermore, if for example plume dispersion is being studied, the results have to be analysed in some way which is meaningful statistically.

The basic form of the lidar signal provided by atmospheric backscatter is shown in Fig. 10.6. Here the laser is firing through a power station plume. The initial rise in the signal is due to the optical arrangement and corresponds to an increasing overlap with distance of the field of view of the telescope with the laser beam. This is soon complete after which the signal decays, principally due to the $1/r^2$ term in Equations (10.1) and (10.2) but also as a result of attenuation of the pulse in its passage through the atmosphere. However, on reaching the plume the signal is significantly enhanced by the presence of particulates. The plume is first encountered at a range of 1.2 km with a small perturbation to the signal followed by a much larger signal at 1.5 km. This section of the plume has a depth of about

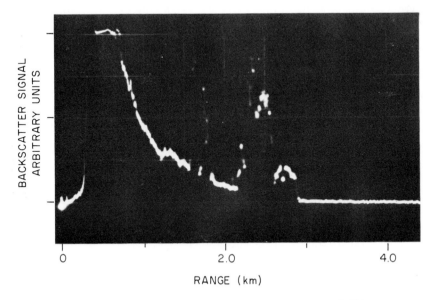

Fig. 10.6   Typical lidar backscatter signal with plume. (Hamilton [15].)

200 m and is followed by a more substantial section starting at a range of 2.2 km and having a depth of 800 m. The plume boundaries are clearly defined and it can be seen from the size of the signal that the pulse has been attenuated in passing through it. As indicated in the figure a plume is not usually homogeneous but exhibits marked internal structure. The object of analysis is to find the position in space of such areas of enhanced backscatter, to follow the change and development of such features with time and where possible to interpret the results quantitatively.

The most straightforward way of examining the return signals is to display signal strength as a function of time on an oscilloscope screen to obtain a record similar to that shown in Fig. 10.6. From such records, with a measurement of the direction in which the laser was fired the position of plume features encountered along the path of the pulse can be determined. The same technique can be used to scan an area of the sky. The laser is fired at a series of vertical angles and by noting the elevation of each shot and the range from which signals are received simple geometry gives the shape, dimensions and position of the area of enhanced backscatter. However, this soon becomes tedious if many shots are used and the first step in improving the analysis is to use the method illustrated in Fig. 10.7. Here the path of the pulse through the atmosphere is represented directly on the screen, a result achieved by coupling a potentiometer to the axis of elevation of the laser and using the voltage to provide the corresponding trajectory of the oscilloscope beam. The backscattered signal is used to modulate the intensity of the beam. Thus a bright region on the oscilloscope represents a cross-section of the area of enhanced backscatter in the atmosphere, in this case a cross-section of the plume. Displays such as this provide vivid, immediate information, a good example of which is shown in Fig. 10.8. This was obtained with the Stanford Research Institute aircraft mounted neodymium-YAG lidar system [22] being used to study the dispersion of a power station plume in hilly terrain. In addition to backscatter from the plume, in this figure shown dark against a light

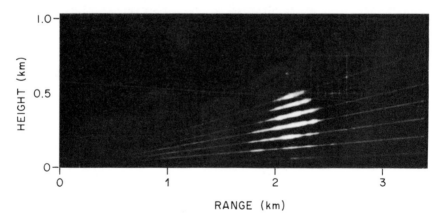

Fig. 10.7 Oscilloscope display of lidar scans of a plume. (Hamilton [15].)

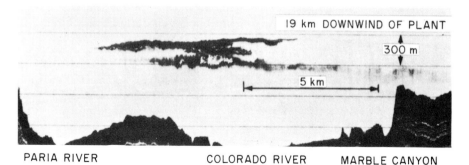

19 km DOWNWIND OF PLANT

300 m

5 km

PARIA RIVER   COLORADO RIVER   MARBLE CANYON

Fig. 10.8 Cross-section of power station plume in hilly terrain: Stanford Research Institute airborne lidar. (Uthe *et al.* [22].)

background, the return signal from the ground is used to provide a cross-section of the terrain being traversed. Figure 10.6 illustrates one of the problems encountered in simple displays of this sort. After the initial rise the signal decreases rapidly, with the result that the accuracy with which the effects of enhanced backscatter can be measured decreases rapidly with range. It also means that if backscatter from different ranges is to be compared then compensation has to be made for the effects of $1/r^2$ and attenuation. One method of doing this is to vary the gain of the photomultiplier. This is achieved by increasing the photomultiplier supply voltage in such a way that the gain increases as the square of the range. Further adjustment can be made to compensate for atmospheric attenuation depending upon the conditions which exist at the time of the experiment. In this way fairly constant signals can be obtained, typically for ranges between about 300 m and 3 km and it is upon this signal that the effects of enhanced backscatter are superimposed.

The techniques described above provide excellent qualitative visualization of return signals. For more detailed analysis it is clear that the examination of oscilloscope displays is unsatisfactory and that considerable advantage is to be gained from digitizing the signal and coupling the system to a computer. An outline of the system used on the Central Electricity Generating Board ruby lidar, is shown in Fig. 10.9. The photomultiplier is used with swept gain and the return signals are digitized in a 20 MHz, 8 bit transient recorder. The unprocessed signals are monitored by direct display on an oscilloscope screen and also stored on floppy disc in a DEC MINC computer. Azimuth and elevation of each laser pulse are also recorded together with the energy of each pulse as it leaves the laser. The signal is sampled every 50 ns, which is equivalent to a range resolution of about 7.5 m, and up to 1000 samples can be taken of each shot. The signals can be displayed immediately in an intensity modulated mode, either as an $X–Y$-signal strength display or as range–time-signal strengths. This is similar to the analogue displays used simply with an oscilloscope. Here however intensity modulation is

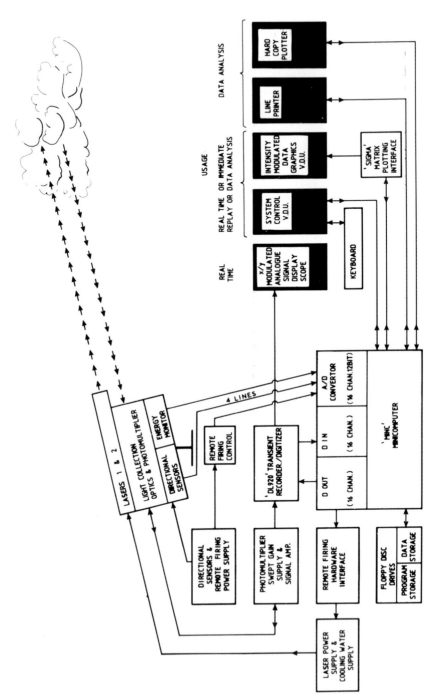

Fig. 10.9   Central Electricity Generating Board single wavelength lidar: schematic of system control and data processing.

achieved with a 16 level grey scale and a first step in processing is achieved by normalizing each signal with respect to the energy in each pulse.

In terms of display this represents little advance over direct analogue techniques, in fact care must be exercised in identifying apparent structure introduced by the discrete nature of the grey scale quantization. It does have the advantage of being able to recall the display at will without recourse to photographs but the great advantage lies in being able to do further analysis. Signals can be normalized and the results of many shots can be averaged. The statistical nature of plume dispersion can be investigated and corrections for variations in background attenuation can be incorporated in the analysis. Also the results are readily available for approaching the problems of making the results quantitative rather than just qualitative. This is difficult, the problems arising principally from the complexity of scattering from particles of unknown shape, composition and size distribution [23]. However, if it is assumed that the character of the particles does not change through the plume and that variations in backscatter are caused only by variations in number density then a quantitative interpretation can be made of the distribution of plume material. An example obtained with the Computer Genetics Corporation lidar is shown in Fig. 10.10 where contours of

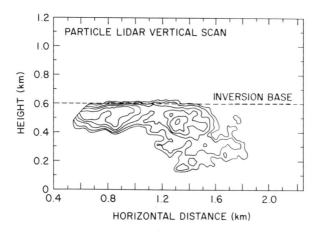

Fig. 10.10   Plume density contours: an example of plume reflection near the base of an inversion. (Computer Genetics Corporation ground based lidar [25].)

the relative density of particles in a plume are plotted. Finally, the form of the contours in the figure clearly show that further analysis is required even when contours have been obtained. A large number of such profiles have to be averaged to obtain statistically significant results for plume height and dispersion. Methods can be developed [24, 25] to perform such analysis once the data has been stored in digital form and results can be obtained rapidly and efficiently.

10.4   Differential lidar

10.4.1   Basic methods

Whereas single wavelength lidar makes a rather generalized measurement of aerosol particles differential lidar, commonly known as DIAL, gives a measurement of specific molecules with their concentration and distribution in the atmosphere. It is becoming routine to measure SO_2 and O_3 and NO_2 can be determined in the same manner. As will be seen later the sensitivity of the technique depends upon several factors such as atmospheric visibility, the spatial resolution required and the distance of the species from the instrument. Typically for SO_2 and O_3 a sensitivity of 20 ppb can be achieved out to a distance of a kilometre or so with a spatial resolution of about 50 m. There is less experience with NO_2 but theory suggests that sensitivity should be similar. Emphasis is on SO_2, NO_2 and O_3 since they are accessible with readily available lasers operating in the ultraviolet and visible regions of the spectrum. Many more molecules could be measured in the infrared but although some success has been achieved with water vapour, ethylene and HCl lack of suitable lasers has limited practical developments.

As with single wavelength lidar DIAL depends on measuring the radiation from a laser pulse backscattered by the atmosphere. The difference is that instead of operating at just one wavelength two or more wavelengths are used such that the difference in the backscattered signal can be related to absorption caused by molecules of interest. Generally two wavelengths are used corresponding to a maximum and minimum in the absorption spectrum, a central feature of the method being that the two wavelengths are sufficiently close for differences in the scattering properties of the atmosphere to be small compared with differences in absorption. This is important because as we have seen in the previous section the magnitude of the backscattered signal does vary with wavelength in a manner which depends upon the shape and size distribution of the atmospheric aerosol and in general this variation is not known. However, if the wavelengths are separated by only a few nanometres or so then these variations are small and differences in backscattered signal can to a good approximation be attributed to absorption. From the absorption spectrum of the molecule being measured the differences in backscattered signal can be interpreted in terms of the concentration of the molecules, spatial distribution being obtained by measuring the difference in the signals as a function of range, as with single wavelength systems.

Detailed treatments of differential lidar can be found in references [13, 14, 16]. Briefly, referring to the lidar Equation (10.2) quoted in the previous section but now written for two wavelengths, on-resonance corresponding to strong absorption and off-resonance to weak absorption, we have

$$P_{on}(r) = \frac{E_{on} T A c \beta_{on}(r)}{2r^2} \exp\left[-2 \int_0^r \alpha_{on}(r')\, dr' \right] \qquad (10.3)$$

$$P_{off}(r) = \frac{E_{off} T A c \beta_{off}(r)}{2r^2} \exp\left[-2 \int_0^r \alpha_{off}(r')\, dr' \right] \qquad (10.4)$$

If it is assumed that the ratio β_{on}/β_{off} does not vary with r, that the difference between α_{on} and α_{off} is dominated by absorption and that each pulse is normalized with respect to initial energy E_0, then

$$\frac{P_{on}(r)}{P_{off}(r)} = \exp\left\{ -2 \int_0^r [\sigma_{on} n(r') - \sigma_{off} n(r')]\, dr' \right\} \tag{10.5}$$

where now absorption is written as the product of the absorption cross-section σ and concentration n.

Differentiating gives

$$n(r) = \frac{1}{2(\sigma_{on} - \sigma_{off})} \frac{d}{dr}\left[\ln \frac{P_{off}(r)}{P_{on}(r)} \right] \tag{10.6}$$

Thus knowing values of σ and measuring the backscattered signal, concentration can be determined as a function of range.

Figure 10.11 shows diagrammatically DIAL return signals and the stages in determining concentration. As with single wavelength lidar the signals show a sharp rise as the field of view of the telescope overlaps the path of the laser beam and then a fall as range increases. Both signals are the same until the region is reached where absorption occurs. There the on-resonance signal decreases more

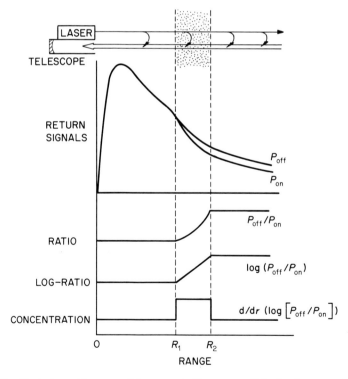

Fig. 10.11   Variation with range of the on and off resonance lidar backscatter signals and stages in determination of pollution concentration.

rapidly until this region is passed and thereafter both signals decrease in a similar manner. The ratio P_{off}/P_{on} increases through the region as more and more of the on-resonance signal is absorbed. The gradient of the ln of the ratio is proportional to the concentration. In the simplified example given in the diagram the concentration is constant so the gradient does not vary. In practice, of course, the concentration is not constant and moreover a region containing absorbing molecules often produces enhanced backscatter due to the presence of increased concentrations of aerosol particles, a power station plume for example. Then both on- and off-resonance signals show a sharp rise at the plume but nevertheless the ratio P_{off}/P_{on} still exhibits the same behaviour and is analysed in the same way to give concentration.

As with single wavelength lidar considerable data processing is required in order to obtain useful results. It is even more important with DIAL because in addition to performing the calculations outlined in Equation (10.6) it is not usually sufficient to measure just two pulses. In practice systems operate close to detection limits and the signal-to-noise ratio on the backscattered signals is small. Consequently a large number of measurements, typically several hundred, have to be averaged in order to obtain an acceptable result. This analysis cannot be done from oscilloscope displays of individual backscatter records, as is sometimes possible with single wavelength lidar, and fast digitization and analysis of signals is essential. In fact computer control is a central feature of any practical system since in addition to data analysis a computer is used to control the tuning of the laser and monitor the energy output, control the sequence of laser firing and digitization and storage of the return signal, interface with any navigation system on the vehicle carrying DIAL and monitor and record results obtained by any other instruments being carried.

10.4.2   Examples of practical systems

In the ultraviolet SO_2 and O_3 can be measured at wavelengths close to 300 nm and in the visible NO_2 at around 450 nm. Systems developed for these measurements usually employ a neodymium-YAG pumped dye laser which provides pulses of high energy, narrow linewidth and short pulse length. The telescope used to collect the backscattered radiation is generally based upon a 0.5 m parabolic reflector and signals are processed via a 10 or 20 MHz, 8 or 10 bit digitizer. The whole system is mounted in a lorry so that it can be moved from one location to another and is preferably capable of making measurements on the move. For such mobile operations the laser is usually fired vertically upwards, scanning of an area of the atmosphere being achieved by movement of the vehicle. Some systems have the facility to change the direction of the laser beam and scan the atmosphere from a fixed position.

An outline of a typical system is shown in Fig. 10.12. The neodymium-YAG laser produces radiation at a wavelength of 1064 nm. This is frequency doubled to give a wavelength of 532 nm which is used as the input to the dye laser. This is tuned to produce alternate pulses of slightly different wavelength around 600 nm which are again frequency doubled to give two wavelengths close to 300 nm. For

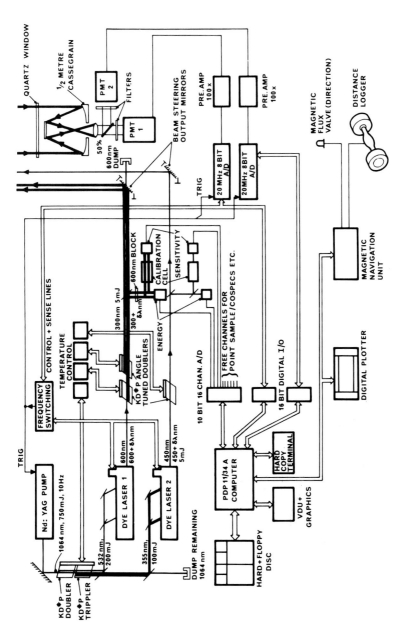

Fig. 10.12 Central Electricity Generating Board differential lidar system – schematic.

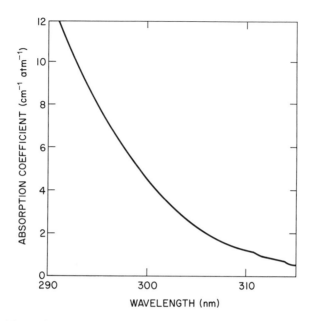

Fig. 10.13 Absorption spectrum of O_3 in the region of 300 nm.

SO_2 it can be seen from the absorption spectrum, Fig. 10.1, that suitable wavelengths for on and off resonance are 300.1 nm and 299.5 nm. The spectrum of O_3 in this region does not have the same well-defined structure, absorption changing more slowly with wavelength as shown in Fig. 10.13. The consequence is that in order to obtain sufficient difference in absorption the two wavelengths must be further apart, typically 20 nm or so. Clearly the presence of one species can interfere with the measurement of the other, particularly the presence of SO_2 in O_3 measurements. Here it is important to choose wavelengths which have the same, preferably low, absorption by SO_2. This can be obtained in practice by incorporating into the system a standard cell containing SO_2 and tuning the two wavelengths to get zero difference. For NO_2 the neodymium-YAG output is tripled to produce a wavelength of 355 nm before entering the dye laser which is tuned to give an output in the region of 450 nm. The pulses which finally emerge from the system have an energy of about 10 mJ and a duration of 10 ns. A small proportion of each pulse is diverted before it leaves the system in order to measure pulse energy. Backscattered radiation is collected by the telescope, passed through a filter to eliminate background radiation at wavelengths other than that of the laser, and measured with the photomultiplier. The signal is then digitized, normalized with respect to the initial energy and analysed as outlined in Section 10.4.1. Pulses are produced at a rate of about 10 per second and several hundred at each wavelength are required for a sensible measurement.

The accuracy and detection limits of differential lidar cannot be defined in the same way as those for a normal point sampler since they depend on factors such as atmospheric visibility, the spatial resolution required, the range at which the measurement is being made and the time resolution, i.e. the number of pulses being averaged. Essentially it is a question of the signal-to-noise ratio in the measured signal which in turn depends on the number of photons received from each measurement. However, it is generally possible to measure SO_2 and O_3 with a range resolution of 50 m and a sensitivity of about 10 ppb out to a range of a few hundred metres and with a sensitivity of 20 ppb or so at a kilometre. Sensitivity for NO_2 is in principle somewhat lower but to date fewer measurements have been made [26–28], most effort being concentrated on SO_2 [29–31] and O_3 [32–34]. A typical result is shown in Fig. 10.14 where there is a clearly defined plume between about 100 m and 550 m with a peak concentration approaching 350 ppb.

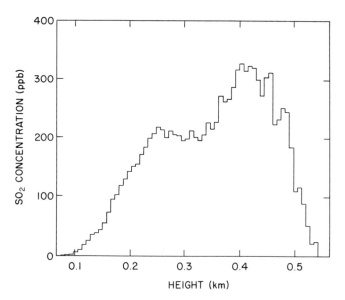

Fig. 10.14 Vertical profile of SO_2 concentration in a power station plume obtained with differential lidar. (Sutton, Central Electricity Generating Board, private communication.)

Systems operating in the ultraviolet and visible have been developed by many laboratories. Figure 10.15 shows a view of the Central Electricity Generating Board system. It is mounted in a lorry with the laser firing vertically. Measurements are made on the move and the vehicle is fitted with a navigation system to enable profiles of SO_2 to be plotted along the route followed by the vehicle. An example of such a route map showing plumes from four power stations is illustrated in Fig. 10.16 with the vertical distribution of the plumes on the same traverse shown in Fig. 10.17. Measurements of ozone made with the

Fig. 10.15   Central Electricity Generating Board differential lidar vehicle. The laser fires
vertically upwards and the receiving telescope is mounted behind the cab.
The interior shows two neodymium-YAG lasers, centre; various data
processing units and a correlation spectrometer, top right.

same system are shown in Fig. 10.18 together with simultaneous results obtained
with an ozone sonde mounted in a radio controlled model aircraft.

The Stanford Research Institute in California have a similar system [35] again
mounted in a vehicle, Fig. 10.19. This operates by scanning a plume from a fixed
site and provides contours of SO_2 concentration.

An aircraft mounted system has been developed by NASA [32] which uses
twin neodymium-YAG lasers. In this case the lasers are fired vertically

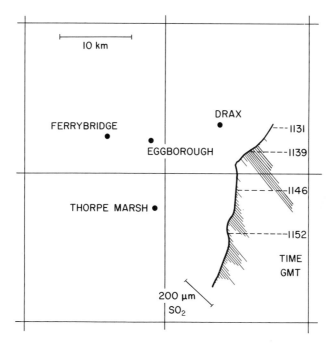

Fig. 10.16 Differential lidar measurements of four power station plumes showing the route followed by the vehicle and the overhead burden of SO_2. (Sutton, Central Electricity Generating Board, private communication.)

downwards so that in addition to the radiation backscattered by the atmosphere large signals are also received by scattering from the ground and can be used to obtain sensitive measurements of the integrated concentration between the aircraft and the ground. This has been used to measure O_3 and water vapour using a wavelength in the region of 720 nm. A sensitivity of about 10 ppb has been obtained for O_3 with a spatial resolution of 200 m. Figure 10.20 shows a comparison of the lidar results with those of a point sampler mounted in another aircraft flying vertical profiles along a track adjacent to the path of the laser pulses. Other neodymium-YAG lorry mounted systems include those developed at the National Physical Laboratory, UK [36], Chalmers University, Sweden [27, 30] and GKSS in Germany [37].

Infrared systems are not so well advanced, in spite of their considerable potential in measuring a wide range of species important in air pollution studies. As mentioned above the main reason is lack of suitable lasers. One possibility is to use a CO_2 laser. This cannot be tuned in the same way as a dye laser to produce radiation over a continuous range of wavelengths but can produce pulses at a number of closely spaced discrete wavelengths in the range 9–11 μm. These can be used to measure molecules with coincident lines in their absorption spectra.

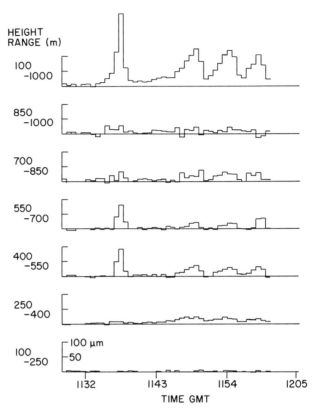

Fig. 10.17   Vertical distribution of SO_2 shown in Fig. 10.16: lower six plots show SO_2 burden in six height ranges up to 1000 m; top graph shows total overhead burden; all four plumes are concentrated mainly between 250 m and 850 m; none is contributing to ground level concentration.

One such system at the University of Munich [19] has been used to measure water vapour above cooling towers and ethylene above an oil refinery, sensitivity for ethylene being about 10 ppb out to a range of about 500 m with a spatial resolution of 100 m. Other CO_2 systems include one developed by NASA [21] for measurements of O_3 and a Stanford Research Institute [20] instrument for measuring water vapour and ethylene. Recently the Stanford Research Institute have also developed an aircraft mounted system using two CO_2 lasers, which was later due to be tested in 1983 [38]. Efforts have been made to extend the tuneability of CO_2 lasers by using high pressure systems, the object being to broaden and overlap the wavelengths at which lasing can be achieved [39]. So far, however, this work has not reached a stage of practical application.

One infrared system which has been used for practical measurements is that developed by GKSS in Germany [40, 41] which operates from an inertial

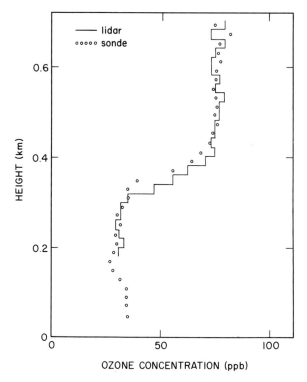

Fig. 10.18 Ozone measurements with ground based differential lidar and comparison with radiosonde measurements. (Ellis and Sutton, Central Electricity Generating Board, private communication.)

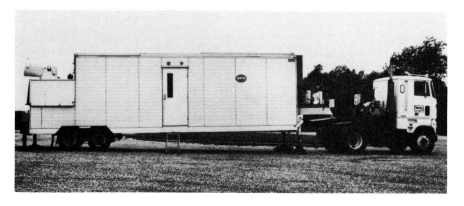

Fig. 10.19 Stanford Research Institute differential lidar vehicle. (Hawley *et al.* [35].)

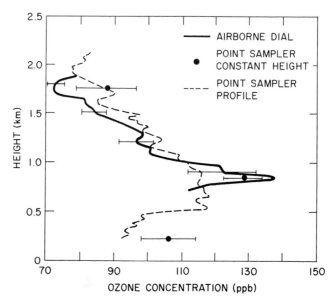

Fig. 10.20   Vertical profile of ozone concentration: comparison between NASA airborne lidar measurements and point sampler. (Browell *et al.* [32].)

platform mounted on a ship (Fig. 10.21). This is based on the DF laser and has been used to measure the dispersion of HCl emissions from a ship in the North Sea incinerating waste products. Again this laser is not tuneable but one of the wavelengths in its emission coincides with an absorption line in the HCl spectrum. The sensitivity is 500 ppb over an 800 m range which is adequate for the high concentrations in the vicinity of the ship but not good enough for more general measurements in the atmosphere where a sensitivity of a few ppb is required.

Continuously tuneable infrared systems are available in the form of diode lasers. These are well developed and available commercially. Unfortunately their output is low, typically a few nJ, which is an order of magnitude too low for DIAL operations. Consequently their use is limited to measurements with retro-reflectors or multipass cells.

10.5 Laser safety

Laser radiation can be harmful. It is the eye which is most sensitive and great care must be taken to avoid risk, particularly with the powerful lasers used in lidar. The effects of very short repetitively pulsed radiation are not well known. The matter is complicated by the wide range of wavelengths, pulse lengths and repetition frequencies of lasers in use and further by effects which occur in the

Fig. 10.21   Infrared differential lidar system mounted on a ship for the measurement of HCl: GKSS. (Weitkamp *et al.* and Herrmann *et al.* [40, 41].)

atmosphere such as focusing by turbulence. Laser users are recommended to consult safety guides such as that published by the British Standards Institution, BS 4803:1983.

10.6   Long pathlength absorption spectroscopy

10.6.1   Differential ultraviolet and visible absorption spectroscopy

Among the most successful examples of the application of long pathlength absorption spectroscopy to the identification and measurement of air pollutants is the rapid scanning spectrometer originally developed by Platt and Perner at the Institute for Chemistry, Jülich, West Germany [42–46]. Over the past several years this system has been used by the Jülich researchers and by Pitts and co-workers at the University of California, Riverside (UCR), to determine the atmospheric concentrations of pollutants such as O_3, SO_2, NO_2, HCHO and, for the first time in the polluted troposphere, the nitrate radical (NO_3) and nitrous acid (HONO) [42–51]. The UCR group has also successfully adapted this technique to in-situ measurements of gaseous species in environmental chambers [52] and automobile exhausts [53] using appropriate multiple reflection cells.

The basic configuration of the differential UV/VIS absorption spectrometer (DUVVS) employed in these studies at various locations in Europe and in California is shown in Fig. 10.22. An essential feature of this system is that the exit slit of the Spex 0.5 m grating monochromator is replaced by a thin metal disc which rotates in the focal plane and into which approximately one hundred 100 μm wide slits are etched. At any given time one such slit serves as the exit slit and an $\sim$ 30 to 40 nm segment of the spectrum dispersed by the monochromator grating is scanned. Since the disc rotates at $\sim$ 1 Hz, about 100 spectral segments are received per second by the photomultiplier (EMI 9656Q). These signals are digitized by a high speed A/D convertor and stored in a DEC PDP-11/23 minicomputer to be signal averaged and further manipulated.

Both Xe high pressure lamps (450 W) and quartz iodine lamps (150–240 W) have been used, with an appropriate collimating mirror, as light sources for the DUVVS system. Optical paths ranging from a few hundred metres up to 17 km have been employed. Since a single scan takes less than 10 ms, atmospheric scintillation has little effect on the signal. Moreover, since typically 10–20 thousand individual scans are recorded over each several minute measurement period (and these spectral records are then signal averaged) even complete interruption of the light beam for a moment has no appreciable effect on the final spectrum. Except under the haziest conditions, stray light effects are negligible due to the narrow field of view of the instrument (~ 0.3–1.1×10^{-4} steradians)

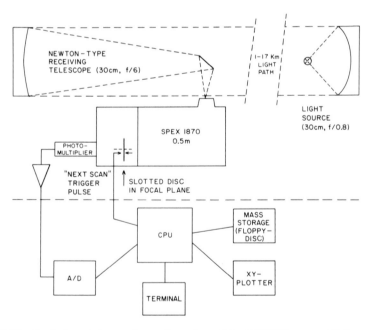

Fig. 10.22   Optical and electronic configuration of the DUVVS system [46].

and in any case can always be normalized by means of appropriate 'stray light' reference spectra.

The concentrations of atmospheric species are derived from Beer's Law but since the light intensity (I_0) in the absence of any absorption cannot be obtained with this technique the 'differential' optical density is employed. Thus, as indicated in Fig. 10.23, only those compounds which have a structured absorption spectrum are accessible and for such compounds

$$\log \frac{I_0}{I} = \alpha C l \qquad (10.7)$$

where

$$I_0 = I(\lambda_1) + [I(\lambda_3) - I(\lambda_1)] \frac{\lambda_2 - \lambda_1}{\lambda_3 - \lambda_1}$$

and α is the 'differential' absorption coefficient at λ_2 for the compound of interest.

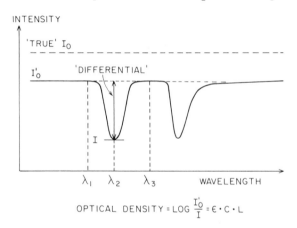

Fig. 10.23 Determination of the 'differential' optical density [46].

In practical applications of the DUVVS system the precision of a measurement is improved by least squares fitting a reference absorption spectrum of the pollutant under investigation to the observed absorption spectrum. In this way all of the usable absorption bands in the scanned spectral region are employed in determining the absorption strength (and hence the concentration) of the air pollutant of interest. This also aids in deconvoluting a complex atmospheric absorption spectrum to which several, or many, air pollutants are contributing.

Although in principle the minimum optical density (OD) obtainable with current versions of the DUVVS system is $(5-10) \times 10^{-5}$ (base 10) for averaging times of several minutes [46], in practice minimum ODs are found to fall in the range $(1-3) \times 10^{-4}$. Some of the atmospheric pollutants observed to date with the DUVVS system and their detection limit for a 10 km optical path and a minimum detectable OD of 10^{-4} are shown in Table 10.1.

Table 10.1   Atmospheric pollutants observed by differential UV/VIS spectroscopy [46]

Substance	Wavelength range (nm)	Differential absorption coefficient[a] (cm² molec⁻¹)	λ (nm)	Detection limits for 10 km light path[b] (ppt)
SO_2	200–230, 290–310	5.7×10^{-19}	300	17
CS_2	200–220, 320–340	4.0×10^{-20}		240
NO	215, 226	2.3×10^{-18}	226	400[c]
NO_2	330–500	1.0×10^{-19}	363	100
NO_3	623–662	1.8×10^{-17}	662	0.5
HNO_2	330–380	4.2×10^{-19}	354	20
O_3	220–330	4.5×10^{-21}	328	2100
HCHO	250–360	7.8×10^{-20}	340	120

[a] 0.3 nm spectral resolution.
[b] Minimum detectable OD $= 10^{-4}$ except as noted.
[c] 1 km light path, minimum detectable OD $= 10^{-3}$.

An example of the application of the DUVVS system to atmospheric measurements in urban atmospheres is shown in Fig. 10.24. In this case the spectrometer was mounted on the rooftop of a building near downtown Los Angeles and two optical paths were employed (~ 1 km and 2.3 km, respectively) over a major highway. Measurable concentrations of HONO were observed after sunset (from spectral features at 354 and 368 nm) and these increased throughout

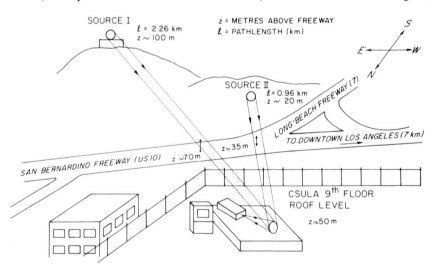

Fig. 10.24   DUUVS field site near downtown Los Angeles with two light sources at different pathlengths [50].

the night to levels as high as 8 ppb just before sunrise. Since (as discussed in Chapter 3) HONO is an initiator of photochemical air pollution via its photolysis, these observations filled an important gap in previous models of air pollution formation which failed to account for the presence of significant concentrations of HONO at sunrise [49].

In similar measurements at several locations in the Los Angeles basin NO_3 radicals were detected at 623 and 662 nm, as shown in Fig. 10.25, yielding time–concentration profiles of the kind shown in Fig. 10.26. As discussed in Chapter 3, NO_3 radical measurements by the DUVVS system have significantly advanced our understanding of air pollution chemistry and acid deposition phenomena.

Recently, in order to conduct measurements over a large geographical area (e.g. the California desert) UCR researchers equipped a mobile van with a DUVVS system and this was used successfully with a light source as much as 17 km distant [51].

It should be noted that Noxon has employed a variation of the DUVVS techniques which is not based on a rapid scanning spectrometer. Thus, he has

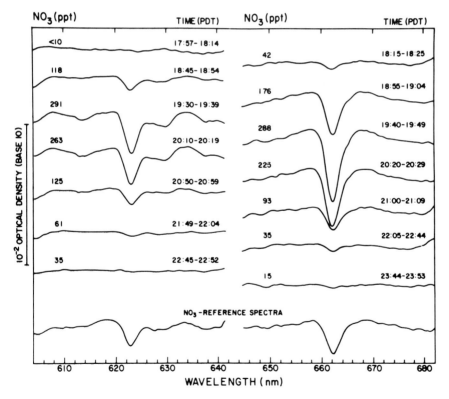

Fig. 10.25 Observation of NO_3 radical absorption bands with DUVVS system in measurements at Riverside, CA [47].

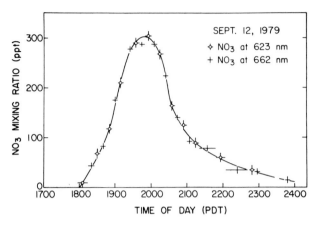

Fig. 10.26   NO$_3$ time–concentration profile observed with DUVVS system at Riverside, CA with 970 m pathlength [47].

measured ambient concentrations of NO$_3$ radicals and NO$_2$ using the sun and moon as light sources at long slant paths through the atmosphere [54–56]. To date, however, these measurements have been confined to clean continental and maritime atmospheres at somewhat elevated altitudes.

10.6.2   Fourier transform infrared spectroscopy

Herget and his colleagues at the US EPA North Carolina laboratories [57–59] developed an FT–IR based 'remote optical sensing of emissions' (ROSE) system. They employed this in two configurations: (1) with a light source located up to 1.5 km distant for making absorption measurements of ambient air pollutants and (2) with an adjustable tracking mirror to make single-ended emission measurements of stack effluents. The optical configuration for the ROSE system is shown in Fig. 10.27 and this system along with its associated computer were placed in a van for transport to measurement sites.

For the long pathlength absorption measurements an $f/5$ telescope of Dall-Kirkham configuration was used with a 60 cm diameter primary mirror. A 1000 W quartz-halogen lamp was employed as the light source. A commercial FT-IR spectrometer (Nicolet 7199) was employed with dual liquid N$_2$-cooled InSb and HgCdTe detectors to cover the spectral region from 600 to 6000 cm^{-1}. The nominal sensitivity of the ROSE system in absorption mode was cited as ~ 1 ppb at 1 km optical path for a number of pollutants accessible in the infrared region and approximately a factor of 10 less in emission mode [58].

Although not strictly 'remote-sensing' instruments the long pathlength FT–IR spectrometer/multiple reflection cell (MRC) systems (see Fig. 10.28) developed originally by Hanst [60, 61] have been important in documenting the atmospheric concentrations of pollutants such as HNO$_3$, HCOOH, HCHO,

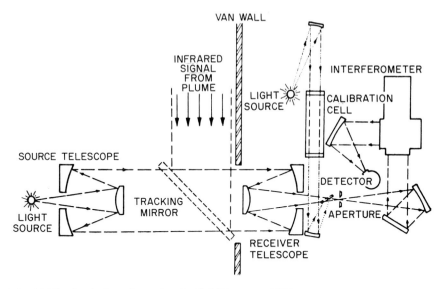

Fig. 10.27 Optical configuration for ROSE system [57].

NH$_3$ and PAN [61–65]. In particular, the eight mirror system shown in Fig. 10.28 was exploited by Tuazon and his co-workers at UCR in measurements over a period of five years in the Los Angeles basin [62–64]. The 22.5 m path MRC shown in Fig. 10.28 afforded total pathlengths in excess of 1 km, yielding the detection limits shown in Table 10.2.

In this system a commercial FT–IR spectrometer (operated at 0.5 cm^{-1} spectral resolution) and liquid N$_2$-cooled InSb and HgCdTe detectors were used. Typically, 30 interferograms obtained over a period of ~ 6 min were co-added permitting simultaneous determinations of several key pollutants with good time resolution. As seen from Table 10.2, a detection limit of ~ 5 ppb was achieved for most pollutants of interest with measurement uncertainties of $\pm(2$–4$)$ ppb.

An example of time–concentration profiles obtained using this kilometre pathlength FT–IR system for O$_3$, PAN, HCHO, HNO$_3$, HCOOH and NH$_3$ over a 36 h period during a severe air pollution episode in the Los Angeles basin is shown in Fig. 10.29 [64].

10.7 Meteorological measurements

10.7.1 Meteorological measurements for pollution surveys

Meteorological measurements important in air pollution surveys are wind speed and direction, stability of the atmosphere and depth of the mixing layer. In studies of the trajectories, dispersion and chemical and physical transformation

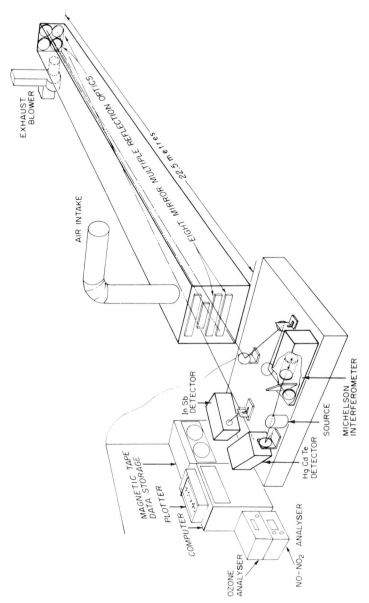

Fig. 10.28   UCR kilometre-pathlength FT–IR system [63].

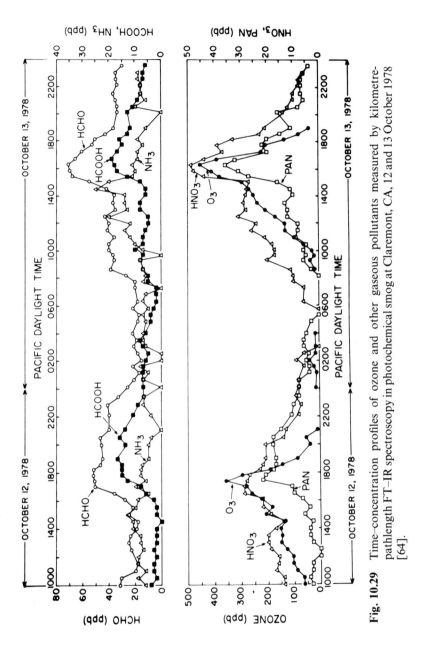

Fig. 10.29   Time–concentration profiles of ozone and other gaseous pollutants measured by kilometre-pathlength FT–IR spectroscopy in photochemical smog at Claremont, CA, 12 and 13 October 1978 [64].

Table 10.2   Calculated detection limits for air pollutants measured with UCR kilometre-pathlength FT–IR system [64]

Compounds	Measurement frequency (cm^{-1})	Absorptivities[a,b] (cm^{-1} atm^{-1})	Resolution (cm^{-1})	Approximate detection limit at 1 km path (ppb)
O$_3$	1055	9.7	1–2	10
PAN	1162	32	2–4	3
NH$_3$	931	27	0.5	4
	967.5	35	0.5	3
	993	21	0.5	4
HNO$_3$	896	12[c]	0.5	6
HCHO	2779	16[c]	0.5	6
	2781.5			
HCOOH	1105	70[c]	0.5	2

[a] $\ln(I_0/I)/pl$.
[b] $\sim 23°$ C, 760 Torr.
[c] Measured from the intensity of the Q branch only.

of emissions such measurements are required up to a height of 1 to 2 km. Traditionally wind measurements are made by releasing small pilot balloons and following their trajectories with theodolites or radar. Atmospheric stability and the depth of the mixing layer are determined by measuring vertical profiles of temperature using sensors mounted on free or tethered balloons. Such measurements are made daily throughout the world from meteorological stations and the results are used extensively in pollution studies. However, in most surveys more detailed information is required. Often the nearest meteorological station is a considerable distance from the site of a survey and since balloon releases are not usually more frequent than once every 6 h information is required on a much shorter time scale. Furthermore these methods are of little use in investigations of atmospheric turbulence which is of central importance in pollution dispersion.

Considerable use is made of remote sensing techniques to make these meteorological measurements. One which is widely used for wind measurements, determination of the mixing layer and estimation of turbulence is sodar. Also used in mixing layer measurements is lidar. Direct measurements of temperature are less well developed but two techniques have been used successfully, one a hybrid of sodar and the other based upon microwave radiometry.

10.7.2   Sodar

Sodar is similar to lidar except that instead of using pulses of light, pulses of sound are employed, hence the acronym *SO*und *D*etection *A*nd *R*anging. When a sound wave is launched into the atmosphere it is backscattered not by aerosol particles as with light but by inhomogeneities in acoustic impedance. Comprehensive treatments of the principles of sodar can be found in references [66, 67] and the theory of the scattering of sound waves is given in detail in references [68–70]. Briefly, the power scattered by the atmosphere and detected by a receiver is given, in a form similar to the lidar Equation (10.2), by

$$P(r) = \frac{E_0 BAc}{2r^2} \sigma(r, \theta)L \tag{10.8}$$

where $P(r)$ is the detected power, E_0 the energy in the transmitted pulse, B the overall efficiency of the system, A the area of the receiver, c the velocity of sound, r the range at which the scattering is taking place and L the attenuation of the pulse in the path to and from the scattering volume. To be strictly analogous to the lidar equation L should appear as an integral over the path of the pulse but since in sodar information is not generally deduced from spatial variation in attenuation it is included as a lumped parameter. An important difference between the acoustic equation and lidar is that the volume scattering coefficient σ is expressed not only as a function of range but also as a function of scattering angle, θ. This is because sodar, unlike lidar, sometimes operates with a receiver separated by a significant distance from the transmitter so the detected signal does not travel back along the same path, $\theta = 180°$, as the transmitted pulse. The inhomogeneities in acoustic impedance which contribute to σ have three components, due to variations in temperature, humidity and wind velocity in the scattering volume. Humidity effects are usually assumed to be small and velocity fluctuations contribute only if $\theta \neq 180°$.

Frequencies used are in the audible range arround 1500 Hz with a corresponding wavelength of 20 cm or so and it is inhomogeneities of about this size which are most effective in producing a backscattered signal.

Sodar is used in two ways. The most straightforward simply measures the backscatter from a pulse as it travels vertically upwards. In situations where there is a well-defined mixing layer this provides results such as that shown in Fig. 10.30 [71] where the backscattered signal is displayed on a facsimile recorder, stronger signals producing a blacker record. Within the mixing layer, turbulence produces much stronger temperature fluctuations than above, hence the clearly marked structure. Much more information can be derived if the Doppler shift of the backscattered signal is measured, since this gives the speed, in the direction of the propagation of the sound pulse, of the volume of air producing the scattering. In general, Doppler systems use three transmit/receive elements (Fig. 10.31), one firing vertically upwards and the others at an angle of 20° or so to the vertical in planes orthogonal to each other. This enables the vertical and horizontal components of wind speed to be calculated as a function of height.

The instrument illustrated in Fig. 10.31 is in effect three separate sodars each employing a single transmit/receive system. Operated in this so-called monostatic mode the received signal returns along the same path as the transmitted pulse and is the result of only temperature variations, i.e. $\theta = 180°$ in Equation (10.8). A

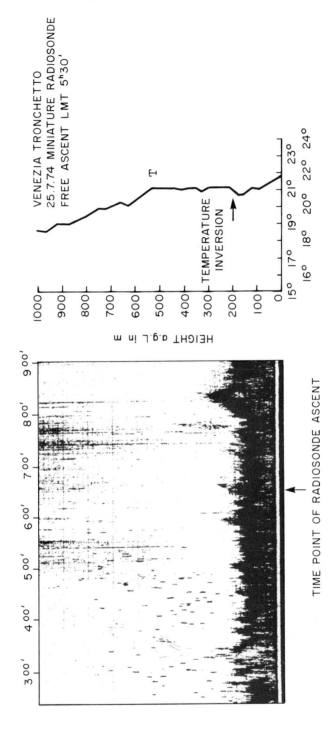

Fig. 10.30   Sodar measurements of the mixing layer and radiosonde temperature profile: Joint Research Centre, Ispra. (Hasenjäger [71].)

Fig. 10.31 Doppler sodar: Bertin trailer mounted system.

larger signal can be obtained if the receiver is offset and $\theta \neq 180°$ but this bistatic mode has the disadvantage that information is received only from the region where the field of view of the receiver overlaps the path of the transmitted pulse. The Doppler system shown in Fig. 10.31 does not have this disadvantage but does suffer from the fact that the components of horizontal velocity used to calculate wind speed and direction are obtained from different regions of the sky. The result is that a degree of horizontal homogeneity is assumed in the analysis of the results.

Comprehensive comparisons have been made between Doppler sodar wind measurements and those made with anemometers mounted on meteorological towers [72]. The height up to which measurements can be made is strongly dependent on meteorological conditions. Mostly it is difficult to go beyond 1 km and regular results can be expected up to about 600 m. Horizontal speed is accurate to a few tenths of a metre per second and vertical speed to within a few centimetres a second. The vertical spatial resolution obtained with the instrument is a few tens of metres and a typical sampling time required to obtain results with the above accuracy is a few minutes. Clearly data analysis with a Doppler system is fairly complicated and requires an on-line computer of some sort. This does, however, have advantages since once such data processing is available it is possible to analyse the results statistically and calculate parameters of atmospheric turbulence. Systems are available with direct outputs not only of wind speed and direction but also of stability categories and inversion heights.

Although the instruments are necessarily large in order to produce a well-collimated sound pulse, collect sufficient of the backscattered signal and exclude extraneous sound, a three element system can be carried on a fairly compact trailer, as shown in the photograph Fig. 10.31, and set up in the field in a short time. Systems are available commercially, through companies such as Remtech in France and AeroVironment in the United States.

10.7.3   Lidar measurements of the mixing layer

In Section 10.3 it was described how lidar signals are enhanced by the presence of aerosol particles in the atmosphere. This can be useful in giving an indication of the stability of the atmosphere and in particular of the depth of the mixing layer. In stable conditions any aerosol tends to be distributed in horizontal layers such as mist in autumn mornings. On the other hand within a mixing layer, aerosols are more uniformly mixed. Ground level emissions, from vehicle exhausts, domestic heating and industrial processes are often trapped within the mixing layer producing a relatively high aerosol concentration above which there is relatively clear air. Lidar measurements often give very clear illustrations of these effects. An example is shown in Fig. 10.32 where the laser was firing vertically from a fixed location. The result from each successive shot is displaced slightly to the right on the display. Compensation has been made for the $1/r^2$ term in the lidar equation and the brightness of the display corresponds to the strength of the return signal. The stable, stratified form of the aerosol distribution is very clear. The figure also shows results obtained later in the day when the ground was warmed by the sun and the stability was disturbed by convection. The display shows the resulting disturbance in the haze layers. Further vivid examples of the use of lidar to measure the development of the mixing layer can be found in references [73, 74] and the use of a mobile lidar to study the effect of an urban

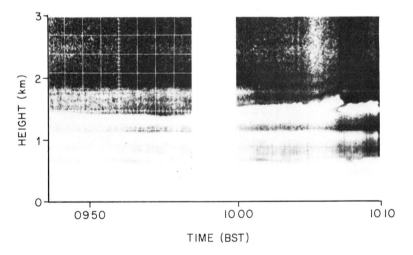

Fig. 10.32   Lidar profiles of the mixing layer.

area on the depth of the mixing layer is illustrated by measurements made on traverses across the city of St Louis [75].

10.7.4   Temperature profiles

Methods for direct remote measurement of temperature profiles in the atmosphere are less well developed than those for mixing layer and wind determination. One technique which has been used successfully is the *R*adar *A*coustic *S*ounding *S*ystem, RASS [76]. With this the progress of a sound pulse through the atmosphere is followed by radar and its speed is measured. Since the speed is proportional to the square root of temperature the measurement gives the variation in temperature along the path of the pulse. The sound pulse is launched from a transmitter such as those used in sodar. This is tracked by an adjacent radar, the return signal being produced by a change of the refractive index in the air caused by pressure variations at the passage of the sound pulse. In general the return is rather weak but by adjusting the frequencies of the sound wave and the radar a resonance between the two waves can be obtained and there is a large enough signal for Doppler analysis. A system used in Italy by the Laboratorio di Cosmogeofisica of Torino [77] uses an acoustic frequency of 360 Hz, wavelength 94 cm, and a radar frequency of 159 MHz corresponding to a wavelength of 188 cm. Results obtained by this system are shown in Fig. 10.33 where they are compared with measurements made at the same time by a conventional balloon borne radiosonde. It can be seen that the agreement is sufficiently close for RASS to be a useful technique. The results commence at a height not much below 200 m since that is where the field of view of the radar starts to overlap the track of the sound pulse. Some degree of signal averaging is required and each measurement takes a few minutes. This is adequate for detailed measurements of the evolution of the mixing layer and for determinations of atmospheric stability required for dispersion studies.

Another technique used to determine temperature profiles is microwave radiometry in which radiation from molecular oxygen is measured at a wavelength in the region of 5 mm, 60 GHz. The radiation from a small volume of the atmosphere, assumed to be in local thermodynamic equilibrium, is temperature dependent and is described by the usual Plank expression. Radiation reaching ground level is the integral of emission from different heights at different temperatures, modified by absorption in the atmosphere between the point of emission and the ground. Radiation reaching the ground is characterized by its brightness temperature T_B and it can be shown [78, 79] that to a good approximation

$$T_B(v) = \int_0^\infty T(h)\alpha(v, h)\exp\left[-\int_0^h \alpha(v, h')\,dh' \right]dh \qquad (10.9)$$

where v is frequency, T temperature, α absorption coefficient and h height. To a first approximation α depends only on frequency and the variation in number density of oxygen molecules resulting from the decrease in pressure with height. The problem is that unlike correlation spectroscopy it is not sufficient to just

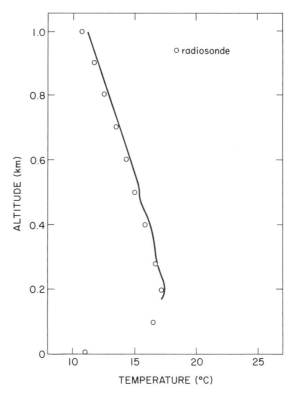

Fig. 10.33   Radio acoustic sounding temperature profile – comparison with radiosonde
measurements: Laboratorio di Cosmogeofisica, Torino. (Bonino *et al.* [77].)

measure the integrated effect of the passage of radiation through the atmosphere,
the whole object of the exercise is to obtain vertical spatial resolution. However,
since the vertical variation of pressure is known an approximation can be made
for the distribution of oxygen molecules in the atmosphere and Equation (10.9)
can be inverted to give temperature profiles. Iterative techniques can then be used
to correct for smaller variations in α due to temperature and the presence of water
vapour. Comprehensive accounts of the methods used are given in references
[80–82]. Two basic experimental techniques are used. One is to measure
radiation at a fixed vertical angle but at different frequencies and the other, most
commonly used in ground based systems, is to keep the frequency fixed but to
vary the vertical angle. The frequency chosen within the absorption band
depends upon the height to which measurements are required. To get good
resolution a large value of α is required but on the other hand a large α limits the
height from which signals can be received.

   One method is to use a frequency of strong absorption for the first few hundred
metres and to use another frequency at lower absorption to extend the

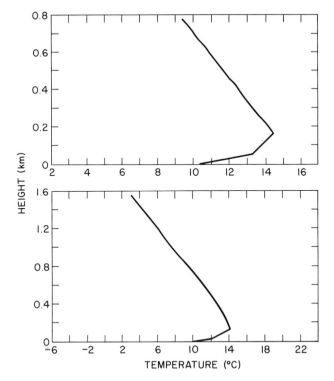

Fig. 10.34 Temperature profile measured by microwave radiometry. (Jeske [83].)

measurements to a greater height. An example is shown in Fig. 10.34 [83] where a frequency of 58 GHz was used for measurements up to 700 m, averaging over a height of 50 m, and 54.5 GHz up to 1500 m averaging over 100 m. Extensive comparisons have been made between radiometer profiles and those obtained with radiosondes and instruments carried in aircraft (see for example [84, 85]), and the radiometer readings are generally accurate to about $\pm 1°$ C up to a height of a kilometre or so. The results in Fig. 10.34 were obtained by making measurements at eight different angles of elevation between $0°$ and $90°$, the calculation of the temperature profile being calculated with a small desk-top computer at the end of each cycle, the whole operation taking about 20 min.

10.8 The use of remote sensing in field studies

10.8.1 Plume rise and dispersion
An example of lidar being used in a relatively straightforward way to determine the trajectory of a plume is illustrated in Fig. 10.35 which shows measurements

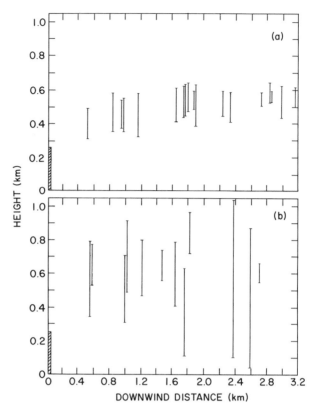

Fig. 10.35 Lidar measurements of plume rise and dispersion: Central Electricity Generating Board. (Scriven [86].)

made at Drax Power Station in England [86]. Two single wavelength lidars were used, one run by the CEGB and the other by the Italian Electricity Authority ENEL. Both operated at 694.3 nm and were used to obtain vertical plume cross-sections by firing at a series of different angles of elevation. They were located such that from a single fixed site, cross-sections could be obtained at distances up to 3 km downwind of the stack. Twenty or thirty shots were used for each cross-section which took less than a minute. Results were recorded on oscilloscope photographs and analysed by hand. No attempt was made to obtain quantitative data on the distribution of material through the plume but the points at which the laser pulse entered and left the plume were usually clearly marked by a sharp change in the backscatter signal. Thus, with simple geometry to allow for the horizontal angle at which the pulse traversed the plume, the horizontal and vertical dimensions of the plume could be obtained. The results in Fig. 10.35a were obtained in the early morning with the plume travelling in stable air above a shallow mixing layer. In Fig. 10.35b, taken in the afternoon of the same hot sunny day, convection produced a deep mixing layer. The difference in dispersion in the two conditions is typical of stable and convective conditions and is shown effectively by the lidar scans.

A sodar was used in the same exercise [86, 87] to measure the depth of the mixing layer. It was a straightforward acoustic sounder rather than a Doppler system and results for the same period as the lidar measurements are shown in Fig. 10.36. The existence in the morning of stable air above about 150 m with convection producing a deepening mixing layer, starting at about 1100 and rising to 700 m by 1300, is consistent with the lidar results. Vertical temperature profiles taken during the same period confirm the early morning stability and subsequent convection activity. The results shown in Fig. 10.36 are typical of those obtained with sodar. Under conditions such as these where there is a well-defined mixing layer capped by a temperature inversion, sodar gives results which can be accurately interpreted. In less well-defined situations, such as the onset of

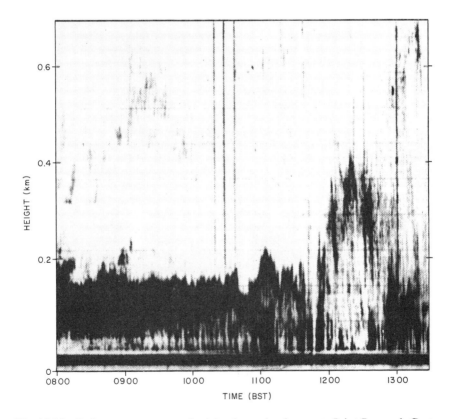

Fig. 10.36   Sodar measurements of mixing layer development: Joint Research Centre, Ispra. (Hasenjäger [87].)

stability in the evening, results are often more difficult to analyse. Nevertheless, the ability to measure the depth of the mixing layer and its variation with time in a continuous manner, together with lidar results, is useful in interpreting plume behaviour and for comparisons with theoretical predictions.

A more recent and comprehensive study of plume rise and dispersion is that organized by the Electric Power Research Institute at the Kincaid Power Station in Illinois [25]. Together with a large array of ground level point samplers, sodar was used to make meteorological measurements and lidar to obtain three-dimensional plume profiles. Two Doppler sodars were employed to measure the depth of the mixing layer and vertical and horizontal components of wind speed. Results were obtained continuously for periods of several months. In all, three lidar systems were deployed at different distances from the stack. The SRI van-mounted DIAL system [35] measuring SO_2 was located at distances up to 3 km. This operated in two modes. The first firing at a fixed vertical angle, making measurements as the plume meandered through the path of the laser pulse, and obtaining 5 min and 1 h averages of SO_2 concentration. In the second it operated at different elevations scanning the plume to give cross-sections of concentration at a rate of six to twelve an hour.

A second, single wavelength, van-mounted lidar (Computer Genetics Corporation) operated at about 5 km from the stack measuring cross-sections of relative aerosol particle concentrations. With this system a cross-section of the plume was obtained every 5 min and from the results the height of the centre of the plume and vertical and horizontal distributions could be obtained either for each cross-section or averaged over a period of an hour.

In situations where the mixing was deep and the plume was being dispersed and brought down to the ground both lidars were operated in a third mode, making horizontal scans at a low level to determine the distribution of the plume at ground level. Here the results can be compared with those of the ground level samplers and reasonable, though not exact, agreement was obtained.

The third lidar [22] was mounted in an aircraft which flew at a height of about 1 km firing vertically downwards as it traversed the plume about 10 km from the stack. These results were processed to give both single plume profiles and 1 h averages. The manner in which the three lidars were deployed is illustrated in Fig. 10.37. Figure 10.38 is another good example, obtained during this exercise, of a plume, initially travelling above the mixing layer, being brought down to the ground as it was entrained when the mixing layer developed during the day. An example of the contours of the aerosol concentration of the plume has already been shown in Fig. 10.10 where the plume was moving not above the mixing layer but trapped within it. It is conditions such as these that, further downwind where the plume reaches the ground, lead to the highest ground level concentrations, so called fumigation episodes. Here the mixing layer is deep enough to trap the plume but not sufficiently deep to allow it to reach its maximum height and dilution.

A further example of fumigation is shown in Fig. 10.39. Here a vehicle measuring SO_2 with a correlation spectrometer and a ground level sensor carried out repeated traverses back and forth along the same route under the plume from

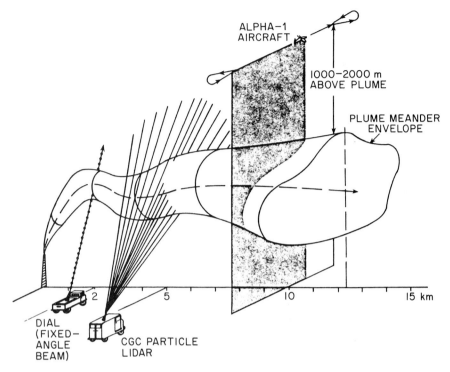

ALPHA–1
AIRCRAFT

1000–2000 m
ABOVE PLUME

PLUME MEANDER
ENVELOPE

DIAL
(FIXED–
ANGLE
BEAM)

CGC PARTICLE
LIDAR

Fig. 10.37   Deployment of lidar for studies of plume dispersion: EPRI Plains Study [25].

the Cordemais Power Station in France near Nantes [88]. The figure shows the
route followed by the vehicle on one side of which is plotted the total overhead
burden of SO_2 measured with the correlation spectrometer and on the other side
the ground level concentration. It also shows the integrals along the route of
burden and ground level concentration for each traverse. It can be seen that,
subject to normal statistical fluctuations, the overhead burden remains the same
throughout the measurements. On the other hand the ground level concentration, practically zero on the first traverse, rises dramatically between the time of
1147 and 1223 and then decreases to about half its maximum value by 1250. This
behaviour corresponds to a deepening mixing layer, similar to that indicated by
the sodar results in Fig. 10.36, in which the plume is first trapped and prevented
from achieving its full rise, giving maximum ground level concentrations, and
then as the mixing layer subsequently deepens further, dilution of the plume
increases and ground level concentrations diminish.

Traverses with a correlation spectrometer such as those shown in Fig. 10.39
can be used to track plumes for considerable distances downwind of a source. An
example is shown in Fig. 10.40 where the track of a plume has been followed for
about 60 km [89]. It shows the overhead burden measured on individual

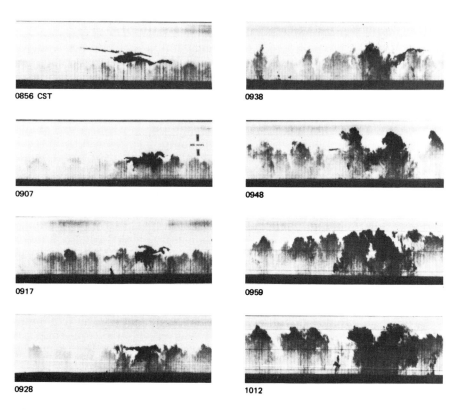

0856 CST

0938

0907

0948

0917

0959

0928

1012

Fig. 10.38   Airborne lidar profiles showing plume entrainment in mixing layer, 10 km downwind of Kincaid Power Plant, 4 May 1980: Stanford Research Institute. (Uthe [38].)

traverses and the interpolated trajectory of the plume. Correlation spectrometers mounted in aircraft have been used to follow plumes over considerably greater distances, up to 1000 km [90, 91].

10.8.2   Measurement of emission fluxes from point sources

Results such as those shown in Figs 10.4 and 10.39 where the profile of overhead burden has been measured across a plume can be used to determine the flux of SO_2 and NO_2 in the plume. The traverse is not in general perpendicular to the direction of plume travel so the first step is to normalize the measurement by multiplying the integral of the burden by the sine of the angle between the traverse and the plume axis. This can be done by approximating the direction of the whole traverse or, if there are many bends in the route, to do it sector by sector. The normalized burden is then multiplied by the wind speed averaged across the

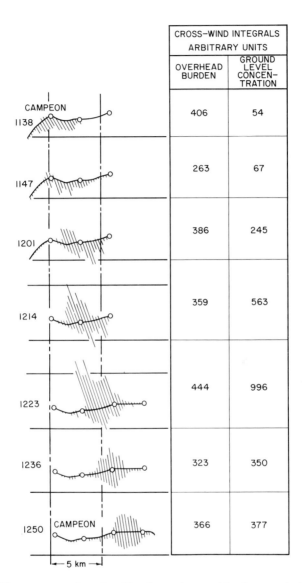

| CROSS–WIND INTEGRALS ARBITRARY UNITS | |
OVERHEAD BURDEN	GROUND LEVEL CONCEN- TRATION
406	54
263	67
386	245
359	563
444	996
323	350
366	377

Fig. 10.39 Measurement of overhead burden and ground level concentration of SO_2 on repeated traverses downwind of a power station during a fumigation episode: EDF. (Dutrannoy and Robé [88]; Hamilton, private communication.)

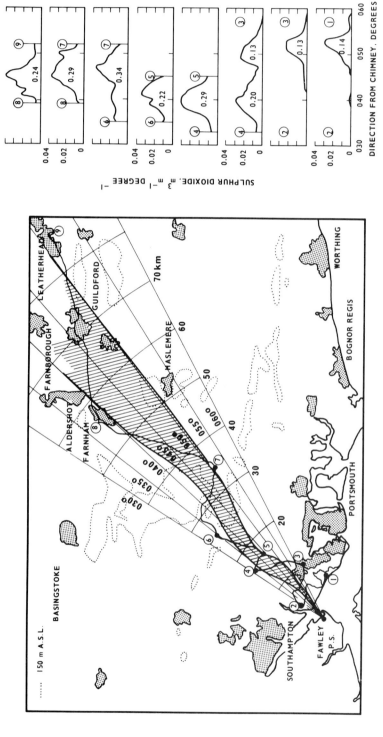

Fig. 10.40 Plume tracking with a van-mounted correlation spectrometer. Graphs of overhead burden on successive traverses and interpolated plume trajectory. (Hamilton [89].)

depth of the plume. Of course the flux determined from a single traverse is not generally accurate due to the statistical nature of dispersion and a number of traverses have to be averaged to obtain a representative result. In an exhaustive study based on 465 traverses Sperling [92] found that an accuracy of 30% could be achieved over a 20 min period with an average of 5 traverses, a major part of the inaccuracy being associated with inherent problems in measuring wind speed. Table 10.3 shows a set of results obtained at a coal burning power station at Drax, Eggborough [86]. SO_2 emissions are estimated from the sulphur content of the coal being burned and the single traverse results show the typical scatter to be

Table 10.3 SO_2 fluxes measured with correlation spectrometers. Single traverses under a plume and averages of several traverses.

Single traverses		Averages			
			Measured arithmetic	*Measured harmonic*	*Number of*
Emission	*Measured*	*Emission*	*mean*	*mean*	*traverses*
14.2	13.5	14.2	44.3[a]	38.2[a]	5
14.2	7.4	14.2	16.5	13.3	6
14.2	22.4	16.2	16.3	12.6	10
14.2	10.6	14.6	17.8	17.0	2
17.5	15.6	18.5	20.1	18.8	26
17.7	21.8	16.3	18.7	14.1	6
17.9	16.0	18.3	44.7[a]	31.7[a]	15
17.9	21.4	18.3	34.0[a]	28.5[a]	13
17.9	27.3	13.0	18.2	16.3	17
17.8	18.5	13.7	16.7	14.8	7
17.8	20.0				
17.8	20.0				
18.1	25.9				
18.1	13.4				
18.4	33.0				
18.7	15.0				
19.0	16.8				
21.0	63.7[a]				
20.8	33.5[a]				
20.1	37.8[a]				
20.1	51.6[a]				
13.0	17.1				
13.0	17.4				
13.0	16.4				
13.0	10.5				
12.9	13.1				
12.9	19.4				
12.9	22.0				
12.9	19.4				

[a] These results were obtained with condensation in the plume.
Units are tonnes SO_2 per hour.

expected from measured fluxes. The table also shows the results of averaging over a number of traverses, each group carried out at the same power station but during different periods. Both arithmetic and harmonic means are shown and it can be seen that the latter generally gives a result closer to the estimated emissions. This is to be expected if the fluctuations in integrated burden are due to variations in wind speed and by the plume direction swinging either towards or away from the direction in which the traverse was made. There are some values of averaged flux which are much higher than the estimated emissions. This is a phenomenon which has been found elsewhere [86] and appears not to be due to the temporary malfunction of a particular instrument since several instruments operating independently on different vehicles carrying out traverses together exhibit the same anomaly at the same time. It occurs when there is condensation in the plume or when the plume is travelling in a dense haze layer. A possible explanation is that light reaching the correlation spectrometer undergoes multiple scattering in the plume and has its effective path length, and hence absorption, increased. Certainly it seems that care is required when there is low cloud, condensation or poor visibility.

Another example of SO_2 flux measurements is shown in Fig. 10.41 [93]. These were made at the oil burning Turbigo Power Station, Italy, during a period when for 3 h the fuel was changed to a type with about half the normal sulphur content. The sequence of traverses show the accompanying decrease in SO_2 flux, displaced by the travel times of the plume from the station to the route along which the traverses were made.

Flux measurements can also be used to examine the oxidation of NO to NO_2 in plumes. This occurs principally through reaction with atmospheric ozone and generally increases from early morning onwards as plume dispersion develops, thus entraining more ozone, and ambient ozone concentrations increase. This effect is masked by plume fluctuations if the flux of NO_2 is measured alone but if SO_2 is measured at the same time the ratio of the burden NO_2/SO_2 is a good

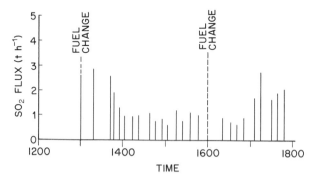

Fig. 10.41 SO_2 flux measurements with a correlation spectrometer 12 km downwind of Turbigo Power Station during a period, 13.00–16.00, when sulphur content of oil being burned was reduced [93]. Each vertical line represents the flux calculated from a single traverse.

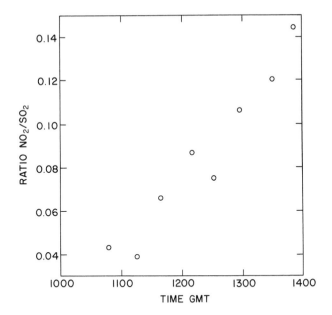

Fig. 10.42   Diurnal variation in the oxidation rate of NO in a power station plume indicated by the changing ratio of the fluxes of NO_2 and SO_2 measured with correlation spectrometers.

indicator of the conversion rate. An example is shown in Fig. 10.42 where the ratio increases from about 0.04 in the morning to 0.14 later in the day. The measurements were made on a day in January, 7 km from a power station, in a wind speed of about 5 m s^{-1}, and correspond to NO conversion between the stack and the traverse of 9% in the morning rising to 31% in the afternoon.

10.8.3   Multisource monitoring in an industrial area

In an industrial area containing a large number of sources emitting material at different heights and rates it is often difficult to interpret ground level concentrations of pollutants in terms of individual sources. It is not always clear what contribution is being made by small, low level local sources compared with larger sources emitting from tall chimneys. Measurements of overhead burden of SO_2 with correlation spectrometers can be useful in studies of this kind, particularly if the instruments are mounted in vehicles so that traverses can be made through the area. In addition to carrying a correlation spectrometer a point sampler would normally be carried to measure ground level concentrations of SO_2. Figure 10.43 shows a traverse on a perimeter road downwind of a predominantly industrial area. The traverse covers a distance of about 15 km and it is evident from both records that a number of sources are contributing to

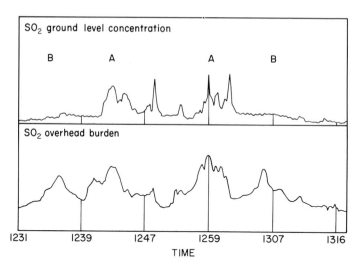

Fig. 10.43   Overhead burden and ground level concentration of SO_2 downwind of an industrial area. Some sources contribute substantially to ground level concentrations, A; while other plumes remain aloft, B.

ground level concentration and overhead burden. There is a degree of correlation between some features in the record of overhead burden and ground level concentration, such as at the points marked A on the figure. It is also clear that plumes from some substantial sources, B, are contributing very little at ground level.

The results shown in Fig. 10.43 were obtained in Ghent, Belgium, during an exercise designed to examine the accuracy with which the flux of SO_2 from an extensive industrial area can be measured. The survey was organized jointly by the EEC and the Belgian Ministry of Science Policy. Measurements in Fig. 10.43 are reported individually in [94–96]. Seven van-mounted correlation spectrometers were used together with an array of fixed point samplers and comprehensive meteorological measurements. The vehicles were deployed along routes around the industrial area in order to make traverses at different distances downwind of the sources and also along a route upwind to measure the flux entering the area. Examples of the results are shown in Fig. 10.44. On the map the values of overhead burden are plotted as a function of position along the routes. Each is the average of several traverses as indicated in the figure where the results of individual traverses and the average is shown for the vehicle which followed the closed loop route in the centre of the map. The increase across the industrial area is clear as are the contributions from some individual sources. However, to evaluate the results quantitatively is more difficult, due principally to uncertainties in assigning a base-line to the measurements. As described in Section 10.2, the base-line of a correlation spectrometer can vary with solar elevation and with

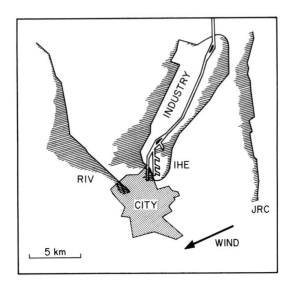

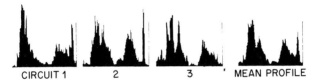

CIRCUIT 1   2   3   MEAN PROFILE

Fig. 10.44 Correlation spectrometer measurements of SO_2 fluxes in an industrial area. Results on map are average of several traverses as shown below for measurements made around the central closed loop [94–96].

meteorological factors such as cloud cover. On a short traverse across a well-defined plume such as that shown in Fig. 10.4 this is not a serious problem but on long traverses through plumes from many sources with no well-defined boundaries it is difficult to identify a base-line. Moreover, a small error integrated across a long traverse can constitute a significant flux. The problem is most acute on upwind traverses where there may be no noticeable plume structure. However, in the example shown in which the instruments on each vehicle were carefully intercalibrated, the measured flux, calculated from on average 10 downwind traverses and 4 upwind, was within about 25% of that estimated from a comprehensive emission inventory for the area. This was one of the better results, the ratios of measured to inventory flux during six separate periods being 2.12, 0.89, 1.36, 1.71, 1.44 and 1.02. Not all the difference should be attributed to the flux measurements – some is due to wind measurement and there may be shortcomings in the inventory. However, the exercise demonstrated that there are difficulties in measuring fluxes accurately in extended multisource areas, due primarily to uncertainties in correlation spectrometer base-lines.

10.9 Conclusions

Advances in the last decade or so have established the usefulness of remote sensing in air pollution studies. Correlation spectroscopy is used regularly to measure SO_2 and NO_2 in industrial plumes and is finding some application in monitoring emissions from extended urban and industrial conurbations. Lidar is effective in measuring plume rise and dispersion and with the use of smoke as a tracer it can be employed to investigate dispersion in situations such as those which occur in the wake of buildings. Differential lidar has been successful in measuring SO_2, NO_2 and ozone. Considerable progress has been achieved in meteorological measurements. Sodar provides three-dimensional wind data together with information on turbulence. Temperature profiles can be obtained with microwave radiometry and with the RASS system. Sodar and lidar give information on the depth of the mixing layer. Most of these techniques have limitations of some sort which restrict their mode of operation. Ground based differential lidar for example cannot operate out to a range much in excess of a kilometre or so for the measurement of SO_2 and NO_2 with the result that, except for measurements very close to a source, systems have to be mounted in vehicles or aircraft. Sodar has a similar restriction in range and is generally limited to measurements within the mixing layer. Correlation spectroscopy is affected by multiple scattering and base-line drift and does not of course operate at night. Nevertheless, within these restrictions remote sensing is proving unique and valuable in measuring important species and meteorological variables in the most important region, the mixing layer.

There are of course some differences between these techniques and the more standard methods of point sampling. Apart from sodar the instruments are not in general used to collect long-term statistical data. Their use has been mainly in intensive studies of particular situations. This is mainly because at their present stage of development the instruments require frequent attention and the measurements sometimes need considerable interpretation before they are as meaningful as the output from a point sampler. The techniques have not reached the stage of becoming standard reference methods for pollution monitoring, one exception perhaps being lidar for the determination of plume opacity [97]. Correlation spectrometers and sodar are available commercially but in general the other remote sensing instruments have been built by the teams who operate them. Even the relatively straightforward vehicle assemblies incorporating point samplers, correlation spectrometers and navigation units, of which there is an increasing number, are put together individually and require considerable resources to incorporate the necessary computer systems and to develop the associated software. There is also a difference in cost. Whereas a standard point sampler is usually available in the price range £5–10K a standard sodar system is likely to be in the region of £50K and a mobile differential lidar in excess of £100K.

As to the future, it seems that remote sensing will be used increasingly both to monitor individual sources and to establish pollutant budgets in urban and industrial areas. Sodar will continue to be developed with perhaps improvements in range on the one hand and on the other higher frequency systems giving more detailed information over shorter ranges. Instruments to measure temperature profiles may become more readily available. Lidar systems will be flown in the Space Shuttle and may find application in pollution measurements. The most fruitful area for development is in the extension of lidar to the infrared where many more species are accessible than in the ultraviolet and visible. Already long path systems using retroreflectors are being developed to monitor emissions in petrochemical plants. However, advances here depend upon the development of more powerful infrared lasers suitable for differential lidar and so far progress in this area has been slow.

Acknowledgements

Many colleagues have provided data for this chapter and their help is gratefully acknowledged. Copyright permission has been granted by several authors and journals. Those involved are quoted with each illustration and their co-operation has been most welcome. Particular thanks are due to D. J. Brassington, P. M. Hamilton, C. Rogers and S. Sutton of the Central Electricity Research Laboratories, Leatherhead, UK. The chapter is published with the permission of the Central Electricity Generating Board.

References

1. *Abstracts of the 10th Int. Laser Radar Conf., October 1980, Silver Spring, Maryland* (abstracts available from the American Meteorological Society, Boston, Mass., USA).
2. *Abstracts of the 11th Int. Laser Radar Conf., June 1982, University of Wisconsin-Madison,* NASA Conf. Publication 2228 (abstracts available from the US Air Force Office of Scientific Research, NASA Langley Research Center, Hampton, Virginia, 23665, USA).
3. Fraser, R. S., Kaufman, Y. J. and Machoney, R. L. (1984) *Atmos. Environ.,* **18,** 2577–78.
4. Browell, E. V., Wilkerson, T. D. and McIlrath, T. J. (1979) *Appl. Opt.,* **18,** 3474–83.
5. Browell, E. V. (1982) *Opt. Eng.,* **21,** 128–32.
6. Millán, M. M. and Hoff, R. M. (1977) *Appl. Opt.,* **16,** 1609–18.
7. Hamilton, P. M., Varey, R. H. and Millán, M. M. (1978) *Atmos. Environ.,* **12,** 127–33.
8. Millán, M. M. and Hoff, R. M. (1977) *Atmos. Environ.,* **11,** 857–60.
9. Millán, M. M. and Hoff, R. M. (1978) *Atmos. Environ.,* **12,** 853–64.
10. Newcombe, G. S. and Millán, M. M. (1970) *IEEE Trans. Geosci. Electron.,* **GE-8,** 149–57.
11. Moffat, A. J., Robbins, J. R. and Barringer, A. R. (1971) *Atmos. Environ.,* **5,** 511–25.
12. Moffat, A. J. and Millán, M. M. (1971) *Atmos. Environ.,* **5,** 677–90.
13. Byer, R. L. (1975) *Opt. Quant. Electron.,* **7,** 147–77.
14. Collis, R. T. H. and Russel, P. B. (1976) in *Laser Monitoring of the Atmosphere* (ed. E. D. Hinkley), Vol. 14, Topics in Applied Physics, Springer-Verlag, Berlin.
15. Hamilton, P. M. (1969) *Phil. Trans. Roy. Soc.,* **265,** 153–72.

16. Byer, R. L. and Garbuny, M. (1973) *Appl. Opt.*, **12**, 1496–505.
17. Van de Hulst, H. C. (1957) *Light Scattering by Small Particles*, Wiley, New York.
18. Middleton, W. E. K. (1952) *Vision Through the Atmosphere*, University of Toronto Press.
19. Rothe, K. W. (1980) *Radio and Electron. Eng.*, **50**, 567–74.
20. Murray, E. R., Hake, R. D., Van der Laan, J. E. and Hawley, J. G. (1976) *Appl. Phys. Lett.*, **28**, 542–3.
21. Stewart, R. W. and Bufton, J. L. (1980) *Opt. Eng.*, **19**, 503–7.
22. Uthe, E. E., Nielson, N. B. and Jimison, W. L. (1980) *Bull. Am. Meteorol. Soc.*, **61**, 1035–43.
23. Uthe, E. E. (1982) *Appl. Opt.*, **21**, 454–9.
24. Uthe, E. E. and Allan, R. J. (1975) *Opt. Quant. Electron.*, **7**, 121–9.
25. Bowne, N. E., Londergan, R. J., Murray, D. R. and Borenstein, H. S. (1983) EPRI Report EA-3074, Electric Power Research Institute, Palo Alto, California.
26. Rothe, K. W., Brinkman, U. and Walther, H. (1974) *Appl. Phys.*, **4**, 181–2.
27. Fredriksson, K., Galle, B., Nyström, K. and Svanberg, S. (1979) *Appl. Opt.*, **18**, 2998–3003.
28. Sugimoto, N., Takeuchi, N. and Okuda, M. (1980) *Abstracts, 10th Int. Laser Radar Conf. Maryland*. See [1, 2].
29. Adrain, R. S., Brassington, D. J., Sutton, S. and Varey, R. H. (1979) *Opt. Quant. Electron.*, **11**, 253–64.
30. Fredriksson, K., Galle, B., Nyström, K. and Svanberg, S. (1981) *Appl. Opt.*, **20**, 4181–9.
31. Baumgartner, R. A. *et al.* (1979) EPRI Report EA-1267, Electric Power Research Institute, Palo Alto, California.
32. Browell, E. V. *et al.* (1983) *Appl. Opt.*, **22**, 522–34.
33. Uchino, O., Maeda, M. and Hirono, M. (1979) *IEEE Trans.*, **QE-15**, 1094–107.
34. Megie, G., Allain, J. Y., Chanin, M. L. and Blamont, J. E. (1977) *Nature*, **270**, 329–31.
35. Hawley, J. G., Fletcher, L. D. and Wallace, G. F. (1982) Technical Digest, Workshop on Optical and Laser Remote Sensing, Monterey California, US Army Research Office.
36. Woods, P. T. and Jolliffe, B. W. (1978) *Opt. Laser Technol.*, **10**, 25–8.
37. Weitkamp, C., Harms, J., Lahmann, W. and Michaelis, W. (1978) *Proc. Soc. Photo-Optics Conf., Utrecht*, Society of Photo-Optical Instrumentation Engineers, Bellingham, WA, USA, pp. 109–18.
38. Uthe, E. (1983) *Abstracts of the 9th Conf. on Aerospace and Aeronautical Meteorology, Omaha, Nebraska*, American Meteorological Society, Boston, USA.
39. Petheram, H. C., Rye, B. J. and Thomas, E. L. (1980) *Abstracts, 10th Int. Laser Radar Conf., Maryland*. See [1, 2].
40. Weitkamp, C., Heinrich, H. J. and Herrmann, W. (1980) *Laser and Electro-Optik (Germany)*, No. 3/1980, 23–7.
41. Herrmann, W., Heinrich, H. J., Michaelis, W. and Weitkamp, C. (1980) *Abstracts, 10th Int. Laser Radar Conf., Maryland*. See [1, 2].
42. Perner, D. and Platt, U. (1979) *Geophys. Res. Lett.*, **6**, 917–20.
43. Platt, U., Perner, D. and Paetz, H. W. (1979) *J. Geophys. Res.*, **84**, 6329–35.
44. Platt, U. and Perner, D. (1980) *J. Geophys. Res.*, **85**, 7453–8.
45. Platt, U., Perner, D., Schröeder, J. *et al.* (1981) *J. Geophys. Res.*, **86**, 11965–70.
46. Platt, U. and Perner, D. (1983) in *Optical and Laser Remote Sensing* (eds D. K. Killinger and A. Mooradian), Springer-Verlag, Berlin, Vol. 39, pp. 97–105.

47. Platt, U., Perner, D., Winer, A. M. *et al.* (1980) *Geophys. Res. Lett.*, **7**, 89–92.
48. Platt, U., Perner, D., Harris, G. W. *et al.* (1980) *Nature*, **285**, 312–14.
49. Harris, G. W., Carter, W. P. L., Winer, A. M. *et al.* (1982) *Environ. Sci. Technol.*, **16**, 414–19.
50. Harris, G. W., Winer, A. M., Pitts, J. N. Jr. *et al.* (1983) in *Optical and Laser Remote Sensing* (eds D. K. Killinger and A. Mooradian), Springer-Verlag, Berlin, Vol. 39, pp. 106–13.
51. Platt, U. F., Winer, A. M., Biermann, H. W. *et al.* (1984) *Environ. Sci. Technol.*, **18**, 365–369.
52. Pitts, J. N., Sanhueza, E., Atkinson, R. *et al.* (1984) *Int. J. Chem. Kinet.*, **16**, 919–939.
53. Pitts, J. N. Jr., Biermann, H. W., Winer, A. M. and Tuazon, E. C. (1984) *Atmos. Environ.*, **18**, 847–854.
54. Noxon, J. F., Norton, R. B. and Marovich, E. (1980) *Geophys. Res. Lett.*, **7**, 125–8.
55. Noxon, J. F. (1981) *Geophys. Res. Lett.*, **8**, 1223–6.
56. Noxon, J. F. (1983) *J. Geophys. Res.*, **88**, 11017–21.
57. Herget, W. F. and Brasher, J. D. (1979) *Appl. Opt.*, **18**, 3404–20.
58. Herget, W. F. and Brasher, J. D. (1980) *Opt. Eng.*, **19**, 508–14.
59. Herget, W. F. (1982) *Appl. Opt.*, **21**, 635–41.
60. Hanst, P. L. (1971) *Adv. Environ. Sci. Technol.*, **2**, 91–213.
61. Hanst, P. L., Wilson, W. E., Patterson, R. K. *et al.* (1975) EPA Publication 650/4-75-006, Environmental Protection Agency, Research Triangle Park, NC.
62. Tuazon, E. C., Graham, R. A., Winer, A. M. (1978) *Atmos. Environ.*, **12**, 865–75.
63. Tuazon, E. C., Winer, A. M., Graham, R. A. (1980) *Adv. Environ. Sci. Technol.*, **10**, 259–300.
64. Tuazon, E. C., Winer, A. M. and Pitts, J. N. Jr. (1981) *Environ. Sci. Technol.*, **15**, 1232–7.
65. Hanst, P. L., Wong, N. W. and Bragin, J. (1982) *Atmos. Environ.*, **16**, 969–81.
66. McAllister, L. G., Pollard, J. R., Mahoney, A. R. and Shaw, P. J. R. (1969) *Proc. IEEE*, **57**, 579–87.
67. Mahoney, A. R., McAllister, L. G. and Pollard, J. R. (1973) *Boundary-Layer Meteorol.*, **4**, 155–67.
68. Takarski, V. I. (1961) *Wave Propagation in a Turbulent Medium*, McGraw-Hill, New York.
69. Monin, A. S. (1962) *Sov. Phys. Acoust.*, **7**, 370–3.
70. Little, C. G. (1969) *Proc. IEEE*, **57**, 571–8.
71. Hasenjäger, H. (1976) CEC Report EUR 5534e, Joint Research Centre, 21020, Ispra, Varese, Italy.
72. MacCready, P. and Warden, J. (1982) EPRI Report EA-2219, Electric Power Research Institute, Palo Alto, California.
73. Collis, R. T. H. and Uthe, E. E. (1972) *Opto-Electronics*, **4**, 87–99.
74. Uthe, E. E. (1972) *Bull. Am. Meteorol. Soc.*, **53**, 358–60.
75. Uthe, E. E. and Russel, P. B. (1974) *Bull. Am. Meteorol. Soc.*, **55**, 115–21.
76. Frankel, M. S. and Peterson, A. M. (1976) *Radio Sci.*, **11**, 157–66.
77. Bonino, G., Lombardini, P. P. and Trivero, P. (1980) *Nuovo Cimento*, **1C**, 207–14.
78. Goody, R. M. (1964) *Atmospheric Radiation*, Oxford University Press, London.
79. Meeks, M. L. and Lilley, A. E. (1963) *J. Geophys. Res.*, **68**, 1683–703.
80. Westwater, E. R. (1972) *Monthly Weather Rev.*, **100**, 15–28.
81. Rodgers, C. D. (1976) *Rev. Geophys. Sp. Phys.*, **14**, 609–24.
82. Schönwald, B. (1978) *Boundary-Layer Meteorol.*, **15**, 453–64.

83. Jeske, H. (1981) Private communication, Meteorologisches Institut der Universität Hamberg.
84. Miner, F. G., Thornton, D. D. and Welch, W. J. (1972) *J. Geophys. Res.*, **77**, 975–91.
85. Hosler, R. C. and Lemmons, T. J. (1972) *J. Appl. Met.*, **11**, 341–8.
86. Scriven, R. A. (1979) CEC Report EUR 6420 EN, Commission of the European Communities.
87. Hasenjäger, H. (1984) Private communication, Joint Research Centre, Ispra, Varese, Italy.
88. Dutrannoy, C. and Robé, M. C. (1978) EDF Report RE 33/78/028, Electricité de France, Chatou, France.
89. Hamilton, P. M. (1974) CEGB Report RD/L/N 131/74, Central Electricity Generating Board, Kelvin Avenue, Leatherhead, UK.
90. Millán, M. M. and Chung, Y. S. (1977) *Atmos. Environ.*, **11**, 939–44.
91. Carras, J. N. and Williams, D. J. (1981) *Atmos. Environ.*, **15**, 2205–17.
92. Sperling, R. B. (1975) EPA Report EPA-600/2-75-077, Environmental Protection Agency, Research Triangle Park, NC.
93. Longhetto, A. *et al.* (1982) *Nuovo Cimento*, **5C**, 299–331.
94. Van der Meulen, A. and van Jaarsveld, J. A. (1982) Report 247604012, National Institute of Public Health, Bilthoven, The Netherlands.
95. Verduyn, G. (1982) Report No. DL/1982/2505/15, Instituut voor Hygiene en Epidemiologie, Brussels, Belgium.
96. Cerutti, C., De Groot, M. and Sandroni, S. (1982) Report EUR 7788 EN, Joint Research Centre, Ispra, Varese, Italy.
97. US Federal Register, 46(208), 1981.

11 Physico-chemical speciation techniques for atmospheric particles

R. M. HARRISON

11.1 Introduction

The term physico-chemical speciation is used to describe the determination of precise physical and chemical forms of environmental trace substances. This is recognized as being of considerable importance for air pollutants, since the speciation influences the toxicity and environmental mobility of a particulate pollutant. For example, a 1 μm diameter aerosol of $PbCl_2$ is highly suited to alveolar deposition and rapid absorption by exposed humans, while 10 μm diameter PbS aerosol is more liable to passage via the gastrointestinal route, but is absorbed only very inefficiently [1]. A simple measurement of lead in air would not discriminate the two forms. It is also readily appreciable that the environmental consequences of sulphate in the form of H_2SO_4 are potentially more damaging than for $(NH_4)_2SO_4$. The other area in which speciation information is valuable is in studies of atmospheric chemistry. Information upon chemical species may provide a valuable insight into chemical processes by which substances are formed or removed from the air [2].

The determination of particle sizes and the factors influencing particle size are considered in Chapter 4. This chapter will thus be restricted to the aspect of deriving information on the chemical speciation of atmospheric particles. The greatest mass of atmospheric particles, particularly those of pollutant origin, occurs in the 'accumulation mode' (Section 4.1). At a diameter of 1 μm, towards the upper end of the 'accumulation' range, there will be about 10^8 particles per cubic metre of air at an atmospheric particle loading of 100 μg m^{-3}. In a poly-disperse aerosol of mass median diameter 1 μm there will be vastly more particles than 10^8 m^{-3} since smaller particles contribute little to mass, but are very abundant numerically.

The great numerical abundance of atmospheric particles has consequences for experimental techniques of speciation. Many methods are microscopic, i.e. they depend upon examination of individual particles to provide chemical information. The scanning electron microscope with X-ray energy spectroscopy (SEM/XES) is the best-known example of such a technique. It has been widely used to characterize airborne particles, but since examination of even a single particle is laborious, it is difficult to generate sufficient data to be representative of the air sample as a whole, unless the aerosol is highly uniform. In ambient air, aerosols are extremely diverse and thus results from microscopic techniques must be treated with some reserve. The use of raster modes to scan a larger area of sample and establish interparticle distributions of several selected elements can give valuable supplementary information upon chemical composition for the sample as a whole, but does not entirely overcome the problem of generation of representative speciation information.

Macroscopic techniques are also available and provide a characterization of the bulk sample, rather than of individual particles. These also have limitations, particularly with regard to selectivity towards particular types of constituent. For example, X-ray powder diffraction (XRD) will identify only crystalline components of a mixture, and thus valuable data upon amorphous or semi-crystalline components is lost when using this method.

11.2   Speciation methods

11.2.1   X-ray diffraction (XRD)

When a collimated monochromatic X-ray beam is focused upon a crystalline material, diffracted X-rays may be detected leaving the sample at angles related to the interplanar spacings within the crystalline sample by the Bragg equation. Hence, after X-ray diffraction examination of a known sample, the interplanar spacings (known as d-spacings) are estimated from the diffraction angles, and the magnitudes and relative intensities of the d-spacings of the sample are compared with known spectra using a search manual*. Complex mixtures may be characterized by this technique as long as there are few overlaps of diffraction lines.

The major drawback of XRD methods in the examination of environmental samples is the requirement for crystalline material. Although many substances exist in the atmosphere as highly crystalline particles, others are present in poorly crystalline or amorphous forms and give little or no response in XRD analysis. The technique is also of limited sensitivity, and is responsive to crystalline components of a sample accounting for in excess of about 1% of a sample of milligram proportions.

Early use of XRD methods was severely hampered by the presence of natural mineral materials such as α-quartz and calcite in air samples, which have strong diffraction patterns which tend to obscure the response of the man-made components of the aerosol. This problem may be overcome by size fractionation of the aerosol to separate coarse and fine particles into several discrete size ranges by impaction, with subsequent XRD analysis of each fraction [3, 4]. Alternatively, use of density fractionation techniques may be productive.

Collection of samples by filtration or impaction inevitably involves use of a substrate material. Certain of these give considerable diffraction patterns in their

---

* Generally the diffraction files of the Joint Committee on Powder Diffraction Standards (JCPDS).

own right, and if the sample is to be examined without prior separation from the substrate, careful selection of the substrate is necessary. O'Connor and Jaklevic [5] have tested cellulose ester, PTFE and polycarbonate and find all to give either discrete diffraction peaks or a diffuse hump, but none was ruled out on these grounds. Davis and Johnson [6] recommend Teflon filters, especially at low particle loadings, although quartz or glass fibre may be used at greater loadings.

In our own work, a number of alternative methods have been developed. One useful technique involves impaction on to non-porous inert FEP Teflon substrate. The impacted particles are then transferred on to a small area of adhesive tape (double-sided Sellotape) which is then analysed by XRD [2]. Secondly, polystyrene (Delbag Luftfilter) filters have been used in a high volume sampler. These are soluble in toluene and after dissolution, the collected particles are filtered on to a cellulose ester filter of far smaller surface area (25 or 47 mm diameter) which is used for the analysis. Fukasawa *et al.* [7] have used dichloromethane as a solvent for Nuclepore and Fuji filters, but it must be recognized that this, and other polar solvents, may dissolve some of the more covalent inorganic compounds. Alternatively, particles may be stripped ultrasonically into hexane from glass fibre filters [3, 4] and collected on a smaller filter. The stripping is not, however, fully efficient and also tends to disintegrate the filter with consequent incorporation of glass fibres in the separated sample. No single best method can be recommended – the most appropriate technique for the task in hand must be selected by trial and error.

Quantification of XRD analysis of atmospheric particles is feasible, but difficult. Davis and Cho [8] and Davis [9] report a method applicable to high volume filter samples in which a thin layer of a reference component is added prior to XRD analysis. The resulting line intensities must then be corrected for the masking effect of the reference component, the absorption of the matrix and the transparency of both reference and sample components. The method requires a rather detailed knowledge of the aerosol composition and is not easily applied to ambient air samples.

Although many workers are limited due to availability of XRD equipment, if choice is available the use of a crystal monochromator or Guinier camera is highly recommended. Both have the advantage of monochromatization which limits problems due to fluorescence, and the Guinier camera is highly suited to examination of the small masses of samples collected in air sampling.

11.2.1.1 *Phases identified in air by XRD*

In order to have some indication of the compounds to be found by XRD methods, some of the more commonly observed phases are listed in Tables 11.1–11.4.

In countries where leaded gasoline is used, the automotive lead compounds in Table 11.1 may be found; some are primary emissions, while others are formed by chemical reactions in urban air. Rather different compounds are encountered as a result of lead smelting (Table 11.1). Metal phases identified in the atmosphere and floor dust of a primary zinc–lead smelter are listed in Table 11.2.

Table 11.1   Lead, zinc and cadmium compounds identified in ambient air and smelter stack effluent using X-ray diffraction

(a) *Automotive lead*	*Ref.*
$PbSO_4 . (NH_4)_2SO_4$	[3, 4, 10, 11]
$PbSO_4$	[4]
$PbBrCl . (NH_4)_2BrCl$	[4]
$\alpha\text{-}2PbBrCl . NH_4Cl$	[4]
$PbBrCl . 2NH_4Cl$	[4]
$PbBrCl$	[4]
(b) *Industrial (smelter) lead and zinc*	
PbS	[12–14]
$PbSO_4$	[12–14]
$PbO . PbSO_4$	[12, 13]
PbO (litharge)	[14]
$Pb°$	[14]
ZnO	[14]
$\alpha\text{-}ZnS$	[14]
CdO	[14]

Table 11.2   Metal phases identified in the internal atmosphere of a primary zinc–lead smelter using X-ray diffraction [15]

PbS	$\beta\text{-}ZnS$	CdO
PbO (litharge)	ZnO	$Cd°{}^a$
PbO (massicot)		$Cd(OH)_2{}^a$
$PbSO_4$		
$Pb°$		
$PbO \cdot PbSO_4$		

a Detected in floor dusts only.

Sulphates are abundant atmospheric compounds and appear also to give strong diffraction patterns, so are readily identified. Some of the reported sulphates are listed in Table 11.3. Natural minerals usually occur in the larger ($>2\ \mu m$) size fractions; Table 11.4 indicates some phases identified in the ambient atmosphere by XRD methods.

11.2.2   Method for sampling and XRD analysis of atmospheric particles (based on [2])

Air samples are collected over a period of 7 days using a high volume Andersen cascade impactor fitted with impaction surfaces of FEP-Teflon and a polystyrene filter back-up (Delbag Luftfilter). Collected particles are removed from the FEP-Teflon film by transfer to adhesive tape (double-sided Sellotape) which is inserted

Table 11.3 Sulphate compounds identified in ambient air using X-ray diffraction

Compound	Ref.
$Fe_2(SO_4)_3 . 3(NH_4)_2SO_4$	[16]
$CaSO_4 . (NH_4)_2SO_4 . H_2O$	[2, 11]
$2CaSO_4 . (NH_4)_2SO_4$	[16]
$PbSO_4 . (NH_4)_2SO_4$	[10, 11, 16]
$PbSO_4$	[15, 16]
$CaSO_4 . 2H_2O$	[16, 17]
Na_2SO_4	[16]
$(NH_4)_2SO_4$	[16]
NH_4HSO_4	[11]
$(NH_4)_2SO_4 . NH_4HSO_4$	[5]
$ZnSO_4 . (NH_4)_2SO_4 . 6H_2O$	[5]
$MgSO_4 . 7H_2O$	[18]
$(NH_4)_2SO_4 . 2NH_4NO_3$	[2, 11]
$(NH_4)_2SO_4 . 3NH_4NO_3$	[2, 11]
$Na_2SO_4 . NaNO_3 . H_2O$	[2]
$Na_2SO_4 . (NH_4)_2SO_4 . 4H_2O$	[2]

Table 11.4 Some natural minerals identified in ambient air by X-ray diffraction

Mineral	Ref.
NaCl (halite)	[3, 11]
Na_2SO_4	[3]
α-SiO_2 (α-quartz)	[3, 7]
$CaCO_3$ (calcite)	[3]
$CaSO_4 . 2H_2O$ (gypsum)	[3, 11, 17]
Biotite	[17]
Muscovite	[17]
Kaolinite	[17]
Plagioclase	[17]
Chlorite	[18]
$MgCO_3 . CaCO_3$ (dolomite)	[11, 17]
Fe_2O_3 (haematite)	[17, 18]
$FeO . Fe_2O_3$ (magnetite)	[17]
$CaSO_4 . \frac{1}{2}H_2O$ (hemihydrate gypsum)	[18]
$MgSO_4 . 7H_2O$ (Epsom salt)	[18]

directly into the powder diffractometer. The polystyrene back-up is dissolved in toluene, and the suspended particles filtered on to a 25 mm diameter membrane filter (Millipore type HAWP). Any storage of impacted or filter samples or of adhesive tape was in a desiccator over silica gel.

In the reported work, XRD analysis was carried out on a Philips PW1050/1720 powder diffractometer with crystal monochromator and Cu K_α radiation, scanning at 0.25 degrees per minute over 2θ angles of 3–70°. Compounds were identified by d-spacing and intensity matching both with the JCPDS diffraction files and with standard compounds run on the same instrument, when available. Some examples of the diffraction patterns encountered appear in Fig. 11.1.

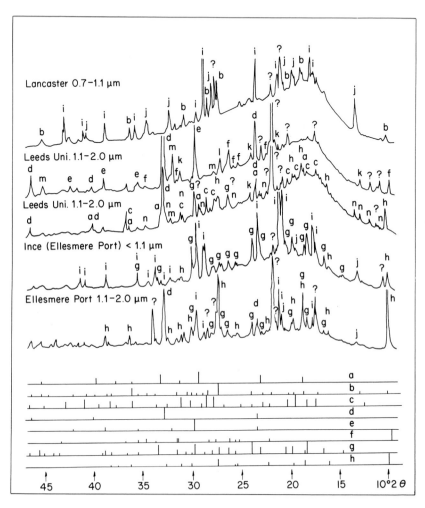

Fig. 11.1   Selected X-ray diffractograms and standard spectra (below) as 2θ (Cu K_α) versus intensity. The phases identified by letters are: (a) NH_4NO_3; (b) $(NH_4)_2SO_4 . 2NH_4NO_3$; (c) $(NH_4)_2SO_4 . 3NH_4NO_3$; (d) NH_4Cl; (e) $NaNO_3$; (f) $Na_2SO_4 . NaNO_3 . H_2O$; (g) $Na_2SO_4 . (NH_4)_2SO_4 . 4H_2O$; (h) $CaSO_4 . (NH_4)_2SO_4 . H_2O$; (i) $(NH_4)_2SO_4$; (j) $PbSO_4 . (NH_4)_2SO_4$; (k) $CaSO_4 . 2H_2O$; (l) α-SiO_2; (m) NaCl; (n) possibly $PbBrCl . 2(NH_4)_2BrCl$, or similar.

11.2.3   Single particle techniques

These come in two main groupings: electron beam methods and laser techniques. The former group has been widely used and includes the electron microprobe, the scanning electron microscope (SEM) and the transmission and scanning transmission electron microscopes (TEM and STEM). X-rays, stimulated by the electron beam, give information upon the elemental composition of the sample. While the electron microprobe is designed primarily for generation of such chemical data, the electron microscope methods give size and morphology data, with elemental information as a secondary benefit.

The most commonly used method is the combination of scanning electron microscopy with X-ray energy spectroscopy (SEM/XES). It is limited to elements of atomic number $\geqslant 11$, and gives only semi-quantitative elemental data. Thus, while interesting data upon approximate particle composition are attainable, no precise identification of chemical compounds is possible. Butler and co-workers [19] used the SEM/XES to characterize airborne particles collected in an urban area, and reported that particles in the 0.5–1.0 μm range could be focused and analysed with a minimum detection limit of 0.9% for the transition elements. Particles individually rich in the elements Ti, Zn and Pb were reported. Hamilton and Adie [20] examined particles of > 3 μm and found similar size distribution patterns for zinc and copper, while lead was appreciably different.

The major limitations upon SEM/XES work are the applicability only to particles of > 0.5 μm, the semi-quantitative and incomplete nature of the elemental analysis data and the restriction on the number of particles analysed. These render the method of very limited value for speciation studies.

Better results may be obtained with transmission electron microscopy. Smaller particles, down to < 0.1 μm may be examined and selective area electron diffraction patterns collected, allowing firm structural assignments for some crystalline materials. Bloch and co-workers [21] have obtained some useful results, summarized in Table 11.5 using this method.

The laser microprobe mass spectrometer (LAMMA) has recently been applied to the analysis of airborne particles. A high intensity laser pulse with a spatial resolution of *c.* 1 μm is focused upon an individual particle collected on a filter or impaction substrate and is used to volatilize and ionize the particle. Analysis of the ion fragments by a time-of-flight mass spectrometer allows generation of both positive and negative ion mass spectra, which may allow characterization of the chemical nature of the particle [22, 23]. This method often generates very complex mass spectra (many particles are aggregates), but shows promise as a research technique, at least.

11.2.3.1   *Transmission electron microscope method for atmospheric particles* (based on [21])

Thin Formvar films are made by immersing microscope object glasses in a 0.25% solution of Formvar in chloroform. After drying, the film is immediately floated off from the glass surface on to distilled dust-free water. A 200 mesh, 3 mm

Table 11.5   Particles detected in aerosol samples by STEM (based on [21])

Identification	Elements detected and relative concentration		Crystal structure	Mean projected diameter (μm)
$(NH_4)_2SO_4$	S(1.00)		$(NH_4)_2SO_4$	1.1
Na_2SO_4	Na(0.17)	S(1.00)	Na_2SO_4	1.7
K_2SO_4	K(1.00)	S(0.33)	K_2SO_4	1.9
$CaSO_4$	Ca(1.00)	S(0.83)	$CaSO_4 . 2H_2O$	1.4
NaCl	Na(0.54)	Cl(1.00)	NaCl	4.6
Fly ash	Al(0.67)	Si(1.00)	Amorphous	2.5
	S(0.26)	K(0.26)		
	Ca(0.07)	Fe(0.72)		
Fe-oxide spheres	Fe(1.00)		Amorphous	1.2
Goethite	Fe(1.00)		Goethite	2.0
TiO_2	Ti(1.00)		Amorphous	0.8
Condensation aggregates	Al(0.07)	Si(0.16)	Amorphous	5.1
	S(0.81)	K(0.12)		
	Ca(0.25)	Fe(0.33)		
	Pb(1.00)			
S-containing particle	S(1.00)			
$PbBr_xCl_y$	Pb(1.00)		Crystalline	1.7
	Br(0.27)	Cl(0.10)		
Quartz	Si(1.00)		Quartz	1.6
Calcite	Ca(1.00)		Calcite	2.2

diameter TEM grid of copper is placed upon the film and the whole is removed from the water surface with a ribbon of parafilm. The Formvar grids are mounted with adhesive tape at one edge on the collector plates of a 5-stage Battelle cascade impactor with a 1 litre min^{-1} flow rate. Short air sampling times (5 min–2 h) are used to limit collection of particles to a thin layer. After air sampling, the samples are coated with a thin conductive carbon layer on the Formvar film side by vacuum evaporation and may be directly mounted into the TEM (in this case a Philips FM-300 with Kevex Si(Li) X-ray detector and Link Systems electronics and data processing system). The computer has a peak integration facility with automatic background subtraction. Interference between peaks is accommodated by a deconvolution program, and quantitative analytical data are obtained from the X-ray intensities by a thin film model which neglects radiation absorption. Selected area electron diffraction data in the form of a spot array ring pattern are also generated.

11.2.4   Speciation of sulphuric acid and other particulate sulphates

In most polluted atmospheres the major part of the sulphate comprises one or more of the three compounds H_2SO_4, NH_4HSO_4 and $(NH_4)_2SO_4$ which represent differing degrees of ammonia neutralization of sulphuric acid formed from SO_2 oxidation processes. Since the environmental effects of the three compounds differ considerably, much effort has been expended upon quantification of the different forms. While XRD is useful in characterization of $(NH_4)_2SO_4$ in aerosols, it is difficult to quantify, and application of XRD to NH_4HSO_4 and H_2SO_4 is rendered impossible by their existence as solution droplets and not crystalline solids (except at very low relative humidity for NH_4HSO_4).

Solvent extraction has been used considerably in the speciation of sulphates [24]. Sulphuric acid is extracted selectively into benzaldehyde, subsequent extraction with iso-propanol extracts NH_4HSO_4, while $(NH_4)_2SO_4$ is dissolved finally into water. Doubts have been cast, however, upon the efficiency [25] and specificity [26] of the benzaldehyde extractant.

Since H_2SO_4 is normally the only particulate strong acid in air, micro-titration of an aqueous extract of atmospheric particles using the Gran plot method can be used to quantify the concentration of H^+ in the extract and thus H_2SO_4 in the aerosol [27, 28]. The main limitations are due to chemical interactions within the extract (acid-base reaction processes) and the possible presence of weak acids.

A glass fibre filter impregnated with 2-perimidinylammonium bromide will react selectively with H_2SO_4 to form the sulphate. Subsequent pyrolysis releases stoichiometric amounts of SO_2, which are quantified [29, 30]. Alternatively, the 2-perimidinylammonium sulphate is quantified colorimetrically by reaction with HNO_3 to form 2-amino,4,6,9-trinitroperimidine [31].

Sulphuric acid will react also with secondary amines, and may be reacted with diethylamine vapour on a filter surface subsequent to collection. The absorbed diethylamine is determined colorimetrically [32]. Alternatively, if ^{14}C-labelled trimethylamine is used, the absorbed amine may be assayed by β-counting, hence giving a measure of the sulphuric acid collected [33]. This is an elegant method, but rather demanding in terms of cost and facilities.

In a sophisticated instrumental design, D'Ottavio *et al.* and Tanner *et al.* [34, 35] use a flame photometric analyser as an ambient aerosol sulphur monitor, after removal of gaseous SO_2 in a diffusion denuder. By heating the air stream, the Na_2CO_3-coated denuder is made to selectively remove H_2SO_4, but not other sulphates. Thus by cyclic addition of ammonia to the gas stream prior to the denuder, a measure of H_2SO_4 concentrations may be achieved. This is very much a research method, rather than a routine procedure, however.

11.2.4.1 *Solvent extraction method for speciation of H_2SO_4, NH_4HSO_4 and $(NH_4)_2SO_4$ in ambient air* (based on [24])

Air is sampled through a Teflon, or acid-washed quartz filter. The filter is divided in portions, and these are analysed for H_2SO_4 by benzaldehyde extraction, $H_2SO_4 + NH_4HSO_4$ by isopropanol extraction and total sulphate after water extraction.

H_2SO_4 measurements

A portion of exposed filter is placed in a 50 ml polypropylene tube and dried overnight in a vacuum desiccator. After returning the desiccator to atmospheric pressure with dry nitrogen, freshly vacuum-distilled benzaldehyde (15 ml) is added under nitrogen, the tube is capped and agitated for 2 h on a mechanical shaker or in an ultrasonic bath. The liquid is then decanted into a 15 ml centrifuge tube and centrifuged at 3000 rev/min for 10 min. Centrifuged extract (10 ml) is transferred by pipette into a 25 ml tube and 10 ml water added. The tube is capped and shaken, causing transfer of extracted H_2SO_4 to the aqueous phase. The mixture is centrifuged for 5 min, after which 8 ml of the aqueous phase is removed and analysed for sulphate by barium chloride turbidimetry (see Section 6.4.1.1; reagent quantities are scaled down appropriately). Analysis may be by alternative methods such as ion chromatography, but u.v. absorption methods such as the barium chloranilate procedure are not suitable due to high and variable blanks caused by extracted benzaldehyde.

$H_2SO_4 + NH_4HSO_4$

A second portion of exposed filter is placed in a 50 ml centrifuge tube and dried overnight in a desiccator. Isopropanol (15 ml) is added, the tube is capped and then agitated for 2 h on a mechanical shaker or ultrasonic bath. The tube is centrifuged at 3000 rev/min for 30 min, and a precise volume of solution (1.5 ml) is transferred by micropipette to a 50 ml flask containing 6.5 ml isopropanol and 2 ml of distilled water. The resultant solution is analysed for sulphate by the barium chloranilate method as follows. To the 10 ml of solution is added Zeocarb 225 Hydrogen Resin (0.5 g; fully swollen and dried), the flask is sealed and agitated mechanically for 15 min. The supernatant liquid is decanted carefully into another 50 ml flask containing barium chloranilate (25 mg; purified by washing 25 g solid with 200 ml of 80:20 isopropanol/water, filtering under vacuum, washing with ethanol and drying). The mixture is mechanically agitated for 30 min, transferred to a centrifuge tube and centrifuged for 10 min at 3000 rev/min. The absorbance of the clear solution is measured at 314 nm against a blank of 80:20 isopropanol/water treated in the same manner. The spectrophotometer is calibrated by means of standard sulphate solutions processed as above within the range 0–10 μg SO_4^{2-} ml^{-1} of final solution.

Total water-soluble sulphate

A third portion of exposed filter is placed in a 50 ml tube, dried and distilled water (15 ml) is added. The tube is capped and agitated for 2 h on a mechanical shaker, or ultrasonic bath. After centrifuging for 10 min at 3000 rev/min, an aliquot (2 ml) is withdrawn and added to isopropanol (8 ml) in a centrifuge tube. The resultant solution is then analysed by the barium chloranilate method as above.

References

1. O'Neill, I. K., Harrison, R. M. and Williams, C. R. (1982) *Trans. Inst. Min. Metall.*, **91**, C84.
2. Harrison, R. M. and Sturges, W. T. (1984) *Atmos. Environ.*, **18**, 1829.

3. Biggins, P. D. E. and Harrison, R. M. (1979) *Atmos. Environ.*, **13**, 1213.
4. Biggins, P. D. E. and Harrison, R. M. (1979) *Environ. Sci. Technol.*, **13**, 558.
5. O'Connor, B. H. and Jaklevic, J. M. (1980) *Atmos. Environ.*, **15**, 19.
6. Davis, B. L. and Johnson, L. R. (1982) *Atmos. Environ.*, **16**, 273.
7. Fukasawa, T., Iwatsuki, M. and Tillekeratne, S. P. (1983) *Environ. Sci. Technol.*, **17**, 596.
8. Davis, B. L. and Cho, N. K. (1977) *Atmos. Environ.*, **11**, 73.
9. Davis, B. L. (1978) *Atmos. Environ.*, **12**, 2403.
10. O'Connor, B. H. and Jaklevic, J. M. (1981) *Atmos. Environ.*, **15**, 1681.
11. Tani, B., Siegel, S., Johnson, S. A. and Kumar, R. (1983) *Atmos. Environ.*, **17**, 2227.
12. Foster, R. L. and Lott, P. F. (1980) *Environ. Sci. Technol.*, **14**, 1240.
13. Lott, P. F. and Foster, R. L. (1977) *Nat. Bur. Stand. (US), Spec. Publ.*, **464**, 351.
14. Harrison, R. M. and Williams, C. R. (1983) *Sci. Tot. Environ.*, **31**, 129.
15. Harrison, R. M., Williams, C. R. and O'Neill, I. K. (1981) *Environ. Sci. Technol.*, **15**, 1197.
16. Biggins, P. D. E. and Harrison, R. M. (1979) *J. Air Pollut. Control Assoc.*, **29**, 838.
17. Davis, B. L. (1981) *Atmos. Environ.*, **15**, 613.
18. Fukasawa, T., Iwatsuki, M., Kawakubo, S. and Mayasaki, K. (1980) *Anal. Chem.*, **52**, 1184.
19. Butler, J. D., MacMurdo, S. D. and Stewart, C. J. (1976) *Int. J. Environ. Stud.*, **9**, 93.
20. Hamilton, R. and Adie, G. (1982) *Sci. Tot. Environ.*, **23**, 393.
21. Bloch, P., Adams, F., Van Landuyt, J. and Van Goethem, L. (1979) *Physico-Chemical Behaviour of Atmospheric Pollutants*, Commission of the European Communities, Luxembourg, 307.
22. Denoyer, E., Van Grieken, R., Adams, F. and Natusch, D. F. S. (1982) *Anal. Chem.*, **54**, 26A–32A.
23. Adams, F., Bloch, P., Natusch, D. F. S. and Surkyn, P. (1981) *Proc. Int. Conf. Environ. Pollut., Thessaloniki, Greece*, 122.
24. Leahy, D., Siegel, R., Klotz, P. and Newman, L. (1975) *Atmos. Environ.*, **9**, 219.
25. Appel, B. R., Wall, S. M., Haik, M. *et al.* (1980) *Atmos. Environ.*, **14**, 559.
26. Eatough, D. J., Izatt, S., Ryder, J. and Hansen, L. D. (1978) *Environ. Sci. Technol.*, **12**, 1276.
27. Brosset, C. (1978) *Atmos. Environ.*, **12**, 25.
28. Brosset, C. and Ferm, M. (1978) *Atmos. Environ.*, **12**, 909.
29. Maddalone, R. F., Thomas, R. L. and West, P. W. (1979) *Environ. Sci. Technol.*, **10**, 162.
30. Thomas, R. L., Dharmarajan, V., Lundquist, G. L. and West, P. W. (1976) *Anal. Chem.*, **48**, 639.
31. Dasgupta, P. K., Lundquist, G. L. and West, P. W. (1979) *Atmos. Environ.*, **13**, 767.
32. Huygen, C. (1975) *Atmos. Environ.*, **9**, 315.
33. Dzubay, T. G., Snyder, G. K., Reutter, D. J. and Stevens, R. K. (1979) *Atmos. Environ.*, **13**, 1209.
34. D'Ottavio, T., Garber, R., Tanner, R. L. and Newman, L. (1981) *Atmos. Environ.*, **15**, 197.
35. Tanner, R. L., D'Ottavio, T., Garber, R. and Newman, L. (1980) *Atmos. Environ.*, **14**, 121.

12 Analysis of precipitation

J. R. KRAMER

12.1 Introduction

The implementation of a precipitation chemistry measurement programme can be an extremely complicated and expensive task. It is necessary that the measurement programme be well designed so that the resulting data will be useful and the programme cost effective. Mancy and Allen [1] suggest eight key steps in the design of a monitoring system:

1. Objectives of measurement: Why are the measurements needed?
2. Models of measurement systems: What is the best type(s) of model(s) to represent the information needed?
3. Determinands/parameters: What are the key determinands; which sampling sites should be selected; and with what frequency should they be sampled?
4. Method of analyses: How will the sampling and measurements be conducted? How will the quality be assured and controlled during the programme?
5. Data processing: How will the data be gathered, stored, confirmed, retrieved and displayed? Will the data and methods be readily available for others to use?
6. Information assimilation: How will the information be assimilated relative to Steps 1 and 2? Is there sufficient or excess information relative to the objectives and conceptual models?
7. Cost analysis: Are the important determinands being measured within budgetary constraints? Is the programme feasible? Should the objectives (Step 1) and conceptual models (Step 2) be modified?
8. Information dissemination: How will the information be disseminated in order to be useful to others?

There are various feedback relationships among the steps; often changes in the steps will occur as programmes evolve. Specifically Steps 1–3 are interrelated and dependent upon Step 7. It is also common that objectives become clearer and change as projects evolve, and it is quite common that methods of analysis change (at least in detail) as the study proceeds. If one wishes to determine spatial or time trends, one is cautioned against change after a programme commences. On the other hand, changes are generally implicit due to technological developments in methodology and information gained from the programme. If a change in the project is made, there should be (a) a careful *overall* assessment of the new programme using the above eight steps as a guide, and (b) an additional study to obtain differences in results due to changes in the programme. Unfortunately, changes in procedures are often not assessed as to comparability of results.

Objectives of precipitation studies generally consider either (a) calibration of models or (b) monitoring. In the first case, often short time (precipitation events) and space (10 km) scales are considered whereas the second case often considers longer term trends in concentrations and fluxes.

There is generally a more precise connection among objectives, models and determinands for calibration studies and a less well defined relationship for monitoring. But it is just as important to have clearly defined objectives for monitoring networks. Models for monitoring networks design often consider (a) emission–transport–deposition processes, (b) biogeochemical cycles and (c) chemical–meteorological factors.

Each project has individual characteristics requiring different specifications. It is important to consider all requirements of analysis not just analysis of precipitation [2]. Other analytical aspects, many of which are discussed elsewhere in this handbook, include meteorological measurements, analysis of soil compared with analyses of entrained dust, and physical and chemical measurements of the watershed.

Analysis of precipitation conveniently falls into three categories: (a) major ions, (b) trace metals, and (c) organic compounds. Major ions generally include Na^+, K^+, Ca^{2+}, Mg^{2+}, NH_4^+, H^+, HCO_3^-, Cl^-, SO_4^{2-}, NO_3^-, and sometimes F^-. These parameters are generally related to geochemical mass balance studies. The cations (Na^+, K^+, Ca^{2+}, Mg^{2+}), H^+ and HCO_3^- are generally related to weathering reactions (dust reactions) involving common minerals; H^+, Cl^-, SO_4^{2-} ions are related to natural and anthropogenic sources; and NH_4^+ and NO_3^- ions are associated with the nitrogen cycle as well as anthropogenic sources [3]. Phosphate is sometimes analysed if aquatic/terrestrial biological productivity is being studied.

Trace metals are often studied in order to compare natural geochemical cycling with anthropogenic analyses. Those trace metals which often show larger anthropogenic than natural emissions are Ag, Cl, Cu, (Hg), Pb, Sb, Se, Zn [4, 5]. These studies are often related to trace metal accumulation profiles in soils, lake sediments and vegetation as well as to toxicity to flora and fauna [6].

Trace organic studies generally consider the atmospheric transport and fate of synthetic volatile organic chemicals such as DDTs and PCBs. Typically these analyses are related to mass budget estimates for a watershed and assessment of chemical reactivity [7].

12.2 Sampling

Sampling aspects to consider are siting and sampler design. Both aspects are interconnected and are related to the objectives of the study.

12.2.1 Siting

Precipitation samples are taken at elevations in the atmosphere using towers [8] or aircraft [9], ships [10], or more commonly at about 1 m elevation on land. Land sites are most common but there are often sampling interferences due to dust re-entrainment, foliage interception/emission effects, land use effects such as agriculture (ploughing, fertilizing), transportation (exhaust emissions, coal dust) and biological effects (insect swarms, etc.). As a working hypothesis, one should always consider that a land-based sample produces a bias reflecting the sampler location. Therefore experiments involving alternate siting, analysis of seasonal data, and different types of sampling need to be carried out on a sampling network. A simple and cost-effective test is to carry out manual sampling at different localities and at different periods of a precipitation event to compare with results of the central sampler. Local ground effects can be assessed by parallel sampling on land and on rafts in lakes, by creating an artificial air/land interface around the sampler (using synthetic 'astroturf', etc.), and by carrying out measurements at different elevations. Comparison of results at different elevations, however, involves effects due to differing scavenging as well as soil entrainment. Thus it is not uncommon to find increased concentrations and apparent increased fluxes at higher elevations!

Precipitation sampling on a ship eliminates much of the effect due to soil re-entrainment, but attention must be paid to contamination from the ship and ship emissions. This is particularly important for analysis of trace metals and organics. A common procedure is to carry out sampling at the bow while the ship is proceeding slowly into the wind; however, many precipitation events are associated with storm fronts resulting in possible contamination from spray. Therefore it is always important to include one or more parameters indicative of spray contamination.

Remote islands, peninsulas and buoys have been used as alternative sampling sites to ships. In addition to problems of spray, samplers in these locations often offer a site for concentration of insects and birds which can lead to elevated results in nutrients.

Sample sites can be considered as manned or unmanned. Generally a manned site is thought to give the highest quality and more representative results, but it is also the most expensive. But this assessment may often be false. Manned sites generally imply access by road or air, and there are other activities which can result in local elevated concentrations in precipitation. This local contamination has to be considered a major problem especially when background concentration studies are carried out in localities with difficult access or severe climate (e.g. Arctic), because the severity of the climate will maximize local emissions at the same time as minimum concentrations are being measured.

Whereas isolated unmanned sites may eliminate local habitation effects, they may be prohibitive due to the type of sampling or cost of access. Unmanned sites can be used for multiple sampling by employing batteries, compressed air, or mechanical energy devices, but they are virtually impossible to use for multiple sampling of snow deposition due to the demand of large amounts of clean thermal energy for melting of samples. A truly isolated site may require individual 'on foot' access thus involving large expenses. Alternatively volunteers may be employed to operate remote sites, or aircraft may be employed.

It is not uncommon for unmanned sites to be the subject of pranks resulting in interesting chemical analyses. Therefore it is worthwhile to include an explanatory sign with unmanned sites as well as to make the samplers as inconspicuous as possible.

As inexpensive and rugged solid state electronics become more and more available, one may expect the increased use of unmanned sites or satellite sampling locations within the proximity of a central manned facility.

It has been common to develop monitoring networks in conjunction with an agency that has other prime mandates. This procedure is invariably mandated due to unacceptable costs related to an independent study system. However, this mandate invariably involves sampling at manned sites such as airports, agricultural study sites or academic laboratories where there is a high probability that there will be local contamination effects especially for trace substances. In these instances, it is advisable that (a) site personnel receive a minimum of training, (b) that sampler site handling be a prescribed duty, not an extra, (c) that a detailed protocol be established for operation of a site and (d) that each site undergo a series of parallel experiments by central laboratory personnel to evaluate sampler location and procedures [11].

There are now a number of precipitation chemistry study programmes going on throughout the world. It is of particular value to choose common sampling sites for different programmes. In this way statistical variation as well as sampling and analytical bias can be ascertained for different programmes. In addition differences due to sampling period and design of sampler can be assessed. It is desirable to have between 10 and 15% of sampling sites common to two studies. This is the same magnitude of overlap as prescribed for replicate and blank sampling at one location [12, 13].

Siting of samplers and associated handling and shipping can be quite expensive even in small-scale programmes. In general it is best to employ the simplest samplers initially; this often means that a clean collector is manually located for each precipitation event or at specific time intervals for bulk solutions. Perhaps the best sources of skilled manpower for this type of programme are volunteer students and teachers in science courses. In this situation, it is worthwhile to develop a simple textbook and laboratory programme in conjunction with the sampling protocol.

In summary, siting and handling of precipitation samplers and samples is a most important and very difficult task in analysis of precipitation chemistry. Untrained manpower is often used; detailed protocols are required, and field experiments should be carried out by central laboratory personnel to assess various aspects of siting and manning. Too often the failure of a programme results from the analyst not venturing into the field. Granat [14] suggests some worthwhile field experiments.

12.2.2   Samplers

A classification of samplers for precipitation chemistry is:

(a) Bulk samplers: collectors continuously open for a period of time (weekly to monthly) collecting precipitation, impacted gas and particulate.

(b) Wet samplers: samplers that open (and close) during each precipitation event.

    (i) Cumulative: samples of various precipitation events are accumulated over some time period (weekly to monthly).

    (ii) Event: each rainfall event is collected in a separate container or the sample container is changed after each event.

    (iii) Sequential: individual samples are taken during a precipitation event. The sampling interval is either constant time or constant volume.

(c) Wet and dry samplers: a sample (cumulative or event) of precipitation and one of non-precipitation are obtained. These samples are normally obtained by alternately covering one container or another; the cover lid is actuated by a rain sensor. Sometimes the 'dry' collector will have some distilled water in it.

Generally sampler design depends upon objectives of study, cost and location (power requirements, access). Ship collectors are often manual versions of automated field collectors because personnel are available. Plastic scoops [15] have been designed and mounted to aircraft for precipitation sampling.

Snow collection presents a special problem for precipitation analysis. Relatively large area collectors are required to obtain enough melt water for analysis. In addition, standard precipitation collectors are inefficient devices for snow; and snow or freezing rain may immobilize a collector if it is not designed properly. Shields have been used to increase efficiency of snow and rain collection. Alternatively large surfaces (plastic sheet beds or garbage cans) have been employed manually on an event basis. In addition, precipitation sensors may not function well for snow whereas they are quite suitable for rainfall.

Other studies have sampled and analysed the fallen snowpack rather than snow fall, but these studies do not necessarily assess changes in chemistry due to post-depositional processes. Generally the more reliable results appear to come from 'bed frame' plastic sheet samplers operated manually on an event basis.

Bulk samplers are generally simple devices ranging from 'funnel and bottle' to cylindrical 'garbage can'. For special studies requiring large volumes, large diameter funnels or inclined troughs with large collection areas are readily constructed. Bulk samplers are quite suitable when samplers can be manned on an event basis. The sampler is then 'opened' just before a precipitation event. For this procedure, it is convenient to have a supply of distilled water available to wash the unit out just prior to collection; it is also worthwhile to submit some of the washings for analysis.

Most bulk samplers have a construction so that evaporation is minimal if the sample must stay in the sampler over a period of time.

Bulk collectors are convenient to use because they can be constructed at low cost and located with minimal additional requirements. They may also reflect total precipitation and dry deposition. But, in many cases, they may seriously give overestimates due to contamination from soil entrainment and bird/insect activities. Where bulk collectors are contemplated for cumulative measurements, they should only be considered as a means for preliminary estimate of commonly occurring chemical substances or as an indicator of exotic substances. Otherwise they can be used on an event basis if proper handling is carried out.

Wet and dry samplers consist of a movable cover over the collector, a sensor switch to activate the cover mechanism, and a power supply. The cover is typically a lid which either slides or is pivoted off the sampler with the use of arms. For electrical power supplies, the lid is moved by a motor. Either the lid or arms are spring loaded so that a tight seal is obtained on the container in order to avoid dust contamination during dry periods.

The sensor switch is best constructed of two parallel plates of clean stainless steel separated by a distance equal to about one-half the average diameter of a raindrop (the spacers need to be changed if used for snow). The outer plate should be slotted. The specific conductance of some 10 μS cm^{-1} of rain closes the circuit activating the motor and lid. If ample electrical power is available, a small tape-type heater can be used under the lower contact plate to evaporate raindrops, thus enabling the sampler to close shortly after precipitation ceases. It is often common to use two or more sensor switches in parallel so that the probability of actuation is very high for the first raindrop. Sensors are generally extended out from sampling instruments, and the sensor surface is slightly inclined.

Electric motors are commonly used to move lids but other devices can also be employed. Compressed air 'motors' can be used for remote installations. Alternatively springs can be employed. In one circular 'lazy susan' mode, multiple sampler containers are deployed on a spring-wound table [16]. In this case the actual weight of the sample acts as a mechanical switch to advance the table one notch to a new container. For event remote samplers, lids may be spring loaded, and the rain sensor device can activate a solenoid to release the lid. Alternatively polyvinyl water soluble paper [17, 18] can substitute for sensor and solenoid. In the two latter examples, the collector has to be serviced after each rain event.

It is important that samplers be designed to eliminate contamination from moving parts or from splash. Lids may accumulate dust during dry periods, and initial precipitation may wash the dust into the collector during opening. The incorporation of flat surface in the sampler design can lead to serious contamination. These surfaces act as areas for dust accumulation and container contamination due to splash when the sampler is open.

De Pena *et al.* [19] have carried out a comprehensive intercomparison study of three precipitation samplers. Their design of sampler assessment can be extended to other samplers.

Samplers for organic chemicals incorporate additional features in order to concentrate the sample and to fix the sample to avoid volatilization and reaction. For example, Strachan and Huneault [20] used a Teflon coated square funnel (46 cm × 46 cm) and a XAD-2 or 7 resin column to sample trace organics. Kawamura and Kaplan [18], however, used sections of galvanized metal cans to obtain manually operated event samples which were immediately processed.

12.2.2.1   *Container material*

Material in contact with sampler will vary depending upon substances analysed. Typically quality linear polyethylene, polypropylene or Teflon are used for common ions and trace metals whereas Pyrex glass or stainless steel are used for sample collection of organics (see Table 12.1). All material should be checked for contamination by carrying out experiments in the field with blanks.

Table 12.1   Containers, preservatives and storage times for different chemical parameters [22]

Parameter	Container[a]	Preservative[b]	Maximum holding time[c]
Inorganic tests:			
Acidity	P	Cool, 4° C	14 days
Alkalinity	P	Cool, 4° C	14 days
Ammonia	P,G	Cool, 4° C H_2SO_4 to pH < 2	28 days
Bromide	P	None required	28 days
Calcium	P,G	Cool, 4° C	28 days
Chemical oxygen demand	P,G	Cool, 4° C H_2SO_4 to pH < 2	28 days
Chloride	P	None required	28 days
Chlorine, total	P,G	None required	Analyse immediately
Chromium VI	P	Cool, 4° C	24 h
Colour	P,G	Cool, 4° C	48 h
Cyanide, total and amendable to chlorination	P,G	Cool, 4° C NaOH to pH > 12 0.6 g ascorbic acid[f]	14 days[i]
Fluoride	P	None required	28 days
Hardness (calcium, magnesium)	P	Cool, 4° C	28 days
Hydrogen ion (pH)	P	None required	Analyse immediately
Magnesium	P,G	Cool, 4° C	28 days
Mercury		In collector HNO_3 to pH < 2	28 days
Metals, heavy[d] except chromium	P	HNO_3 to pH < 2	6 months
Nitrate	P,G	Cool, 4° C	48 h
Nitrate–nitrite	P,G	Cool, 4° C H_2SO_4 to pH < 2	28 days
Nitrite	P,G	Cool, 4° C	48 h
Nitrogen, Kjeldahl and organic	P,G	H_2SO_4 to pH < 2	28 days
Organic carbon	P,G	Cool, 4° C HCl or H_2SO_4 to pH < 2	28 days
Orthophosphate	P,G	Filter immediately. Cool, 4° C	48 h

Table 12.1   *Continued.*

Parameter	Container[a]	Preservative[b]	Maximum holding time[c]
Oxygen, dissolved probe	G bottle and top	None required	Analyse immediately
Phenols	G only	Cool, 4° C H₂SO₄ to pH < 2	28 days
Phosphorus (elemental)	G	Cool, 4° C	48 h
Phosphorus, total	P,G	Cool, 4° C H₂SO₄ to pH < 2	28 days
Potassium	P	Cool, 4° C	28 days
Residue, total	P,G	Cool, 4° C	7 days
Residue, filterable	P,G	Cool, 4° C	7 days
Residue, non-filterable (TSS)	P,G	Cool, 4° C	7 days
Residue, volatile	P,G	Cool, 4° C	7 days
Silica	P	Cool, 4° C	28 days
Sodium	P	Cool, 4° C	28 days
Specific conductance	P	Cool, 4° C	7 days
Sulphate	P	Cool, 4° C	28 days
Sulphide	P,G	Cool, 4° C, add zinc acetate plus sodium hydroxide to pH > 9	7 days
Sulphite	P,G	Cool, 4° C	Analyse immediately
Surfactants	P,G	Cool, 4° C	48 h
Temperature	P,G	None required	Analyse immediately
Turbidity	P,G	Cool, 4° C	48 h
Organic tests [e,k]			
Acrolein[j] and acrylonitrile	G, Teflon-lined septum	Cool, 4° C 0.008% Na₂S₂O₃[f] Adjust pH to 4–5[11]	14 days
Benzidines	G, Teflon-lined cap	Cool, 4° C 0.008% Na₂S₂O₃[f]	[l]
Chlorinated hydrocarbons	G, Teflon-lined cap	Cool, 4° C	[l]
Haloethers	G, Teflon-lined cap	Cool, 4° C 0.008% Na₂S₂O₃[f]	[l]
Nitroaromatics and isophorone	G, Teflon-lined cap	Cool, 4° C	[l]
Nitrosamines[g]	G, Teflon-lined cap	Cool, 4° C Store in dark 0.008% Na₂S₂O₃[f]	[l]
PCBs	G, Teflon-lined cap	Cool, 4° C[h] pH 5–9	[l]
Phenols	G, Teflon-lined cap	Cool, 4° C 0.008% Na₂S₂O₃[f]	[l]

Table 12.1 *Continued.*

Parameter	Container[a]	Preservative[b]	Maximum holding time[c]
Phthalate esters	G, Teflon-lined cap	Cool, 4° C	[l]
Polynuclear aromatic hydrocarbons	G, Teflon-lined cap	Cool, 4° C 0.008% $Na_2S_2O_3$[f] Store in dark	[l]
Purgeable aromatics	G, Teflon-lined septum	Cool, 4° C 0.008% $Na_2S_2O_3$[f] HCl to pH < 2	14 days
Purgeable halocarbons	G, Teflon-lined septum	Cool, 4° C 0.008% $Na_2S_2O_3$[f]	14 days
TCDD	G, Teflon-lined cap	Cool, 4° C 0.008% $Na_2S_2O_3$[f]	[l]
Pesticides tests			
Pesticides	G, Teflon-lined cap	Cool, 4° C pH 5–9	[l]
Radiological tests			
Alpha, beta and radium	P,G	HNO_3 to pH < 2	6 months

[a] Polyethylene (P), or Pyrex glass (G) or equivalent.
[b] Sample preservation should be performed immediately upon sample collection. For composite samples, each aliquot should be preserved at the time of collection. When use of an automated sampler makes it impossible to preserve each aliquot, then samples may be preserved by maintaining at 4° C until compositing and sample splitting is completed.
[c] Samples should be analysed as soon as possible after collection. The times listed are the maximum times that samples may be held before analysis and still considered valid.
[d] Samples should be filtered immediately on site before adding preservative for dissolved metals.
[e] Guidance applies to samples to be analysed by GC, LC or GC/MS for specific compounds.
[f] Should only be used in the presence of residual chlorine.
[g] For the analysis of diphenylnitrosamine, add 0.008% $Na_2S_2O_3$[f] and adjust pH to 7–10 with NaOH within 24 h of sampling.
[h] The pH adjustment may be performed upon receipt at the laboratory and may be omitted if the samples are extracted within 72 h of collection. For the analysis of aldrin, add 0.008% $Na_2S_2O_3$.
[i] Maximum holding time is 24 h when sulphide is present.
[j] Samples receiving no pH adjustment must be analysed within 7 days of sampling.
[k] Samples for acrolein receiving no pH adjustment must be analysed within 3 days of sampling.
[l] 7 days until extraction, 40 days after extraction.

12.2.2.2 *Sample preservatives*

Generally no preservative is used for common ions, although phosphorus, ammonia and nitrate may react. Hakkarinen [21] found that accumulated wet-only samples had higher pH and higher Ca and Mg than individual event samples representing the same sample period. This is either due to contamination or the result of slow neutralization of acid by mineral particles in the sample.

Sampling for trace metals necessitates the prior addition of high purity HNO_3 acid so that the final sample pH is about 1.

Samples obtained for organic analysis are often filtered through a clean 0.4 μm filter to eliminate biological activity. Alternatively they may be fixed with $HgCl_2$.

Sample container and preservation vary with the substances analysed. Table 12.1 [22–24] summarizes containers and preservatives for some commonly analysed substances.

12.2.3 Field procedures

It is critical that precise field procedures with quality control be carried out for the field aspect of a precipitation chemistry programme. These procedures are especially important when using personnel with a varying range of knowledge and experience. Simplicity, precise instructions, redundancy of measurement, and quality control are the four important aspects of a satisfactory field operation.

There should be a minimum number of field operations, and the operations should be written out in detail and preferably enclosed with each container shipment. Sampling containers should not be stored at field facilities. If possible, in order to avoid unknown effects at field stations, they should be shipped in time to reach the site only a few days before use. They are best sealed, with instructions to break the seal only at time of change. A sample report form is desirable which also requests information about the site (e.g. construction), sampler operation, and variations from standard procedure. Simple yes/no answers are best. Field analytical measurements should be minimal. Generally amount of precipitation, kind of precipitation, and possibly pH and specific conductance are measured by field personnel at the site. If pH and specific conductance are measured, solutions and electrodes should be systematically rotated, and reference solutions and test solutions are best shipped and replaced with each shipment. Also other chemicals, such as distilled water, along with beakers, etc., should be replaced each time. Thus the responsibility for sampling, calibration and quality control rests with the laboratory.

Measurement of pH, specific conductance and amount of precipitation can be checked against lab pH, lab specific conductance and container volume respectively. If the pH is near 4 but changes from field to laboratory, there should also be a change in specific conductance due to the large equivalent conductance of H^+ ion. Container volume if markedly different from precipitation gauge amount generally indicates misfunction of sampler or mishandling of sample container.

There are some simple and worthwhile experiments to check field operations. Sample containers can be processed throughout the complete field operation with a blank or known concentration. The sampler would not be operated for this experiment. An operational mode experiment would consider addition of an exotic tracer known not to be common in precipitation. Other experiments would involve rinsing of the container with distilled water in the field followed by collection of a second rinse for analysis. It is recommended that 10–15% of analyses be of a quality control type.

It is important that all sample containers and apparatus be numbered so that the history of use can be ascertained. This is particularly important for networks which exhibit a wide range in amount and type of parameters.

Field programmes become more difficult in the sequence common major ions, trace metals, and trace organic chemicals. The number of field sites should decrease with increasing difficulty of estimate. It is most desirable to use only qualified laboratory personnel in the estimate of trace metals and trace organics. The requirements in quality control are very stringent, and for acceptable analyses, it is best to have a small laboratory at each site to carry out spiking, blank analysis, preservation, and special handling. This procedure results in a decrease in the number of analyses but an increase in the confidence of the final results.

Details of field procedures will vary depending upon the programme purpose and master variables to be measured. The quality control programme for the field operations of the US National Atmospheric Deposition Program (NADP) [25] and the respective field manual [26] develops specific points relating field operations to laboratory analysis. Headings consider programme purpose, site selection, types of samples, observer's instructions, completion of field forms, routine maintenance, and quality assurance tests. There are also appendices relating to each piece of equipment.

12.3 Analysis

The following considers analytical procedures for major ions, trace metals, organics, and potentiometric titrations. Major ions are considered in the most detail with special emphasis on protolytes and pH.

12.3.1 Filtration

Samples may be filtered prior to analysis. All precipitation samples may be assumed to contain particulates. These particulates may be very important for ionic concentrations especially Ca^{2+} ion and H^+ ion (alkalinity), trace metal concentrations, and substrates for non-polar organic concentrations. On the one hand samples may be filtered so that chemical principles relating to a single phase can be applied (e.g. homogeneity, ion balance), or to better define a reactive phase (e.g. trace metal), or to determine distribution coefficients between solids and aqueous phases, or to determine the chemical and physical nature of the solid phase; on the other hand, filtration can introduce complications due to contamination especially with respect to trace substances or due to the filter and accumulated particles acting as a chemical substrate for removal of reactive substances. Filter paper, filtration equipment, containers and tubing must be investigated for possible contamination and surface effects. The best approach, if feasible, is to use a labelled substance. Flexible tubing is known to absorb trace metals, sulphate, and sulphur dioxide; fritted glass filter bases adsorb trace metals and organics; it is common to use Teflon filtering material and coarse mesh plastic filtering bases for all but organic substances [27]. It is common to acid wash/distilled water rinse all filters prior to use for inorganic substances. Selected samples of cleansed filters should be checked for blanks by analysing distilled water filtered through the medium.

Glass filter units and glass fibre filters are often used for analysis of organics. Filtering apparatus is cleansed in chromic acid, and glass fibre filters are baked at about 600° C for 24 h to remove impurities. Glass and stainless steel tubing are used for connections.

Filtration is often carried out at the analytical laboratory. If filtration is to be carried out in the field, individual sealed package units should be supplied by the central laboratory for each sample. The packaging should be preassembled so as to require minimum handling.

12.3.2   Major ions

There are now many precipitation chemistry programmes which analyse for major ions. This group of ions (with concentrations ranges) generally consists of, H^+ (~ 0–1000 μeq l^{-1}), Ca^{2+} (~ 0–300 μeq l^{-1}), Mg^{2+} (~ 0–100 μeq l^{-1}), Na^+ (~ 0–50 μeq l^{-1}), K^+ (~ 0–50 μeq l^{-1}), NH_4^+ (~ 0–100 μeq l^{-1}), SO_4^{2-} (~ 0–300 μeq l^{-1}), NO_3^- (~ 0–40 μeq l^{-1}) and Cl^- (~ 0–500 μeq l^{-1}). In addition, it is worthwhile to analyse for specific conductance, k (~ 10–300 $\mu S/cm$).

Typical analyses give results in the low 10 s μeq l^{-1} with H^+, NH_4^+ and Ca^{2+} ions often being the dominant cations and SO_4^{2-}, and NO_3^- ions often being the dominant anions. A sample with high values of Ca^{2+} alkalinity (HCO_3^-) and high pH (>5) generally is indicative of dust contamination, whereas elevations of concentrations of Na^+, K^+, Ca^{2+}, Mg^{2+}, Cl^- and SO_4^{2-} indicate a marine source. Concentration data for samples are often adjusted assuming that Cl^- ion comes entirely from sea-spray and that the composition of sea-spray is the same as sea-water. Thus the following factors are multiplied by Cl^- ion concentrations (μeq l^{-1}) to determine the sea-spray content. The amount of ion (μeq l^{-1}) in precipitation due to sea-spray, assuming constant sea-water composition, is as follows (multiply by Cl ion concentration (μeq l^{-1}) – assumed to be from sea only): Na^+, 0.858; K^+, 0.019; Ca^{2+}, 0.038; Mg^{2+}, 0.193; SO_4^{2-}, 0.104; HCO_3^-,* 0.004.

Nitrate and ammonia are assumed to be negligible in sea-water. Other ions, especially Sr^{2+}, Br^-, F^-, and B may be included in the sea-salt factor. Various studies have confirmed and negated [28–30] the assumption of constant sea-water composition for precipitation studies.

Chloride ion is often assumed to be solely of marine origin, but it can originate from other sources such as coal combustion and industrial emissions. In this case either sodium or magnesium ion concentrations may be assumed to reflect mostly a marine source. Since chloride, sodium and magnesium are the three major constituents of sea-water, it is desirable to always assess the $[Na^+]/[Cl^-]$ and $[Mg^{2+}]/[Cl^-]$ ratios relative to the above values.

There are many methods for the analysis of major ions in precipitation

---

* Subtract from H^+ if pH <5.

[36–40]. Typically flame photometric, flame/flameless photometric, colorimetric, potentiometric, ion chromatographic and titrimetric techniques are used. Thus atomic absorption spectrometers, colorimeters (automated), pH meters, ion chromatographs and titration apparatus are equipment required.

Details of methods of analysis can be found in many publications. Reference books to automated colorimetric techniques [31], ion chromatography [32], and specific ion electrodes [31, 33] should be available in most laboratories. Ion chromatography, with use of the single column (plus guard column) has rapidly replaced colorimetric methods for anions (SO_4^{2-}, (SO_2 aqueous), NO_3^-, Cl^-) because of the general lack of interferences in precipitation samples, good detection limits and sensitivities, instrument ruggedness, and simplicity of operation permitting use even in field environments. Short columns allow complete analysis in 5 min, and coupled to a microprocessor could be used in a continuous in situ analysis mode. Detection limits for Cl^-, SO_4^{2-}, and NO_3^- using a 100 μl sample are typically 0.1, 1 and 0.2 $\mu eq\ l^{-1}$.

12.3.2.1 *pH and other protolytes*

H^+ ion concentration is often the major contributor of positive ions to a precipitation sample. Therefore it is important that it be measured carefully and precisely. Unfortunately the pH scale tends to suppress the important difference of a small error in reading at pH 4 compared with a pH of 6. For example, an error of 0.03 units at pH 4 is about 7 $\mu eq\ l^{-1}$ whereas the same error at pH 6 is 0.1 $\mu eq\ l^{-1}$. In order to avoid this pitfall one should use concentration units ($\mu eq\ l^{-1}$) routinely.

pH is measured electrometrically using a good quality research meter readable to 0.001 pH units or 0.1 mV; generally a quality combination electrode is employed. Due to the low ionic strength of the precipitation sample, junction potentials and streaming potentials may give rise to large errors. Streaming potential can be held constant by always stirring at a constant rate when taking measurements; junction potentials can be held constant by careful selection of (reference) electrode, by adding a neutral salt to increase the ionic strength, by decreasing the concentration of KCl in the reference electrode, and by a constant stirring rate [34].

A series of experiments are required on a planned basis to check out the pH meter and electrode system.

The pH meter can be routinely checked by:

(a) Shorting out the electrode input and checking the EMF (0 mV) or pH (7). Using the pH mode or relative EMF mode one should be able to zero the shorted input.
(b) Apply a known potential of about 60 mV. At 25° C, the pH should increase/decrease by about 1 unit or by the equivalent voltage change.
(c) An electrode in a pH buffer of 7 should be adjustable to 0 mV or a pH 7 reading. Inability to obtain the correct isopotential reading, may be due to an electrode fault. Try a second electrode.

Electrodes may be checked by:

(a) Confirming readings in buffer solutions of pH 4 and 7.
(b) Titrating a weak NaCl solution (0.0005 N) with acid titrant or base titrant. The base titration should be carried out in a closed system or under an inert gas (CO_2-scrubbed argon or nitrogen) blanket to avoid atmospheric CO_2 contamination. This titration not only checks the electrode response but also allows determination of acid/base impurities in the salt and distilled water, and it allows overall evaluation of the system response.

The following is recommended for electrode calibration, as well as routine H^+ ion concentration determination.

Use a sample of about 100 ml, a microburette of 2 ml capacity and standardized hydrochloric acid (against Na_2CO_3) of about 0.1 N. For titration, first zero the acid reading; then draw the acid into the burette tip. Measure the apparent pH while stirring; add about 1 ml of 0.05 N NaCl and measure the pH again. Add a small aliquot of acid (concentration C_a eq l^{-1} of a volume v_a l) and record the apparent pH (pH′) or EMF (E). Note the time required to obtain a stable reading. Continue addition of acid to an apparent pH of 3–3.5. The electrode should respond in a Nernstian fashion

$$pH' \text{ or } E = a + b \log [H^+] \tag{12.1a}$$

where $[H^+]$ is the H^+ ion concentration in a constant ionic medium, and a and b are coefficients.

Substituting in measurements

$$pH' \text{ or } E = a + b \log \left[\frac{v_a C_a}{V + v_a} \right] \tag{12.1b}$$

where V is the sample volume. a and b can be determined by least squares analysis. Equation (12.1b) can be used to check the response of the electrode as well as to calibrate the electrode. Similarly a base titration can be used for experiments at higher pHs.

Any effect of the salt (plus distilled water) can be assessed by a Gran analysis of the blank titration [35]. For pHs less than about 5, $F_1 = (V + v_a) 10^{-pH'}$ (or $(V + v_a) 10^{-E}$) is linear with v_a so that

$$F_1 = c + d v_a \tag{12.2a}$$

and v_e, the equivalence point volume, is

$$v_e = -c/d \tag{12.2b}$$

and the resultant acid (negative) or base (positive) of the blank (A) (eq l^{-1}) is

$$A = \frac{v_e C_a}{V} \tag{12.2c}$$

where c and d are linear least squares coefficients. The value of A should be less than ± 2 μeq l^{-1} for the blank (NaCl) titration. Note that in the above analysis EMF may be used throughout.

Many precipitation samples have apparent pHs less than 5. In this range, H^+ ion would normally be the only important protolyte. Thus the blank procedure outlined above can be used on a precipitation sample to determine the H^+ ion concentration. This value, in turn, can be compared with a direct estimate. In essence this is the same as a standard addition technique. The resultant A will be negative and from

$$A \simeq -[H^+] \qquad (12.3a)$$

$$pH \simeq -\log A \qquad (12.3b)$$

Thus one may use a number of data points to determine the pH.

Electrodes more often than not are the major source of problems in potentiometric measurements; and the reference electrode often gives the greatest difficulty. Electrodes are best maintained by storing them in solutions similar to the solutions to be measured; they often are best maintained while being stirred. More than one electrode is advisable when measurements are being made over a wide range of H^+ ion concentrations.

H^+ ion is often the major protolyte in precipitation samples, but there are also H_2CO_3–HCO_3^-–CO_3^{2-} and NH_3–NH_4^+ systems. Seymour (discussed in Kramer [35]) has developed the titration functions to determine strong acid $[H^+]$, total carbonate ($[H_2CO_3]+[HCO_3^-]+[CO_3^{2-}]$) and total ammonia ($[NH_4^+]+[NH_3]$). Titration using KOH is carried out for an acidic sample in a closed system at constant ionic strength (KCl). Four linear interdependent functions are derived for the equivalence points; volumes and values are obtained in an iterative fashion.

12.3.2.2 *Calcium, magnesium, potassium and sodium*

Calcium, magnesium, potassium and sodium are almost universally analysed by flame atomic absorption (FAA) analysis. Standard references can be referred to for details. Calibration curves (including a blank) are determined for the cations, and the concentration of the sample is determined from the calibration curve. Either a linear calibration curve or a slightly curved modification [41] is fitted to the data.

It is important to suppress interferences in the FAA analysis. This is achieved by adding a 'releasing agent' to obtain a uniform matrix. Lanthanum chloride is often added for calcium and magnesium analyses, and caesium chloride is added for sodium and potassium analyses.

Normally FAA analyses are carried out on samples that are acidified to a pH of about 2.

12.3.2.3 *Ammonia*

Ammonia is normally analysed colorimetrically but it can be determined using a specific ion electrode [42] with standard addition. If a complete base titration is carried out as outlined in Section 12.3.2.1, that result can be compared to those from more conventional methods.

The automated phenate colorimetric method is commonly used for the low levels (~ 0–100 μeq l^{-1}) [43]. A blue coloured complex is measured from the reaction of alkaline phenol and hypochlorite with ammonia. Tartrate and citrate are added to suppress metal oxy-hydroxide formation. The method performs suitably except in the rare case of coloured samples. A blank solution including all reagents should routinely be run to check for possible contamination of reagents.

The ammonia specific ion electrode consists of a pH electrode system and a gas permeable membrane. The solution is adjusted to a pH of 11+ so that the ammonia exists as NH_3. NH_3 equilibrates between sample and inner solution, and NH_3 in the inner solution equilibrates with water to form NH_4^+ and OH^-, and the change in pH is read. The minimum concentration level for analysis is about 10 μeq l^{-1}; the major problem in analysis is the equilibration across the membrane; generally a standard addition technique is recommended; the method, however, has been applied to a number of aqueous systems [44]. The value of the ammonia electrode is that it can be used in the field in situ. This would be valuable for cases where oxidants may alter the ammonia concentration.

12.3.2.4   *Sulphate*

Sulphate can be determined colorimetrically using methyl-thymol blue [45], by titration with thorin [38, 46], or by ion chromatography [47]. All methods have been used, with ion chromatography being the method presently preferred. The automated methyl-thymol blue (MTB) method has been used more commonly in the past perhaps because it can be automated. However, this method requires adjusting the sample to high pH for colour development of Ba-MTB where organics may react to interfere with the signal. The MTB method relies upon the reaction of and colour change of excess barium ions with MTB after first forming $BaSO_4$; therefore the concentration matching of sample SO_4^{2-} ion and Ba^{2+} reagent is important for maximum sensitivity. The MTB method should be satisfactory for most precipitation samples. However, the method is best automated as the colour development is not stable in the presence of air.

The thorin method depends upon the change in colour of thorin and the Ba-thorin complex. Barium is titrated in slightly acidic medium to a sample which has been passed through a cation exchange column. The colour change is best determined electronically through 20–50 mm of solution using a filter and photocell arrangement.

The ion chromatographic method is rapid, requires a small sample, and is apparently free of analytical interferences. Single column ion chromatographs perform as well as double column ion chromatographs. Units can be constructed from available commercial parts which are totally suitable for analysis [48] (see also nitrate and chloride). No sample preparation is needed. A sample of 100 μl gives a detection limit of 2 μeq l^{-1}.

12.3.2.5   *Nitrate*

Nitrate can be analysed colorimetrically, by ion chromatography, and by specific ion electrode.

The colorimetric method can be implemented manually [49] or in an automated flow through version [36, 50]. The technique actually determines nitrate and nitrite ions; nitrite ions are assumed to be negligible. The procedure consists of reducing nitrate ions to nitrite ions using a copper–cadmium reduction column which reacts with a dye solution to produce a reddish-purple colour. The procedure is relatively free of interferences.

The ion chromatographic method can be carried out for a single column or double column instrument using an anion column. No sample preparation is required. A sample of 100 μl gives a detection limit of about 1 μeq l^{-1}. The method yields discrete signals for chloride, nitrate and sulphate [51].

The nitrate electrode is a liquid-membrane probe using an organophilic membrane [52]. The detection level, however, for Nerstian linearity is only about 10^{-4} M, making it suitable for samples with elevated nitrate concentrations. Some anions will cause interferences or excessive drift due to poisoning of the membrane.

12.3.2.6 *Chloride*

Chloride can be measured colorimetrically, by ion chromatography, titrimetrically, and by specific ion electrode.

The common colorimetric method uses the reaction of mercuric thiocyanate with chloride to liberate thiocyanate which is measured as ferric thiocyanate. The detection limit is about 2 μeq l^{-1}.

Chloride is readily determined using a specific ion electrode. The chloride ion electrode consists of a silver chloride–silver sulphide solid membrane which has a detection limit of about 50 μeq l^{-1} [31, 33]. A mercuric sulphide–mercurous chloride electrode has a detection limit of 0.5 μeq l^{-1} [53]. Other halides, Br and I will be measured as chloride; the electrodes can be poisoned by precipitation of insoluble salts (sulphides, oxides) on the surfaces. However, they can be used in a continuous analysis mode.

The titrimetric method uses a Ag–HgCl electrode (made by electroplating a piece of Hg wire in KCl) in an acetic acid buffer solution; silver is the titrant [31]; the equivalence point may be determined by a Grans plot [31, 33] (see Section 12.3.2.1).

Chloride is readily determined by ion chromatography along with nitrate and sulphate. No sample preparation is necessary. A 100 μl sample gives a detection limit of about 0.1 μeq l^{-1}.

12.3.2.7 *Specific conductance*

Specific conductance is measured using a conductance cell connected to a high frequency AC bridge circuit. Values measured are typically less than 100 μS cm^{-1} at 25° C.

The conductivity meter is calibrated against KCl solutions. At 25° C, the relation between specific conductance (k) of KCl (μS cm^{-1}) and concentration (C, molar) for dilute solution (<0.002 M) is given by

$$k_{\text{KCl},25°} = 0.367 + 1.467 \times 10^5 C$$

Major factors in specific conductance measurements are contamination of electrode surface and temperature.

Since specific conductance is a relatively simple measurement, it is worthwhile to calibrate electrodes frequently, and for laboratory measurements, estimates at constant temperature using a water bath for control are desirable since there is a relatively large temperature effect. Otherwise the temperature should be measured and the solution can be corrected to 25° C by comparison with the value for KCl at the ambient temperature.

$$k(\mu S \text{ cm}^{-1}, 25° \text{ C}) = \frac{k_{KCl, 25} k_t}{k_{KCl}}$$

where $k_{KCl, 25}$, k_t and k_{KCl} are the specific conductances of a known KCl solution at 25° C (e.g. for 0.0005 M KCl, $k_{KCl, 25} = 73.9 \ \mu S \text{ cm}^{-1}$) for sample and reference KCl solutions at temperature t.

The above equation should not be used to extrapolate temperatures more than 5° C since the temperature coefficient of specific ions vary somewhat thus making the temperature effect of a sample dependent upon its composition [31, 54].

12.3.2.8   Consistency checks for major ions

Consistency calculations for major ions consist of ion balance, comparison of calculated and measured specific conductance, and comparison of measured pH and pH from Gran titration. These checks should be routinely applied to analyses of major ions.

The ion balance check is generally reported as the ratio (R) of the sum of cations (eq l^{-1}) to the sum of anions (eq l^{-1}) or

$$R = \frac{\Sigma[+]}{\Sigma[-]}$$

Data are generally questioned if $0.85 < R > 1.15$, and data are considered consistent for $0.95 < R > 1.05$. As the total ion concentration decreases, one should expect R to deviate increasingly from one.

The value of R depends upon the assumption of a homogeneous aqueous phase, the completeness of inclusion of major ions, and the analytical error. Thus calculation and assessment of R on unfiltered solutions can be deceptive due to the partial incorporation of solid phase in some of the major ion analyses. The major ions making up precipitation are usually quite constant, but consistent measurement of values less than or greater than unity may be indicative of another important ion, or a consistent bias in analysis. Generally the ion ratio, R, however, is used to indicate suspect analyses. It is worthwhile to determine if this flag coincides with the specific conductance check.

Calculation of specific conductance is made using concentration of individual major ions. The calculation is based upon the Kohlrausch rule which states that the equivalent conductance of a solution at infinite dilution is equal to the sum of the equivalent conductances of individual ions. Assuming this rule holds at low concentrations, and using the definitions of equivalent conductance and specific conductance

$$k_c(25° \text{ C}) = \Sigma \lambda_i C_i$$

where k_c (25° C) is the specific conductance in μS cm^{-1}, λ_i is the equivalent conductance adjusted for units and C_i is the ion concentration (μeq l^{-1}) for the ith ion.

Values of λ_i for the specified units are [55]:

	λ_i
H$^+$	0.3498
Na$^+$	0.0501 1
K$^+$	0.0735 2
Ca^{2+}	0.0595
Mg^{2+}	0.0530 6
Positive alkalinity (HCO$_3^-$)	0.0445
SO$_4^{2-}$	0.0800
Cl$^-$	0.0763 5
NH$_4^+$	0.0734
NO$_3^-$	0.0714

The equivalence conductance of H$^+$ ion is many times greater than that of other ions. Hence the measurement of H$^+$ ion will have a great influence on the calculation of specific conductance for most precipitation samples. Normally alkalinity will be negative for precipitation samples, and in this case, the absolute value of alkalinity can be used for the H$^+$ ion concentration.

pH is a master variable for most precipitation chemistry studies. Thus it is important that the H$^+$ ion concentration be measured precisely. Therefore it is worthwhile to always compare the direct measurement of pH to the Gran analysis of alkalinity (Section 12.3.2.1) when alkalinity is less than about -10 μeq l^{-1}. This comparison reveals the overall precision of the measurement and, perhaps more important, reveals consistent analytical bias.

12.3.3 Trace metals

Certain trace metals have been measured in precipitation because the atmosphere may often be the more important flux to the environment. Secondly due to the typical low pH of precipitation and general lack of chemical complexing agents, a large portion of a given trace metal may be chemically more reactive and potentially toxic than from other sources such as runoff.

Various estimates have been made for trends in the concentrations of trace metals in the atmosphere [4, 5, 56, 57]. Some of the metal transport may be due to natural processes, but attention is more often given to increases in concentrations due to anthropogenic processes. One estimate suggests Ag, As, Cd, Cu, Hg, Pb, Sb, Se and Zn are due mainly to anthropogenic emissions, but this list may change depending upon locale and better data.

Chemical speciation details of trace metals in precipitation are poorly known.

Often only total concentrations are estimated and sometimes filtered portions are included. In general, the filtered fraction increases as particulate concentration and pH decrease and at pHs less than 4 may account for almost all of the trace metal. It is important to remember that almost all natural particles are elevated in trace metals compared with precipitation. Complexing capacity techniques have been used to estimate the 'labile' form of trace metals in a few cases [58, 59], and gas chromatography–vapour atomic absorption techniques have been developed to measure certain metal–organic complexes [60, 61]. Various operationally defined chemical speciation methods [62] applied to surface waters could be applied to speciation of trace metals in precipitation.

Sampling and preservation of samples for trace metal analyses is important and difficult; therefore, analysis of trace metals should generally be considered separately from other studies. Sampling containers (see Table 12.1) are generally linear polyethylene or polypropylene. Many trace metals adsorb on the surface walls of the container unless an (acid) preservative is used. Volatile metals, especially mercury, require a special preservative, and for these metals, sampling for the volatile form as well as the aqueous form is desirable. The use of preservatives is designed for total analysis; thus the addition of a preservative does not permit speciation or filtration analysis. To separate portions of a trace metal, either individual sampling is required, or the sample must be processed (filtered, sequestered) before the preservative is added. Alternatively discrete samples may be taken for (INAA) analysis for total metal estimates where possible. Due to container reactivity, it is imperative to carry out sample processing in the field under carefully controlled conditions.

The possible programmes for trace metal analysis consist of:

(a) Total metal analysis. Samples taken with preservative specific to groups of metals, or a discrete sample is taken, and analysed. Volatile trace metals must be preserved in the field.

(b) Trace metal fraction. (i) A separate sample may be filtered, preferably immediately after collection, or (ii) a sample is reacted with a reagent column (exchange resin, complexing medium), preferably during sample collection.

Since trace metals occur at low concentration levels, it is very easy to introduce an artifact or contamination into the sampling, processing and analysis steps. It is important, therefore, to carry out experiments using blank, known solution, and spiked solutions. It is worthwhile to use radioisotopes in these studies.

Trace metals are generally estimated by flame (FAA) or furnace atomic absorption (GFAA). When facilities are available, instrumental neutron activation analysis (INAA) or proton-induced X-ray analysis (PIXE) offer sensitive non-destructive methods [63]. Table 12.2 lists detection limits for various elements by FAA, GFAA, PIXE, ICP and INAA techniques.

Inductively coupled plasma analysis (ICP) [31] is becoming more popular for the analysis of trace elements because of the ability to measure many parameters simultaneously and the lack of matrix problems. However, the detection limit is marginal for many of the trace elements of interest in precipitation (Table 12.2).

Table 12.2 Approximate detection limit[a] (DL) and sensitivity[b] (SE) for furnace atomic absorption (GFAA) [64], instrumental neutron activation (INAA) [65, 66], proton-induced X-ray emission (PIXE) [65, 66] and inductivity coupled plasma (ICP) [31] analyses. Units in $ng\ g^{-1}$ unless otherwise indicated

Element	GFAA DL	SE	INAA DL	PIXE DL	ICP DL
Aluminium	0.01	0.04	1	4	10
Antimony	0.08	0.08			
Arsenic	0.08	0.12	0.2	0.6	
Barium	0.04	0.04			
Beryllium	0.003	0.01			5
Cadmium	0.0002	0.002	18	2	
Calcium (mg l^{-1})	0.001	0.001	0.060	0.011	2
Chromium	0.004	0.04	5	2	8
Cobalt	0.008	0.08	1.5	0.5	6
Copper	0.005	0.04			6
Gallium	0.01	0.05			
Iron	0.01	0.03	500	0.9	50
Lead	0.007	0.04		0.5	50
Lithium	0.01	0.04			10
Magnesium (mg l^{-1})	10^{-6}	10^{-4}	0.08	0.05	0.1
Manganese	0.001	0.01	0.2	1	5
Mercury	0.2	0.4			
Molybdenum	0.03	0.1			5
Nickel	0.05	0.2	4.4	0.3	10
Osmium	2	3			
Palladium	0.05	0.2			
Potassium (mg l^{-1})	0.004	0.004	0.01		0.5
Silver	0.001	0.005			2
Sodium (mg l^{-1})	0.004	0.004	0.01	0.1	0.5
Strontium	0.01	0.02	10	0.2	5
Tin	0.03	0.07			20
Zinc	0.001	0.003	34	0.2	20

[a] Concentration of analyte that results in a signal twice the standard deviation of the baseline noise.
[b] Sensitivity is concentration (or mass) yielding 1% absorption (0.0044 absorbance units).

12.3.3.1 *Atomic absorption analysis*

Flame atomic absorption analysis after chelation with APDC (ammonium pyrolidine dithiocarbamate) and extraction with MIBK (methyl iso-butyl ketone) or a similar system has been the standard technique for measuring low concentrations of many of the trace metals [31]. Generally direct flame analysis is not recommended due to poor detection levels and variable matrix effects. Chelation and extraction result in low detection limits and a constant matrix for analysis. The technique is time consuming, however.

Furnace atomic absorption methods have become more popular in the past ten years because of need for a small sample, no chemical preparation, and low detection limits [67–69].

There are some major problems that arise with GFAA. Calibration curves will vary as the matrix changes. Therefore the method of standard additions is required for many analyses. Other substances (especially organics) may be thermally emitted, which changes the background. Some metals (e.g. Pb) may partially volatilize over a wide range of temperature unless matrix and furnace modifications are made. Small differences in sample positioning in the furnace may result in change in optical signals and gas type, gas flow, and graphite coatings greatly affect sensitivity [67–69]. To overcome these difficulties automated systems have been developed for detailed temperature programming of the furnace, auto-pipetting and injection, and provisions for standard addition and matrix manipulation techniques. It is highly probable that furnace atomic absorption analysis will be, with various modifications, the major analytical method for analysis of trace metals in precipitation.

Certain volatile elements, especially mercury, represent special cases in sampling/preservation and analysis. Mercury typically is present at less than $0.05\ \mu g\ l^{-1}$ concentration, requiring very special handling, preservation and analysis. High purity (sub-boiling distillation) HNO_3 acid is maintained in the sampler; linear polyethylene is a suitable container material, and cold vapour atomic absorption analysis using a long absorption cell (~ 50 mm) is required. The analysis for total mercury consists of oxidation of organomercury compounds with sulphuric acid, potassium permanganate and potassium persulphate. The oxidized mercury is then reduced to elemental mercury with stannous sulphate in a hydroxylamine solution. The elemental mercury is separated from solution with air into the optical cell [31]. Alternatively using similar sample preparation, the mercury is concentrated on to a (gold) substrate and then liberated into an optical cell [70, 71]. Both techniques are usually automated.

Arsenic is also measured as a gas, arsine, but in an optical tube furnace at $800°$ C. Organic arsenic compounds are decomposed by oxidation using a sulphuric acid/persulphate solution. The arsenic is reduced to arsine using sodium borohydroxide, and the arsine is stripped from the solution with nitrogen and decomposed to arsenic in the optical path furnace. The detection limit of $1\ \mu g\ l^{-1}$ is obtained for a 100 mm optical path [72].

12.3.3.2   *Instrumental neutron activation analyses*

Instrumental neutron activation analysis (INAA) offers an excellent analytical tool for many trace elements and major parameters (Table 12.2 [73]). This technique may not be employed universally due to the need for a reactor, but selected samples in any trace metal study should be submitted for INAA comparison if the INAA technique is suitable for the elements in question. Thus the INAA technique should be considered a primary reference method. When INAA techniques are not suitable (e.g. Fe, Ni, Cd, Pb) proton-induced X-ray emission (PIXE) may be used if a source of protons is available.

12.3.4   Organics

Many organic compounds are found in the atmosphere due to their volatility; they are generally adsorbed on particulates and they are scavenged by rainfall [74–76]. This may be a major pathway of trace organics to water-bodies [7]. However, there may be as a result, a reverse emission from water/solid to the atmosphere as well. Thus the more volatile fractions represent an interesting case of transportation by air/water saltation.

Organics should be sampled independently of other parameters. Usually Teflon, glass or stainless steel are used for container linings. Organics are usually sequestered on to resins and later extracted [20, 77, 78]. They are then analysed by gas chromatography or gas chromatography–mass spectrometric techniques [77, 78].

Most analyses for organics have concentrated upon anthropogenic sources and toxic forms. The recent analyses of large amounts of di-methyl sulphide from the ocean surface [79] adds a new perspective to the sulphur cycle and raises interesting questions regarding processes and precursors.

12.3.5   Other analyses

Specific analyses may be carried out to better define sources and processes. Some discussion was given to trace metal speciation, for example. In addition, particle identification [80] and stable isotope studies [81–85] help define provenance and process. Particles can be analysed as to local endemic soil mineral material and exotic material and anthropogenic material by X-ray diffraction analysis and scanning and transmission electron microscopy. Electron microscopy (with elemental analytical capability) is often needed due to the small particles being X-ray amorphous.

Stable sulphur and oxygen isotopes are useful in separating marine sources and other sources. Sea-water has a constant sulphur and oxygen isotope value in the SO_4^{2-} ion; other values would be fixed due to the sulphur source, oxidation/reduction of sulphur and the oxygen isotopic value at the time of formation of the SO_4^{2} ion. For example, a simple study would be to compare the isotope values in SO_4^{2-} in precipitation to those in the water of a small headwater catchment to discern if there are other sources or chemical modifications of the atmospheric sulphate in the catchment.

12.4   Concluding comment

There are now large amounts of precipitation chemistry data particularly for major ions. Much of the data are probably questionable from a sampling and analytical perspective. Perhaps more important is, what do the data represent relative to precipitation? In designing and carrying out a precipitation monitoring system, one must always remember that the atmosphere–land–water interface is dynamic and constantly changing. Ground level sampling will reflect the terrain. Secondly, the elemental buffering intensity [86] of a particle is many times that of most aqueous solutions. Thus the multitude of mechanisms for concentration of particles in a precipitation sample near ground level will often have an important influence on the precipitation chemistry.

References

1. Mancy, K. H. and Allen, H. E. (1982) In *Examination of Water for Pollution Control* (ed. M. J. Suess), Pergamon Press, Oxford, ch. 1.
2. Campbell, S. A. (ed.) (1984) *Sampling and Analysis of Rain*, No. STP 823, American Society for Testing and Materials, Philadelphia, Pa.
3. Kramer, J. R. and Tessier, A. (1982) *Environ. Sci. Technol.*, **16**, 606A–15A.
4. Nriagu, J. (1979) *Nature*, **279**, 409–11.
5. Galloway, J. N., Thornton, S. D., Norton, S. A. *et al.* (1982) *Atmos. Environ.*, **16**, 1677–700.
6. Van Loon, J. C. and Beamish, R. J. (1977) *J. Fish Res. Bd Can.*, **34**, 899–906.
7. Eisenreich, S. J., Capel, P. D. and Loney, B. B. (1982) In *Physical Behavior of PCBs in the Great Lakes* (ed. D. Mackay), Ann Arbor Science, Michigan, pp. 181–212.
8. Acres Consulting Services Limited, Applied Earth Science Consultants Inc. (1975) Report for Canada Cen. for Inland Waters, *Atmospheric Loading of the Upper Great Lakes*, Vols 1–3, Burlington, Ontario.
9. Semonin, R. G. and Beadle, R. W. (1977) *Precipitation Scavenging*, ERDA Conf. 741003, National Technical Information Service, US Dept of Commerce, Springfield, Va 22161.
10. Sievering, H., Dave, M., Dolske, D. A. *et al.* (1979) *An Experimental Study of Lake Loading by Aerosol Transport and Dry Deposition in the Southern Lake Michigan Basin*, EPA-905/4-79-016, US Environmental Protection Agency, Region V, Chicago 60604.
11. Heintzenberg, J., Bischof, W., Odh, S. A. and Moberg, B. (1983) *The Investigation of Possible Sites for a Background Monitoring Station in the European Arctic*, Report AP-35, Department of Meteorology, University of Stockholm.
12. US Environmental Protection Agency (1979) *Handbook for Quality Control in Water and Wastewater Laboratories*, EPA-600/4-79-019, USEPA, Cincinnati, Ohio.
13. Laurence, J., Chau, A. S. Y. and Aspila, K. I. (1982) *Can. Res.*, 35–37.
14. Granat, L. (1977) In *Precipitation Chemistry (1974)* (eds R. G. Semonin and R. W. Beadle), ERDA Conf. 741003, National Technical Information Service, US Dept of Commerce, Springfield, Va 22161, pp. 531–51.
15. Wisniewski, J. and Cotton, W. R. (1977) In *Precipitation Scavenging* (eds R. G. Semonin and R. W. Beadle), ERDA Conf. 741003, National Technical Information Service, US Dept of Commerce, Springfield, Va 22161, pp. 611–24.
16. Gascoyne, M. (1977) *Atmos. Environ.*, **11**, 397–400.
17. Liljestrand, H. H. and Morgan, J. J. (1978) *Environ. Sci. Technol.* **12**, 1271–3.
18. Kawamura, K. and Kaplan, I. R. (1983) *Environ. Sci. Technol.*, **17**, 497–501.
19. de Pena, R. G., Pena, J. A. and Bowers, V. C. (1980) *Precipitation collectors intercomparison study*, Department of Meteorology, Pennsylvania State University.
20. Strachan, W. M. J. and Huneault, H. (1984) *Environ. Sci. Technol.*, **18**, 127–30.
21. Hakkarinen, C. (1983) Comparing data from six North American precipitation chemistry networks. *Air Pollut. Control Assoc. Spec. Symp. on Atmospheric Tracers, Detroit, Mich.*

22. USEPA (1982) *Handbook for Sample Preservation of Water and Wastewater*, EPA-600/4-82-029, USEPA, Washington, DC.
23. Environment Canada (1983) *Sampling for Water Quality*, Water Quality Branch, Inland Waters Directorate, Ottawa.
24. Marenthal, E. J. and Becker, D. A. (1976) *Interface*, **5**, 49–62.
25. Stensland, G. J., Peden, M. E. and Bowerson, V. C. (1983) NADP quality control procedures for salt deposition sample collection and field measurements. *Air Pollut. Control Assoc. 76th Ann. Mtg, Atlanta*, no. 83-32.2.
26. National Atmos. Deposition Program (1982) *NADP Instruction Manual: Site Operation*, Colorado State University, Fort Collins, 80523.
27. Danielsson, L. G. (1982) *Water Res.*, **16**, 179–82.
28. Chesellet, R., Morelli, J. and Buat-Menard, P. (1972) *J. Geophys. Res.*, **77**, 5116–31.
29. Kroopnick, P. (1977) *Pac. Sci.*, **31**, 91–106.
30. Bonsang, B., Nguyen, B. C., Gaudry, A. and Lambert, G. (1980) *J. Geophys. Res.*, **85**, 7410–16.
31. Goulden, P. D. (1978) *Environmental Pollution Analysis*, Heyden, London.
32. Fritz, J., Gjerde, D. and Pohlandt, C. (1982) *Ion Chromatography*, A. Huethig, New York.
33. Orion Research (1982) *Handbook of Electrode Technology*, Orion Research, Cambridge, Mass.
34. Ingman, F., Johannson, A., Johannson, S. and Karlson, R. (1973) *Anal. Chim. Acta*, **64**, 113–20.
35. Kramer, J. R. (1982) In *Water Analysis. Part 1 Inorganic Species*, Vol. 1 (eds R. A. Minear and L. H. Keith), Academic Press, New York, pp. 85–134.
36. Peden, M. E., Showron, L. M., McGurk, F. M. (1979) *Precipitation Sample Handling, Analysis, and Storage Procedures*, Illinois State Water Survey COO-1197-57, University of Illinois, Urbana.
37. Environment Canada (1974) *Analytical Methods Manual*, Water Quality Branch, Inland Waters Directorate, Ottawa.
38. APHA, AWWA and WPCF (1981) *Standard Methods for the Examination of Water and Wastewater*, 15th edn, American Public Health Association, Washington.
39. US Environmental Protection Agency (1979) *Methods for Chemical Analysis and Waters and Wastes*, EPA-600/4-79-020, USEPA, Cincinnati, Ohio.
40. US Geological Survey (1979) *Methods for Determination of Inorganic Substances in Water and Fluvial Substances*, Vol. 5, ch. Al, USGS, Washington.
41. Limbek, B. E., Rowe, C. J., Wilkinson, J. and Routh, M. W. (1979) *Am. Lab.*, 89–98.
42. US Environmental Protection Agency (1979) Methods 305.3 and 351.4 in *Methods for Chemical Analysis and Waters and Wastes*, EPA-600/4-79-020, EPA, Cincinnati, Ohio.
43. Tetlow, J. A. and Wilson, A. L. (1965) *Analyst*, **89**, 453.
44. Shebata, N. (1976) *Anal Chim. Acta*, **29**, 299.
45. Lazrus, A. L., Hell, K. C. and Lodge, J. P. (1965) In *Automation in Analytical Chemistry: Technicon Symposium, Medead, 1966*, pp. 291–3.
46. Tragardh, C. and Granat, L. (1982) *Automatic Spectrophotometric Determination of Sulfate with the Thorin Method*, Report AC-37, Dept of Meteorology, University of Stockholm.
47. Tyree, S. Y., Stouffer, J. M. and Bollinger, M. (1979) In *Ion Chromatographic Analysis of Environmental Pollutants*, Vol. 2 (eds J. D. Mulek and E. Sawicki), Ann Arbor Science, Michigan, pp. 295–304.

48. Willison, M. J. and Clarke, A. J. (1984) *Anal. Chem.*, **56**, 1037–9.
49. APHA, AWWA and WPCF (1981) Method 213B in *Standard Methods for the Examination of Water and Wastewater*, 15th edn, American Public Health Association, Washington.
50. Environment Canada (1974) Method 07105 in *Analytical Methods Manual*, Water Quality Branch, Inland Waters Directorate, Ottawa.
51. The Wescan Ion Analyzer, Winter 1983. Wescan Instruments, 3018 Scott Blvd Santa Clara, Calif. 95050 (1983), p. 5–6.
52. Potterton, S. and Shults, W. D. (1967) *Anal. Lett.*, **1**, 11.
53. Sekerka, I., Lechner, J. F. and Wales, R. (1975) *Water Res.*, **9**, 663–5.
54. Hem, J. D. (1982) In *Water Analysis. Part 1 Inorganic Species*, Vol. 1 (eds R. A. Minear and L. H. Keith), Academic Press, New York, pp. 137–60.
55. Harned, H. S. and Owen, B. B. (1958) *The Physical Chemistry of Electrolytic Solutions*, 3rd edn, Reinhold, New York.
56. Kramer, J. R. (1978) In *Sulfur*, Vol. 1 (ed. J. Nriagu) Wiley Interscience, New York, p. 325.
57. Jeffries, D. S. and Snyder, W. R. (1981) *Water, Air Soil Pollut.*, **15**, 127–52.
58. Goulden, P. D. (1978) *Environmental Pollution Analysis*, Heyden, London, pp. 63–6.
59. Noubecker, T. A. and Allen, H. E. (1983) *Water Res.*, **17**, 1–14.
60. Chau, Y. K., Wong, P. T. S. and Goulden, P. D. (1975) A gas chromatographic–atomic absorption spectrophotometer system for the determination of volatile alkyl lead and selenium compounds. In *Proc. Int. Conf. Heavy Metals in the Environment* (ed. T. C. Hutchison), Univ. of Toronto, Ontario, Canada, Vol. 1, pp. 295–301.
61. Chau, Y. K., Wong, P. T. S. and Goulden, P. D. (1976) *Anal. Chim. Acta*, **85**, 421–4.
62. Florence, T. M. and Batley, G. C. (1980) *CRC Crit. Rev. Anal. Chem.*, **9**, 219–96.
63. Cawse, P. A. and Peirson, D. H. (1974) *An Analytical Study of Trace Elements in the Atmospheric Environment*, Report AERE-R-7134, Atomic Energy Research Establishment, Harwell.
64. Instrumental Laboratories (1981) *Guide to Analytical Values for IL Spectrometers*, Report 91, IL, Boston, Mass.
65. Landsberger, S., Jervis, R. E., Kajrys, G. and Monaro, S. (1983) *Int. J. Environ. Anal. Chem.*, **16**, 95–130.
66. Landsberger, S., Jervis, R. E. and Monaro, S. (1984) In *Trace Analysis*, Vol. 4 (ed. J. F. Lawrence), Academic Press, New York.
67. Cooksey, M. and Barnett, W. B. (1979) *At. Absorpt. Newsl.*, **18**, 10.
68. Slavin, W., Manning, D. C. and Carnick, G. R. (1981) *Anal. Chem.*, **53**, 1504.
69. Fernandez, F. J., Beaty, M. M. and Barnett, W. B. (1981) *At. Spectrosc.* **2**, 16.
70. Yoshida, Z. and Motojima, K. (1979) *Anal. Chim. Acta*, **106**, 405–10.
71. Goulden, P. D. and Anthony, D. H. J. (1980) *Anal Chim. Acta*, **120**, 129–39.
72. Fishman, M. and Spencer, R. (1977) *Anal. Chem.*, **49**, 1599–602.
73. Jervis, C. E., Landsberger, S., Lecomte, R. *et al.* (1982) *Nucl. Instrum. Methods*, **193**, 323–9.
74. Barkenbus, R. D., MacDougall, C. S. and Griest, W. H. (1983) *Atmos. Environ.*, **17**, 1537–43.
75. Meyers, P. A. and Hites, R. A. (1982) *Atmos. Environ.*, **16**, 2169–75.
76. Gether, J., Gjøs, N. and Lunde, G. (1976) Isolation and characterisation in precipitation of organic micro-pollutants associated with long-range transport in air. No. 8, *SNSF, NISK*, 1432, Aas-NLH, Norway.
77. Lunde, G., Gether, J., Gjøs, G and Lande, M. B. S. (1976) Organic pollutants in precipitation in Norway. No. 9, *SNSF, NISK*, 1432, Aas-NLH, Norway.

78. Murphy, T. J. (1983) In *Toxic Contaminants in the Great Lakes* (eds J. Nriagu and M. S. Simmons), Ann Arbor Science, Michigan, pp. 53–80.
79. Andreae, M. O. and Raemdonck, H. (1983) *Science*, **221**, 744–7.
80. Cunningham, P. T., Johnson, S. A. and Yang, R. T. (1979) *Environ. Sci. Technol.*, **8**, 131–5.
81. Mizutani, Y. and Rafter, T. A. (1969) *N.Z. J. Sci.*, **12**, 69–80.
82. Ludweig, F. L. (1976) *Tellus*, **28**, 427–33.
83. Nielsen, H. (1974) *Tellus*, **26**, 312–21.
84. Holt, B. D., Engelkemeir, A. G. and Venters, A. (1972) *Environ. Sci. Technol.*, **6**, 338–41.
85. Holt, B. D., Cunningham, P. T. and Kumar, R. (1981) *Environ. Sci. Technol.*, **15**, 804–8.
86. Stumm, W. and Morgan, J. J. (1981) *Aquatic Chemistry*, 2nd edn, Wiley Interscience, New York, p. 561.

13 Low-cost methods for air pollution analysis

H. W. DE KONING

13.1 Introduction

The term 'low cost' tends to mean different things to different people. Some will identify it with primitive and ineffectual methods while others see it as appropriate technologies. Appropriate technologies are considered superior to the primitive technology of bygone ages but at the same time much simpler and cheaper than the super technologies of the rich. One can call it self-help technology – a technology to which everybody can gain access and for this reason is generally low cost [1, 2].

In a given circumstance a technology is low cost when it is decided that it appears to be the only one applicable to the local community. It will be subject to the availability of resources needed for its application such as trained manpower, capital outlay, production and maintenance facilities, budgetary constraints, local availability of materials, etc. For these reasons low-cost technologies may vary from one country to another but in the field of air pollution the variations are probably relatively minor.

In this chapter, a number of principles and procedures are described which will permit the implementation of relatively simple and low-cost air pollution monitoring systems. The procedures given are for the most common types of air pollution, i.e. sulphur dioxide, nitrogen dioxide, carbon monoxide, oxidants and suspended particulate matter. Where possible, references are given to more complete descriptions and also to intercomparison studies which indicate certain operating characteristics that may be helpful in the selection of the correct procedure.

The methods selected, although considered low cost, are in no sense of the word inferior to more high cost technology. Generally speaking, the higher technology affords more automation of the procedures, both for the sampling and analysis of the pollutants as well as for the recording of the data produced. Such technologies also generally involve fairly complex operational, calibration and maintenance procedures, thereby increasing the cost considerably. The point to be made here is that even if at some time in the future it would be considered necessary to install more automatic air monitoring equipment it would be prudent to start with a relatively simple system to determine the exact need and size before proceeding.

13.2   General considerations

This section deals with various aspects which the reader should note when considering adopting a low-cost procedure for measuring air pollution.

13.2.1   An air monitoring network

A monitoring network most commonly consists of a number of sampling stations spread out over the area to be surveyed and a laboratory which has the dual role of serving as a centre from which the monitoring operation is conducted and where the analyses on the samples collected are carried out [3]. Usually there are facilities to store, repair and test equipment as well as office facilities to keep records and carry out other administrative duties such as keeping inventories of spare parts, reordering supplies, keeping suppliers' catalogues, etc. in the laboratory. To keep the network operating smoothly, transportation is required to travel to and from the monitoring stations. It is important that there is free access to the monitoring stations at all times and that the technician who services them can get to the monitoring instruments without difficulty. Figure 13.1 illustrates the various components of an air monitoring operation.

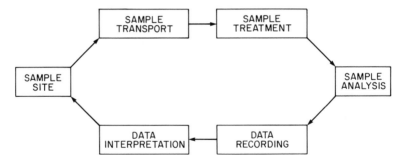

Fig. 13.1   Air monitoring system concept.

13.2.2   Operating conditions

These refer to a variety of factors that may have an effect on the sampling or analytical procedure. For example, if the monitoring operation is carried out in a tropical climate, care must be taken that bubblers do not run dry or that water is added in time. Also, if water is to be kept at the bottom of a dustfall bucket, periodic checks should be made in dry countries, the frequency of which will vary according to climatic conditions. In cold countries, on the other hand, antifreeze may have to be added to prevent the water from turning to ice. High humidity may cause filters to tear and invalidate the results.

   The ambient temperature will have an effect on the collection efficiency of a gas in a liquid medium. This factor should not be overlooked when operating a particular procedure in a tropical region. It is recommended under these conditions to run two bubblers in series and to determine the percentage collection efficiency in each.

Collection efficiency is the ratio between the amount of pollutant collected from a certain sampled volume of the ambient air and the amount of the pollutant that is actually contained in this volume of air. The collection efficiency (CE) is given as a percentage and can be expressed as follows:

$$CE(\%) = \frac{\text{pollutant conc. in inlet air} - \text{pollutant conc. in outlet air} \times 100}{\text{pollutant conc. in inlet air}}$$

Usually when a pollutant is removed from the ambient air, there is some loss, which means that the collection efficiency is seldom 100%. The fact that chemicals in the gaseous state form a true solution with the air makes it more difficult to separate them from the air than suspended particulate matter.

A high collection efficiency (95% or above) is desirable but not absolutely essential for obtaining acceptably reliable results. The important point is that the collection efficiency for a certain procedure be known and reproducible, thus allowing corrections to be made. However, collection efficiencies should not be less than 80%.

The speed with which the air passes through the collecting medium exerts a major effect on the collection efficiency, since it determines the time during which the air stays in contact with the collecting medium. A flow rate higher than that recommended would have the disadvantage that a certain amount of the gas or vapour to be collected would pass through the collecting medium without being retained.

If the procedure adopted does not already indicate a flow rate, it is recommended to test the collection efficiency, using known pollutant concentrations with different flow rates, in order to establish the most suitable one. If necessary, two bubblers should be run in series all the time if operating conditions are such that around 15 to 20% or more is collected in the second bubbler.

Climatic factors can also affect the stability of a sample. High temperatures generally tend to reduce their lifetime and even if samples are known to be stable for relatively long periods, they should be brought to the laboratory without undue delay and properly stored. Insulated containers of the type used for picnics are useful to prevent samples from heating up too much during transport.

Another point to consider in operating air pollution monitoring equipment is the deterioration of materials. Climatic conditions can simply cause hoses to develop pinholes, which results in air leakage. Deterioration can also occur in gaskets, electrical parts of the pump, etc., causing the system as a whole not to function perfectly. Most of these problems can be alleviated through constant vigilance and maintenance and the experienced technician will easily learn how to deal with this sort of problem.

As a last point, care should be taken when substituting one piece of equipment for another and generally a test should be run to see that its operating characteristics are comparable to the one it is replacing. This also holds if certain pieces of equipment have been repaired or manufactured locally. Care should be exercised when changing from one batch of chemicals to another or when changing suppliers for a given chemical or set of chemicals.

13.2.3   What to look for when selecting a method

The following characteristics are important:

13.2.3.1   *Sensitivity*

This is a function of the quantity of air needed to be sampled to provide a detectable amount of the air pollutant. To measure fluctuation in pollution concentrations during short-term fumigations it is necessary to use methods capable of detecting small quantities of the gaseous pollutant. When integrated values over a longer time interval are desired, a less sensitive method could be employed.

13.2.3.2   *Specificity*

The effect of interfering materials present in the concentration range anticipated is an important consideration in the selection of suitable air monitoring methods. In many instances it is possible to eliminate interfering materials by selective absorption or by other chemical reactions and thus avoid errors in analysis. At times it may be desirable to monitor with a non-specific method that measures a group of compounds that have similar chemical properties, e.g. the acidimetric method.

13.2.3.3   *Precision*

This is a measure of the reliability and stability of the equipment, reagents and technique employed in the method. Good precision does not, however, necessarily indicate relatively high accuracy. Accuracy, which is a measure of the deviation from a true value, is determined by measuring known amounts of a material and observing the correspondence between the measured and the true values. In air analysis wherein many methods rely upon the removal of the pollutant from the air mass, accuracy is a function of collection or separation efficiency, flow metering, analytical variation and other sources of error. Figure 13.2 illustrates examples of precision and accuracy in target shooting.

13.2.3.4   *Stability of reagents*

Reagents that have comparatively long shelf lives are advantageous. Their use eliminates the need for frequent reagent preparation and calibration. Absorbents that are easily susceptible to changes in composition and reactivity owing to light, temperature, turbidity, or air oxidation, may produce serious errors in the analysis.

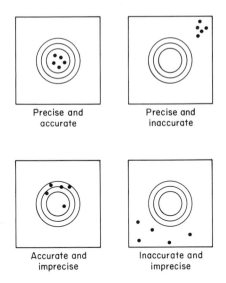

Fig. 13.2 Illustration of precision and accuracy.

13.2.3.5 *Calibration*

Ease of calibration is one important factor in selecting a method for measuring an air pollutant. The calibration is not only concerned with the analytical procedure but also with the air flow or air volume measuring device and the pump. A careful record of the calibrations carried out should be kept. Such a record will not only indicate the frequency of the need for recalibration but will also assist in detecting sources of error if it is suspected that the equipment or the analytical procedure is not functioning properly.

13.2.4 Sampling train

A sampling train has four essential elements: first, a sample line through which the air is sampled; next a device by which the pollutant under study is collected from the air sample for analysis (e.g. a bubbler or a filter), followed by a means to measure the air volume (or flow); finally, a pump required to make the air move through the system. These four elements are usually connected together with tubing [4, 5]. Figures 13.3 and 13.4 illustrate fairly typical sampling trains.

The material of construction of the sampling line must be such that absorption or diffusion errors are not introduced. Nylon and polypropylene are most suitable materials for sample lines in most cases, although Teflon is the only material satisfactory for ozone sampling lines. Glass tubing may also be used, with joints constructed by inserting the ends of the two pieces into a short piece of plastic tubing and butting them together, so that the air sample only comes into contact with glass.

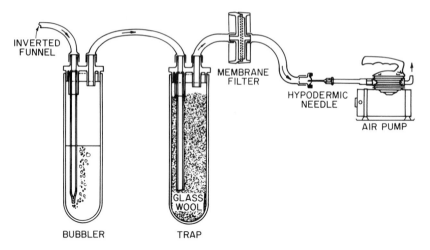

Fig. 13.3 Sampling train: the glass wool trap serves to take moisture out of the air which would damage the pump, the filter is to avoid dust entering the hypodermic needle and the needle itself acts to control the air flow at a predetermined rate depending on its size.

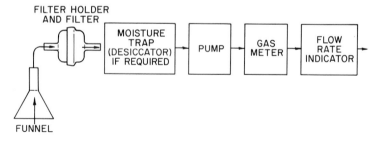

Fig. 13.4 Schematic diagram of sampling train for measuring suspended particulate matter using a filter.

Flasks used for air sampling are called bubblers. Although there exist many types, they consist basically of a glass flask with a tight-fitting stopper through which pass (1) an air inlet, which is a tube extending to the lower portion of the flask, and (2) an air outlet which is a short glass tube usually just through the stopper. In this way the air drawn through the flask passes through the collecting medium where the gas under examination is retained. Bubblers come in different shapes and sizes; a simple home-made version is shown in Fig. 13.5.

The total air volume that passes through the bubbler can be measured either

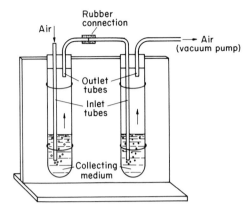

Fig. 13.5 'Home made' simple bubblers in series.

direct with a gas meter or indirectly by measuring the flow rate and the running time and computing the air volume from this.

There is a wide variety of suitable electrically driven pumps. The pump flow capacity required is determined by the total volume of air to be moved, together with the restrictions and obstructions in the sample train. Diaphragm pumps are most often used because of the wide range of capacities available, their low cost and easy maintenance. However, sampling trains which include filters that develop particularly high resistances may require some form of carbon vane pump.

The various elements of a sampling train are described in much greater detail in [4]. The point to be made, however, is that the various components are probably locally available in one form or another and can be made to fit together to produce an adequate low-cost instrument. Care must be taken to ensure that the sampling train does not have leaks since this will cause a significant error in the result.

13.2.5 Total volume of air to be measured

For a certain flow rate, the time during which the sample is collected will determine the total volume of the sample. Depending on the range of concentrations expected, a minimum volume of air is required. Since the total volume of air sampled depends on the flow rate and duration of sampling, these two factors must be considered when determining how long to sample. If a low range of concentrations is expected, the collection of a small volume of air is not enough, unless the analytical method is extremely sensitive. It is generally recommended to undertake some tests to determine either the optimum flow rate or sampling duration to obtain pollutant concentrations which fall somewhere well above the minimum detectable level for the procedure used.

13.3 Selected methods for measuring air pollutants

13.3.1 Sulphur dioxide

13.3.1.1 *Lead sulphation candle or plate*

This technique depends on the reaction of the SO_2 in the ambient air with lead oxide to form lead sulphate. The lead oxide in paste form is painted in a thin layer on a gauze-wrapped cylinder (called a lead candle), a glass plate or other support and allowed to dry. During exposure to atmospheric SO_2, the lead dioxide layer is converted to lead sulphate. The sulphate content is qualitatively determined following conversion to insoluble $BaSO_4$.

Sulphation candles and plates can be quite readily made and analysed with minimal laboratory facilities. Procedures for preparation are given in [6, 7]. Recent innovations, such as the preparation of PbO_2 plates by electrodeposition to assure a more uniform and reproducible application of reagent, may serve to improve precision and reliability of this method in the future.

The duration of exposure is determined by the quantity of sulphate needed for analysis and the ambient SO_2 concentration. Typical exposures are for a period of 30 days, but prevailing conditions may mandate the use of either shorter or longer periods. It is assumed that if the lead dioxide reduced to lead sulphate is less than 15%, the reactivity of the lead dioxide layer is constant and, therefore, the sulphation rate is proportional to the ambient SO_2 concentration. The result is reported as the sulphation rate expressed in units such as mg SO_3 per 100 cm^2 per month. Conversion of the sulphation rate to other units such as μg SO_2 m^{-3} of air is generally not recommended.

13.3.1.2 *Acidimetric method*

Sulphur dioxide in ambient air is oxidized to sulphuric acid by absorption in a dilute solution of hydrogen peroxide. The acidity of this solution is determined by titration with standard alkali and the result calculated as SO_2. The method is applicable to 24-h sampling periods.

Fairly standard laboratory equipment is used for the sampling train and laboratory analysis. The collection efficiency of the impinger should be checked, particularly in warm climates. The efficiency should be of the order 92 to 96%. The indicator solution used for the titration is a special, commercially available, mixed indicator solution (0.06 g bromocresol green and 0.04 g methyl red in 100 ml of methanol). Detailed instructions for this method can be found in [4, 8].

The acidimetric method measures net gaseous acidity of the air samples to within an accuracy of 10% for concentrations greater than 100 μg m^{-3}. If the SO_2 concentration is expected to be less than 100 μg m^{-3}, it is possible that interference from ammonia gas will be appreciable. If this is anticipated, the ammonia content of an aliquot of the absorbing solution should be determined (using Nessler's reagent) and a corrected value for the SO_2 concentrations calculated.

13.3.2   Nitrogen dioxide

13.3.2.1   *Sodium arsenite method*

The NO_2 in ambient air is collected by bubbling it through a solution of sodium hydroxide and sodium arsenite. The overall collection efficiency is 85% in the range 50 to 750 $\mu g \, m^{-3}$. The concentration of the nitrite ion formed during sampling is determined colorimetrically. The method can be used to collect 24-h samples.

The sodium arsenite method requires fairly standard equipment for the sampling train. The analysis of the samples requires sulphanilamide, *N*-(1-naphthyl)-ethylenediamine dihydrochloride, hydrogen peroxide (30%) and phosphoric acid (85%). The final colorimetric determination requires a spectrophotometer capable of measuring absorbance at 540 nm.

Nitric oxide is a positive and carbon dioxide a negative interferent. The average error resulting from normal ambient concentrations of NO and CO_2 is small for most monitoring situations. The range of the method is from 9 to 750 $\mu g \, m^{-3}$. The method is described in more detail in [9, 10] and in Section 6.3.5.1.

13.3.3   Carbon monoxide

Carbon monoxide cannot very conveniently be measured with a wet-chemical manual procedure [11]. Two methods are described, the detector tube procedure which is mostly used for spot sampling, and a relatively low-cost instrumental procedure using the Ecolyzer instrument*.

13.3.3.1   *Detector tube method*

Detector tubes are glass tubes, approximately 125 mm long and 7 mm in diameter, sealed at both ends, with a calibration scale painted on the outside of the tube (Fig. 13.6).

The tubes contain a reagent on a solid support capable of reacting with the carbon monoxide to produce a coloured stain. Just before using the tube the tips are broken off and a prescribed volume of air is drawn through using a squeeze bulb or hand pump. The CO concentration is determined by reading the length of the stain produced directly against the scale marked on the tube. Care should be taken to follow the manufacturer's instructions such as the direction of the air flow. Detector tubes as well as the hand pumps are commercially available from various manufacturers.

Detector tubes have been primarily designed for use in industrial hygiene investigations, where higher CO concentrations may be experienced than those occurring in ambient air. Carbon monoxide detector tubes, however, can also be used to measure ambient concentrations above 6 mg CO per m^3 and are

* Recommendation only, no endorsement intended.

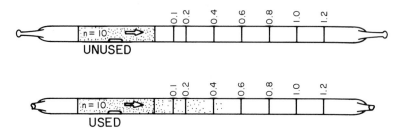

Fig. 13.6 Detector tubes.

particularly useful in motor vehicle traffic surveys for measuring instantaneous concentrations of carbon monoxide [4].

13.3.3.2 *Instrumental method*

An instrument known as the Ecolyzer has been used over the past 10 years as a relatively low-cost method for making ambient CO measurements. This instrument measures CO continuously and uses an electrochemical detection principle. For ambient air measurements a range from 0 to 55 mg CO per m^3 is used. The size of the instrument is about $20 \times 20 \times 35$ cm and a battery operated version is also available. According to the manufacturer's technical information, only certain hydrocarbons interfere but this should pose no problem in most monitoring situations. The basic cost for this instrument is around US$2500. The instrument requires calibration with a CO calibrating gas contained in a cylinder.

13.3.4 Oxidant

13.3.4.1 *Neutral buffered potassium iodide method*

Oxidants can be produced in a polluted atmosphere as a result of photochemical reactions taking place in the presence of strong sunlight. Oxidants are substances other than oxygen, which exhibit oxidizing properties. Ozone is the most important oxidant present in polluted urban air and oxidant measurement results are commonly loosely equated with ozone measurements [12].

In the neutral buffered KI method, air is passed through a bubbler containing a 1% solution of potassium iodide buffered at pH 6.8 ± 0.2. The liberated iodine is measured colorimetrically at 350 nm and is directly related to the concentration of oxidant (expressed as ozone) present in the air, the iodine concentration being determined from a calibration curve prepared with standard iodine solutions. The method is described in greater detail in [4, 5].

The method is capable of oxidant measurements in the range from about 20 to 20 000 $\mu g\ m^{-3}$. Sulphur dioxide produces an interference which can be removed by placing a chromium trioxide scrubber upstream from the bubbler. This method is suitable for sampling periods up to 30 min. Absorbance must be measured 30–60 min after sampling.

13.3.5   Suspended particulate matter

13.3.5.1   *Dustfall*

Free-falling dust can be measured by locating an open collector on a suitable stand in the area to be studied (see Fig. 13.7).

Collector design and specifications vary widely from country to country. As a rule of thumb the height of the container should be at least three times the diameter to ensure that the wind will not cause dust already in the container to be blown out again. All containers used in a given survey should of course be the same to ensure comparability of results.

Fig. 13.7   Dustfall gauge.

As an added precaution to prevent dust being blown out of the container, some water is poured in the bottom. Sometimes an algicide is added to the water to prevent the growth of algae.

The collectors are usually left outside for one month, after which they are returned to the laboratory. The dust is washed out of the collectors and, after screening out leaves, twigs, insects and other non-dust materials, the following parameters are determined

Total weight of dried insoluble material
Weight of material dissolved in water

Dustfall measurements are commonly used to indicate the readily settleable portion of suspended particulate matter. The results are either expressed in $mg/m^2/day$ or tons/mile/month. Typical values for a city are:

Residential area   65–100 $mg\ m^{-2}\ day^{-1}$
Light industrial area   100–200 $mg\ m^{-2}\ day^{-1}$
Heavy industrial area   150–350 $mg\ m^{-2}\ day^{-1}$

This method of measuring suspended particulate matter is not very accurate but it does give an indication of the overall dirtiness of an area, and it is useful for studying the distribution of suspended particulate matter over an area and for determining trends; it is described in greater detail in [5].

13.3.5.2   *High-volume (Hi-Vol) sampling method*

There is a large variety of methods for measuring suspended particulate matter in air. The most common methods use filters which are analysed either by weighing them before and after exposure; or by measuring the density of the resulting stain on the filter paper with a photoelectric reflectometer following exposure. There are several possible variations of each of these methods [13].

Aside from the differences in analytical procedures, there are differences in the speed with which the air passes through the filter. Generally speaking, if the air speed is low (~ 2 m³ air per 24 h, low-volume sampling) only particles up to about 10 μm are collected on the filter. Again many variations exist. The low-volume method probably imitates somewhat better the action of a human respiratory system whereas the high-volume procedure is somewhat more reliable and produces results that are directly comparable from one location to another. For further information, the reader is referred to [4, 5, 8].

Because of the relative simplicity, both of the type of equipment involved and the analytical procedure, the high-volume method is described below.

The sampler consists of a motor and blower enclosed in a shelter (Fig. 13.8). The filter surface is arranged horizontally, facing upwards and is protected by a roof that keeps out rain and generally prevents the collection of particles larger than 100 μm. Filters are made of glass or synthetic organic fibre. The amount of suspended particulate matter is calculated by dividing the net weight of the particulate matter (weighed to the nearest milligram) by the total air volume sampled.

Air flow is measured with a rotameter and, on the assumption that the flow rate decreases uniformly during the sampling period (because of the build-up of particulate matter on the filter), the average of the initial and final flow rates is

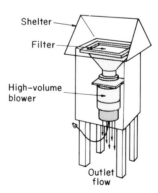

Fig. 13.8   Assembled sampler and shelter.

used to calculate the volume of air sampled. The normal sampling period is 24 h, but in some very dry countries where particulate loadings are very high because of naturally occurring dust, shorter sampling periods may have to be considered.

In this case care should be taken to avoid introducing a bias. For example, if suspended particulate matter loadings are high during the day because of certain meteorological conditions, care should be taken not to represent the results obtained as daily averages but to be specific about the time period during which the samples were obtained. Also, if it is suspected that the suspended particulate matter mass on the filter contains a high percentage of wind-blown dust, an effort should be made to estimate its percentage by microscopic examination.

13.4 Additional considerations for selecting a low-cost air pollution measurement method

There are additional factors when considering the implementation of an air monitoring survey at minimal cost. These factors, although not directly related to the selection of the sampling and analytical procedures, nevertheless have an important bearing on the development of a low-cost monitoring scheme as a whole. These factors are described below.

13.4.1 Equipment requirement

When selecting a low-cost method, a careful look should be taken at the equipment that is required. This concerns not only the glassware and instrumentation that is required for sampling but also the analysis. Often, for example, a spectrophotometer is required for colorimetric analysis but in some cases a much cheaper colorimeter will suffice. Similarly, a balance may be required to weigh chemicals or filters. Again, the required precision for the balance should be known before making a purchase. In most instances a refrigerator will be required to store samples, stock solutions, etc., and in tropical countries air conditioning of the laboratory is recommended. The main reason for this is that the reaction rate of chemicals during analysis varies with temperature and an effort should be made to keep the conditions under which the analyses are carried out approximately constant, preferably around 20–25° C.

There is often the need for substituting prescribed pieces of equipment with what is available locally. Such equipment may either have been manufactured locally or regionally or may have been adapted for a particular application.

Similarly, chemicals may have been supplied from a local source. There are advantages in this in that the supply may be easily accessible and at lower cost. Most low-cost procedures lend themselves to substitution of certain elements or even to modifications to suit local circumstances. It should be stressed, however, that any change from a prescribed procedure should be tested very rigorously to observe whether any differences occur because of the change made. The same holds true if changes are made after the air monitoring survey has been in progress. For example, if a switch is made to a cheaper source of a particular dye, this should be carefully checked out by running parallel tests using the old dye and the new one.

Another example of substitution has been reported in WHO Offset Publication 57 [14] where the comparison of the results obtained with equipment supplied by WHO is compared with data obtained with similar equipment developed by the National Environmental Engineering Research Institute in Nagpur, India. The equipment involved was for the measurement of suspended particulate matter and SO_2. The results of this study indicated some systematic differences among the results obtained with the two sets of equipment but did not in any way diminish the usefulness of the locally manufactured equipment for application in air pollution monitoring survey work.

13.4.2   Calibration

Calibration is an important part of any monitoring survey since it relates directly to the validity of the results obtained. No matter what the results of the survey are going to be used for, the questions will sooner or later be asked, 'How good are these data?' or 'How much confidence can one have in these data?'. Only with good calibration procedures can such a question be answered with confidence.

Calibration refers not only to the analytical procedure but also to the sampling procedure. During sampling it is important to know what the air flow (or air volume) is. It is also important to know, and this should be checked regularly, that all the air that is moved by the pump is going through the collection device and is not sucked in, for example, through leaks in the lines, or the pump itself. Particularly if the resistance in the sampling system is high, this may become a problem. For the calibration of the analytical part, it is important that the procedures are simple and can be carried out with ease. There are static and dynamic ways of calibration. The latter involves producing gas mixtures using bottles or permeation devices. For the procedures recommended in the previous section, mostly static calibration procedures can be used which involve preparing a series of solutions. Such a series represents a number of known concentrations of the pollutant in question and affords the preparation of a calibration curve. A new curve needs to be prepared from time to time.

Aside from the validity of the results obtained in a survey, calibration is important for their comparability. Most countries make air quality measurements and in order to be able to compare the results obtained in different countries, it is important that acceptable calibration procedures are used [15]. The comparing of results is not only important from a scientific point of view but also from a more international standpoint in that countries now wish to show that environmental conditions are within acceptable limits. In this connection it is useful to note that there are several international monitoring programmes carried out within the framework of UNEP's Global Environmental Monitoring System (GEMS) which is involved in international intercalibration studies among laboratories in different countries [16–19].

13.4.3   Record keeping

Records should be kept of all results obtained for air sampling and analysis, since these serve as a basis for further evaluation and, more important, for a determination and proof of compliance with guidelines or standards for the protection of human health or the environment in general. A record or log book for a monitoring survey should be set up as a long-term operation and should, in addition to providing space and columns for entering dates and results obtained, allow space for entering comments and unusual events (the weather for example), a change of instrument (replacement of a pump), late pick-up of a sample or any other event or difficulty that may have a bearing on the interpretation of the results obtained. There should also be space in the record book where calibration dates and results can be entered. Copies of old calibration graphs should be kept in case they are needed at a later date. It is generally best to keep track of all possible relevant information since it is never known beforehand what may be required some years later.

A second major reason for keeping an adequate log book (or possible two log books if it appears complicated to enter all information into one) is for maintenance and stock-keeping. These two activities have a very important effect on the smooth continuous operation of the survey as well as its cost.

Maintenance activities should be recorded in the log book since they relate to the performance of the air monitoring operation but, more important, after a while, as experience is gained, it will become clear how much periodic maintenance is required to prevent breakdowns or other unexpected stoppages, thereby avoiding loss of data and optimizing the lifetime of the equipment used. These rules apply both to monitoring and to laboratory equipment. Examples of periodic maintenance are the replacement of brushes in pump motors, lubrication of moving parts in motors according to manufacturer's specifications or cleaning of sampling lines which collect dust from the atmosphere, etc.

Consideration should be given to the number and type of spare parts which must be held in stock to maintain a programme efficiently. The first decision that must be made is whether a replacement for any particular part is likely to be needed, and if so the expected life of the part in question. For example, there is no need to have in stock a spare case for an instrument – instrument cases are not likely to break. Pump diaphragms have an expected life of over 12 months, so a limited stock only of these need be carried. Under dusty conditions, a line filter may last a few days at the most, so a larger stock is needed. Other considerations include the availability of the spare part, its cost, and whether bulk discounts are available. If a part is readily available from a store in the city, there is no need to carry a large stock of these: enough to satisfy an immediate need is all that is required. If the part must be imported, or is on a long delivery time, a large enough stock must be carried to ensure that the monitoring programme is not interrupted through instrument failure. For example, the rubber gasket of a high-volume sampler requires periodical replacement. This is a specialized item, hence it is usually ordered from the instrument manufacturer, not from a local store. On the other hand, items such as fuses are so cheap that a larger stock can be carried.

References

1. Darrow, K., Keller, K. and Pam, R. (1981) *Appropriate Technology Source Book – A Guide to Practical Books and Plans for Village and Small Community Technology*, Vols 1 and 2, Volunteers in Asia, Box 4543, Stanford, California 94305.
2. Longdon, R. J. (ed.) (1977) *Introduction to Appropriate Technology toward a Simpler Lifestyle*, Rodale Press, Emmaus, Pa.
3. World Health Organization (1977) *Air Monitoring Programme Design for Urban and Industrial Areas*, WHO Offset Publication 33, WHO, Geneva.
4. World Health Organization (1976) *Selected Methods of Measuring Air Pollutants*, WHO Offset Publication 24, WHO, Geneva.
5. Lawrence Berkeley Laboratory (1975) *Instrumentation for Environmental Monitoring, Air*, University of California, Berkeley.
6. American Society for Testing and Materials (1971) D-2010-65, Standard Method for Evaluation of Total Sulfation in Atmosphere by Lead Peroxide Candle. In *Annual Book of ASTM Standards*, Part 23, ASTM, Race St, Philadelphia, Pa, pp. 514–17.
7. Intersociety Committee (1970) *Health Lab. Sci.*, **7**, 164.
8. Organisation for Economic Co-operation and Development (1964) *Methods of Measuring Air Pollution*, OECD, Paris.
9. Margeson, J. H., Beard, M. E. and Suggs, J. S. (1977) *J. Air Pollut. Control Assoc.*, **27**, 553–8.
10. Christie, A. A., Lidzey, R. G. and Radford, D. W. F. (1970) *Analyst*, **95**, 519.
11. World Health Organization (1979) *Carbon Monoxide*, WHO Environmental Health Criteria Document 13, WHO, Geneva.
12. World Health Organization (1978) *Photochemical Oxidants*, WHO Environmental Health Criteria Document 7, WHO, Geneva.
13. World Health Organization (1979) *Sulfur Oxides and Suspended Particulate Matter*, WHO Environmental Health Criteria Document 8, WHO, Geneva.
14. World Health Organization (1980) *Air Quality in Selected Urban Areas, 1977–1978*, WHO Offset Publication 57, WHO, Geneva.
15. Schneider, T., de Koning, H. W. and Brasser, L. J. (1978) *Air Pollution Reference Measurement Methods and Systems (Studies in Environmental Science 2)*, Elsevier, Amsterdam.
16. de Koning, H. W. and Köhler, A. (1978) *Environ. Sci. Technol.*, **12**, 885–9.
17. Kirov, N. Y. (1979) *Asian Environ.*, **2**, 22–30.
18. World Health Organization (1983) *Air Quality in Selected Urban Areas, 1979–1980*, WHO Offset Publication 76, WHO, Geneva.
19. World Health Organization (1980) *Analysing and Interpreting Air Monitoring Data*, WHO Offset Publication 51, WHO, Geneva.

14 Planning and execution of an air pollution study

D. J. MOORE

14.1. Introduction

Pollution monitoring is an expensive business and it should not be undertaken lightly. In a world of limited resources, any monitoring programme will probably have taken priority over some other socially useful exercise. It will therefore be assumed in the following discussion that data are being acquired because:

(a) Some undesirable effect has been noted, or
(b) There is reason to suppose that if certain levels of contamination of the air, or precipitation from the air, are reached some harmful effect will occur, and
(c) There is some possibility of remedial actions being taken, should this be found desirable.

Figure 14.1 illustrates the features and functions of any air pollution monitoring exercise which has been designed in accordance with the above assumptions.

The first row of boxes represents the path along which the effluent travels from its source(s) to the receptor(s) where the harm may possibly be done (the source-to-sink history). The second row shows the various types of monitoring corresponding to each of these stages (i.e. the monitoring network). Some of these stages may be absent in some instances (see Section 14.3).

If the information acquired by the monitoring network is to be of any use then it must be transmitted (row 3), probably with some processing en route, either in 'real time' or after sufficient data had been collected (row 4) to those responsible for interpreting it in terms of what plant control and/or receptor protection measures are necessary (row 5).

Additional information would normally be available in the form of historical analogues and/or mathematical/physical models to help in any decision-making. Experience might also show that the scale of monitoring could be varied according to plant-operating or meteorological conditions and so some monitoring control procedure might be desirable. If it is established that no problems exist, then the monitoring could be terminated.

The way in which the data from any monitoring exercise are handled will, of course, depend very much on the objectives of the survey. These, in turn, together

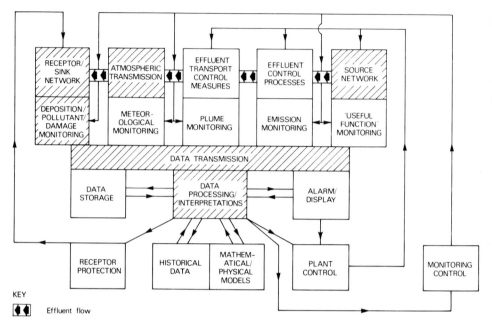

Fig. 14.1 Features and functions of a useful air pollution monitoring exercise. The hatched sections will always be present.

with any prior knowledge of the likely source-to-receptor history, should have determined the form of the monitoring network, the quantity of data to be collected and the rate of acquisition of data, within overall limits set by the money, manpower, instrumentation and resources available.

It is therefore necessary that any discussion of data handling and analysis of results should include likely source-to-receptor histories, monitoring networks and also objectives.

In some states and countries, legislation has already been enacted or is being prepared, to lay down maximum allowable concentrations (MACs) or some equivalent terms for some pollutants for various types of exposure. Where this is the case, it might be argued that the sole purpose of a monitoring programme is to ensure that the law is being complied with and that the historical data available to those responsible for control measures are reflected in the legislation. This view could be correct if the laws have been carefully framed, but where the law includes such details as the dispersion models to be used in calculating ground-level concentrations (GLCs) and the acceptable patterns of monitoring sites, the legislation has probably been made too inflexible to be efficient. This may well have been done for reasons of bureaucratic convenience.

In the following discussion it will be assumed that a flexible approach is possible, that each situation is to be evaluated on its own merits and that no Procrustean regulations exist.

14.2   Objectives of the monitoring programme

14.2.1   General

Assuming that a monitoring programme is contemplated for cases (a) or (b) of Section 14.1, in case (a) the object of the monitoring exercise will be:

  (i)   To identify the pollutant (if this is not already known)
 (ii)   To identify the source of the pollutant (if this is not already known)
(iii)   To assess the effectiveness of any control measures taken.

In case (b), the object will be:

  (i)   To determine existing dosage levels of the pollutant, so that an objective assessment may be made as to whether control or receptor protection measures are necessary now or in the future
 (ii)   If source control measures are taken, to assess their effectiveness.

14.2.2   Pollutant identification

Where damage has been observed, the only certain way to identify the pollutant responsible is to establish a direct causal relation between the pollutant and the damage.

This is best done by identifying the pollutant at the site of the damage, e.g. asbestos fibres in carcinomas, or by reproducing the damage in controlled experiments, which will also establish dosage thresholds for damage.

An alternative is to establish statistical correlations between damage and dosage. This is much less satisfactory because of the many correlations existing between pollution dosage and many other human activities (e.g. pollution and urban living) likely to cause damage. The initiation of high-cost pollution control programmes on the basis of correlations alone may well result in a waste of resources.

14.2.3   Source identification

When it has been recognized that it is desirable to reduce the dosage levels for a particular pollutant or group of pollutants, it will be necessary to establish the major source(s). The assumption that sources are responsible in ratio to their total emissions is another pitfall which could lead to serious misapplication of resources, if expensive control measures were imposed irrespective of source type. The dosage levels observed from a given source will be a complicated function of emission rate, the location of the source with respect to the receptor, the topographical features of the area, the prevailing meteorological conditions and the height of the plume above the receptor (see Chapter 2).

Procedures for source identification are discussed in Section 14.7.

14.2.4   Economic assessment of damage versus control

If it has been established that certain sources cause damage, before control measures are taken, a proper assessment of the relative costs of the damage and control procedures needs to be made. The costs of the control procedure should include the effects of any change in the emission due to the control process and any effects on ground or water of its products.

It may therefore be necessary to monitor changes in minor, but possibly more toxic, components of the emission resulting from operation of the control equipment, as well as the major pollutant for which the control process was initiated.

Other side-effects of the control process which will be important are changes in the physical nature of the emission (e.g. reduction in temperature or addition of liquid droplets) which could adversely affect the plume rise and so reduce the dilution of the residual effluent at ground level, in addition to making the plume more visible.

14.2.5   On-line plant control

If plant operation leads to acceptable dosages under most weather conditions, it may be necessary to operate some form of control under meteorological conditions which result in abnormally low dilution at the receptor. An example of such a control measure would be the substitution of low-pollutant fuel, too expensive for continuous use.

The monitoring system providing the information leading to initiation of such control might consist of

(i) Meteorological instruments, e.g. to measure wind direction and mixing depth to ensure correct source identification
(ii) Fixed dosage monitoring at sensitive or representative receptor points
(iii) A mobile survey unit, or
(iv) Combinations of two or all of these systems.

Alternatively, if control equipment is installed, monitoring will be necessary to detect its failure. This would normally involve effluent monitoring, but some back-up from field monitoring would help in deciding whether plant shut-down is necessary following control equipment failure.

14.2.6   Control of future developments

Provided acceptable dosages from pollutants have been established and existing levels are found to be satisfactory, a monitoring programme might be designed to assess changes in the general level of pollution likely to result from future developments. The objective here would be to assist with decisions about the need for restrictions on such development in certain areas and/or the introduction of control measures at some time in the future. The validation of plume rise and dispersion models for tall stack emissions, so that these models could be used to determine suitable stack heights for future large installations, is an example of an objective in this category.

14.2.7   Receptor protection

If control measures fail or are too expensive, or there is an accident, receptor protection may be the only effective way of dealing with pollution. This would be initiated either as a result of plant function monitoring, effluent monitoring or field monitoring.

Receptor protection may also be necessary as a result of long-term exposure to pollution. An example is the liming of lakes which have been exposed to inflow of acid waters.

14.2.8   Detection of long-term trends

It has only fairly recently been established (see, for example, Junge [1] and Lovelock [2]) that the atmosphere as it exists today is largely a mixture of organic effluents and very different from the inorganic atmospheres to be expected in lifeless worlds in a physically similar situation to our own. There is no reason to suppose that the atmosphere is naturally in some state of delicate equilibrium, beneficial to all existing creatures and likely to be upset only by the unnatural activities of man. The composition of the atmosphere is continually changing, and these changes have led to the replacement of many plant and animal species. Man's days are therefore probably numbered if he does not interfere with the atmosphere. If he does, then he may be reversing a trend that would have led to his destruction or he may be accelerating or even initiating such a trend. In any case, long-term monitoring of atmospheric constituents at background stations appears to be prudent. Some interesting facts have already come to light as a result of such monitoring, e.g. in recent decades the long-term trend of reducing CO_2 levels due to growth of vegetation has been reversed, part of which is probably due to emissions from fossil-fuel burning [3].

Changes in the dissolved material in precipitation may have important effects on soils, e.g. control of particulate emissions could lead to increased acidity in rain as a side-effect.

In general the requirements for such monitoring will be the opposite to those for source identification or source study networks: here it is the background that is to be measured, so the sites should be chosen for minimal effect from individual sources.

14.2.9   Monitoring control

Because adequate monitoring is usually expensive, one object of monitoring should be to establish or to confirm that certain pollutants are *not* likely to be harmful at the levels found and consequently that further monitoring is unnecessary. Alternatively, it may be established that pollutant monitoring will be necessary only in certain meteorological conditions, so continuous meterological monitoring, supported by a pollutant monitoring facility for use as required, may be sufficient.

14.3   Effluent history from source to receptor

14.3.1   General

Effluent control processes and/or effluent transport control measures (Sections 14.3.3 and 14.3.4) may or may not be present in any particular instance, but in any useful monitoring exercise there will always be a source network, and atmospheric transmissions to a receptor/sink network.

The dosage received by a receptor will depend on the rate of emission of the effluent into the atmosphere, the effluent transport control measures taken, the dilution and attenuation during the atmospheric transmission and its own ability to take up or reject the pollutant (Section 14.3.6). In general it will be possible to avoid adverse effects by source control, effluent control, effluent transport control or receptor protection.

14.3.2   Source network

The source network may consist of a single tall chimney or a large number of small sources distributed over a large area (e.g. motor vehicles or houses in a city). In the case of secondary pollutants the source (or sources) is the plume (or plumes) some distance from the source(s) of the primary pollutants.

Provided the sources of the primary pollutants are known, information on their characteristics and locations should be included in the historical data available to assist the data analysis and decision-making processes.

14.3.3   Effluent control processes

A control process may be defined as the modification, for other than economic reasons, of the effluent produced as a by-product of the primary function of a plant or process. The object of this modification will be the improvement of the environment.

It is important that, in assessing the true value of the improvement, a full assessment of the environmental impact of the control processes be made, e.g. a sulphur-removal process will not, in general, affect the total addition of sulphur to the environment, unless the control processes produce a useful product and the production of this useful product by existing methods is consequently reduced.

The installation of a control process means that monitoring the useful function will no longer be sufficient to determine the emission, unless detailed information on the control processes is available and the process itself is very reliable.

14.3.4   Effluent transport control procedures

The plant should be constructed in such a way that the maximum possible dilution of plume material is achieved before it diffuses down to the surface. It must be understood that whatever the height of release, in the absence of precipitation, a substantial proportion of emitted material will travel many hundreds of kilometres before it is deposited, whatever its effective height of release, in all but the most stagnant meteorological conditions. The adverse effects on long-range transport of high level compared with low level releases are second order, while the risks associated with relying on control equipment alone to achieve near-field control, when very substantial emissions are involved, are not easily determinable.

Procedures for achieving high dilution factors were discussed in Chapter 2.

14.3.5   Atmospheric transmission

The wind transports the effluent material from source to receptor and impacts it on protuberances from the surface. Atmospheric turbulence dilutes it on the way to the receptor and diffuses it into or through the laminar flow layer on the receptor surfaces. Precipitation and cloud water are contaminated and effect the wet removal of pollutants.

The design of an effective monitoring network and the realistic interpretation of the data from it therefore depend primarily on a detailed understanding of the atmospheric transmission processes. They were discussed in detail in Chapter 2.

14.3.6   The receptor/sink network

In addition to the dilution by atmospheric turbulence as it travels from source to receptor, the pollutant may be attenuated by deposition on the surface or modified by reactions with other atmospheric gases or aerosols.

Distant receptors may therefore be protected from the effluent by this attenuation process as well as by the atmospheric dispersion. Conversely, successful effluent transmission control procedures, by eliminating high ground-level concentrations and high deposition rates near the source, may result in some increase in the concentrations (which for an individual source will be small anyway) at very long distances where the pollution has become well mixed through the boundary layer.

Receptors may be regarded as active if they ingest air or passive if the air passes over or through them under the action of the wind. In the former case, if the ventilation rate is independent of wind speed, the dosage they receive will be a product of pollutant concentration, ventilation rate, exposure time and the efficiency with which the receptor retains the pollutant. In the second case, it will be the product of the ground-level concentration, exposed surface area and a deposition velocity. The deposition velocity will be a function of the wind speed and turbulence, the physical characteristics of the receptor and the chemical or biological resistance of the receptor to the pollutant [4]. In the case of aerosols the problem is complicated by the inertia of the particles, large particles tending to impinge on projections into the air stream, while the smaller particles follow the streamlines round the projection. Impaction on the surface does not necessarily mean retention by it, as the particle may bounce off or be re-entrained by the wind (saltation). Particle fall velocity will also be important in determining the dosage received by horizontal surfaces if the particles are large enough.

Rain or snow may be regarded as receptors. The degree to which they are contaminated depends in a complicated way on the total amount of pollutant in the volume of air that the falling raindrop or snowflake swept out on its way to the ground coupled with its ability to take up pollutant material from the atmosphere.

A form of deposition which is receiving increasing attention at the present time is the deposition of cloud droplets on the surface or their impaction on trees, other vegetation and other projections from the surface. This has been termed, perhaps rather unfortunately, occult deposition [5]. Routine measurements of occult depositions have not been reported because they depend on the efficiency with which the receptor removes the cloud droplets.

From the above, it will be inferred that a proper assessment of relations between air concentrations and damage may require a great deal of supplementary information about the properties of the receptor and the pollutant.

Chemical reactions of pollutant with other atmospheric gases or aerosols are also complicated processes. Solar radiation, relative humidity, temperature and the presence of suitable catalysts or materials forming intermediate products can all be important.

14.4   The monitoring network

14.4.1   General

It is possible that a monitoring network would contain no pollutant sensors at all but would consist solely of meteorological instruments, backed up by information on the current functioning of the source and experience in the form of historical data or dispersion models. However, in general the monitoring network will record some or all of the variables listed in Sections 14.4.2 to 14.4.8.

14.4.2   Function monitoring

The effluent is a by-product of some process or function which some people find useful or desirable (e.g. manufacturing, power supply, transport). One way to monitor the rate of emission of effluent is to monitor the useful function. The relation between the rate of emission of effluent and the rate of performance of the useful function will first have to be established. This need not necessarily involve measurements of the effluent. For example, if the effluent is produced by burning fuel and the products of combustion are known, the rate of emission could be calculated from knowledge of the rate of fuel usage and the amount of each component of the fuel retained in the (unemitted) ash.

The units of function monitoring will be quantities like megawatts for electricity generation stations, cars per kilometre or cars per second for roads, kilograms of fuel burnt per second or energy output for heating systems, etc.

14.4.3   Emission monitoring

Continuous, accurate monitoring of the emissions from any plant is a difficult undertaking because of the difficulty of maintaining the instrumentation in good working order under adverse conditions. Although trials of limited duration are usually sufficient to relate emissions and function monitoring where no control equipment is installed, continuous monitoring of a controlled pollutant is much more necessary to ensure that the control equipment is working properly. The physical changes (temperature, liquid carry-over (drift), etc.) produced by the control processes should also be monitored to ensure that the transport control procedures (Section 14.3.4) remain effective in ensuring maximum dilution of the remaining pollutants (and possibly newly introduced pollutants resulting from the control process). One difficulty with emission monitoring is that what is usually recorded is the concentration (or possibly flux) of effluent at some point or points in the ducting or stack, while what is required is the total emission. Conversion of one measurement to another requires an accurate knowledge of the total volume emission of gas from the plant and, if the pollutant and flow rate (in the case of flux measurements) are not uniform across the duct, of the distribution of flux, requiring multipoint sampling. Leakage of air into the duct near the sampling point can seriously affect the results.

A second point, sometimes overlooked, is that if the emission is hot, then if the (gaseous) emission is expressed as a volume it must be corrected to atmospheric temperature when GLC is determined. For this reason it is better to express emission as $kg\ s^{-1}$ rather than vol/vol of flue gas or $m^3\ s^{-1}$.

14.4.4   Plume monitoring

Plume monitoring involves monitoring the effluent between the emission point(s) and its arrival at the receptor network. Remote sensing equipment (lidar, correlation spectrometers, laser spectroscopy, etc.), airborne sensors (using aircraft, including kites, balloons and model aircraft) and instrumented towers have all been used.

All these techniques tend to be expensive and/or difficult to operate over extended periods, so their main function has tended to be the validation or development of plume-rise and dispersion models and/or pollutant-reaction models.

It is possible to use remote sensing techniques to monitor emissions as an alternative, or a supplement, to plant monitoring. Closed-circuit television monitoring of emissions from the stack has been used to detect excessive dust emission. Plume photography is perhaps the least expensive method of monitoring near-field general properties of emissions.

14.4.5 Meteorological monitoring

The most relevant meteorological parameters are wind speed, direction, vertical temperature gradient, where relevant the depth of the mixing layer, and lateral and vertical components of atmospheric turbulence. If practicable, these should be measured over the whole plume depth using tower-borne instruments, a balloon or aircraft. In practice, near-surface observations only are normally available. If dosage monitoring is impracticable on a universal basis, previous experience (including dispersion models) will enable the most disadvantageous meteorological conditions to be defined. If some source or emission control is deemed necessary in those conditions, meteorological monitoring may be a more practicable way of initiating the control measures than dosage monitoring, because of the very large number of monitoring sites required to cover the whole area round the source. However, from large sources with very tall stacks, the highest ground-level concentrations are observed in unstable meteorological conditions with light or moderate winds and a limited mixing height, the plume levelling off at the top of the mixing layer but being wholly retained within it. In such conditions, a slight change in the wind speed and/or mixing layer height can result in the plume escaping from the mixing layer, giving zero or low ground-level concentrations (see, for example, Moore [6]). A mobile sampling unit (Section 14.5.8), fitted with remote sampling equipment in addition to point monitors, which is able to establish whether or not high GLCs are in fact occurring when the meteorological monitoring indicated they were likely, could save a large proportion (possibly 60–75%) of unnecessary control measures in such conditions.

14.4.6 Damage monitoring

In a way damage monitoring – using the word damage in the sense of any interference with normal processes – is effected at all times by the general public. They will complain if a pollutant offends their senses, if they think it is affecting their property or if they can detect some physiological effect.

The purpose of any damage monitoring programme should therefore be:

(a) To quantify any qualitative monitoring by the general public
(b) To detect damage of more subtle forms, not immediately apparent to the public at large.

The quantifying of damage is necessary if the correlations between damage and dosage, necessary to put any economic assessment of control processes on a rational basis, are to be established.

It is, of course, easier to suggest that damage should be quantified than to specify how this should be done, especially in the case of damage to health or amenity. However, we continually have to cost benefits and services in practice in all aspects of life, e.g. with regard to costs of medical treatment, so there is some basis of experience to help with this difficult task.

14.4.7   Dose monitoring

Damage will be related to dose rather than GLC and it may be desirable to use an instrument which simulates the exposure of the receptor, if it is not possible to measure the dose the receptor receives.

The directional dust deposit gauge [7] is a fairly early example of such a device. It is an attempt to simulate the soiling of vertical surfaces or the passage through windows and doors of airborne dust.

The measurement of GLC alone may lead to difficulties in relating damage to a particular source. As an example, if the deposition velocity of a pollutant on a surface were proportional to the wind speed and the pollutant concentration were proportional to source strength, but inversely proportional to the wind speed, one might find little correlation between damage and concentration. This could mean that fluctuations in concentration were due primarily to variable wind speed rather than to variable emission rates. It would be wrong to infer that the pollutant was not responsible for the damage. It is therefore important to ensure that all variables that are likely to affect the dose received by sensitive receptors, as well as ground-level concentration, are monitored.

14.4.8   Monitoring ground-level concentration (GLC)

The results of pollution measurement are usually expressed either as mass fractions (mass/mass of air), volume concentrations (vol/vol of air) or mass concentrations (mass/vol of air). The latter has, unfortunately, lately become the most widely used for gaseous pollutants. I say 'unfortunately' because mass/vol of air is not conserved when the pressure or volume of the air changes, and this can be misleading when one is comparing samples taken at different heights. The use of these units has arisen because sampling rates are usually expressed as volume of air per unit of time while the response of the receptor usually depends on the mass of pollutant sampled.

The conversion to concentration requires a knowledge of the volume sampling rate and the time taken to produce the response. As atmospheric concentrations fluctuate in time and space, the peak concentration reported during a given sampling period will depend in a rather complicated way on the volume sampling rate and the response time of the instrument.

If the sampling instrument is mounted in a vehicle, the movement of the vehicle will also effect some smoothing.

The highest peak readings recorded during any sampling period will therefore be given by the instruments which effect minimum smoothing, i.e. those with a fast response and, in the case of vehicle mounted samplers, a slow speed of traverse of the plume.

The requirements for high sensitivity are that as large a mass as possible of the pollutant should be collected, implying, for a given sensor, that as large a volume as possible should be sampled, i.e. the opposite requirement for fast response.

The response time of an instrument itself may be complicated by factors such as turbulent mixing of the sampled air within the instrument and adsorption and desorption of the pollutant on surfaces within the instrument.

If the instrument records continuously then readings could be taken far more frequently than the performance of the instrument merits, leading to unnecessary expense in data processing and storage.

Whether or not the measurement is expressed as concentration or dosage (concentration × sampling time), it is essential to include the sampling time as additional information. The minimum useful sampling time is about three times the response time of the instrument. The design of pollution monitoring systems will be discussed in Section 14.5.

14.4.9   Optical effects of pollutants

The optical effects of emission constitute one form of damage to the environment. The visible appearance of plumes may cause offence near the source, while some reduction in sunlight and visibility may occur at long distances, even on a global scale.

Measurements to assess optical effects may be made either by remote sensors or by point samplers such as the nephelometer. Such measurements are often regarded as pollution-monitoring rather than damage-monitoring. However, the optical effects of pollutants in the atmosphere are complicated by gas-to-aerosol conversion, the shape, size, refractive index and dielectric constant of the particles, all of which may change with the relative humidity if the particles are hygroscopic. The scattering will be directional: both scattering and absorption will be dependent on the light wavelength for a given particle size. Polarization may also be important.

Although good correlations between nephelometer readings of the light scattering coefficient and mass of suspended particulate have been demonstrated, these relations are necessarily of local application and such readings are therefore to be regarded primarily as an indicator of one of the effects of pollution.

Although most gases are colourless, some like NO_2 absorb visible light. A brown haze may therefore result either from scattering and absorption by particles of sub-micron size and/or from the presence of NO_2.

A very comprehensive collection of papers on plumes and visibility appears in White *et al.* [8].

14.4.10   Monitoring wet deposition

Measurement of rainfall is difficult because rain tends to swirl around the splash into and out of rain gauges (see Chapter 2). Further difficulties arise in the measurement of wet deposition because the pollutant will be dry deposited on the gauge surfaces in between rain events. It is therefore preferable to use a gauge which is sealed off when it is not raining. Such a device, which also exposes a filter paper to collect dry deposition when it is not raining, is described by Pattenden *et al.* [9].

Alternatively, the gauge is washed at intervals when there has been no rain, and the washings measured for deposition. For this procedure to be accurate, the GLC of the pollutant should be monitored as well, so that its deposition velocity can be estimated by dividing the deposition rate by the GLC.

14.5   The design of pollution monitoring systems

14.5.1   General

Having decided that pollution monitoring is necessary, one is faced with the prospect of reconciling:

(a) The monitoring requirements
(b) The funds available
(c) The staff available, and
(d) The instrumentation available.

One must decide how many samplers to purchase, whether they are to be installed in fixed patterns, in a vehicle or vehicles, or to be transported from one pattern to another in different conditions or at different stages in the monitoring programme.

The duration of sampling should also be anticipated and the minimum averaging period (see Section 14.5.2) specified.

14.5.2   Choice of minimum averaging period

Assuming that there is a choice of samplers, one of the most important factors determining the final selection will be the shortest averaging period that is thought to be relevant to the problem under investigation. In general, a faster response instrument is more difficult to operate, more expensive and produces more data to handle than a corresponding slow response instrument. On the other hand, if the peak concentrations over periods of a few seconds or minutes are of paramount importance, daily or hourly average readings may have little relevance. It is small consolation to the condemned man to tell him that the daily average HCN concentration in the gas chamber is well below the MAC (maximum allowable concentration). The averaging period should therefore be included in the definition of each MAC. If MACs have been defined, then the minimum averaging period included in their definitions could determine the minimum averaging period for a monitoring programme.

It is difficult to lay down any general guidelines, but the following give some indication of requirements for different applications:

Order of minimum averaging period	*Type of survey*
10 s	Mobile sensors, acute respiratory effects; studies of puffs
3 min	Useful for studying acute health effects if faster response not available
1 h	Time average concentrations; dispersion studies, diurnal changes, discrete source studies; damage to plants
24 h	Effects of weather systems; chronic health effects; area source studies
1 month	Seasonal and annual variation; long-term effects from global source

A faster response instrument than that specified will be more than adequate for the task indicated, *provided* the sensitivity and zero stability are adequate (see Section 14.5.3).

14.5.3   Choice of instruments

Having fixed the minimum sampling period one may be faced with a choice of instrument. If possible, discussion with groups already monitoring in the same field is desirable, but progress in instrument development is so rapid that this is often not possible.

The following factors must all be considered carefully:

(a) Is the response time short enough?

(b) Is the response specific to the pollutant considered; if not, what interference can be expected from other pollutants?

(c) Cost of the instruments

(d) Estimated running cost

(e) Quality and number of staff required to operate the network and provide first-line maintenance

(f) Back-up available from manufacturers in the event of serious malfunction (whether replacement instruments are available if some have to be taken out of service)

(g) Power requirements

(h) Is the output suitable for the anticipated data transmission/recording system?

(i) Size and weight

(j) Housing requirements

(k) Zero stability and zero checking facility

(l) Sensitivity, calibration stability and calibration check (preferably at more than one point)

(m) Has the instrument an established record of trouble-free running in the field?

(n) Are the weather conditions to which the proposed network is to be exposed likely to create difficulties not experienced in previous use?

Continuously recording instruments are generally not the best for monitoring low background concentrations of pollutants. In this case it is better to pass a large volume of air through some suitable collector (filter, scrubber, etc.) and then determine the total quantity of pollutant collected by a suitable analysis technique. It is important that the pollutant be fixed on collection, so that it does not evaporate or disappear in some other way later in the sampling period.

The efficiency of filters may change as they become clogged with collected material – a factor which may lead to mis-estimation of particulate matter collected in clean air [10].

14.5.4 Choice of mobile, fixed, transportable or combined sampling system

The development of compact, fast response instruments with modest power requirements has made the choice of a mobile sampling facility increasingly attractive in recent years. If the facility also includes remote sensing equipment, such as the correlation spectrometer, or dual frequency lidar, differentiation between effects from large, high-level sources and adjacent low-level sources becomes much easier. Data transmission difficulties are avoided and far fewer sampling instruments are needed – a great advantage since the cost of the samplers is probably at least an order of magnitude greater than that of older, slow response instruments.

The disadvantage of mobile sampling is that, unless aircraft are used, sampling is restricted by the available road network. Interference from pollution produced by vehicles or other adjacent small sources (or even the sampling vehicle itself) may be difficult to quantify if point samplers alone are employed. The influence of individual small sources upwind on ground-level concentration measurements may be very important, e.g. if SO_2 is being sampled, readings in the wake of a single house 50 m upwind, a hamlet of 100 homes 1 km upwind and town of 10^4 houses 10 km upwind could all be comparable if coal or some other sulphur-containing fuel were being used. Sheltering effects of hedges, copses, etc. can reduce readings by a factor of 2 or more for pollutants which are taken up by vegetation. Undulating country produces variability, reduced ventilation in valleys may trap pollution from local sources and shelter samplers from the effects of distant sources. Increased turbulence on ridges and on lee slopes may bring down pollution from elevated plumes which has been passing over lower ground nearer the source.

These effects will tend to produce random fluctuations in the readings of GLC from a mobile sampler and systematic differences between samplers in a fixed network. The success of a fixed sampling system will therefore be very dependent on the care with which the sites are chosen in regard to the prime purpose of the monitoring programme.

A transportable system has the advantage that sites can be changed if they are found to be unsuitable, or the network can be varied to suit changing conditions: e.g. it can be set out each day in what is anticipated to be the downwind direction. The power requirements, size and weight of the instruments available would be important factors in such surveys. Data reduction could also present problems if other than average values over the period of exposure were required, as individual records (autographic or possibly magnetic tape from cassette recorders) would have to be synchronized and processed after the event. On-line use of the data would be difficult.

A mixed system, with a mobile facility backed up by carefully selected continuous recording at a few selected sites, is an attractive option if sufficient funds and staff are available.

14.5.5   How many pollutants should be monitored?

If all sources produced pollutants in the same ratios, if all subsequent depletion processes were the same and if there were no chemical reactions in the atmosphere, monitoring the concentrations of one pollutant would be sufficient to quantify the behaviour of all the others. This situation may be approached fairly closely in the vicinity of a single source, provided there is no appreciable background, but in any comprehensive area or distant-source survey, full understanding of the dispersion and modification processes may require monitoring of several secondary as well as primary pollutants. Once again, it is difficult to lay down any general guidelines.

Measurements of a second pollutant may provide useful information on the origins of the first. For example, if smoke is emitted only from low-level sources and SO_2 from both high and low levels, then occurrence of high SO_2 GLCs only in association with high smoke readings would be a strong indication that most of the observed SO_2 came from low-level sources as well (see also Section 14.7.4.6).

The interest in the incidence of acid rain in recent years makes it imperative to obtain a complete ion balance in rainfall composition measurements, if this is at all possible. Organic, as well as inorganic, ions usually play a role of some significance [11].

14.5.6   Height and exposure of samplers

In general there will be a change of pollutant concentration with height. Near individual sources this will be due to the location of the source. For example, with ground-level sources the concentration will decrease with height; with elevated sources the concentration will increase with height until the axis of the plume is reached. If the pollutant is absorbed by the surface or projections from it, there will be an increase in concentration with height due to the depletion of the pollutant in the surface layers, once the gradients due to initial source location have been sufficiently smoothed out.

In general, the more turbulent the flow, the less gradient there will be near the ground, and conversely in stagnant conditions or very smooth flow, pollution from distant sources will tend to disappear near the surface. This is one of the reasons for the spatial variability in pollution already mentioned in Section 14.5.4.

It is therefore desirable to site sampling inlets at a reasonably constant height not too close to the surface (say 2 m), to avoid obviously sheltered locations where absorption can begin above the sampling height (e.g. orchards, near hedges, woods, high walls, etc.), especially if the pollutant is expected to be taken up at their surfaces. In undulating country, it should be recognized that location may be an important factor and that full effects of a distant source are unlikely to be experienced in valleys and possibly on low ground upstream of a ridge. On the other hand local low-level emissions may be trapped and give locally high concentrations at sheltered locations.

When monitoring within area sources such as a city, the important consideration is the spatial variability of pollution within the area. For a given emission density, as the area becomes larger and the general level of pollution rises, the effect of nearby individual small sources becomes relatively less important compared with the combined effect of all the distant sources, but may still be predominant in unfavourable locations. If we choose ideal sites, with no local sources (e.g. the centres of parks), these will not be typical. Pollution levels may in fact be unrealistically low compared with inhabited areas, as the development of stable layers over the relatively cool open spaces may effectively cut them off from the sources at times. In these conditions, GLCs may fall dramatically if the clear area is a sink for the pollutant(s) considered. It would therefore seem to be better to choose representative sites with an average density of development for the part of the city which the site is intended to represent.

If information of the vertical profile of pollutants is required, it may be possible to site samplers on suitable towers. An alternative is to locate the intakes on the tower and the sampler at ground level [12]. Careful assessment of the likely losses, time delay and time smoothing effect of the connecting tube is necessary with this arrangement.

14.5.7 Layout and spacing of instruments in fixed surveys

14.5.7.1 *General*

A wind direction measured at one height at one location over a relatively short period ($\leqslant 1$ h) is generally an imprecise indicator of the direction of a plume from a particular source in the vicinity of the wind recorder. Marked changes of wind with height (sea-breezes, etc., see Chapter 2) aside, it is probably safe to conclude that the best estimate of plume direction deduced from such a wind direction measurement would fix the plume as being within a 45° sector centred about that direction, except in convective conditions with light winds, where the uncertainty could be greater. The only way one could be reasonably sure of measuring a concentration near the axis (peak value for the distance considered) from such a plume would be to have a fairly dense network of instruments as discussed in Section 14.5.7.2.

On the other hand if one wishes to arrive at the frequency distributions in, for example, different wind directions, or at different distances from the source, much of the information from such a dense network would be redundant, so different spacing criteria are necessary (Section 14.5.7.3).

The number of sites that are operated in a network within an area source depends largely on the resources available for, and the objects of, the survey (Section 14.5.7.4).

14.5.7.2 *Discrete source surveys – verification of dispersion models*

The main requirement in the verification of dispersion models is accurate location of the plume axis. As wind direction measurement does not generally enable one to do this (Section 14.5.7.1) we have to arrange the instruments in arcs (of at least three instruments) in such a way that the average reading of the instrument giving the highest reading is not less than some large fraction (say 0.9) of the axial value (C_A).

Assuming that a Gaussian distribution of material across wind (i.e. concentration C) at a point θ radians off the plume axis is given by:

$$C = C_A \exp - \theta^2/(2\theta_y^2)$$

where θ_y is the standard deviation of the cross-wind distribution in radians), the above criterion means that

$$\frac{2}{\theta_s} \int_0^{\theta_s/2} \exp - (\theta^2/(2\theta_y^2)) \, d\theta > 0.9$$

i.e.

$$\int_0^{t=\theta_s/(2\theta_y)} \exp - (t^2/2) \, dt > 0.9(\theta_s/(2\theta_y)) \tag{14.1}$$

where θ_s is the angular spacing between samplers. Expression (14.1) implies

$$\theta_s/(2\theta_y) < 0.8$$

Taking 0.08 radians as a typical value of θ_y [13] over the 1-h sampling period, we have

$$\theta_s < 0.128 \text{ radians}$$

i.e.

$$\theta_s < \text{about } 8°$$

so the maximum angular spacing of samplers should be about 7.5°.

This separation also permits estimation of θ_y as well as C_A.

The *distance* of maximum concentration from a single emitter will be determined by the height of the plume axis and the dispersive properties of the atmosphere.

If the vertical spread of the plume increases as some fractional power ($1/n$) of the distance downwind and the cross-wind spread is linear with distance, then the distance of maximum concentration will increase as (plume height)^n. For tall chimneys the average distance of the maximum appears to be about $H^2/12$, where H is the plume height at the average wind speed (Chapter 2), but individual maxima occur over a wide range of distances.

The variation of GLC with distance is relatively flat in the vicinity of the concentration maximum, so the distance does not appear to be very critical. Figure 14.2 shows one of several types of GLC pattern observed in the Plains Site Project (Section 14.8.2.1).

Minimum requirements for the sampling pattern are: arcs of at least three instruments at three distances or at two distances supported by single instruments nearer to and further from the source than the arcs and if possible between them. One of the arcs should be at the distance where the maximum GLC is expected to occur most frequently, calculated as indicated above. A second arc should be about half this distance as high concentrations occur most frequently at about half the average distance. See Chapter 2 for details of calculations of distance of maximum GLC.

14.5.7.3 *Discrete source surveys – statistics of incidence of various levels of pollution*

In this case the principal object would be to see how the frequency of occurrence of various levels of pollution from the discrete source(s) being investigated varies with direction and/or distance from the source.

To assess the variation with direction, the sites would normally be set out in a ring or concentric rings centred at the source. The spacing requirement is the opposite to that of Section 14.5.7.2: what is now required is that, when one site is in the plume, the adjacent sites are normally unaffected. If we regard unaffected as recording $<10\%$ of the average recorded at the 'in-plume' sampler, again assuming a Gaussian distribution we have

$$0.1 \times (2) \int_{0}^{\theta_s/2} \exp(-\theta^2/2\theta_y^2) \, d\theta > \int_{\theta_s/2}^{3\theta_s/2} \exp(-\theta^2/2\theta_y) \, d\theta$$

as the criterion in this case.

Taking $\theta_y = 0.08$ radians, this criterion is satisfied if

$$\theta_s > 0.22$$

i.e. the separation of the sites should be *at least* $13°$; $22.5°$ appears to be adequate for most locations [15].

If resources are sufficient for only one ring, this should be set out as close as possible to the estimated distance of maximum ground-level concentration (Section 14.5.7.2). It would be an advantage, in this case, to have additional sites at around half and at twice the ring radius in the direction of the wind which is expected to give the most frequent and/or highest ground-level concentrations.

If important topographical features are present, sites should be chosen where the maximum effects from these features are expected, in addition to, or possibly in place of, the regular pattern.

14.5.7.4 *Area surveys (i.e. sites within the source area)*

It is more difficult to specify the number of spacing of sites in an area survey. In general it will depend on the resources available for, and objects of, the

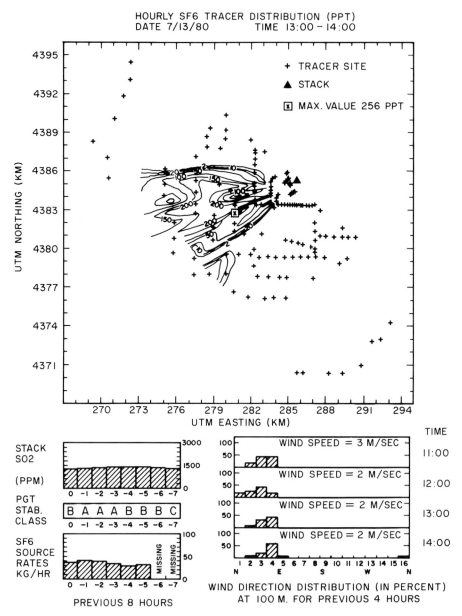

Fig. 14.2   Example of Coherent type of GLC (in pptv) observed at the Plains Site. From Bowne and Londergran [14]. Copyright © 1983, Electric Power Research Institute. EPRI EA-3074, *Overview, Results, and Conclusions for the EPRI Plume Model Validation and Development Project: Plains Site.* Reprinted with permission.

monitoring exercise. The correct order of density seems to be something like $\sqrt{N}$ where N is the population in tens of thousands for cities over 250 000 (i.e. a minimum of five sites) or R, where R is the radius in kilometres, where the situation is not complicated by topography. In the latter case, instruments should be sited in areas which are expected to be specially difficult.

A number of objective methods for calculating the spacing of sites have appeared in the literature but they all presuppose some knowledge of the spatial variability of pollution and the correlations between readings at different sites, neither of which would be known until some measurements have been made. Objective methods include principal components and cluster analysis. For a brief discussion and further references see Munn [16].

In general the centre of GLC will be displaced from the centre of emission by an amount which depends on the effective source height of the emissions. Dilution factors will be much greater for high-level emissions than for low-level emissions and the meteorological conditions resulting in the minimum atmospheric dispersion will also be different (Chapter 2). It is therefore important to site samplers near the centre of emission and downwind in the prevailing direction for stable, light wind conditions (for low-level emissions) and downwind in the prevailing directions for strong winds, and for convective conditions to observe maximum effects from high- or medium-level emissions.

14.5.7.5   *Distant source surveys*

Most of the requirements for siting instruments for this type of survey were mentioned in Section 14.5.6. To recapitulate:

(a) Effects of local, low-level sources should be avoided
(b) Sheltering effects by absorbing features (hedges, copses, etc.) should be avoided
(c) Sheltering effects resulting from reduced atmospheric dispersion (siting in valleys, etc.) should be recognized and considered in relation to the overall objectives of the survey.

14.5.7.6   *Global effects surveys*

If a pollutant is to have a global effect, it is likely to have a long life in the atmosphere and therefore it is unlikely that sheltering effects will be important. The major consideration will therefore be the choice of sites in which fluctuations caused by individual sources are a minimum, i.e. as remote as possible from human activity.

14.5.7.7   *Multi-purpose surveys*

Although a survey may have been designed for a specific purpose, e.g. monitoring a point source, it may also provide much useful information on other features of the pollution in the locality. It may be possible to maximize this subsidiary function at little additional expense by consideration of such possibilities at the planning stage. However, one should avoid the pitfall of trying to use information from existing surveys for some purpose other than that for which it was intended, if there are serious deficiencies in the data for that purpose. For example, although it may be reasonable to estimate the likely distribution of 24-h readings from a dispersion model predicting hourly average concentration distributions, it would be unwise to attempt to modify the dispersion model in light of any discrepancies which arose between observed and predicted 24-h readings. Comparison between a wide range of observed and predicted 1-h readings would be the only reliable way to do this.

14.5.8 Mobile monitoring

14.5.8.1 *Surface systems*

The location of the vehicle at any particular time must be recorded if the measurements made by a mobile system mounted in a motor vehicle are to be meaningful. The simplest way of doing this is simply to mark autographic recordings of the monitored pollutant(s) when recognizable features (intersections, etc.) are passed. The time the vehicle was at any point can then be deduced from the route map by interpolation. However, if the object of the survey is to produce measurements, at, for example, direction intervals of 1° referred to an origin at some prominent source, the reduction of such records can be a time-consuming business. Automatic recording of distance travelled and bearing, using a gyro or corrected magnetic compass, greatly facilitates such analysis. Data logging and if possible some computing facility should be available in the vehicle to support the navigational equipment.

Systematic traversing of the plume at one or more distances downwind is probably the most fruitful procedure in single-source monitoring. However, if the standard deviation of the cross-wind distribution of plume material is 5° (about 0.08 radians), a traverse which takes the vehicle out of the measurable plume at both sides will probably be at least 5 × 5°, i.e. 4 km at 10 km downwind. Allowing for road orientation not being ideal and turn-round at the end of each traverse, one is hard pressed to approach a total of ten traverses an hour. Four or five is a more realistic average figure, especially if a change of distance is involved between traverses.

Meandering of the plume tends to distort the pattern and so each traverse gives only an approximation to the instantaneous cross-wind spread of the plume, therefore several hours sampling will be required to define the plume envelope in a way that could be related to patterns observed from fixed networks.

14.5.8.2 *Airborne systems*

Aircraft are generally expensive to operate but have the advantage of greater freedom of movement, three-dimensional capability and higher speed of traverse.

Location now involves height as well and it is not possible to use the dead-reckoning method of position-fixing because of the effect of wind drift. On-line presentation of processed data to the observer is more difficult than in a motor vehicle; especially if a light aircraft is involved.

Aircraft surveys are therefore more applicable to specialist investigations, especially those involving chemical reactions in the atmosphere, rather than obtaining comprehensive statistics on dispersion. However, a few interesting data on cross-wind distributions after several hundred kilometres of travel over water have been obtained [17].

Figures 14.3–14.5 illustrate this.

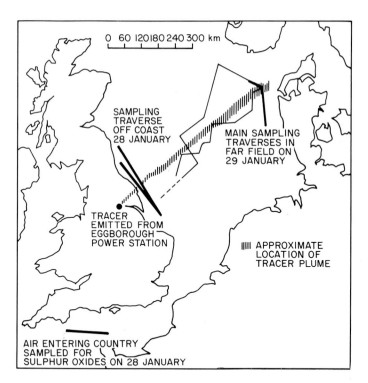

Fig. 14.3 Flight path along main sampling traverses on 28 and 29 January 1981. (Reproduced from an internal CEGB report.) Copyright © 1983, Electric Power Research Institute. EPRI EA-3217, *The Fate of Atmospheric Emissions Along Plume Trajectories Over the North Sea.* Reprinted with permission.

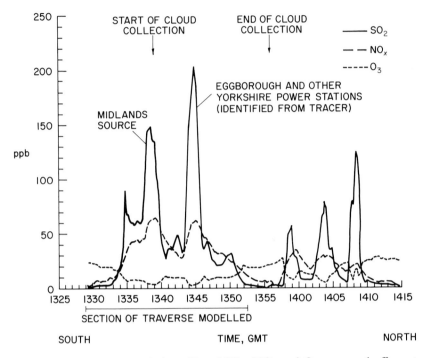

Fig. 14.4 Horizontal cross-wind profiles of SO_2, NO_x and O_3 measured off coast at 300 m altitude, 120 km from Eggborough. (Reproduced from an internal CEGB report.) Copyright © 1983, Electric Power Research Institute. EPRI EA-3217, *The Fate of Atmospheric Emissions Along Plume Trajectories Over the North Sea*. Reprinted with permission.

14.6   Data handling

14.6.1   Data transmission

If the information from the monitoring network is to be used on a real time basis then transmission to some control point will be essential. Even if the data are not to be used on-line for control purposes, data logging and inspection of the performance of the monitoring network are facilitated by such transmission. In general, each sensor will transmit at some predetermined interval and it is desirable that the value of the variable transmitted be averaged over the period between transmissions. Radio or telephone lines may be used.

In the absence of any control facility, the signal will still have to be transmitted from the sensor to some suitable recording device located nearby and in the case of electrical outputs it is important that the connecting cable be protected from any interference.

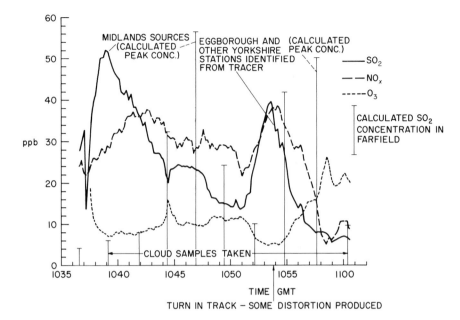

Fig. 14.5 SO$_2$, NO$_x$ and O$_3$ profiles measured at 390 m altitude, 660 km from Eggborough. (Reproduced from an internal CEGB report.) Copyright © 1983, Electric Power Research Institute. EPRI EA-3217, *The Fate of Atmospheric Emissions Along Plume Trajectories Over the North Sea*. Reprinted with permission.

14.6.2 Data storage

For on-site recording, magnetic tape, either in a cartridge or cassette, is now preferred to autographic recording. Accurate time marking is essential if the data are to be compared with variables recorded at other sites. More expensive pieces of equipment, e.g. sodars, may have micro-processor or computer-based on-site data reduction, with subsequent storage on floppy disc.

A central data logging facility would probably use hard disc rather than tape. Data formatting is usually a trade-off between immediate comprehension (data stored e.g. as ASCII characters) and high density (e.g. binary characters).

In choosing the interval between readings for digital recording the following factors are important:

(a) The response time of the instrument
(b) The expected minimum time for an effect from the pollutant on the receptors of interest to be noticed
(c) Limitations to data storage and expenditure or computing time.

On-site autographic recording had many advantages, including back-up for the digital system and quick on-site assessment of instrument malfunction. However, equipment reliability has improved considerably in recent years and it is now very easy to create graphs of the data on site with the aid of a portable microcomputer. It is now, therefore, hardly necessary to run autographic recorders in parallel with digital recording equipment.

14.6.3 On-line alarm/display systems

The alarm indicates to the plant control engineer when a situation requiring remedial action has arisen. It may be audible, visible or both.

The network data may be displayed on a map, either digitally or by dials, or by a print-out. The print-out may be produced at intervals automatically or by interrogation. If the central logging system includes a computer, the display could include averages, peaks, standard deviations, etc. over longer periods (e.g. over 1 h if the data were recorded every 1–5 min). In this case the alarm might operate when a given time average value was exceeded or a level was exceeded for more than a given number of scans.

Presentation for subsequent assimilation is dealt with in Section 14.7.

14.6.4 On-line recognition of defective readings

It is obviously undesirable to initiate expensive control measures on the basis of high readings of pollutant concentration which subsequently are shown to have been erroneous. However, the elimination of erroneous data is often a difficult business because one is constantly seeking the abnormal, and elimination of extreme readings as suspect could defeat the object of the exercise. Familiarity with the instrumentation will normally expose fairly well-defined failure patterns and the bulk of defective readings can usually be eliminated by setting up suitable criteria to cater for these. The elimination process is greatly aided if there is some redundancy in the network, and/or built in calibration facilities and zero checks.

14.7 Analysis of results

14.7.1 General

In addition to the data available from the monitoring network, there will be background information which is of relevance (historical data, for example (Section 14.7.2)) and possibly concurrent monitoring of relevance to the investigation by other organizations (e.g. meteorological data by the national weather service). To some extent the choice of an area for specialized monitoring may be influenced by the existence of useful parallel monitoring or historical data.

Correction of defective readings on-line was mentioned in Section 14.6.4; retrospective correction is discussed in Section 14.7.3.

The presentation of valid data is discussed in general terms in Section 14.7.4 and physical models in Section 14.7.5. Examples follow in Section 14.8.

14.7.2 Availability of historical data

14.7.2.1 *Source inventories and characteristics*

It is possible that some emission monitoring (or function monitoring) will permit hourly estimates to be made of emissions from some or all of the sources affecting the sampling area. However, it is more likely that only general historical information, such as total sales of fuel to householders, traffic density in selected census periods, etc. will be available. Even so, it should be possible to build up a reasonably accurate estimate of emission from the various types of source at different times of the day and year and in different weather conditions. In addition to estimates of the emission from each source, an estimate of the effective height of release should be made. This will vary with meteorological conditions and to some extent with rate of emission. In many cases it is likely that the plumes from small sources with short chimneys will be subject to interference from the wakes of adjacent buildings, so the effective height in moderate or strong winds may be less than the actual height of release.

The location of sources with respect to any marked topographical features (Section 14.7.2.3) is also important. See Chapter 2 for a description of methods of estimating plume rise and the effects of buildings and topography.

14.7.2.2 *Climatological data*

The following information is likely to be of use:

(a) Frequency of occurrence of different wind speeds at different times of day in different direction ranges (if available over range of effective source heights) including local winds such as sea breezes, etc.
(b) Frequency of occurrence of different mixing depth (in association with (a) if possible)
(c) Strength and extent of stable layers
(d) Rainfall and humidity
(e) Sunshine and cloud cover
(f) Atmospheric pressure.

14.7.2.3 *Topographical information*

The existence of any marked topographical features which prevent the spreading out of pollution, disrupt and prevent the proper rise of plumes, or reduce the effective height of plumes passing overhead, should be noted. In a well-designed survey they should have been taken into account when the location of sampling points was decided (Section 14.5). Important features include changes of surface characteristics at urban boundaries, woods and forests, coasts, rivers, etc.; and hills, mountains, valleys, etc.

14.7.2.4   *Experience of similar source networks or other information (e.g. epidemiological) on damage/dosage relations for the pollutants under investigation*

Such information as is available from earlier monitoring in approximately similar locations with similar sources would provide valuable assistance, both in planning the network and in deciding if the readings, when obtained, are reasonable. In the absence of proven dispersion or of other physical models, developing analogue prediction models with the aid of any additional available data may be the most practicable course (see also Section 14.7.4.6).

Where dosage/emission relations have been established, dosage/damage data from earlier field or laboratory studies will help determine or may have already established whether the observed pollutants levels are cause for concern.

14.7.3   Elimination of erroneous readings

Introduction of errors when samples are taken from field monitoring points to a central laboratory – often with a time lapse of several weeks – has been discussed by Paterson [18] who used a correlation technique for deciding whether or not observations were reliable.

Continuously recording instruments are normally not subject to common errors, except as a result of bad design or calibration technique.

Where there has been some change in zero or sensitivity between calibrations, it is usually possible to decide at which point the observations became unreliable if there is some redundancy in the network, e.g. if the readings are normally correlated with those from adjacent sites and the correlation pattern changes markedly. If more than one pollutant is being monitored, correlations between pollutants may become anomalous. However, some faults, e.g. volume sampling rate errors, may be common to all pollutants.

Calibration checks should, if possible, involve at least one intermediate point in addition to checking zero and full-scale readings, otherwise changes in response characteristics (e.g. linearity) may not be detected.

14.7.4   Statistics

14.7.4.1   *General*

The display of the collected results is probably the most important part of any monitoring exercise, if it is to be of any general use. It must also be borne in mind that many of those who may need to make decisions based on such information are not experts in chemical analysis, meteorology or statistics so the presentation of the results in an easily assimilable manner is very important.

14.7.4.2   *Mean values over specified averaging periods*

Because the concentration of pollutants fluctuates in time, the highest concentration measured during a given sampling period decreases as the length of the averaging period increases.

The mean of all the 3-min periods in an hour will, of course, be the hourly mean, but we might, for example, compare the highest 3-min mean during an hour with the hourly mean value and express this ratio of the highest mean over the shortest averaging period divided by the mean over the whole sampling period as a peak/mean ratio.

In general the peak/mean ratio *decreases* as the mean increases, because a high mean value usually implies that pollutant plume directions have remained steady during the sampling period, so there is less variability in short-period average concentrations, and also that the sampling point is near the centre rather than near the edge of the pollutant cloud distribution over the sampling period.

It is therefore important to qualify information on peak/mean ratios by such statements as 'at the point of maximum (hourly average) concentration' or 'for occasions when the hourly average GLC exceeded 1 pphm', if they are to be interpreted and applied correctly. (1 pphm = 1 part per hundred million by volume.)

14.7.4.3   *Frequency distributions*

The frequencies with which given concentrations, for given averaging periods, are exceeded at each sampler, together with the absolute maximum concentration, are probably the most important statistical data provided by the survey.

Separate frequency distributions may be prepared for different periods of the day or year (e.g. night/day, summer/winter), for different wind directions or weather situations, etc., if sufficient data are available. Data from several sites may be combined in some or all of these distributions.

Other useful information is the distribution of periods for which a given concentration is exceeded during a day, year or similar interval. The data are sometimes plotted on log probability paper, which has the advantage that if the frequency distribution is log normal the cumulative frequency distribution appears as a straight line, the 50% value is the geometric mean and the slope of the line gives the geometric standard deviation.

14.7.4.4   *Diurnal or annual variations*

The readings over specific periods of the day, if readings have continued over several days, or specific parts of the year, if readings have continued for several years, may be averaged to investigate diurnal or annual variations. Some measure of the scatter of the individual readings about the mean values should also be obtained to enable one to decide whether any cyclic variations which appear are significant. It may be convenient to subdivide the data according to wind speed and/or cloud cover if diurnal effects are being investigated.

14.7.4.5   *Mean values in different weather situations and/or wind directions*

Both the emission and the dispersion of the effluent material before it returns to ground level may be affected by weather conditions. Subdivision of data by time of day or year (Section 14.7.4.4) will, to some extent, take account of this, but further investigation of the effect of temperature, vertical temperature gradient, wind direction and/or speed may be desirable.

Any marked effects should be apparent for mean values but the scatter of the individual readings from which the means were deduced is also important in assessing significance.

A convenient method of displaying wind direction effects is the pollution rose (see Section 14.8).

14.7.4.6   *Correlations and regression analysis*

Correlations between concentrations of the same pollutant and any of the following measured over the same or different time intervals may be of interest:

   (a) Meteorological variables
   (b) Other pollutant concentrations at the same site
   (c) Damage
   (d) The same or other pollutant concentrations at other sites.

Correlations between successive measurements of one pollutant at the same site at different time intervals (auto-correlation) are also useful (e.g. in deciding the distance of the principal sources affecting the site).

If multiple regression techniques are used it must be remembered that relations with no physical basis are not likely to be applicable elsewhere unless data from other locations (Section 14.7.2.4) have been included in their development. If possible, the relations developed should have their basis in some simple physical model. It must also be remembered that relations developed by statistical techniques only are not sufficient to establish causal relations between variables especially where dosage/damage relations are concerned. The existence of a third, common, unmonitored factor should always be suspected if it is difficult to establish the causal relation independently.

Developments of multiple regression techniques include eigenvector or principal components analysis and the closely related factor analysis. These involve the creation of uncorrelated ranked factors which account for ever-decreasing fractions of the variance of the pollution data.

These factors may then be related to particular types of source, meteorological variables, etc.: see Benarie [19] for further discussion of statistical methods and Parekh and Husain [20] for an example of the application of factor analysis to identify source types responsible for episodes of high aerosol concentration in New York State.

14.7.5 Evaluation of physical models

14.7.5.1 *General*

Physical models were discussed in Chapter 2. Surveys may be designed specifically to test such models, or the data from surveys whose principal purpose was to provide information, i.e. statistics, may be used in model evaluation. Since it is assumed in the model development that the ground-level patterns of concentration and deposition are determined by independent variables, it is a requirement that these, as well as the pollutants, should have been measured if the models are to be assessed satisfactorily. It is also essential that the pattern measured was actually the result of emissions from the source or source area under investigation. This involves knowledge of the background from sources which have not been considered or the addition of some unique tracer to the effluent. The surveys may be divided into typical ranges of source receptor separation distance as indicated in Sections 14.7.5.2 to 14.7.5.4. Some examples of surveys which have been used to test or evaluate physical models are given in Section 14.8.

14.7.5.2 *Near field (up to 25 km)*

Single source studies

The evaluation of physical models using data from near field surveys depends to a large extent on the expectations from the modelling.

In the limit, one might have made such extensive meteorological measurements that only the equation of continuity was being tested.

However, it is unlikely that such detailed atmospheric structure could be predicted in advance of the event, so to be useful it seems reasonable to use in model evaluation only those independent variables which are likely to be forecastable.

If this is done it appears that the weather, on an hour-to-hour basis, may be subdivided roughly into 3 (or at most 6) stability categories, 3 (or at most 9) wind speed classes and 8 (or at most 16) wind direction classes. The ensemble average maximum GLC for such a category should, as we saw in Chapter 2, be predictable to within perhaps $\pm 10\%$ and the highest observed value within the data set to within about $\pm 20\%$. However, the maximum on a given occasion can be within a range of more than twice the ensemble average down to zero and its location may often be outside the area predicted to be affected by the plume.

Therefore, a model may appear satisfactory in that it is capable of locating the ensemble maximum and predicting its magnitude with reasonable accuracy. However, if the r.m.s. error in predicting the hourly mean plume direction is about 0.1 radians, i.e. about 6 degrees, there is no way in which accurate predictions of GLC at a fixed site can be made on an hour-to-hour basis.

Multiple source studies

In this case at least some of the samplers will be within the source network. This will certainly be the case where urban models are being evaluated. In these cases only real emissions can be investigated, although tracers could be used to evaluate the dispersion characteristics of particular source types. Accurate source inventories are therefore very important.

Background concentrations from out-of-town sources may still be of some significance, especially where secondary pollutants are involved.

Removal processes are of importance in stagnant meteorological conditions and will also affect the vertical profiles of pollutants which are deposited on the surface.

14.7.5.3   *Medium range (20–250 km)*

The effects of removal processes will become increasingly important as distance downwind increases. However, plumes from individual sources, or small groups of sources, can still be identified and ratios of concentrations and fluxes of tracer compared with those of reactive pollutants can help in the determination of removal or conversion rates. If there are emissions between those sources whose behaviour is being modelled and the survey sites, it is essential that their strengths and locations are known. One problem is the role of material, particularly OH radicals and ozone, already present above the mixing layer and brought into it by subsidence or penetrative convection.

14.7.5.4   *Long range (>250 km)*

The use of field data in this range of distance from the source usually consists of:

(a) Checking the validity of trajectory calculations, and
(b) Checking and tuning the values of conversion and removal rate parameters.

The evaluation will be very sensitive to the accurate estimation of the effects of near field sources. It follows that surveys where there are no emissions between the sources under investigation and the survey sites are most likely to give accurate model evaluation.

As far as modelling wet deposition is concerned, accurate three-dimensional back trajectories for air arriving over the survey sites while it is raining are needed. This can be checked to some extent by identifying pollutants characteristic of each major source area. This means that sampling intervals should be short enough to ensure that this can be achieved.

At very long ranges, the residual pollution in air which has passed through the rain systems will return to the surface. The trajectories required to model this behaviour will be even more difficult to construct than those required to model the wet deposition.

14.8 Examples of monitoring networks and data presentations

14.8.1 General

The principal features of some examples of the various types of survey listed in Sections 14.5.7 and 14.5.8 are given in Sections 14.8.2–14.8.6. Multi-purpose surveys (Section 14.5.7.7) are not included separately, but where a network has been used to provide information on subjects other than the major purpose of the investigation, these are mentioned.

The discussions are restricted to major programmes mostly occupying periods of a year or more, but are not intended to be comprehensive.

14.8.2 Discrete source surveys

14.8.2.1 *US power plant studies*

Following the comprehensive LAPPES [21] study funded by the US Government the Electric Power Research Institute (EPRI) initiated the Plume Model and Validation (PMV and D) project. This falls into three parts, concerned with (a) flat, (b) moderately complex and (c) mountainous terrain.

It is intended to provide definitive information on the behaviour of hot plumes from tall chimneys, and to provide sufficient concurrent source and meteorological data in three dimensions to enable mathematical plume models to be fully evaluated. A hands-off evaluation of existing models is to be followed by refinement and development where necessary. The flat terrain (Plains Site) experiment has already been evaluated [14].

The field measurements included:

(a) 28 fixed, 2 mobile and 1 airborne aerometric stations
(b) Up to 200 transportable tracer (SF_6) samplers
(c) 1 mobile particle, 1 mobile SO_2 and 1 airborne particle LIDAR for measuring plume cross-sections
(d) 2 acoustic Doppler sounders, 1 double theodolite T-sonde system and a 100-m meteorological mast
(e) Source sampling included efflux velocities and temperatures and emission rates for SO_2, NO_x and SF_6.

The plume validation was regarded as unsatisfacotory, probably because of the optimistic expectation that individual hour-to-hour peak concentrations could be predicted accurately (Fig. 14.6).

14.8.2.2 *CEGB Midlands Region studies of SO_2 around power stations*

Continuous ground-level recording (autographic) of SO_2 (minimum averaging period: 3 min) with patterns of up to twenty-two recorders at West Burton (Fig. 14.7) and High Marnham. These patterns conform to the spacing criteria for statistical point source surveys given in Section 14.5.7.3 [12, 15].

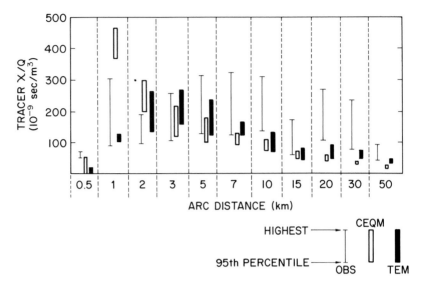

Fig. 14.6   Top observed 5% values of GLC/emission rate for SF$_6$ at each arc distance at Kinkaid Power Plant (Plains Project) compared with predictions by two Gaussian plume models (CEQM and TEM). From Bowne and Londergan [14]. Copyright © 1983, Electric Power Research Institute. EPRI EA-3074, *Overview, Results, and Conclusions for the EPRI Plume Model Validation and Development Project: Plains Site.* Reprinted with permission.

Figure 14.7 also illustrates one of the forms of data display-pollution roses (Section 14.7.4.5). The variability of average concentrations at instruments in similar locations, even with careful siting, is noticeable in Fig. 14.7. This is probably due to local sheltering (and possibly source) effects. The high readings with pollution coming from the north-west to south-west direction illustrate the dominance of distant sources over the large power station emissions.

Methods of identifying the power station contribution were: (a) recognition of a characteristic peaky trace above the relatively steady background contribution, and (b) comparing the results from adjacent recorders, subtending an angle of about 90° at the source, and subtracting the mean reading of the outer two as background from the readings of the inner three as plume. Plume readings ($\geqslant 1$ pphm), however, rarely showed on more than one meter at a time [15].

Wind speed and direction were obtained from instruments on a 45 m tower at the power station site and supplementary data on lapse rate and 450 m wind were estimated by the Meteorological Office.

Frequencies of occurrence of 3 min or hourly average concentrations were displayed graphically as functions of wind speed. Other presentations of the data were frequencies of the occurrence of various concentration ranges for different averaging times. Some of these were subdivided by temperature lapse conditions, with further subdivisions by wind speed ranges and between winter and summer.

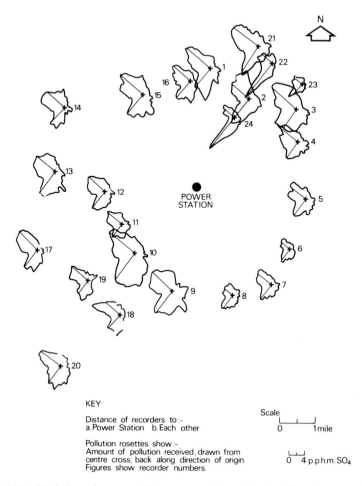

Fig. 14.7 Variation of average SO_2 concentration with wind direction at each recorder site at High Marnham.

Data display was mostly tabular, but log-probability plots were also used (Fig. 14.8). Occasions with exceptionally high GLCs were described in some detail. Variation with distance from the source was discussed by Martin and Barber [22].

Concentration distributions were compared with values calculated from dispersion equations, and background concentration distributions, mostly from distant sources in this rural area, were also tabulated. So, although the principal objective of the survey was acquisition of statistics of GLCs around power stations, it could also be regarded as a multi-purpose survey (Section 14.5.7.7) because of the important applications to dispersion models and distant sources.

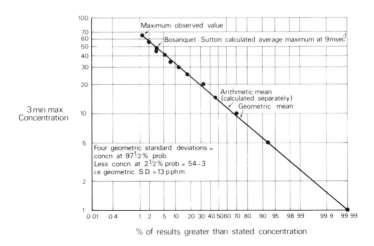

% of results greater than stated concentration

3 min maxima

Concn. (pphm)	No. of recorder-hours in which given 3 min max. was recorded. Recorders							
	12	11, 15	10, 14	7, 8, 9, 13.	4, 5 16, 17	1, 6, 18, 19, 20, 22	2, 23	3, 21 24
	Distance from source							
	$\frac{1}{2}$	3	$5\frac{1}{2}$	7	10	$12\frac{1}{2}$	14	17 (miles)
	0.8	5	9	11	16	20	23	28 (km)
0	173	322	258	419	452	635	409	285
1–4	27	10	19	62	61	91	22	22
5–9	46	36	44	115	118	67	16	16
10–14	13	35	25	82	36	18	5	1
15–19	8	17	9	30	7	9	0	1
20–24	6	26	7	9	6	5	2	1
25–29	8	8	6	6	2	1		
30–34	6	8	3	1	1			
35–39	5	4	0		0			
40–44	2	3	1		1			
45–49	1	2						
50–54	1	0						
55–59	2	1						
60–64	0	0						
65–69	2	2						
70, 71, 86, 90	1 each							
	131	152	114	305	232	191	45	41

Frequency of occurrence of given concentration ranges W 65/6 and S 66.

Fig. 14.8 Tabular display of 3 min maxima of SO_2 from High Marnham at different distances. Upper diagram shows distribution at 5 km plotted on log probability paper. (Reproduced with permission of Pergamon Press.)

The CEGB Tilbury/Northfleet and Eggborough/Drax surveys, which were model evaluation and combined evaluation/statistical surveys respectively were discussed in Chapter 2.

14.8.3   Area surveys

14.8.3.1   *Urban surveys*

The British Petroleum (BP) Reading survey [23–25] was a study designed to test dispersion and empirical models for a town of 120 000 inhabitants, spread over an area of about 100 km^2. There were forty monitoring sites for SO_2 (Fig. 14.9) – far more than would normally be installed in a routine survey (Section 14.5.7.4). Concentrations were averaged over 6 h.

Supporting meteorological data were made at a height of about 13 m from a fixed and a transportable mast and from a balloon sonde.

Detailed source inventories were prepared and estimates of emission from low-level sources were made from historical data (Section 14.7.2.1). Six-hourly fuel consumption figures were available for some of the larger sources.

Data were displayed on contour maps (Fig. 14.9 is an example) for all wind directions and for each direction octant.

Empirical expressions gave better correlations with the data than physically based dispersion equations. However, Pasquill [26] demonstrated that the physical models gave a good representation of the overall average value and frequency distribution. There were large discrepancies between individual observed and predicted values due, Pasquill suggests, to the practical uncertainties in specifying meteorological conditions precisely at all points and times.

An interesting report giving guidance on network design in urban areas based on sampling requirements and information available from the results of past surveys (e.g. at Nashville and Sheffield) is given by Blokker [27].

14.8.3.2   *Regional surveys*

The growth of interest in secondary pollutants in the last decade because of their involvement in visibility degradation and dissolved material in rain is reflected in the recently completed SURE experiment (Sulphate regional experiment). This was another EPRI funded project involving 54 ground-based aerometric stations, up to five aircraft and special sonde ascents to supplement the available standard meteorological data. In addition an up-to-date source inventory for the whole of the eastern USA.

The major objectives of the project were:

(a) Compilation of factual information on the transport and fate of south and north compounds originating in the eastern USA.
(b) Reliable description, based on these data, of the concentrations of compounds such as SO_2 and SO_4^{2-}.

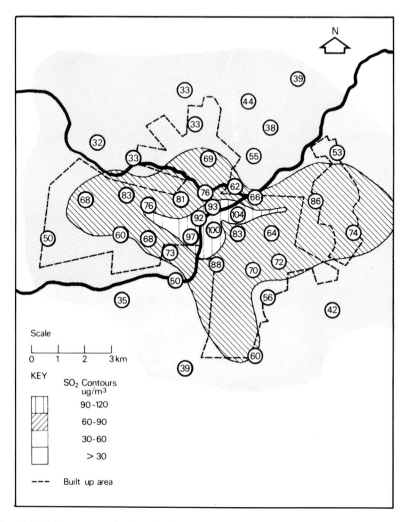

Fig. 14.9 SO$_2$ averages for the whole year (all wind directions) at Reading. (Reproduced by permission of Pergamon Press.)

(c) The development of models to predict these concentrations from source and meteorological data.

There are three volumes covering the findings of the investigation [28]. Some of the most interesting were:

(a) Annual average background concentrations of SO$_4^{2-}$ in the north-eastern USA are <8 μg m^{-3}, representing 50–75% of the SO$_4^{2-}$ in urban air.

(b) Episodes of $>40\ \mu g\ m^{-3}\ SO_4^{2-}$ were associated, with major source areas up to 300 and occasionally 500 km away, where stagnant meteorological conditions were followed by advection in a shallow mixing layer. Such conditions were usually predictable.

14.8.4 Distant sources and global effects

The effect of distant sources on air quality has been under investigation for many years. The work which began under the Long Range Transport of Air Pollutants (LRTAP) Project is continuing under the co-operative programme for monitoring and evaluation of long-range transmission of air pollutants (EMEP). This [29] includes an updated source inventory for 150 × 150 km squares of the earlier study (Fig. 14.10, from [30]) which included measurements of chemicals in

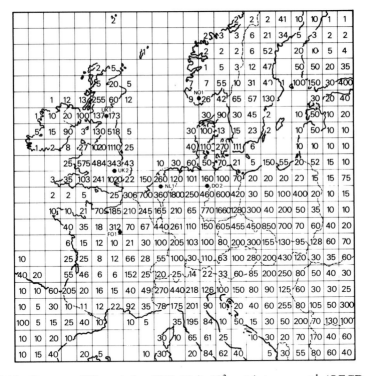

Fig. 14.10 European SO_2 emission 1972. Unit: 10^3 metric tonnes y^{-1}. (OECD project Long Range Transport of Air Pollutants (LRTAP) preliminary report). Better information has now been received for many countries. For instance, total 1973 emissions of SO_2 for the UK were about 5.7 million metric tonnes, which is 12.3% less than the 6.5 million tonnes indicated in Fig. 14.10. Similar uncertainties may of course apply to the estimated emissions of other countries. (Reproduced with permission of Pergamon Press.)

Estimated values of SO_2 decay and transformation rates, etc.

Samp. Station	Date of 1st day (1973)	No. of days	k_0 10^{-5} s^{-1}	k_t 10^{-6} s^{-1}	h_0 10^3 m	R_{SO_2}	R_{SO_4}	$\bar{a}_d$ µg m^{-3}	$\bar{q}_d$ µg m^{-3}	S_a µg m^{-3}	S_q µg m^{-3}	$\bar{b}_d$ µg m^{-3}	S_b µg m^{-3}
DO2	1 April	62	1.3	0.78	2.2	0.514	0.466	14.4	13.1	12.6	8.4	2.5	2.0
FO1	1 April	62	2.2	1.9	0.57	0.537	0.538	24.9	25.9	26.8	12.9	5.9	8.4
NO1	1 Dec.	59	1.4	1.2	1.2	0.527	0.551	6.4	5.1	7.5	5.0	3.0	3.2
NO1	15 Dec.	58	0.72	1.6	1.8	0.469	0.602	8.4	7.0	10.1	6.1	4.5	3.7
NL1	1 June	61	2.5	4.8	2.0	0.602	0.532	10.2	11.0	10.0	4.4	10.5	9.4
NL1	1 Aug.	60	1.6	2.2	1.7	0.722	0.370	17.3	19.5	16.8	8.5	11.0	9.7
NL1	1 Oct.	61	1.2	3.1	1.0	0.564	0.673	25.5	27.8	26.9	14.1	11.1	10.8
UK1	1 April	60	3.6	1.2	0.95	0.455	0.469	28.6	28.2	18.9	10.9	6.2	4.1
UK1	1 Oct.	61	3.9	1.5	0.64	0.437	0.389	36.5	41.1	18.7	11.5	9.7	7.2
UK2	1 Aug.	62	2.7	3.0	1.0	0.678	0.688	8.2	8.1	10.1	6.7	5.5	6.8
UK2	1 Oct.	53	3.6	2.3	0.65	0.527	0.551	12.8	11.8	13.1	8.8	5.0	5.8

Symbols (mean value over period designated by a bar): k_0 estimated SO_2 decay rate; k_t estimated transformation rate SO_2–SO_4; h_0 estimated mixing height parameter; R_{SO_2} empirical correlation coefficient between observed and computed SO_2 concentrations; R_{SO_4} empirical correlation coefficient between observed and computed SO_4 concentrations; $\bar{a}_d$ mean value of observed SO_2 concentrations; $\bar{q}_d$ mean value of computed SO_2 concentrations; S_a empirical S.D. of observed SO_2 concentrations; S_q empirical S.D. of computed SO_2 concentrations; $\bar{b}_d$ mean value of SO_4 concentrations; S_b empirical S.D. of observed SO_4 concentrations.

precipitation and SO_2 and particulate sulphate in air. Figure 14.10 shows the location of some of the sampling stations and some interesting historical data relevant to the project, the 1972 emissions of SO_2 in 127 km squares over Europe. Decay rates for SO_2 and transformation rates of SO_2 into particulate sulphate were derived from the data (Table on p. 618). The sampling period for the measurements was 24 h. Part of the work involved trajectory calculations and reference is made in the paper to the method used to calculate them.

An alternative approach to the same problem is to follow pollution over the North Sea using aircraft. The work described by Kallend [17] follows earlier work described in a paper by Smith and Jeffrey [31]. Vertical cross-sections of SO_2 and sulphate were measured by a sampling aircraft flying approximately normal to the wind direction over the North Sea. SO_2 emission was calculated for 20 km side squares over the UK and trajectories from source to sampling area were calculated, with estimates of the uncertainty in the calculated paths.

Measurements of atmospheric dust concentrations over the North and South Atlantic using a mesh technique are described by Parkin *et al.* (1972). They found that concentrations of Saharan dust decreased by an order of magnitude over a distance of about 3000 km but that the dust was still present in quantities of around 10 μg m^{-3} over the West Indies, three orders of magnitude higher than concentrations of dust over the western approaches of the British Isles, despite the large industrial emissions in north-eastern USA.

Acknowledgements

Similar diagrams to Figs 14.3–14.5 appear in the EPRI report [17] and to Fig. 14.7 in the *Journal of the Institute of Fuel* and are reproduced by courtesy of the publishers.

References

1. Junge, C. (1972) *Q. J. R. Meteorol. Soc.*, **98**, 711–29.
2. Lovelock, J. E. (1972) *Atmos. Environ.*, **6**, 579–80.
3. Liss, P. S. and Crane, A. J. (1983) *Man-made Carbon Dioxide and Climatic Change*, Glo Books, Norwich, pp. 3–8; 109–11.
4. Chamberlain, A. C. (1966) *Proc. R. Soc. London, Ser. A*, **290**, 236–65.
5. Dollard, G. J. and Unsworth, M. H. (1983) *Atmos. Environ.*, **12**, 775–80.
6. Moore, D. J. (1975) *Proc. Inst. Mech. Eng.*, **189**, 33–43.
7. Lucas, D. H. and Moore, D. J. (1964) *Int. J. Air Water Pollut.*, **8**, 441–54.
8. White, W. H., Moore, D. J. and Lodge, J. P. (eds) (1981) *Atmos. Environ.*, **15**, 1785–2646.
9. Pattenden, N. J., Branson, J. R. and Fisher, E. M. R. (1982) Trace element measurements in wet and dry deposition and urban particulates at an urban site. In *Deposition of Atmospheric Pollutants* (eds H. W. Georgii and J. Pankrath), Reidel, Dordrecht.
10. Biles, B. and Ellison, J. McK. (1975) *Atmos. Environ.*, **9**, 1030–2.

11. Skeffington, R. (1984) *Atmos. Environ.*, **18,** 1683–1704.
12. Martin, A. and Barber, F. R. (1973) *Atmos. Environ.*, **7,** 17–38.
13. Clarke, R. H. (1979) *A Model for Short and Medium Range Dispersion of Radionuclides Released to the Atmosphere*, National Radiological Protection Board, Harwell.
14. Bowne, N. E. and Londergan, R. J. (1983) *Overview Results and Conclusions from the EPRI Plume Model Validation and Development Project: Plains Site*, EPRI EA-3074 Project 1616-1, Electric Power Research Institute, Palo Alto.
15. Martin, A. and Barber, F. R. (1966) *J. Inst. Fuel*, **39,** 294–307.
16. Munn, R. E. (1981) *Air Pollution Problems – 2. The Design of Air Quality Monitoring Networks*, Macmillan, London, pp. 55–6.
17. Kallend, A. S. (1983) *The Fate of Atmospheric Emissions along Plume Trajectories over the North Sea*, Summary Report EPRI EA-3217, Electric Power Research Institute, Palo Alto.
18. Paterson, M. P. (1975) *The Atmospheric Transport of Natural and Man-made Substances*, PhD Thesis, University of London.
19. Benarie, M. (1980) *Air Pollution Problems – 1. Urban Air Pollution Modelling*, Macmillan, London, pp. 204–28.
20. Parekh, P. P. and Husain, L. (1981) *Atmos. Environ.*, **9,** 1717–25.
21. Schiermeur, F. A. and Niemeyer, L. E. (1970) *Large Power Plant Effluent Study (LAPPES)* (3 vols): Vols 1 and 2, US Department of Health, Education and Welfare, 1970; Vol. 3, US Environmental Protection Agency, 1972.
22. Martin, A. and Barber, F. R. (1967) *Atmos. Environ.*, **1,** 655–78.
23. Marsh, K. J. and Foster, M. D. (1967) *Atmos. Environ.*, **1,** 527–50.
24. Marsh, K. J., Bishop, K. A. and Foster, M. D. (1967) *Atmos. Environ.*, **1,** 551–60.
25. Marsh, K. J. and Withers, V. R. (1969) *Atmos. Environ.*, **3,** 281–302.
26. Pasquill, F. (1972) *Q. J. R. Meteorol. Soc.*, **98,** 469–94.
27. Blokker, P. C. (1973) *Major Aspects in Air Pollution Monitoring in Urban and Industrial Areas*, Report 7/73, Stichting CONCAWE, The Hague.
28. Mueller, P. K. and Hidy, G. M. (1983) *Regional Experiment: Report of Findings, EPRI EA-1901*, Electric Power Research Institute, Palo Alto.
29. NILU (1980) *Summary Report from the First Phase of EMEP*, Norwegian Institute for Air Research, Oslo.
30. Eliasson, A. and Saltbones, J. (1975) *Atmos. Environ.*, **9,** 431–6.
31. Smith, F. B. and Jeffrey, G. H. (1975) *Atmos. Environ.*, **9,** 643–60.
32. Parkin, D. W., Phillips, D. R., Sullivan, R. A. L. and Johnson, L. R. (1972) *Q. J. R. Meteorol. Soc.*, **98,** 798–808.

15 Quality assurance in air pollution monitoring

A. APLING

15.1 Quality and quality assurance

Quality assurance is a well-developed discipline applied in the industrial production of goods and the provision of commercial services [1]. The consistent application of quality assurance to the measurement of air pollution is most evident in countries where large-scale monitoring networks are a major feature of the control of air pollution and chief among these is the USA [2–5]. There is now a considerable literature concerning the quality assurance of air pollution monitoring and the purpose of this short chapter is to bring some of the more important items to the attention of those about to embark on monitoring programmes. This will be done by citing references in the context of a brief analysis of the various elements of a monitoring chain that can usefully be identified and subjected to quality assurance routines. The treatment throughout assumes that the monitoring is aimed at gaseous species which can be uniquely identified chemically. In principle, however, the techniques can be applied to all forms of air pollutant.

15.2 Definitions

The British Standards Institution Handbook 22 [6] gives the following definitions:

> Quality: The totality of features and characteristics of a product or service that bear on its ability to satisfy a given need.
> Quality assurance: All activities and functions concerned with the attainment of quality.

The products of air pollution monitoring are measurements of air pollutant concentrations. The quality resides in the confidence that can be placed in them in terms of the sampling region of which they are representative and their potential numerical error. The ability in principle of an air monitoring system to supply data of sufficient quality to satisfy given needs is therefore dependent on careful initial selection of the monitoring sites and of instruments capable of supplying data of the required specificity, accuracy and precision [7, 8]. Quality assurance is the sum of all activities aimed at ensuring a continued understanding of the character of sampling sites and maintaining the performance of instruments.

Many terms can be used to express and describe the estimated error of a numerical result but for clarity only two will be used here: accuracy and precision [9, 10, 11].

Accuracy error

This is taken as expressing the potential deviation of the central estimate of the measured quantity from the true value. For most practical purposes it is associated with *calibration procedures* or drift in the *sensitivity* of a measuring instrument and can be expressed as a percentage error applicable to all measurements within the instrument's usable range.

Precision error

This is taken as a measure of the ability to produce the same result for the same measured quantity on successive occasions. It is principally associated with the *intrinsic noise* of an instrument, the ability to establish the *zero* of the measuring scale, and *drift in the base line*. It can be expressed as potential error in measured units.

15.3   Elements of the monitoring chain

Four basic elements of a monitoring system can be identified, each of which can be the subject of a set of quality assurance procedures. They are:

(a) Site location and character
(b) Sampling line integrity
(c) Instrument performance
(d) Calibration.

These four elements will be treated in turn, the main quality assurance functions indicated and principle references given.

15.4   Site location and character

Measurements are useful only if we know to what they apply. It is therefore necessary that when a measuring site is located, the character of the region for which it is supposed to be representative is initially well defined. This may vary from a site which is intended to represent a very limited volume, such as a kerbside site aimed at peak levels of traffic pollution, through to a rural site aimed at being representative of a wide geographic region [3, 8].

Quality assurance encompasses those activities designed to check routinely the continued character of the site. In the case of a kerbside site, for example, it will be necessary to ensure that traffic density and mode of operation remain constant or at least that any changes are well monitored. For sites intended to be representative of larger regions, the influence of local sources is an obvious point to keep under surveillance. Factors which can influence the pollutant concentration at a given site may not be immediately obvious. The construction of a building which alters the topography of the local area can, for example, be important [12–14]. In general, the exercise of defining the location and character of a site when it is first selected will indicate the potential difficulties in assuring that these parameters have not changed.

15.5 Sampling line integrity

A sampling line has two tasks: to conduct a representative sample of the atmosphere to the measuring instrument and, while doing so, be effectively invisible to both sample and instrument. Quality assurance of sampling lines will involve regular cleaning at a frequency which will depend on the general quality of the sample atmosphere. Any filters installed to protect the instruments or main sampling line from the build-up of particulate material need to be renewed before the collected material affects significantly the concentration of sample gases or the back pressure in the sampling line. Heating systems designed to avoid condensation in the line must be monitored for correct functioning. This could be done automatically by monitoring suitably located temperature sensors. Finally, but most important, checks should be carried out frequently to ensure that sample air is being drawn through the line at the specified rate.

Unfortunately, careful investigations of sampling line performance and the effect of factors such as filter loading are rare. In specifying a quality assurance schedule it will probably be necessary to measure and assess many of these factors as part of the initial establishment of the monitoring system [3].

15.6 Instrument performance

The main elements of instrument performance that need to be assured are:

(a) Noise
(b) Linearity of response
(c) Response time
(d) Interference.

(Sensitivity and stability are more appropriately dealt with under the heading of calibration below.)

Possible reasons for instruments deviating from their initial specification with regard to these characteristics include ageing and failure of electronic components, reduced performance of pumping systems and the build-up of deposits within the instrument's own sampling line and reaction chambers.

Monitoring of air pollution inevitably involves a measure of faith that between checks the instrument performance characteristics remain essentially unchanged, and the obvious way to minimize uncertainty is to carry out the quality assurance checks as frequently as possible. An ideal solution is to equip the instrument with self-test equipment so that checks on performance can be carried out and the results logged along with the air pollution data. Examples might be continuous logging of photomultiplier housing temperature, response to surrogate signals either optical or electronic, and alarm signals indicating failure of components such as the drive belts of chopper wheels.

More often, however, instrument performance is checked when the instrument is brought back to the laboratory where test atmospheres and equipment are more conveniently located and employed. The quality assurance schedule must in this case be dictated by the known or estimated frequency of deviation from specified performance. Preventive maintenance based on such statistics should be the deciding factor, not simply convenient milestones of normal operation, such as the need to replace chemicals filters or other components of known limited life-time. Finally, complete overhaul and cleaning of instruments should be undertaken on a routine basis at least annually [3].

15.7   Calibration

There is nothing mysterious about calibration, but it can be technically demanding. In essence it consists of preparing standard atmospheres of the pollutant to be detected, including a zero atmosphere, presenting these to the instrument and noting the response [15]. Standards, with the exception of the zero standard, can be either primary or base standards. The former have a concentration estimate based on the measurement of fundamental quantities such as mass, time or length while the latter usually depend on following some prescribed procedure. It may not be convenient or even possible to deploy either form of standard directly in the field, in which case a secondary or transfer standard will be required. For quite a wide range of pollutants such secondary standards can now be contained in specially treated gas pressure cylinders [16].

The calibration schedule also represents a check on the stability of an operating instrument. A sequence of exposures to a test atmosphere will indicate whether or not the sensitivity of the instrument is remaining within specification while the record of response to zero gas will indicate any instability in the base line. As with performance checks, therefore, frequency of calibration leads to increased confidence in the overall operating characteristics of an instrument.

Quality assurance as far as calibration is concerned lies principally in the routine recording and checking of the data necessary to calculate the concentration of the primary or base standards and their use to calibrate transfer or secondary standards, followed by careful monitoring of the instrument's response to calibration.

15.8 Discussion and further checks

The quantitative objective of the quality assurance function in air pollution monitoring is to ensure that when results are reported, estimates of the precision and accuracy errors can always be given based on the calibration and performance check data and on records which show that the monitoring system was operating under the originally specified conditions. One must always be able to write:

The concentration of pollutant A over period T was:

x units $\pm p$ units $\pm a\%$

where p is the estimate of precision error and a the estimate of accuracy error [10, 17].

No system or schedule of quality assurance can be guaranteed to reveal every possible malfunction or deviation from specification and it is always worthwhile considering additional checks that can be applied. Three in particular should be mentioned:

(a) Data continuity
(b) Extreme measurements
(c) Inter-laboratory checks.

Data continuity is simply the process of inspecting the time series of collected data for periods which stand out, either in terms of the levels measured, the base line or the noise, from surrounding data. An essential part of using data continuity is that when a new period of data is added to the existing archives, no attempt should be made to harmonize existing data with that being added. On the contrary, a new set of data should be corrected and verified using solely the parameters of calibration and instrument performance estimated for that period. In this way isolated periods of poor data or gradual trends arising from a progressive malfunction can often be identified. Good quality assurance records for the period can often help to explain the problem and perhaps enable further corrections to be made allowing the retention of the data in the full time series.

By extreme measurements is meant the occurrence of measured values outside the normally expected range. This of course assumes that some statistical examination of an existing data set has been carried out in order to define the normal range. With modern data acquisition systems such checks of measured data against predefined standards can be done automatically and even in real time [18]. It may be, of course, that the extreme value turns out to be verifiable. Good quality assurance records are the key to identifying whether this is so or not.

Finally, inter-laboratory checks are amongst the most powerful techniques for investigating whether the estimates of measurement error, assured by a quality assurance programme, are really justifiable. The techniques employed typically consist of circulating standard material, in this case a test gas atmosphere, to a large number of laboratories [19–22]. In the most comprehensive tests an instrument operating in the field is employed and test gas is introduced to the sampling line under exactly the same conditions as a normal ambient sample. Because sampling lines are very demanding in terms of gas volume, test sample gas is more commonly introduced indirectly to the operating instrument. In this case errors introduced by virtue of malfunctioning of the sampling line will not be detected.

Inter-laboratory tests can be disquieting experiences. One's own primary standards and quality assurance techniques are being directly compared with the independently operated and possibly physically different systems of another laboratory. If there is any discrepancy someone has to be wrong. It is, however, of the essence of quality assurance that procedures, schedules and tests be regularly applied so that only data fulfilling specified standards of validity survive to be reported and interpreted.

References

1. Knowles, R. (1983) No short cuts to quality assurance. *BSI NEWS*, October 1983, 13. (Note: this issue was devoted to the topic of quality assurance.)
2. USEPA (1976) *Quality Assurance Handbook for Air Pollution Measurement Systems: Vol. 1, Principles*, Document EPA-600/9-76-005, US Environmental Protection Agency, North Carolina.
3. USEPA (1977) *Quality Assurance Handbook for Air Pollution Measurement Systems: Vol. 2, Ambient Air Specific Methods*, Document EPA-600/4-77-027a, US Environmental Protection Agency, North Carolina.
4. Hauser, T. R. (1979) *Environ. Sci. Technol.*, **13**, 1356–62.
5. ACS Committee on Environmental Improvement (1980) *Anal. Chem.*, **52**, 2242–9.
6. BSI (1983) *Quality Assurance. BSI Handbook 22: 1983*, British Standards Institution, London.
7. WHO (1977) *Air Monitoring Programme Design for Urban and Industrial Areas*, World Health Organisation, Geneva.
8. Reay, J. S. S. (1979) *Phil. Trans. R. Soc. London, Ser. A*, **290**, 609–23.
9. Eisenhart, C. (1969) Realistic evaluation of the precision and accuracy of instrument calibration systems. In *Precision Measurement and Calibration*, Nat. Bur. Stand. (US), Spec. Publ. 300, **1**, 21–47.
10. Evans, E. G. and Rhodes, R. C. (1981) *Summary of Precision and Accuracy Assessments for the State and Local Air Monitoring Networks*, Document EPA 600/4-84-032, US Environmental Protection Agency, North Carolina.
11. Suggs, J. C. and Curran, T. C. (1984) *Incorporating Measurement Uncertainty into Air Quality Evaluations*, Document EPA 600/D-84-208, US Environmental Protection Agency, North Carolina.
12. Kennedy, I. M. and Kent, J. H. (1977) *Atmos. Environ.*, **11**, 541–7.
13. Wedding, J. B., Lombardi, D. J. and Cermak, J. E. (1977) *J. Air Pollut. Control Assoc.*, **27**, 557–60.
14. Kotake, S. and Sans, T. (1981) *Atmos. Environ.*, **15**, 1001–3.
15. Nelson, G. O. (1971) *Controlled Test Atmospheres*, Ann Arbor Science Publishers, Ann Arbor.

16. Peperstraete, H. J. (1982) Gaseous reference materials – certification – traceability. In *Physico-Chemical Behaviour of Atmospheric Pollutants: Proceedings of the 2nd European Symposium, Varese 29 Sept.–1 Oct. 1981*, D. Reidel, Dordrecht.

17. NAS (1977) *Analytical Studies for The US Environmental Protection Agency: Vol 4, Environmental Monitoring*, National Academy of Sciences, Washington, DC, pp. 68–9.

18. Roosken, A. A. M. (1978) Real time validation of air quality data. In *Air Pollution Reference Measurement Methods and Systems* (eds T. Schneider, H. W. deKoning and L. J. Brasser), Elsevier, Amsterdam.

19. Youden, W. J. and Steiner, E. H. (1975) *Statistical Manual of the Association of Official Analytical Chemists*, Association of Official Analytical Chemists, Washington, DC.

20. Foster, J. F. and Beatty, C. H. (1974) *Interlaboratory Cooperative Study of the Precision and Accuracy of the Measurement of Nitrogen Dioxide Content in the Atmosphere Using ASTM Method D1607*, ASTM Data Series Publication DS 55, American Society for Testing and Materials, Pennsylvania, Pa.

21. Apling, A. J., Peperstraete, H. J. and Rudolf, W. (1982) *CEC Harmonization of Methods for Measurement of NO_2*, Report EUR 7865 EN, Commission of the European Communities, Luxembourg.

22. Muylle, E. P. and Peperstraete, H. J. (1982) *Harmonization of Methods for Measurement of SO_2*, Report EUR 8052 EN, Commission of the European Communities, Luxembourg.

Index

Absorbance, 242
Absorption, 470
  spectra, 465
  spectroscopy, 290
  spectrum, 470, 478
Acetaldehyde, 142
Acetylene, 403
Acidimetric methods, 570, 284
Activated carbons, 394
Accuracy, 622
Adiabatic lapse rate, 2
Adsorbents, 66
Aerosol, 155, 502, 508
  laser, 470
  particle counters, 176
  particles differential lidar, 478
  source sampling procedures, 58
  source sampling procedures, high volume
    stack sampler, 59
  source sampling procedures, mobile sources,
    60
Aerosols, 25
  centrifugal collection, 41
  diffusion sampling, 41
  effects upon sampling, 30
  electrostatic precipitation, 40
  filtration, 34
  impaction sampling, 48
  impingement, 38
  inhalable particles sampling, 45
  Nuclepore filters, sequential filtration, 52
  sampling errors, 30
  source sampling procedure, British Standard
    method, 55
  source sampling procedure, ASTM method,
    55
  thermal precipitation, 39
Air monitoring network, 564
Air volume to be measured, 569
Air quality chromatograph, 398
Air quality standards for particulate matter,
  hydrocarbons and carbon monoxide, 390
Air streams, 112

Airborne systems, 600
Alarm/display systems, 604
Alkanes, 140, 387
Alkenes, 141, 147, 387
Aldehydes, 144, 147, 370, 371, 388
Alkoxy radicals, 140
Alkyl nitrates, 388
Alkyl peroxynitrates, 140
Alundum thimbles, 56
American high volume sampler, 42
Ammonia, 149, 324, 549
Amperometric procedure, 444
Amines, 328
Ammonium salts, 337
Anabatic winds, 114
Analysis of results, 604
Analytical methods, 415
Analytical procedures, 451
Analytical techniques, 387
Anodic stripping voltammetry, 237
Anticyclones, 108
Antimony, 215
Anthropogenic processes, 553
Area surveys, 597
Aromatics, 142, 387
Aromatic hydrocarbons, 148
Arsenic, 215, 556
Arsenite method, 314
Artifact particulate matter, 36
Asbestos, 205
Atomic spectroscopy, 246
Atomic absorption analysis, 555
Atomic absorption detection limits, 248
Atmospheric backscatter, 473
Atmospheric dispersion, 99
Atmospheric stability, 3, 498
Atmospheric transmission, 585
Automatic smoke instruments, 170
Automobile exhaust, 388
Averaging periods, 607

β-radiation monitors, 170
Backscatter, 479

Bag sampling, 75
Base standards, 624
Bead packed absorbers, 68
Benzene, 388
Beryllium, 215, 245, 258
Biacetyl, 143
Bicarbonate-coated glass tube and filter technique, 434
Bismuth, 215
Blast furnaces, 202
Box models, 121
Breakthrough volume, 391
British Standard deposit gauge, 181
British Standard particulate sampler, 41
Bromides, 450
Brominated hydrocarbons, 456
Buffering intensity, 558
Bulk collectors, 82
Butene, 398

Cadmium, 215
Calcium, 549
California 7-mode driving cycle, 19
Calibration, 567, 576, 624, 396
Calibration methods, 409
Capillary restricted flow method, 410
Carbon dioxide lasers, 472
Carbon disulphide, 305
Carbon monoxide, 387, 389
Carbon oxysulphide, 305
Carbonization industry, 201
Carbonyls, metal, 261
Carbosieve, 394
Cascade impactors, 38, 48, 172
Cellulose filters, 35, 163
Cement industry, 202
Centrifugal impactors, 51
Ceramic industry, 202
Charcoal adsorption tubes, 71
Chelation with APDC, 555
Chemical ionization mass spectrometry, 408
Chemiluminescence, 288
    analysers, 139, 327
    detectors, 321
    methods, 249, 361
Chlorinated hydrocarbons, 455
Chromium, 215
Chromotropic acid, 373
Chrysotile, 205
Classification of particulates, 189
Climatological data, 605
Coke ovens, 201
Cobalt, 215
Cold bath solutions, 70
Cold advection, 112
Collection efficiency, 22
Collimated holes structure diffusion battery, 53

Colorimetric methods, 284, 286
Condensation traps, 74
Contamination, 545
Continuous instrumental analysers, 396
Continuous vertical retorts, 201
Convergence, 107
Copper, 215
Correlation analysis, 608
Correlation spectroscopy, 291, 465, 508, 509
Coulometric analysers, 300
Crocidolite, 205
Chromosorb, 101, 105, 395
Cryogenic systems, 392
Cryogenic trapping techniques, 391
Cryostat, 392
Cupolae, 203
Cyclone aerosol samplers, 51

Dosage relations, 606
Dose monitoring, 588, 589
Data handling, 602
    requirements, 10
Data presentations, 611
Data quality, 557
Data storage, 603
Data transmission, 602
Data continuity, 625
Density gradient liquids, 189
Density gradient separation, 189
Deposition, 124
Deposition velocity, 180
Deseaming mills, 203
Detector tube method, 571
DF laser, 488
DIAL, 478, 508
Dichotomous sampler, 47, 174
Differential lidar, 472, 478
Differential ultraviolet and visible absorption spectroscopy, 489, 490
Diffusion battery, 41, 53, 177
Dimethyl sulphide, 146, 303
Dimethylmercury, 264
Dinitrogen pentoxide, 138
Diode lasers, 488
Directional sampler, 177
Discrete source surveys, 597, 611
Distant source surveys, 599
Dispersion, 95
Dispersion staining, 195
Dithizone, 243
Diurnal variations, 607
Domestic fires, 200
Doppler sodar, 501, 508
Double paper tape sampler method, 434
Drift, 622
Driving cycles, 18
Dry deposition, 125, 156

Dry test gas meter, 23
Durapak, 401
Dust identification table, 203
Dust reference library, 199
Dustfall, 156, 573
   sampling, 179
Dye laser, 480, 485
Dynamic methods, 330, 409

Eddy diffusivity models, 120
Effective height, 119
Efficiency of filters, 165
Effluent control processes, 584
Effluent transport, 585
Eight-port sampling instrument, 169
Electric arc furnaces, 203
Electrical conductivity analysers, 300
Electrochemical, 418
Electrode, 551
Electron capture detectors, 401
Electron capture GC, 378
Electron microscopy, 205
Elemental carbon, 167
Emission factors for particulate matter, 157
Emission fluxes, 510
Emission monitoring, 587
Emission spectrography, 228
Equipment requirement, 575
Error, 622
Erroneous readings, 606
Ethyl mercaptan, 303
Ethylene, 387, 478, 486
European ECE driving cycle, 20
Exponential dilution method, 332
Exposure of samplers, 594
Extraction and clean-up procedures, 414, 555

FEP-Teflon sampling bags, 410, 411
Field experiments, 544
Filter treatment, 37
Filter paper, 545
Filter tubes, 56
Filters, blank concentrations, 36
Filters, cleansing, 546
Filters, head loss, 36
Filters, membrane, 216
Filter materials, impurities, 217
Filtration, 545
Filtration efficiency, 35, 53
Fine particles, 46
Flame Atomic Absorption, 554
Flame ionization detector, 396
Flow measurement, 23
Fluoride species, 425
Fluorides, 425
Fluorocarbons, 453
Fluorometry, 241

Fluorescence, 290
Fog sampling, 82
Forced convection, 96
Formaldehyde, 133, 139, 142, 373
Formic acid, 139
Fourier transform, 370
Fourier transform infrared spectroscopy, 383, 494
Frequency distributions, 607
Fritted glass absorbers, 68
Fumigation, 508
Function monitoring, 586

Gas absorption bottles, 69
Gas chromatographic phase adsorption, 72
Gas chromatography, 292, 350, 389, 418, 557
Gas chromatography/Mass spectrometry, 404, 407, 418
Gas-liquid chromatography, 399, 417
Gas sampling, 65
   adsorption, 66, 67
   condensation, 70
   determination, 15
   efficiency, 68, 73
   errors, 65
   flexible container, 74
   grab, 71
   passive, 76
   rigid container, 76
   syringe, 75
   systems, 395
Gas source sampling
   mobile sources, 78
   stationary sources, 81
Gaussian models, 119, 120, 127
Geochemical cycling, 536
Glass fibre filters, 35, 164
Global effects surveys, 599
Glyoxal, 143
Grab sampling, 71
Gran analysis, 548
Graphitized carbon black, 404
Graticules, 209
Gravimetric techniques, 168
Greenburg–Smith impinger, 38, 56, 434
Ground-level concentration, 589

Halogenated hydrocarbons, 452
High-performance liquid chromatography, 416
High-volume sampling method, 168, 414, 574
Historical data, 605
Horizontal retorts, 201
Humidity, 4
Hunt formula, 7
Hydrocarbons, 387
Hydrocarbon fraction or airborne particulate matter, 413

Hydrogen chloride, 445, 478, 488
Hydrogen peroxide, 137, 369
Hydrogen sulphide, 303, 306
Hydroperoxyl radicals, 135
Hydroxyl radicals, 133, 135

Impaction sampling, 48
Indophenol-blue method, 325
Inductively coupled plasma analysis, 554
Inertial impactors, 48, 50
    wall losses, 50
Inhalable particle matter, 46
Inorganic oxyacids, 346
Instrumental neutron activation analysis, 224
Instrument performance, 623
Intake and transfer component, 21
Intake performance, 21
Interference, 623
Inter-laboratory checks, 625
Intermittent vertical retorts, 201
Iodide method, 359
Ion balance check, 552
Ion chromatography, 550, 551
Ion electrode, 551
Ionic stability of rainwater samples, 83
Ion exchange, 434
Ion-selective electrode, 430, 437
Iron, 215
Iron and steel industry, 202
Iron carbonyl, 263
Isopropyl benzene, 412

Japanese driving cycle, 20

Katabatic winds, 114
Ketones, 141

Land breezes, 113
Laser linewidth, 472
Laser microprobe mass spectrometer, 529
Laser Raman spectroscopic techniques, 399
Laser safety, 488
Layout of instruments, 595
Lead, 215
    alkyls, 267
    particulate, 270
Lead peroxide candle method, 77
Lead sulphation candle or plate, 570
Lidar, 464, 469, 473, 498, 502, 505, 508
Lidar equation, 470, 471, 499
Light reflectance, 166
Linearity, 623
Log book, 577
Long path absorption spectroscopy, 464
Long-term trends, 583
Low temperature ashing, 218
Low-cost air pollution monitoring systems, 563

Magnesium, 549
Manganese, 215
Martin's diameter, 209
Mass median aerodynamic diameter, 173
Mass spectroscopy, 389, 407
Mass spectrometry, spark source, 239
Membrane filters, 35, 164
Medium range, 610
Mercaptans, 311
Mercury, 556
    dialkyl, 264
    elemental, 266
    monoalkyl, 264
Metal pollutants, 215
Metal recovery, 218
Meteorological measurements, 100, 495
Meteorological monitoring, 588
Methane, 387, 398
Methyl mercaptan, 303
Methyl orange method, 444
Methyl-thymol blue, 550
Methylcyclohexane, 403
Methylglyoxal, 143
Methylmercury, 264
Microdiffusion technique, 436
Microscopic techniques, 197
Microwave radiometry, 498, 503
Midget impinger, 38
Mie scattering, 470
Mists, 215
Mixing depth, 106
Mixing layer, 95, 498, 495, 500, 502, 508
Molybdenum, 215
Monitoring
    exercise, 580
    control, 583
    networks, 611
Monoterpenes, 146
Muffle furnace, 218
Multiple scattering, 467, 514
Multiple source studies, 610
Multi-purpose surveys, 599
Multisource monitoring, 515

National Atmospheric Deposition Program, 545
Ndir, 418
Neodymium-YAG laser, 471
Neodymium-YAG lidar, 474
Nephelometer, integrating, 175
Network sites, 5
Neutral buffered potassium iodide method, 355, 572
Neutron activation analysis, 224
Nickel, 215
Nickel carbonyl, 262
Nitrate, 550
Nitrate particulate, 149

Nitric acid, 133, 137, 139, 382
Nitric oxide, 279, 313, 514
Nitrate radical, 133, 136, 139, 489, 493, 494
Nitrogen dioxide, 139, 279, 313, 465, 466, 468, 469, 478, 480, 483, 489, 494, 510, 514
Nitrous acid, 133, 135, 137, 138, 139, 382, 489, 492
Noise, 622, 623
Non-dispersive infra red spectroscopy, 390
Non-methane hydrocarbon monitors, 396
Nuclepore filters, 164

Occult deposition, 125
OECD project Long Range Transport of Air Pollutants, 617
Oil fired boilers, 201
Open hearth furnaces, 203
Optical effects, 590
Organic lifetimes, 146
Organic particulate, 150
Organics, 557
Organo-sulphur compounds, 310
Orsat analysis, 10
Overhead burden, 468, 485, 509
Oxidants, 346
Oxides of nitrogen, 133, 288
Oxides of sulphur, 133
Oxyacids, 381
Oxygenated compounds, 370
Ozone, 134, 139, 144, 359, 478, 480, 483, 489
Ozone generator, 368

Pararosaniline, 294
Particle diameter, 211
Particle identification, 557
Particle (proton) induced X-ray emission (PIXE), 222, 544, 556
Particulates, 473
Particulate chloride, 445
Particulate matter, 387
Particulate pollutants, 155
Permeation tubes, 330, 396, 409
Peroxyacetyl nitrate, 139, 140, 144, 495, 275, 280, 343, 383, 388
Peroxynitric acid, 137
Pesticides, 455
Phases identified in air by X-ray diffraction, 525
Photochemical smog, 344
Photoionization detection, 401
Piezoelectric mass monitors, 171
Pilot balloons, 498
Pitot tubes, 24
Plume density contours, 477
Plume monitoring, 587
Plume rise and dispersion, 115, 505, 508
Plume rise formulae, 116

Polarography, 235
Pollutant identification, 581
Polychlorinated biphenyls, 455
Polycyclic aromatic hydrocarbon, 389
Polyester bags, 74
Porous polymers, 394
Portable mass spectrometry, 408
Potassium, 549
Power station plume, 473
Power stations, 611
Precipitation, 3, 535, 536, 540, 541
Precision, 566, 622
Pretreatment for particulates, 429
Primary standards, 624
Pulverized fuel boilers, 200
Pumps, 25

Quality assurance, 621

Radar Acoustic Sounding System, 503
Radioactivation methods, 223
Radiosonde, 500
Rainwater sequential samplers, 82, 83
Rayleigh scattering, 470
Recognition of defective readings, 604
Refrigerants, 70
Receptor protection, 583
Receptor/sink network, 585
Regional surveys, 615
Removal of interfering ions, 434
Remote optical sensing of emissions, 494
Remote sensing, 463, 498
Response time, 623
Retroreflector, 464
Ring oven methods, 231
Rotameter, 24
Ruby laser, 471
Ruby lidar, 472, 475

S pitot tube, 17
Saltzman analysis, 6
Sample traps, 411
Sampler, Hi-Vol, 255
Samplers
  bulk, 539
  wet only, 539
Sampling
  anisokinetic, 27
  directional, 30
  duration, 5
  isokinetic, 27
  line, 623
  network, 5
  periodicity, 6
  precipitation, 536, 538, 544, 537
  precision, 7

Sampling—*contd.*
  procedures, 391, 451
  random, 7
  scheduling, 5
  site criteria, 4
  system, 593
  train, 567
Saran, 395
Satellite borne instruments, 464
Scanning electron microscopy, 523, 529
Scattering, 470, 477
Screen type diffusion batter, 54
Sea breezes, 113
Secondary standard, 624
Sensitivity, 566, 624
Shell type boilers, 200
Sieve techniques, 208
Signal processing, 473
Silica gel, 394
Single source studies, 609
Sintering plant, 202
Site location, 622
Siting, multiple programmes, 538
Size distribution, 156
Size selective inlets, 46
Skylight, 465, 467
Smoke, 156
Smoke sampler, 41
Snow, 539
Snowpack, 539
Sodar, 464, 498, 507, 508
Sodium, 549
Sodium arsenite method, 571
Solar radiation, 464, 465
Solid adsorbents, 391, 394
Solvent extraction method, 531
Source
  height, 97
  identification, 581
  inventories, 605
  network, 584
  sampling
    location, 10
    mobile, 18
    site selection, 10
    stationary, 8
Soxhlet apparatus, 395
Space Shuttle, 464
Spare parts stock, 577
Speciation methods, 524
Specific conductance, 544, 551, 553
Specific ion electrode, 549
Specificity, 566
Spectroscopic methods, 352
Spectrophotometric methods, 241, 438
Spiral absorbers, 68
Spread of a plume, 126

Stability, 624
  categories, 128
  of reagents, 566
Stack sampling, 10
  moisture determination, 11
  temperature determination, 15
  velocity determination, 15
Stacked filters sampling, 52
Standard atmosphere, 624
Standard gas mixtures, calibration, 329
Standard smoke, 166
Standard pitot tube, 16
Static methods, 329
Static procedures, 409
Statistics, 606
Steam distillation, 429, 435
Stoker fired boilers, 200
Subsidence, 107
Sulphate, 333, 549
  particulate, 149
  speciation, 531
Sulphur dioxide, 149, 293, 465, 466, 468, 469,
    478, 480, 483, 489, 508, 509, 510
  sampling of, 74
Sulphuric acid, speciation, 530
Support coated open tubular columns, 401
Surface pollution pattern, 129
Surface systems, 600
Synoptic weather map, 110

Tedlar bags, 396
Teflon bags, 75
Temperature, 102
Tenax GC, 394, 395, 412, 455
Tenax TA, 395
Test atmosphere, 22
Tetrachloromercurate, 294
Tetra-alkyl lead, 450, 267, 268
Tetraethyl lead, 267
Tetramethyl lead, 267
Thermal precipitator, 40
Thermal desorption, 394, 404, 405
Thermal radiation, 103
Thermo-desorption cold trap injector, 405
Thin-layer chromatography, 415
Thoracic particle fraction, 46
Thorin, 550
Time series, 625
Tin, 215
TISAB, 437
Titanium, 215
Titrimetric method, 438
Toluene, 142, 388
Topographic targets, 464
Total oxidants, 353

Total suspended particulate matter, 215
Total suspended particulates, 156
Total suspended particles (TSP) measurement, 42
Trace Metals, 545, 553
Trajectories, 124
Transfer standard, 624
Transmission electron microscope, 529
Transmittance, 242
Triethanolamine method, 319
Trimethylbenzene, 412
TSP measurement, flow controllers, 45
TSP measurement, passive deposition, 44
Tubing, 545
Turbulence, 95, 105
Two-stage thermal desorption, 405

Ultraviolet absorption, 363
Unsettled cyclonic, 112
Upslope winds, 114
Urban surveys, 615
US Federal driving cycle, 19
US Environmental Protection Agency method, 58
US national air quality objectives, 5
US National air quality standard reference methods, 6

Vacuum sublimation, 415
Vanadium, 215
Vehicles, exhaust sampling, 18
Venturi meter, 23
Vinyl chloride, 456
Virtual impactor, 51, 173
Visibility degradation, 149
Volatile hydrocarbons, 387
Volume backscatter coefficient, 470
Volume extinction coefficient, 470
Volume meters, 23
Volume scattering coefficient, 499

Warm advection, 111
Water vapour, 478, 485, 486
West Gaeke method, 294
Wet deposition, 125
Wind directions, 608
Wind speed, 101

Xylenes, 143, 412
X-ray diffraction, 205, 524, 557
X-ray diffractograms, 528
X-ray emission analysis, 219
X-ray fluorescence, 220
X-ray energy spectroscopy, 523, 529

Zinc, 215